A Course on Many-Body Theory
Applied to Solid-State Physics

Lecture Notes in Physics — Vol. 11

A Course on Many-Body Theory Applied to Solid-State Physics

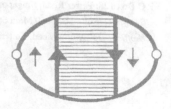

Charles P. Enz

Department of Theoretical Physics
University of Geneva

World Scientific
Singapore • New Jersey • London • Hong Kong

Published by

World Scientific Publishing Co. Pte. Ltd.
P O Box 128, Farrer Road, Singapore 9128
USA office: Suite 1B, 1060 Main Street, River Edge, NJ 07661
UK office: 73 Lynton Mead, Totteridge, London N20 8DH

Library of Congress Cataloging-in-Publication Data

Enz, Charles P. (Charles Paul), 1925–
    A course on many-body theory applied to solid-state physics /
Charles P. Enz
        p.   cm.
    Includes bibliographical references and index.
    ISBN 99715033360. -- ISBN 9971503379 (pbk.)
    1. Solid state physics.   2. Many-body problem.   I. Title
QC176.E58   1992
530.4'1--dc20   92-30895
                                                        CIP

Copyright © 1992 by World Scientific Publishing Co. Pte. Ltd.

Printed by Singapore National Printers Ltd.

PREFACE

This book is based on a course on many-body theory I had been asked to give in the Postgraduate Teaching Program of Physics in French Switzerland (Troisième Cycle de Physique en Suisse Romande) at the Federal Institute of Technology (Ecole Polytechnique Fédérale) in Lausanne during the summer terms of 1982 and of 1985. But some of the materials had already been presented at earlier invitations, 1962 in Leysin (IVe Cours de Perfectionnement de l'Association Vaudoise des Chercheurs en Physique), 1965 in Lausanne (Troisième Cycle, notes by Michel Romério) and in Geneva (University, notes by Fabio Barblan), 1967 in Trieste (International Centre for Theoretical Physics), 1968/69 in Lausanne (Troisième Cycle, notes by Michel Droz), 1969 in Majorca (International School of Physics) and 1971 in Geneva (Troisième Cycle, notes by Dionys Baeriswyl). In comparing these different courses the evolution, both in scope and in sophistication appears striking to me. Interestingly, this evolution is both, my own and that of physics itself. The last decades have indeed brought exciting discoveries and interpretations in ever faster succession. For this reason I think it is a good time now to write such a book, in spite of the large number of already existing texts on many-body theory. On a more personal level this timing also means a certain retrospective.

My ambition and satisfaction in writing this book was to reach a self-contained unity by giving as complete explanations as possible. This of course is a time-consuming and often frustrating task. But when I think of the disappointments in reading certain theoretical reviews which just repeat formulas without explaining them I become convinced that I did the right thing. But the question now is how well these explanations withstand criticism. Fortunately, in several instances I benefited from clarifying discussions with Daniel Loss, Ching Zhou and Ora Entin-Wohlman which I acknowledge gratefully here.

Wanting to "leave nothing unexplained" of course sets other limits. So I had to leave out many exciting topics or cover them only with few words instead of detailed expositions. This, however, is the author's privilege which is comparable to that of a conductor's: it is he who selects the music to be performed. But he also works with the musicians to realise his interpretation, which may or may not be applauded in the concert hall. The performers in the orchestra, on the other hand, have to work their daily exercises at home if they wish to belong to the orchestra or even become soloists. For the students of this book there is also a selection of exercises, all with detailed solutions and many of a similar kind just as the scales of the musicians all resemble each other. In this sense the book is indeed a course, as also in the exposition of the two introductory chapters, in particular the combinatorial toccata leading to Wick's theorems for a complex time path which is my composition.

The material of the remaining three chapters, however, is selected more according to its research interest. Topics like localization by disorder or mesoscopic transport had greatly fascinated me because of their lack of intuitive evidence. Then of course came high-T_c superconductivity, re-sounding like a huge Richard Strauss orchestra; it still rings in the ears. But the topic of most concern to me had always been the problem of magnetism. I even had doubts of ever grasping the subject as a unity and I did not find much help in books or reviews. But what came out of a consider-able effort as Chapter 5 reassures me that there may indeed be some unity and intelligibility in the matter. In all these efforts I was helped by Francine Gennai-Nicole in typing the manuscript in TeX and by Jean-Gabriel Bosch in drawing the figures in MacDraw. The numerous computer problems I ran into were all solved with great competence and patience by Andreas Malaspinas. To all three I address my sincere thanks for their expert help.

Many readers may find my style tiresome because of the obsession of "leaving nothing unexplained". I have no excuse for this because it is both personal and eminently Swiss. Indeed, it is the attitude that made Swiss watch-makers famous and let the people of this country be concerned with the security of the Alpine passes ever since the foundation of the Confœderatio Helvetica 700 years ago. Since this book is ready just for Lady Helvetia's big celebration it gives me pleasure and pride to dedicate it to Her 700th anniversary.

Geneva, summer 1991   Charles P. Enz

CONTENTS

Chapter 5. MAGNETISM 257

Chapter 1

ELECTRONS AND PHONONS AND THEIR QUANTIZATION

The subject of electrons and phonons in solids is treated in many books of which we wish to mention particularly those by Ashcroft and Mermin[1] and by Ziman[2]. In this introductory chapter only ideal crystals and their excitations, electrons and phonons, are considered. Invariance under translations of the lattice then implies the fundamental property of Bloch functions for the electrons, enunciated by Bloch in 1928[3]. On the other hand, if many phonons or electrons are present, they have to be described by totally symmetric or totally antisymmetric wavefunctions since, as was shown in detail by Pauli (Ref. 4, Section 14), these are the only acceptable permutation symmetries that the states describing many identical particles can have. In this spirit, second quantization is introduced in a systematic way and forms, together with the description of electrons and phonons and their interactions, the basis for the subsequent chapters.

1. Electron and Phonon Modes in Perfect Crystals

Except for the discussion of fluids and disordered solids in Chapter 3, the physical systems considered in this course are *perfect crystals* in 3-dimensional **r**-space built up from a *periodic lattice* of unit cells of given point group symmetry. The lattice vectors

$$\mathbf{R} = \sum_{i=1}^{3} m_i \mathbf{a}_i \; ; \quad m_i = 0, \pm 1, \pm 2, \dots \tag{1.1}$$

are generated by the primitive translations $\mathbf{a}_i$ which span the *unit cell C* centered at $\mathbf{R} = 0$ whose volume is $v = \mathbf{a}_1 \cdot (\mathbf{a}_2 \times \mathbf{a}_3)$.

Periodicity in **r**-space generates a *reciprocal lattice* in **k**-space through

1

the basic relation

$$e^{i\mathbf{K}\cdot\mathbf{R}} = 1 \tag{1.2}$$

where the reciprocal lattice vectors $\mathbf{K}$ may be expressed in terms of primitive translations $\mathbf{b}_j$ in analogy to (1.1),

$$\mathbf{K} = \sum_{j=1}^{3} n_j \mathbf{b}_j \; ; \quad n_j = 0, \pm 1, \pm 2, \dots . \tag{1.3}$$

It is easily seen that the basic relation (1.2) is satisfied by choosing the $\mathbf{b}_j$ according to the conditions

$$\mathbf{a}_i \cdot \mathbf{b}_j = 2\pi \delta_{ij} . \tag{1.4}$$

These conditions are satisfied with $\mathbf{b}_1 = 2\pi \mathbf{a}_2 \times \mathbf{a}_3 / v$, etc.

Excitations in a perfect crystal may be described by a *field on the lattice*[5], $f(\mathbf{R})$, which has a Fourier representation

$$f(\mathbf{R}) = \sum_{\mathbf{k} \in Z} g(\mathbf{k}) e^{i\mathbf{k}\cdot\mathbf{R}} . \tag{1.5}$$

In writing this equation *discrete wavevectors* were assumed which is most convenient for calculations. Also, because of the basic relation (1.2) the $\mathbf{k}$-sum may be restricted to the unit cell Z of the reciprocal lattice, centered at $\mathbf{K} = 0$. This domain is called the *first Brillouin zone* or *reduced zone*, and $\mathbf{k} \in Z$ is called a *crystal momentum*[1,6] divided by $\hbar$ ($h = 2\pi\hbar$ is Planck's constant). Explicitly, Z is defined as the semi-open domain

$$\mathbf{k} = \sum_{j=1}^{3} \kappa_j \mathbf{b}_j \; ; \quad -\frac{1}{2} < \kappa_j \leq +\frac{1}{2} \tag{1.6}$$

whose boundary ∂Z is formed by the median planes of the rays from $\mathbf{K} = 0$ to all neighboring $\mathbf{K}$-vectors necessary to close Z.

Discreteness of the $\mathbf{k}$-vectors in Eq. (1.5) is achieved by imposing on the crystal the *periodic boundary conditions*

$$f(\mathbf{R} + 2L_i \mathbf{a}_i) = f(\mathbf{R}) \; ; \quad i = 1, 2, 3 \tag{1.7}$$

where L_i are large positive integers. This means that the *crystal volume* is $V = Lv$ with $L = 8L_1 L_2 L_3$. Equations (1.5), (1.7) imply that $\mathbf{k}_i \cdot L_i \mathbf{a}_i / \pi = \nu_i$ are integers, and with Eqs. (1.4), (1.6) one finds that

$$\kappa_i = \frac{\nu_i}{2L_i} \; ; \quad -L_i + 1 \leq \nu_i \leq L_i \; ; \quad i = 1, 2, 3 . \tag{1.8}$$

This shows that L is also the number of $\mathbf{k}$-points contained in Z. Combining these explicit values (1.8) with Eqs. (1.1), (1.4) and (1.6), it is easy to derive the two relations (Problem 1.1)[1,2,6]

$$\frac{1}{L} \sum_{\mathbf{R} \in V} e^{i\mathbf{k}\cdot\mathbf{R}} = \sum_{\mathbf{K}} \delta_{\mathbf{k},\mathbf{K}} \tag{1.9}$$

where $\mathbf{k}$ is not restricted to Z, and [1,2]

$$\frac{1}{L} \sum_{\mathbf{k} \in Z} e^{i\mathbf{k}\cdot\mathbf{R}} = \delta_{\mathbf{R},0} \tag{1.10}$$

where $\mathbf{R}$ is not restricted to V. These relations express, respectively, orthonormality modulo a vector $\mathbf{K}$ and closure in Z of the plane wave fields. Applied to Eq. (1.5), relation (1.9) yields the Fourier amplitude

$$g(\mathbf{k}) = \frac{1}{L} \sum_{\mathbf{R} \in V} f(\mathbf{R}) e^{-i\mathbf{k}\cdot\mathbf{R}} = g(\mathbf{k} + \mathbf{K}) \tag{1.11}$$

which is seen to have the periodicity of the reciprocal lattice.

Free *conduction electrons* or *holes* in a perfect crystal are described by a *one-body Hamiltonian*

$$H^{\mathrm{el}} = \frac{\mathbf{p}^2}{2m} + U(\mathbf{r}) \tag{1.12}$$

with a *periodic potential* $U(\mathbf{r}) = U(\mathbf{r} + \mathbf{R})$, $\mathbf{p}$ being the electron's momentum and m its mass. Thus H^{el} is invariant under the group of translations by $\mathbf{R}$. Since this group is abelian, i.e. all its elements commute with each other, the wavefunctions ψ in the Schrödinger equation $H^{\mathrm{el}}\psi = \varepsilon\psi$ form a one-dimensional unitary representation of this group which may be parametrized by a wavevector $\mathbf{k} \in Z$. Thus the action of the translation by $\mathbf{R}$ is

$$\psi_{\mathbf{k}}(\mathbf{r} + \mathbf{R}) = e^{i\mathbf{k}\cdot\mathbf{R}} \psi_{\mathbf{k}}(\mathbf{r}) . \tag{1.13}$$

From this representation it is easily seen that $e^{-i\mathbf{k}\cdot\mathbf{r}}\psi_{\mathbf{k}}(\mathbf{r})$ is a periodic function so that *Bloch's theorem*[1]

$$\psi_{\mathbf{k}}(\mathbf{r}) = e^{i\mathbf{k}\cdot\mathbf{r}} u_{\mathbf{k}}(\mathbf{r}) ; \quad u_{\mathbf{k}}(\mathbf{r} + \mathbf{R}) = u_{\mathbf{k}}(\mathbf{r}) \tag{1.14}$$

holds.

Insertion of (1.14) into the Schrödinger equation shows that the periodic part $u_{\mathbf{k}}(\mathbf{r})$ satisfies a modified Schrödinger equation $H_{\mathbf{k}}^{\mathrm{el}} u_{\mathbf{k}} = \varepsilon_{\mathbf{k}} u_{\mathbf{k}}$ with $H_{\mathbf{k}}^{\mathrm{el}} = (\mathbf{p} + \hbar\mathbf{k})^2/2m + U(\mathbf{r})$ which may be solved in the unit cell C by imposing the periodic boundary conditions (1.14). Quantum mechanics then tells us that there is a countably infinite set of solutions which may be labeled by a *band index* n. Since the electron has spin $1/2$ there is an additional degree of freedom $\sigma = +, -$, indicating the spin direction up or down and which is represented by the column vectors

$$|+\rangle = \begin{pmatrix} 1 \\ 0 \end{pmatrix} \; ; \; |-\rangle = \begin{pmatrix} 0 \\ 1 \end{pmatrix} . \tag{1.15}$$

Thus the Schrödinger equation may finally be written in the form

$$H^{\text{el}}|k\rangle = \varepsilon_k|k\rangle \tag{1.16}$$

where k is a composite *mode index* such that $\pm k \equiv (n, \pm\mathbf{k}, \sigma)$. The eigenstates are $|k\rangle = \psi_{n\mathbf{k}}|\sigma\rangle$ where $\psi_{n\mathbf{k}}$ are *Bloch functions* of the form (1.14) and the eigenvalues $\varepsilon_k = \varepsilon_{n\mathbf{k}}$ are the *energy bands*. Orthonormality in the crystal volume V,

$$\langle k|k'\rangle = \delta_{kk'} , \tag{1.17}$$

combined with Eqs. (1.9), (1.13) and (1.14) implies the orthonormality in the unit cell C,

$$L \int_C d^3 r \, u_{n\mathbf{k}}^*(\mathbf{r}) u_{n'\mathbf{k}}(\mathbf{r}) = \delta_{nn'} . \tag{1.18}$$

While the Bloch functions are extended over the whole crystal volume V, the *Wannier functions*[1]

$$w_n(\mathbf{r} + \mathbf{R}) = \sum_{\mathbf{k} \in Z} \psi_{n\mathbf{k}}(\mathbf{r} + \mathbf{R}) \tag{1.19}$$

are localized around $\mathbf{r} = -\mathbf{R}$. They may be viewed as fields (1.5) with internal degrees of freedom n and $\mathbf{r} \in C$ (they are not, however, observables in the quantum mechanical sense). We may thus calculate their Fourier amplitude according to (1.11) and find, with use of Eqs. (1.9) and (1.13), that

$$\psi_{n\mathbf{k}}(\mathbf{r}) = \frac{1}{L} \sum_{\mathbf{R} \in V} w_n(\mathbf{r} + \mathbf{R}) e^{-i\mathbf{k}\cdot\mathbf{R}} = \psi_{n\mathbf{k}+\mathbf{K}}(\mathbf{r}) . \tag{1.20}$$

This periodicity with the reciprocal lattice is of course also shared by the energy bands $\varepsilon_{n\mathbf{k}}$.

Since the Hamiltonian (1.12) is a real operator, the Schrödinger equation is *time-reversal invariant* which implies *Kramers' theorem*[7] $u_{n-\mathbf{k}}^*$ $= u_{n\mathbf{k}}$, $\varepsilon_{n-\mathbf{k}} = \varepsilon_{n\mathbf{k}}$. Combining this reflection symmetry of $\varepsilon_{n\mathbf{k}}$ with the periodicity $\varepsilon_{n\mathbf{k}+\mathbf{K}} = \varepsilon_{n\mathbf{k}}$ one easily finds that two bands $n \neq m$ crossing at a border point of the reduced zone Z have symmetric slopes,

$$\frac{\partial \varepsilon_{n\mathbf{k}}}{\partial \mathbf{k}} = -\frac{\partial \varepsilon_{m\mathbf{k}}}{\partial \mathbf{k}} \; ; \; \varepsilon_{n\mathbf{k}} = \varepsilon_{m\mathbf{k}} \; ; \; \mathbf{k} \in \partial Z , \tag{1.21}$$

while for a single band, $n = m$, the slope is zero.

The final remark concerning free electrons is that often relativistic corrections to the Hamiltonian (1.12) cannot be neglected, the most important being the *spin-orbit interaction*[7,8]

$$U_{\text{s.o.}} = \frac{\hbar}{4m^2c^2}(\vec{\sigma} \times \nabla U) \cdot \mathbf{p} \tag{1.22}$$

where $\vec{\sigma} = (\sigma^x, \sigma^y, \sigma^z)$ are the Pauli spin matrices acting on the states (1.15). Since $m\mathbf{v} = \mathbf{p} = \hbar\nabla/i$, $U_{\text{s.o.}}$ is of the order $(v/c)^2U$ (c is the light velocity). The main effect of $U_{\text{s.o.}}$ is to mix band and spin states, so that now $|k\rangle = \sum a_{n\sigma, n'\sigma'}(\mathbf{k})\psi_{n'\mathbf{k}}|\sigma'\rangle$. This gives rise to *double group* representations of the crystal symmetry[7,8].

Phonons in a perfect crystal are described by the displacement vectors $\mathbf{u}_\nu(\mathbf{R})$ of the basis atoms (or ions in the case of a conductor) $\nu = 1, \ldots, B$ from their equilibrium position $\mathbf{R} + \mathbf{r}_\nu$ in the unit cell at $\mathbf{R}$ and by the conjugate momenta $\mathbf{p}_\nu(\mathbf{R}) \equiv M_\nu\dot{\mathbf{u}}_\nu(\mathbf{R})$, M_ν being the mass of the atomic species ν. Thus the cartesian components $u_{\nu i}(\mathbf{R})$ and $p_{\nu i}(\mathbf{R})$ are *fields*[5] with $3B$ internal degrees of freedom. *Free phonons* are defined by the harmonic approximation of the interatomic potential, i.e. by the Hamiltonian

$$\mathcal{H}^{\text{ph}} = \sum_{\mathbf{R}\in V,\nu} \left\{ \frac{\mathbf{p}_\nu^2(\mathbf{R})}{2M_\nu} + \frac{1}{2}\sum_{\mathbf{R}'\in V,\nu'} \mathbf{u}_\nu(\mathbf{R}) \right.$$
$$\left. \times \mathbf{C}_{\nu\nu'}(\mathbf{R}-\mathbf{R}')\mathbf{u}_{\nu'}(\mathbf{R}') \right\} \equiv \sum_{\mathbf{R},\nu} H_{\mathbf{R},\nu}^{\text{ph}} \tag{1.23}$$

where the $\mathbf{C}_{\nu,\nu'}$ are real 3×3 *force-constant matrices* with the crystal periodicity (1.7) and $H_{\mathbf{R},\nu}^{\text{ph}}$ is the one-body Hamiltonian of the atom with equilibrium position $\mathbf{R} + \mathbf{r}_\nu$. The Hamiltonian (1.23) is diagonalized by transforming to *normal-mode amplitudes* or *phonon coordinates* Q_q, P_q according to

$$\mathbf{u}_\nu(\mathbf{R}) = \hbar\sum_q (LBM_\nu\omega_q)^{-1/2}\mathbf{e}_q(\nu)e^{i\mathbf{q}\cdot\mathbf{R}}Q_q$$
$$\mathbf{p}_\nu(\mathbf{R}) = \sum_q (M_\nu\omega_q/LB)^{1/2}\mathbf{e}_q(\nu)e^{i\mathbf{q}\cdot\mathbf{R}}P_{-q} \tag{1.24}$$

where $\mathbf{q}$ is a crystal momentum, $\mathbf{q} \in Z$. Here q is a composite *mode index* such that $\pm q = (\mu, \pm\mathbf{q}, j)$. $\mu = 1, \ldots, B$ is the *branch index*, $j = 1, 2, 3$ the *polarization index*, $\omega_q/\hbar$ the (real) *phonon frequency* and $\mathbf{e}_q(\nu)$ is a (complex) *polarization vector*. In Eq. (1.24) reality of $\mathbf{u}_\nu$ and $\mathbf{p}_\nu$ implies

$$Q_{-q} = Q_q^*, \; P_{-q} = P_q^*, \; \omega_{-q} = \omega_q, \; \mathbf{e}_{-q} = \mathbf{e}_q^* . \tag{1.25}$$

The cartesian components $e_q(\nu, i)$ of the polarization vectors $\mathbf{e}_q(\nu)$ are eigenfunctions of the *dynamical matrix*[1,2,5]

$$\mathbf{D}_{\nu\nu'}(\mathbf{q}) = \hbar^2 \sum_{\mathbf{R} \in V} (M_\nu M_{\nu'})^{-1/2} \mathbf{C}_{\nu\nu'}(\mathbf{R}) e^{-i\mathbf{q} \cdot \mathbf{R}} \tag{1.26}$$

with eigenvalues ω_q^2,

$$\sum_{\nu', i'} D_{\nu i, \nu' i'}(\mathbf{q}) e_q(\nu', i') = \omega_q^2 e_q(\nu, i) . \tag{1.27}$$

This *eigenvalue equation* for the phonons is the analog of the Schrödinger equation (1.16) for the electrons. However, the eigenfunctions $e_q(\nu, i)$ are elements of a $3B$-dimensional vector space while $\psi_{n\mathbf{k}}(\mathbf{r})$ are elements of a *Hilbert space*. This analogy also extends to the orthonormality relation (1.18) which in the phonon case takes the form

$$\frac{1}{B} \sum_\nu \mathbf{e}^*_{\mu \mathbf{q} j}(\nu) \cdot \mathbf{e}_{\mu' \mathbf{q} j'}(\nu) = \delta_{\mu\mu'} \delta_{jj'} . \tag{1.28}$$

Using Eqs. (1.24)–(1.28) and crystal momentum conservation (1.9), one shows that, after some algebra, the phonon Hamiltonian (1.23) simplifies to a sum of harmonic oscillator Hamiltonians (Problem 1.2),

$$\mathcal{H}^{\mathrm{ph}} = \frac{1}{2} \sum_q \omega_q (P_q^* P_q + Q_q^* Q_q) . \tag{1.29}$$

Since P_q and Q_q are canonically conjugate variables (which have been defined dimensionless), the quantization condition is

$$i[P_q, Q_{q'}] = \delta_{qq'} \tag{1.30}$$

where the bracket means the commutator $[A, B] \equiv AB - BA$.

Equation (1.30) is, in fact, a *mode quantization* of the fields $\mathbf{p}_\nu(\mathbf{R})$, $\mathbf{u}_\nu(\mathbf{R})$ which therefore become *observables* in the quantum mechanical sense. Introducing *creation and annihilation operators* b_q^+ and b_q by the prescription

$$Q_q = (b_q + b_{-q}^+)/\sqrt{2}, \quad P_{-q} = -i(b_q - b_{-q}^+)/\sqrt{2} \tag{1.31}$$

which, together with (1.30), implies the usual commutation relations

$$[b_q, b_{q'}^+] = \delta_{qq'} , \tag{1.32}$$

all other commutators being zero, the *phonon number*

$$N_q = b_q^+ b_q \tag{1.33}$$

also becomes an observable. With Eqs. (1.31), (1.33) the Hamiltonian (1.29) goes over into the quantized form

$$\mathcal{H}^{\mathrm{ph}} = \sum_q \omega_q (N_q + \frac{1}{2}) \tag{1.34}$$

where the term $\omega_q/2$ represents the *zero-point energy* of mode q. This *canonical field quantization* was first introduced by Dirac to quantize electrodynamics[9]. The case of phonons is seen to be completely analogous.

2. Many-body Description of One- and Two-body Forces

One-body forces are produced by an *external field* $F(\mathbf{r}, t)$ which may depend on time and which acts on an observable O_j of particle $j = 1, \ldots, N$ (electron or ion) with position $\mathbf{r}_j$ and momentum $\mathbf{p}_j$. These forces are mediated by an N-body interaction energy

$$\delta \mathcal{H}_t = \sum_{j=1}^{N} \frac{1}{2} \{F(\mathbf{r}_j, t), O_j\} = \int_V d^3r F(\mathbf{r}, t)\tilde{d}(\mathbf{r}) = \frac{1}{V} \sum_q \tilde{F}(\mathbf{q}, t)d(-\mathbf{q}) \ . \tag{2.1}$$

Here the anticommutator $\{A, B\} \equiv AB + BA$ takes care of the symmetrization which is necessary in case O_j does not commute with $\mathbf{r}_j$ (see the examples below), and

$$\tilde{d}(\mathbf{r}) = \sum_{j=1}^{N} \frac{1}{2} \{\delta(\mathbf{r} - \mathbf{r}_j), O_j\} \tag{2.2}$$

is the *density* associated to the observable O_j[5]. In the last expression (2.1) the Fourier-transformed density

$$d(\mathbf{q}) = \int_V d^3r \tilde{d}(\mathbf{r})e^{-i\mathbf{q}\cdot\mathbf{r}} = \sum_{j=1}^{N} \frac{1}{2} \{e^{-i\mathbf{q}\cdot\mathbf{r}_j}, O_j\} \tag{2.3}$$

is introduced, and the external field is assumed to have the inverse Fourier representation

$$F(\mathbf{r}, t) = \frac{1}{V} \sum_q \tilde{F}(\mathbf{q}, t)e^{i\mathbf{q}\cdot\mathbf{r}} \ . \tag{2.4}$$

It should be noted that because of the continuity of $\mathbf{r}$, the discrete $\mathbf{q}$-vectors are here defined by Eqs. (1.6), (1.8) with *unrestricted* values of the integers ν_i. As a consequence, the periodic boundary conditions (1.7) are generalized to

$$F(\mathbf{r} + 2L_i\mathbf{a}_i) = F(\mathbf{r}) \ ; \quad i = 1, 2, 3 \tag{2.5}$$

and Eqs. (1.9), (1.10) are replaced, respectively, by

$$\frac{1}{V} \int_V d^3r e^{i\mathbf{q}\cdot\mathbf{r}} = \delta_{\mathbf{q},o} \tag{2.6}$$

and

$$\frac{1}{V} \sum_{\mathbf{q}} e^{i\mathbf{q}\cdot\mathbf{r}} = \delta(\mathbf{r}) \tag{2.7}$$

where the last expression is the Dirac function.

The most important examples of electronic densities are obtained by inserting into Eq. (2.2) the observables $O_j = 1$, p_j/m, $\sigma^j/2$ where $\mathbf{p}$ is the electron momentum and $\vec{\sigma}$ are the Pauli spin matrices introduced in Eq. (1.22),

$$\sigma^+ \equiv \frac{1}{2}(\sigma^x + i\sigma^y) = \begin{pmatrix} 0 & 1 \\ 0 & 0 \end{pmatrix} ; \quad \sigma^- \equiv \frac{1}{2}(\sigma^x - i\sigma^y) = \begin{pmatrix} 0 & 0 \\ 1 & 0 \end{pmatrix} ;$$

$$\sigma^z = \begin{pmatrix} 1 & 0 \\ 0 & -1 \end{pmatrix} . \tag{2.8}$$

The corresponding expressions (2.3) and fields $F(\mathbf{r}, t)$ are[1]: the number density

$$n(\mathbf{q}) = \sum_j e^{-i\mathbf{q}\cdot\mathbf{r}_j} \tag{2.9}$$

which couples to an electric potential $eU(\mathbf{r}, t)$ (e is the elementary charge), the current density

$$\mathbf{j}_n(\mathbf{q}) = \sum_j \frac{1}{2m}\{e^{-i\mathbf{q}\cdot\mathbf{r}_j}, \mathbf{p}_j\} \tag{2.10}$$

which couples to a vector potential $(e/c)\mathbf{A}(\mathbf{r}, t)$, and the spin density

$$\mathbf{s}(\mathbf{q}) = \sum_j \frac{1}{2}\vec{\sigma}_j e^{-i\mathbf{q}\cdot\mathbf{r}_j} \tag{2.11}$$

which couples to a magnetic field $g\mu_B \mathbf{H}(\mathbf{r}, t)$, ($g \cong 2$ is the spectroscopic splitting factor and $\mu_B = e\hbar/2mc$ the Bohr magneton).

This description by densities $\tilde{n}(\mathbf{r})$ and $\tilde{\mathbf{j}}_n(\mathbf{r})$ leads naturally to the notion of an *electron fluid* described by hydrodynamic equations (see Section 13). The surprising fact which recently emerged from an analysis of size effects in ohmic conduction is that the electron fluid in metals like copper has a viscosity comparable to water (for copper $\eta(300K) = 0.6cP)$[10].

In the case of the atoms or ions forming a perfect crystal the $\mathbf{r}_j$ are the positions $\mathbf{R}+\mathbf{r}_\nu +\mathbf{u}_\nu(\mathbf{R})$, $\nu = 1, \ldots, B$. The number and current densities of the atomic species ν then are

$$n_\nu(\mathbf{q}) = \sum_{\mathbf{R}} e^{-i\mathbf{q}\cdot(\mathbf{R}+\mathbf{r}_\nu+\mathbf{u}_\nu(\mathbf{R}))} \tag{2.12}$$

and

$$\mathbf{j}_\nu(\mathbf{q}) = \sum_{\mathbf{R}} \frac{1}{2M_\nu} \{ e^{-i\mathbf{q}\cdot(\mathbf{R}+\mathbf{r}_\nu+\mathbf{u}_\nu(\mathbf{R}))}, \mathbf{p}_\nu(\mathbf{R}) \} . \tag{2.13}$$

$n_\nu(\mathbf{q})$ may be measured by X-ray or by elastic neutron scattering (see Section 15). On the other hand, $\sum_\nu M_\nu \mathbf{j}_\nu(0) = \sum_{\mathbf{R},\nu} \mathbf{p}_\nu(\mathbf{R}) = L \sum_\nu M_\nu \mathbf{c}$ where $\mathbf{c}$ is the center-of-mass velocity of the crystal. Because of the periodic boundary conditions (2.5), $\mathbf{c} = 0$ which means that with these boundary conditions recoils of the crystal as a whole are not described[5].

By taking for O_j the one-body Hamiltonians H_j^{el} and $H_{\mathbf{R},\nu}^{\mathrm{ph}}$ of Eqs. (1.12) and (1.23), respectively, one obtains the energy densities of electrons and phonons,

$$h^{\mathrm{el}}(\mathbf{q}) = \sum_j \frac{1}{2} \{ e^{-i\mathbf{q}\cdot\mathbf{r}_j}, H_j^{\mathrm{el}} \} \tag{2.14}$$

and[5]

$$h^{\mathrm{ph}}(\mathbf{q}) = \sum_{\mathbf{R},\nu} \frac{1}{2} \{ e^{-i\mathbf{q}\cdot(\mathbf{R}+\mathbf{r}_\nu+\mathbf{u}_\nu(\mathbf{R}))}, H_{\mathbf{R},\nu}^{\mathrm{ph}} \} , \tag{2.15}$$

respectively. The limit $\mathbf{q} = 0$ in (2.14) and (2.15) produces the many-body Hamiltonians of free electrons and free phonons,

$$\mathcal{H}^{\mathrm{el}} = h^{\mathrm{el}}(0) = \sum_j H_j^{\mathrm{el}} \tag{2.16}$$

and

$$\mathcal{H}^{\mathrm{ph}} = h^{\mathrm{ph}}(0) = \sum_{\mathbf{R},\nu} H_{\mathbf{R},\nu}^{\mathrm{ph}} . \tag{2.17}$$

The last relation is nothing else than Eq. (1.23) which is thus recognized as already being a many-body Hamiltonian with respect to the atoms. This is so because (1.23) is a function of the *phonon fields* $\mathbf{p}_\nu(\mathbf{R})$ and $\mathbf{u}_\nu(\mathbf{R})$. Since, on the other hand the *electron fields*, namely the Wannier functions (1.19), are not observables, the many-body Hamiltonian of the free electrons must be built up from the observables H_j^{el} of individual *particles* (electrons).

Two-body forces among electrons and ions are, in essence, shielded Coulomb forces in *Born-Oppenheimer approximation* [6]. This is an adiabatic approximation in which the electrons follow instantaneously the ionic motion, which is justified since $v_{\mathrm{ion}}/v_{\mathrm{el}} \sim \sqrt{m/M_\nu} \sim 0.01$. The two-body forces among the conduction electrons are described by the two-body interaction Hamiltonian

$$\mathcal{H}^{\text{el-el}} = \frac{1}{2} \sum_{j,l=1}^{N} (j \neq l)\phi(\mathbf{r}_j - \mathbf{r}_l) \tag{2.18}$$

where $\phi(\mathbf{r})$ is a shielded Coulomb potential. Similarly, the two-body interactions between electrons and ions are described by a shielded Coulomb potential which, however, depends on the atomic species ν,

$$\mathcal{H}^{\text{el-ion}} = \sum_{\mathbf{R} \in V, \nu, j} \{\phi_\nu(\mathbf{R} + \mathbf{r}_\nu + \mathbf{u}_\nu(\mathbf{R}) - \mathbf{r}_j) - \phi_\nu(\mathbf{R} + \mathbf{r}_\nu - \mathbf{r}_j)\} . \tag{2.19}$$

Developing this expression in powers of the displacements $\mathbf{u}_\nu$ and retaining only the first term, one obtains the usual *electron-phonon interaction*[2,6,7]. Making use of Eqs. (1.9), (1.24) and (2.9) it may be cast into the form (Problem 1.3)[6]

$$\mathcal{H}^{\text{el-ph}} = \sum_{\mathbf{K}} \sum_{\mathbf{q}'} \sum_{q} \delta_{\mathbf{q}+\mathbf{q}',\mathbf{K}}\gamma_q(\mathbf{q}')n(\mathbf{q}')Q_q \tag{2.20}$$

where

$$\gamma_q(\mathbf{q}') = \frac{i\hbar}{v} \sum_\nu (LBM_\nu\omega_q)^{-1/2}\mathbf{q}' \cdot \mathbf{e}_q(\nu)\tilde{\phi}_\nu(\mathbf{q}')e^{i\mathbf{q}'\cdot\mathbf{r}_\nu} = \gamma_{-q}^*(-\mathbf{q}') \tag{2.21}$$

is the coupling function and $\tilde{\phi}_\nu(\mathbf{q})$ is the Fourier transform of the potential $\phi_\nu(\mathbf{r})$ defined as in Eq. (2.4). In Eq. (2.20) $\mathbf{q} \in Z$ but $\mathbf{q}'$ is unrestricted so that *momentum conservation* expressed by the Kronecker-δ may require a $\mathbf{K} \neq 0$. Such a momentum transfer to the lattice is called an *Umklapp process*[1,2,6,7] (German for flip over). This non-rigorous conservation modulo a $\mathbf{K}$ reflects the discreteness of the translation group of a perfect crystal which is a *broken symmetry* as compared to the continuous translation group of a fluid.

Two-body elastic forces among the atoms (or ions) are contained in the potential energy term of the Hamiltonian (1.23). Terms of higher order in the displacements $\mathbf{u}_\nu$ may also play a role, in particular in insulators. Such *anharmonic terms* involving 3 or more phonons give rise to phonon-phonon interactions[2,5,6,7]

$$\mathcal{H}_n^{\text{ph-ph}} = \frac{1}{n!} \sum_{\mathbf{K}} \sum_{q_1 \ldots q_n} \delta_{\mathbf{q}_1 + \ldots + \mathbf{q}_n, \mathbf{K}} C_{q_1 \ldots q_n}^{(n)} Q_{q_1} \ldots Q_{q_n} ; \quad n \geq 3 . \tag{2.22}$$

Here $\mathbf{q}_1 \ldots \mathbf{q}_n$ are crystal momenta and, as in Eq. (2.20), the Kronecker-δ expresses momentum conservation modulo a $\mathbf{K}$. Note that Umklapp processes $\mathbf{K} \neq 0$ are only possible for $n \geq 3$; i.e. for interaction terms.

Anharmonic effects play a role in the thermal properties of dielectric crystals such as thermal expansion, heat conduction or second sound[1,5]. They are particularly important for light atoms for which M_ν is of the order of the proton mass, since in this case the displacements u_ν are comparable to the lattice spacings, even at low temperatures. This is the case of *quantum solids*[1] where tunneling of the atoms to vacant sites takes place.

In the case of the electron-ion interaction (2.19), higher order terms in the displacements u_ν have also been considered in the literature. Such *multi-phonon processes*[2,11] were once invoked to explain some temperature anomaly in the carrier mobility of semiconductors (see Section 16); they certainly play a role in the optical spectra of semiconductors[7]. Very recently they have been invoked in the problem of high-temperature superconductivity (see Chapter 4) where anharmonicity (e.g. double-well potentials) play an important role (see, e.g. Ref. 12).

3. Second Quantization

The objects of many-body theory, as of field theory, are processes in which excitations of the system appear or disappear. This means that the *number of excitations* of a given mode is an *observable* in the quantum mechanical sense, and the changes are effected by *creation and annihilation operators*. For phonons these operators followed naturally from canonical field quantization in Section 1 which was trivial once the normal-mode form (1.29) of the atomic many-body Hamiltonian (1.23) was established.

In trying to apply a similar procedure to electrons one encounters the following dilemma: Should one pursue the analogy with the phonons or the one with the atoms? The first is based on the correspondence between the eigenvalue equations (1.16) and (1.27), the second on the correspondence between the many-body Hamiltonians (2.16) and (2.17). The first analogy proceeds by decomposition of the electron's wavefunction ψ into normal modes $|k\rangle$ and subsequent mode quantization. This is the method of canonical field quantization. The second analogy proceeds by construction of creation and annihilation operators for excitations defined by any basis of one-electron states. This method is called *second quantization*[4,13] since the one-electron states are already solutions of a Schrödinger equation. The two methods are of course equivalent[4], at least in a non-relativistic framework, since in the method of field quantization it is also possible to transform from normal modes to any other basis.

The idea of second quantization goes back to Jordan and Klein for *bosons* and to Jordan and Wigner for *fermions* (see Ref. 4). Since there is no extra work involved, we introduce second quantization both for fermions and for bosons although we shall use it only for electrons, which are fermions. The common feature shared by bosonic and fermionic particles is their quantum nature of being identical and *indistinguishable* within each species which means that permutations among them have no observable effect. The difference between bosons and fermions lies in the *fundamental property* that *many-boson states* are *even* and *many-fermion states* are *odd* under the exchange of two particles[4].

If $f_1(i), \cdots, f_N(i)$ represent any one-particle states of the particle labeled $i = 1, \cdots, N$, then indistinguishability requires the N-particle state built up from products of these $f_j(i)$ to be completely symmetrized with respect to the particle label i. Taking into account the fundamental property of many-boson/fermion states, the N-particle state takes the form[4]

$$|f_1 \cdots f_N\rangle = S_{\pm}|f_1(1) \otimes \cdots \otimes f_N(N)\rangle \qquad (3.1)$$

where $f_1(1) \otimes \cdots \otimes f_N(N)$ is the tensor product,

$$S_{\pm} = \frac{1}{N!} \sum_P (\pm)^{\eta_P} P \qquad (3.2)$$

is the *symmetrizing operator* which selects the totally symmetric/antisymmetric combination of the $N!$ permutations $P = (1 \cdots N \rightarrow i_1 \cdots i_N)$ of the particle labels, i.e.,

$$P|f_1(1) \otimes \cdots \otimes f_N(N)\rangle = |f_1(i_1) \otimes \cdots \otimes f_N(i_N)\rangle \qquad (3.3)$$

and η_P is the number of transpositions necessary to generate P. It is easy to see that $S_{\pm}$ is a projector, $S_{\pm}^2 = S_{\pm}$, and that it is hermitean, $S_{\pm}^+ = S_{\pm}$ (Problem 1.4). In addition, the states (3.1) satisfy the fundamental property

$$P|f_1 \cdots f_N\rangle = (\pm)^{\eta_P}|f_1 \cdots f_N\rangle . \qquad (3.4)$$

It is also evident that the antisymmetric states are just *Slater determinants*. It may also be shown that the scalar product between two N-boson/fermion states is given by (Problem 1.5)

$$\langle g_1 \cdots g_N|f_1 \cdots f_N\rangle = \frac{1}{N!} \sum_P (\pm)^{\eta_P} \langle g_1|f_{i_1}\rangle \cdots \langle g_N|f_{i_N}\rangle . \qquad (3.5)$$

We now define the creation operator $a^+(f_o)$ which adds a particle in state f_o to the state (3.1), by the $(N+1)$-particle state

$$a^+(f_o)|f_1 \cdots f_N\rangle = \sqrt{N+1}|f_o f_1 \cdots f_N\rangle . \qquad (3.6)$$

Defining the annihilation operator $a(f_o)$ as hermitian conjugate of $a^+(f_o)$, its action may be calculated from the general matrix element $\langle f_1 \cdots f_N | a(f_o) | g_o g_1 \cdots g_N \rangle$ using (3.6) and hermiticity. The result for boson/fermion states is (Problem 1.6)

$$a(f_o)|g_o g_1 \cdots g_N\rangle = \frac{1}{\sqrt{N+1}} \sum_{j=0}^{N} (\pm)^j \langle f_o|g_j\rangle |g_o g_1 \cdots g_{j-1} g_{j+1} \cdots g_N\rangle .$$

$$(3.7)$$

It follows from (3.6) and (3.7) that a^+ is linear and a is antilinear in its argument,

$$a^+(\lambda f + \mu g) = \lambda a^+(f) + \mu a^+(g) ,$$
$$a(\lambda f + \mu g) = \lambda^* a(f) + \mu^* a(g) ,$$

$$(3.8)$$

and also that

$$a^+(g)a(f)|g_1 \cdots g_N\rangle = \sum_{j=1}^{N} \langle f|g_j\rangle |g_1 \cdots g_{j-1} g_{j+1} \cdots g_N\rangle .$$

$$(3.9)$$

The most important properties of these operators, however, are the commutation/anticommutation relations in the case of bosons/fermions (Problem 1.7),

$$[a(f), a^+(g)]_{\mp} = \langle f|g\rangle ,$$
$$[a(f), a(g)]_{\mp} = [a^+(f), a^+(g)]_{\mp} = 0 .$$

$$(3.10)$$

Here we introduced the notation $[A, B]_{\mp} \equiv AB \mp BA$ for the commutator $[\,]$ and for the anticommutator $\{\,\}$, respectively.

The definitions (3.6) and (3.7) are still too general to be useful. The crucial step is to define the creation and annihilation operators for a complete system of orthonormal one-particle states $|\varphi_n\rangle$ satisfying

$$\langle \varphi_n|\varphi_m\rangle = \delta_{nm}, \quad \sum_n |\varphi_n\rangle\langle\varphi_n| = 1 \qquad (3.11)$$

by

$$a_n^+ \equiv a^+(\varphi_n), \quad a_n \equiv a(\varphi_n) . \qquad (3.12)$$

The relations (3.10) then become those of field quantization[4,13)],

$$[a_n, a_m^+]_{\mp} = \delta_{nm} ,$$
$$[a_n, a_m]_{\mp} = [a_n^+, a_m^+]_{\mp} = 0 .$$

$$(3.13)$$

Here the last relation implies, in particular that for fermions $(a_n^+)^2 = 0$. This expresses the *Pauli exclusion principle* which says that no one-fermion state may be occupied by more than one particle[4]. Writing the ν_n-fold occupation of a one-particle state $|\varphi_n\rangle$ as

$$\underbrace{|\varphi_n \cdots \varphi_n\rangle}_{\nu_n \text{ times}} = |\nu_n\rangle \tag{3.14}$$

where for fermions $\nu_n = 0$ or 1, one reads from Eqs. (3.6) and (3.7) that the usual relations[4]

$$\begin{aligned}
a_n^+|\nu_n\rangle &= \sqrt{\nu_n + 1}|\nu_n + 1\rangle \ , \\
a_n|\nu_n\rangle &= \sqrt{\nu_n}|\nu_n - 1\rangle
\end{aligned} \tag{3.15}$$

hold. And from (3.9) it is easily seen that

$$N_n = a_n^+ a_n \tag{3.16}$$

is the *number operator* for which (3.14) are the eigenstates with eigenvalues ν_n.

The closure relation in (3.11) implies for any *one-particle operator* O_1 and any one-particle state $f(1)$ that

$$O_1|f\rangle = \sum_{nm} |\varphi_n\rangle O_{nm}\langle\varphi_m|f\rangle \tag{3.17}$$

where

$$O_{nm} \equiv \langle\varphi_n|O_1|\varphi_m\rangle \ . \tag{3.18}$$

Defining now the *second-quantized one-particle operator* $O^{(1)}$ by

$$O^{(1)} \equiv \sum_{nm} O_{nm} a_n^+ a_m \ , \tag{3.19}$$

it is readily shown with the help of Eqs. (3.9) and (3.17) that for any N-particle state (Problem 1.8)

$$O^{(1)}|f_1 \cdots f_N\rangle = \sum_{j=1}^{N} O_j|f_1 \cdots f_N\rangle \tag{3.20}$$

where O_j acts only on the state $f(j)$. Thus second quantization of a one-particle operator O_1 is the prescription $\sum_{j=1}^{N} O_j \rightarrow O^{(1)}$ contained in Eqs. (3.18), (3.19).

The one-electron operators of Section 2 are easily calculated in a *plane-wave basis* within the crystal volume V and with spin functions (1.15),

$$|\mathbf{k}\sigma\rangle = V^{-1/2} e^{i\mathbf{k}\cdot\mathbf{r}}|\sigma\rangle \ . \tag{3.21}$$

Making use of relation (2.6) the result for the number , current and spin densities (2.9), (2.10) and (2.11) is, respectively

$$n(\mathbf{q}) = \sum_{\mathbf{k},\sigma} a^+_{\mathbf{k}-\mathbf{q}/2,\sigma} a_{\mathbf{k}+\mathbf{q}/2,\sigma} \; , \tag{3.22}$$

$$\mathbf{j}_n(\mathbf{q}) = \frac{\hbar}{m} \sum_{\mathbf{k},\sigma} \mathbf{k} a^+_{\mathbf{k}-\mathbf{q}/2,\sigma} a_{\mathbf{k}+\mathbf{q}/2,\sigma} \tag{3.23}$$

and

$$s_{\pm}(\mathbf{q}) = \frac{1}{2} \sum_{\mathbf{k}} a^+_{\mathbf{k}-\mathbf{q}/2,\pm} a_{\mathbf{k}+\mathbf{q}/2,\mp} \; , \tag{3.24}$$

$$s_z(\mathbf{q}) = \frac{1}{2} \sum_{\mathbf{k},\sigma} \sigma a^+_{\mathbf{k}-\mathbf{q}/2,\sigma} a_{\mathbf{k}+\mathbf{q}/2,\sigma} \; . \tag{3.25}$$

In the last two equations use was made of the definitions (2.8) and of the relation

$$\langle \tau' | \vec{\sigma} | \tau \rangle = \vec{\sigma}_{\tau',\tau} \tag{3.26}$$

which follows from (1.15). As an application, the second-quantized electron-phonon interaction $\mathcal{H}^{\text{el-ph}}$ is obtained by insertion of the density (3.22) into Eq. (2.20). It is represented, for $\mathbf{K} = 0$, by the *electron-phonon vertex* in Fig. 1.1 where an emerging, converging and wavy line represents, respectively, an emitted electron, $a^+_{\mathbf{k}+\mathbf{q}/2,\sigma}$, an absorbed electron, $a_{\mathbf{k}-\mathbf{q}/2,\sigma}$, and a phonon, $Q_{\mu\mathbf{q}j}$.

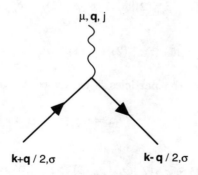

Fig. 1.1: The electron-phonon vertex representing Eqs. (2.20), (3.22) with $\mathbf{K} = 0$.

The plane-wave basis (3.21) is inconvenient for the second-quantized form of the free-electron Hamiltonian (2.16), for which the natural basis are of course the eigenmodes defined by Eq. (1.16) so that

$$\mathcal{H}_o^{el} = \sum_k \varepsilon_k a_k^+ a_k . \tag{3.27}$$

However, in this physical basis the densities (3.22)–(3.25) become more complicated. The number density, for instance, takes the form

$$n(\mathbf{q}) = \sum_{\mathbf{K}} \sum_{kk'} \delta_{\mathbf{k}-\mathbf{k}-\mathbf{q},\mathbf{K}} n_{kk'}(\mathbf{K}) a_k^+ a_{k'} . \tag{3.28}$$

Here use was made of Eqs. (1.9) and (1.14), and the matrix element (3.18) is

$$n_{kk'}(\mathbf{K}) = L \int_C d^3 r u_{n\mathbf{k}}^*(\mathbf{r}) e^{i\mathbf{K}\cdot\mathbf{r}} u_{n'\mathbf{k}'}(\mathbf{r}) \delta_{\sigma\sigma'} . \tag{3.29}$$

Fortunately, the plane-wave basis (3.21) is often a sufficiently accurate representation of the eigenmodes, provided that no interband transitions and no Umklapp processes are involved.

Second quantization of *two-particle operators* O_{12} proceeds along the line described above for one-particle operators. The problem is complicated by the fact that the labels 1 and 2 in O_{12} designate operations on *distinguished* one-particle states $f(1)$, $g(2)$ in the non-symmetrized product space $|f(1) \otimes g(2)\rangle$, see Eq. (3.1). Now it is easy to show that the one-particle basis (3.11) also furnishes a basis $|\varphi_n(1) \otimes \varphi_{n'}(2)\rangle$ in this product space with orthonormality

$$\langle \varphi_m(1) \otimes \varphi_{m'}(2) | \varphi_n(1) \otimes \varphi_{n'}(2)\rangle = \delta_{mn}\delta_{m'n'} \tag{3.30}$$

and closure

$$\sum_{nn'} |\varphi_n(1) \otimes \varphi_{n'}(2)\rangle\langle \varphi_n(1) \otimes \varphi_{n'}(2)| = 1 . \tag{3.31}$$

Indistinguishability of the particles is expressed by the symmetry in the particle labels,

$$O_{12} = O_{21} \tag{3.32}$$

which implies that

$$O_{nn',mm'} \equiv \langle \varphi_n(1) \otimes \varphi_{n'}(2) | O_{12} | \varphi_m(1) \otimes \varphi_{m'}(2)\rangle = O_{n'n,m'm} . \tag{3.33}$$

In analogy to Eq. (3.17) we now want to express $O_{12}|fg\rangle$ by making use of the closure (3.31). Here $|fg\rangle$ is a symmetrized two-particle state (3.1), that is $|fg\rangle = (1/2)(|f(1)\otimes g(2)\rangle \pm |g(1)\otimes f(2)\rangle)$, and $\langle \varphi_m(1)\otimes\varphi_{m'}(2)|fg\rangle = \langle \varphi_m \varphi_{m'}|fg\rangle$. Therefore, and because of the symmetry (3.33), application of

(3.31) yields

$$O_{12}|fg\rangle = \sum_{nn'mm'} |\varphi_n\varphi_{n'}\rangle O_{nn',mm'} \langle\varphi_m\varphi_{m'}|fg\rangle \qquad (3.34)$$

where all the states, except those in the definition (3.33), are now symmetrized. Defining the *second-quantized two-particle operator* $O^{(2)}$ by

$$O^{(2)} \equiv \frac{1}{2} \sum_{nn'mm'} O_{nn',mm'} a_n^+ a_{n'}^+ a_{m'} a_m , \qquad (3.35)$$

a rather lengthy calculation making use of Eqs. (3.4), (3.9), (3.13), (3.33) and (3.34) yields, for any N-particle state, the connection (Problem 1.9)

$$O^{(2)}|f_1 \cdots f_N\rangle = \frac{1}{2} \sum_{j,l=1}^{N} (j \neq l) O_{jl}|f_1 \cdots f_N\rangle \qquad (3.36)$$

where O_{jl} acts with label j on f_j and with label l on f_l only. Thus second quantization of a two-particle operator O_{12} is the prescription $\frac{1}{2}\sum_{j,l=1}^{N}(j \neq l)O_{jl} \to O^{(2)}$ contained in Eqs. (3.33), (3.35).

Let us apply the result (3.35) to the two-electron interaction (2.18) in the plane-wave representation (3.21). The first step is to calculate the matrix element (3.33),

$$\phi_{\mathbf{k}\sigma\mathbf{k}'\sigma',\mathbf{p}\tau\mathbf{p}'\tau'} = \frac{1}{V^2} \int_V d^3r_1 \int_V d^3r_2 e^{-i\mathbf{k}\cdot\mathbf{r}_1} e^{-i\mathbf{k}'\cdot\mathbf{r}_2} \phi(\mathbf{r}_1 - \mathbf{r}_2)$$

$$\times e^{i\mathbf{p}\cdot\mathbf{r}_1} e^{i\mathbf{p}'\cdot\mathbf{r}_2} \langle\sigma|\tau\rangle\langle\sigma'|\tau'\rangle = \frac{1}{V}\tilde{\phi}(\mathbf{k} - \mathbf{p})\delta_{\mathbf{k}+\mathbf{k}',\mathbf{p}+\mathbf{p}'}\delta_{\sigma\tau}\delta_{\sigma'\tau'} . \qquad (3.37)$$

Here the double integral over $V \otimes V$ is best performed in the center of mass and relative coordinates $\mathbf{x} = (\mathbf{r}_1 + \mathbf{r}_2)/2$ and $\mathbf{y} = \mathbf{r}_1 - \mathbf{r}_2$. As in Problem 1.2 (see Fig. 1.4 below) it is important to note that, due to the periodicity (2.5), the domain of $\mathbf{x}\otimes\mathbf{y}$ may again be transformed into $V\otimes V$. Hence Eqs. (2.6) and (2.3) apply in all rigour to the $\mathbf{x}$ and $\mathbf{y}$ integrations, respectively, $\tilde{\phi}$ being the Fourier transform of ϕ. Insertion of (3.37) into (3.35) yields the second-quantized form of the interaction (2.18),

$$\mathcal{H}^{el-el} = \frac{1}{2V} \sum_{\mathbf{k}\sigma\mathbf{k}'\sigma'\mathbf{q}} \tilde{\phi}(\mathbf{q}) a_{\mathbf{k}\sigma}^+ a_{\mathbf{k}'\sigma'}^+ a_{\mathbf{k}'+\mathbf{q},\sigma'} a_{\mathbf{k}-\mathbf{q},\sigma} . \qquad (3.38)$$

It is represented by the *electron-electron vertex* in Fig. 1.2 where emerging and converging lines represent, respectively, creation and annihilation operators.

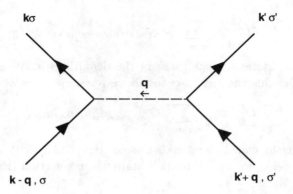

Fig. 1.2: The electron-electron vertex representing Eq. (3.38).

For densities $\tilde{d}(\mathbf{r})$ a property of particular importance is their *translation covariance*

$$\tilde{d}(\mathbf{r}) = e^{i\mathbf{P}\cdot\mathbf{r}}\tilde{d}(0)e^{-i\mathbf{P}\cdot\mathbf{r}} \tag{3.39}$$

where $\hbar\mathbf{P}$ is the total momentum operator. By differentiation of (3.39) one obtains the infinitesimal form which after Fourier transformation (2.3) reads

$$[\mathbf{P}, d(\mathbf{q})]_- = \mathbf{q}\, d(\mathbf{q}) . \tag{3.40}$$

For electrons one easily finds, using Eqs. (2.10) and (3.23), that the total momentum is

$$\hbar\mathbf{P}_{\mathrm{el}} = \sum_j \mathbf{P}_j = m\mathbf{j}(0) = \hbar \sum_k \mathbf{k} a_k^+ a_k . \tag{3.41}$$

In order to verify the covariance (3.40) for one- and two-particle densities we calculate successively, making use of the anticommutation relations (3.13) and of the identities

$$[AB, C]_- = A[B, C]_\mp \pm [A, C]_\mp B , \tag{3.42}$$

the following commutators:

$$[\mathbf{P}_{\mathrm{el}}, a_k]_- = -\mathbf{k} a_k, \quad [\mathbf{P}_{\mathrm{el}}, a_k^+]_- = +\mathbf{k} a_k^+ \tag{3.43}$$

$$[\mathbf{P}_{\mathrm{el}}, a_{\mathbf{k}\sigma}^+ a_{\mathbf{k}-\mathbf{q},\tau}]_- = \mathbf{q} a_{\mathbf{k}\sigma}^+ a_{\mathbf{k}-\mathbf{q},\tau} \tag{3.44}$$

$$[\mathbf{P}_{\mathrm{el}}, a_{\mathbf{k}\sigma}^+ a_{\mathbf{k}-\mathbf{p},\tau} a_{\mathbf{k}'\sigma'}^+ a_{\mathbf{k}'+\mathbf{p}-\mathbf{q},\sigma'} a_{\mathbf{k}'\tau'}]_-$$
$$= \mathbf{q} a_{\mathbf{k}\sigma}^+ a_{\mathbf{k}-\mathbf{p},\tau} a_{\mathbf{k}'\sigma'}^+ a_{\mathbf{k}'+\mathbf{p}-\mathbf{q},\tau'} . \tag{3.45}$$

By taking linear combinations with fixed $\mathbf{q}$ of the expressions in Eqs. (3.44) and (3.45), any one- and two-particle densities $d(\mathbf{q})$ of the form (3.19) and (3.35), respectively, may be constructed which all satisfy the Fourier-transformed infinitesimal translation covariance (3.40). In particular $N_{el} = n(0)$ defined by (3.22) and any electron Hamiltonian (concerning Umklapp processes see below) are of the form $d(0)$ and hence commute with $\mathbf{P}_{el}$.

For phonons, identification of the total momentum with the analogous expression $\sum_{\mathbf{R},\nu} \mathbf{p}_\nu(\mathbf{R}) = \sum_\nu M_\nu \mathbf{j}_\nu(0)$ defined by Eq. (2.13) is not possible because, as discussed after that equation, the center of mass of the crystal is fixed. The correct analogy here is with creation and annihilation operators,

$$\mathbf{P}_{ph} = \sum_q \mathbf{q}\, b_q^+ b_q \tag{3.46}$$

so that with (1.31) one finds

$$[\mathbf{P}_{ph}, P_q] = \mathbf{q} P_q \; ; \quad [\mathbf{P}_{ph}, Q_q] = -\mathbf{q} Q_q \; . \tag{3.47}$$

Hence, any linear combination of products $\prod_{i=1}^n Q_{q_i}$ with $\sum_{i=1}^n \mathbf{q}_i = \mathbf{q}$ yields a density $d(\mathbf{q})$ with the correct translation covariance (3.40). In particular, $d(0)$ describes normal n-phonon processes (2.22) and commutes with $\mathbf{P}_{ph}$, as does $N_{ph} = \sum_q b_q^+ b_q$.

As to Umklapp processes, they occur with electronic or phononic densities $d(\mathbf{q})$ with $\mathbf{q} = \mathbf{K} \neq 0$. According to (3.40) such a density does not commute with $\mathbf{P} = \mathbf{P}_{el} + \mathbf{P}_{ph}$. However, using Eqs. (2.3), (1.2) and (3.39) one derives the weaker invariance under discrete lattice translations $\mathbf{R}$,

$$e^{i\mathbf{P}\cdot\mathbf{R}} d(\mathbf{K}) e^{-i\mathbf{P}\cdot\mathbf{R}} = \int_V d^3r\, \tilde{d}(\mathbf{r}+\mathbf{R}) e^{-i\mathbf{K}\cdot(\mathbf{r}+\mathbf{R})} = d(\mathbf{K}) \tag{3.48}$$

where in the last equality the periodic boundary condition (2.5) was invoked. The invariance (3.48) expresses what in Ref. 6 (p. 742) is called *multiplicative momentum conservation* which, when cast into additive form, holds only modulo a reciprocal lattice vector $\mathbf{K}$.

4. Groundstate Energy in Hartree-Fock Approximation

Matter consists of many-fermion systems, namely nucleons in nuclei and electrons in atoms, molecules and solids. In first quantization the Hamiltonian of a system of N fermions is typically of the form

$$\mathcal{H} = \sum_{j=1}^N H_j + \frac{1}{2} \sum_{j,l=1}^N (j \neq l) V_{jl} \tag{4.1}$$

where H_1 describes free particles and V_{12} the interaction between them. Since the problems are in general too complicated for an analytical solution, approximate methods have been developed to determine the groundstate and, hopefully, the lowest excited states. The most important of these methods is based on the idea of a *self-consistent field* U_1^{H} introduced by Hartree[14]. U_1^{H} replaces the two-fermion term V_{12} in such a way that one-particle functions φ_k are determined by the eigenvalue equation

$$(H_1 + U_1^{\mathrm{H}})|\varphi_k\rangle = \varepsilon_k^{\mathrm{H}}|\varphi_k\rangle \tag{4.2}$$

and by the condition

$$U_1^{\mathrm{H}} = \sum_l \langle\varphi_l(2)|V_{12}|\varphi_l(2)\rangle f_l \ . \tag{4.3}$$

Here

$$f_k = \begin{cases} 1, & \text{if } \varphi_k \text{ is occupied in } |\phi\rangle \ , \\ 0, & \text{if } \varphi_k \text{ is empty in } |\phi\rangle \end{cases} \tag{4.4}$$

and $|\phi\rangle$ is the groundstate of the N-fermion system which in Hartree's method is a simple product of N functions φ_k. The exclusion principle is only partially satisfied by choosing all φ_k different but not applying the antisymmetrization (3.2).

Since U_1^{H} is a functional of all the φ_k occupied in $|\phi\rangle$, (4.2) is a complicated nonlinear integro-differential equation which has to be solved by iteration. However, as first noticed by Fock and by Slater[15], Hartree's method is equivalent to a *variational principle*[16] in which $|\phi\rangle$ is determined such that

$$E(\phi) = \langle\phi|\mathcal{H}|\phi\rangle = \text{minimum} \ . \tag{4.5}$$

In addition, Fock and Slater have corrected the deficiency in the application of the exclusion principle by expressing $|\phi\rangle$ as a *Slater determinant* of the occupied φ_k. It is obvious that in this form second quantization is the appropriate framework for the Hartree-Fock method[17].

We follow here the exposition of Thouless[18], writing the Hamiltonian (4.1) as $\mathcal{H} = \mathcal{H}_o + \mathcal{H}_{\mathrm{int}}$ where, according to Eqs. (3.19) and (3.35),

$$\mathcal{H}_o = \sum_{kl} H_{kl} a_k^+ a_l \tag{4.6}$$

and

$$\mathcal{H}_{\mathrm{int}} = \frac{1}{2} \sum_{kk'll'} V_{kk',ll'} a_k^+ a_{k'}^+ a_{l'} a_l \ . \tag{4.7}$$

Here $a_k^+ \equiv a^+(\varphi_k)$ and, as seen from Eq. (4.2), $\mathcal{H}_o$ is in general not diagonal in the basis φ_k. The Hartree-Fock groundstate simply is

$$|\phi\rangle = a_{k_1}^+ \cdots a_{k_N}^+ |0\rangle \tag{4.8}$$

where all the labels k_j are different and such that the one-particle states with lowest energy $\varepsilon_k^{\mathrm{H}}$ are filled up,

$$\sum_k \varepsilon_k^{\mathrm{H}} f_k = \text{minimum} . \tag{4.9}$$

The variational principle (4.5) is now implemented by varying the basis functions φ_k,

$$\delta\varphi_k = \sum_l \eta_{kl} \varphi_l \tag{4.10}$$

where the orthonormality of the φ_k implies the conditions

$$\eta_{kl} + \eta_{lk}^* = 0 . \tag{4.11}$$

To obtain the variation of the groundstate $|\phi\rangle$,

$$|\delta\phi\rangle = \sum_{j=1}^N a_{k_1}^+ \cdots a_{k_{j-1}}^+ \delta a_{k_j}^+ a_{k_{j+1}}^+ \cdots a_{k_N}^+ |0\rangle , \tag{4.12}$$

we note that, according to Eq. (3.12) and the linearities (3.8) and (4.10),

$$\delta a_n^+ = a^+(\delta\varphi_n) = \sum_m \eta_{nm} a_m^+ . \tag{4.13}$$

Inserting (4.13) into (4.12) and anticommuting a_m^+ to the left of $a_{k_1}^+$, expressing in addition δ_{nk_j} by the anticommutation relation (3.13), we obtain

$$|\delta\phi\rangle = \sum_{j=1}^N (-)^{j-1} \sum_{nm} \eta_{nm} a_m^+$$
$$\times a_{k_1}^+ \cdots a_{k_{j-1}}^+ (a_n a_{k_j}^+ + a_{k_j}^+ a_n) a_{k_{j+1}}^+ \cdots a_{k_N}^+ |0\rangle . \tag{4.14}$$

Since all the k_j are different, a_n in the first term of the bracket may be anticommuted to the left of $a_{k_1}^+$ while a_n in the second term of the bracket may be anticommuted to the right of $a_{k_N}^+$ where it annihilates the vacuum state. Therefore, using (4.8),

$$|\delta\phi\rangle = N \sum_{nm} \eta_{nm} a_m^+ a_n |\phi\rangle , \tag{4.15}$$

and it is easily seen that because of the conditions (4.11),

$$\delta\langle\phi|\phi\rangle = \langle\delta\phi|\phi\rangle + \langle\phi|\delta\phi\rangle = 0 . \tag{4.16}$$

The variational principle (4.5) now becomes

$$\delta E = \langle \delta\phi | \mathcal{H} | \phi \rangle + \langle \phi | \mathcal{H} | \delta\phi \rangle = 0 \tag{4.17}$$

where insertion of (4.16) gives

$$\delta E = N \sum_{nm} \eta_{nm} \langle \phi | [\mathcal{H}, a_m^+ a_n]_- | \phi \rangle . \tag{4.18}$$

To calculate this expression we note that because of the exclusion principle

$$\langle \phi | a_k^+ a_l | \phi \rangle = f_k \delta_{kl} \tag{4.19}$$

and

$$\langle \phi | a_k^+ a_{k'}^+ a_{l'} a_l | \phi \rangle = f_k f_{k'} (\delta_{kl}\delta_{k'l'} - \delta_{kl'}\delta_{k'l}) . \tag{4.20}$$

Evaluating the commutator in (4.18) with the Hamiltonians (4.6) and (4.7), using the anticommutation relations (3.13) and the identities (3.42), application of Eqs. (4.19), (4.20) and of the symmetry (3.33) yields

$$\delta E = N \sum_{nm} \eta_{nm} \{ H_{nm} + U_{nm}^{\mathrm{HF}} \}(f_n - f_m) \tag{4.21}$$

where

$$U_{nm}^{\mathrm{HF}} \equiv \sum_k (V_{nk,mk} - V_{nk,km}) f_k . \tag{4.22}$$

Since only terms with $f_n \neq f_m$, that is in particular, with $n \neq m$, contribute to the sum in (4.21) and since for $n \neq m$ the η_{nm} are not restricted by (4.11), the extremum condition (4.17) implies that

$$H_{nm} + U_{nm}^{\mathrm{HF}} = 0, \text{ if } f_n \neq f_m , \tag{4.23}$$

but undetermined otherwise. Extending Eq. (4.23) to the case $n \neq m$ but $f_n = f_m$, one may write

$$H_{nm} + U_{nm}^{\mathrm{HF}} = \varepsilon_n^{\mathrm{HF}} \delta_{nm} . \tag{4.24}$$

We now define the *Hartree-Fock operator*

$$\mathcal{U}^{\mathrm{HF}} \equiv \sum_{nm} U_{nm}^{\mathrm{HF}} a_n^+ a_m \tag{4.25}$$

and let it act, together with the Hamiltonian (4.6), on the one-particle state $|\varphi_k\rangle$. Applying the anticommutation relations (3.13) and Eq. (4.24) one then easily finds

$$(\mathcal{H}_o + \mathcal{U}^{\mathrm{HF}})|\varphi_k\rangle = \varepsilon_k^{\mathrm{HF}} |\varphi_k\rangle \tag{4.26}$$

which, formally, is the same as Hartree's self-consistent equation (4.2).

According to Eq. (3.33) the two terms in the definition (4.22) of U_{nm}^{HF} may be written in the form

$$U_{nm}^{\text{H}} \equiv \sum_k V_{nk,mk} f_k = \sum_k \langle \varphi_n(1) \otimes \varphi_k(2) | V_{12} | \varphi_m(1) \otimes \varphi_k(2) \rangle f_k$$

$$= \langle \varphi_n(1) | U_1^{\text{H}} | \varphi_m(1) \rangle \qquad (4.27)$$

and

$$U_{nm}^{\text{ex}} \equiv \sum_k V_{nk,km} f_k = \sum_k \langle \varphi_n(1) \otimes \varphi_k(2) | V_{12} | \varphi_k(1) \otimes \varphi_m(2) \rangle f_k$$

$$= \langle \varphi_n(1) | U_{12}^{\text{ex}} | \varphi_m(2) \rangle . \qquad (4.28)$$

Here U_1^{H} is Hartree's self-consistent field (4.3) and

$$U_{12}^{\text{ex}} = \sum_k \langle \varphi_k(2) | V_{12} | \varphi_k(1) \rangle f_k \qquad (4.29)$$

the *exchange field*. In second-quantized form the fields (4.3) and (4.29) combine to the Hartree-Fock operator (4.25),

$$\mathcal{U}^{\text{HF}} = \mathcal{U}^{\text{H}} - \mathcal{U}^{\text{ex}} . \qquad (4.30)$$

The two terms in Eq. (4.30), or equivalently Eqs. (4.27) and (4.28), are represented by the *field diagrams* of Fig. 1.3 where the dashed line represents the vertex of Fig. 1.2. An example of a Hartree-Fock field is the periodic potential $U(\mathbf{r})$ in Eq. (1.12).

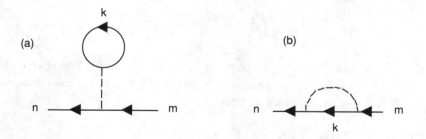

Fig. 1.3: (a) Hartree field (4.27). (b) Exchange field (4.28).

In analogy to Eq. (4.9) the one-particle states φ_k filled up in the groundstate $|\phi\rangle$ are selected by the condition that

$$E^{\text{HF}} = \langle \phi | \mathcal{H}_o + \mathcal{U}^{\text{HF}} | \phi \rangle = \sum_k \varepsilon_k^{\text{HF}} f_k = \text{minimum} . \qquad (4.31)$$

Since from Eqs. (4.7), (4.20) and (4.22) follows with the help of (4.19), (4.25) that

$$\langle\phi|\mathcal{H}_{\text{int}}|\phi\rangle = \frac{1}{2}\langle\phi|\mathcal{U}^{\text{HF}}|\phi\rangle = \frac{1}{2}\sum_k U_{kk}^{\text{HF}} f_k , \qquad (4.32)$$

one finally obtains for the groundstate energy (4.5), with $\mathcal{H} = \mathcal{H}_o + \mathcal{H}_{\text{int}}$,

$$E = \sum_k (H_{kk} + \frac{1}{2}U_{kk}^{\text{HF}})f_k = \frac{1}{2}\sum_k (H_{kk} + \varepsilon_k^{\text{HF}})f_k . \qquad (4.33)$$

The largest of the filled energies $\varepsilon_k^{\text{HF}}$ is called the *Fermi energy* ε_F, and the points k in wavenumber space satisfying $\varepsilon_k^{\text{HF}} = \varepsilon_F$ form the *Fermi surface*. Since the presence of an interaction $\mathcal{H}_{\text{int}}$ does in general preserve a discontinuity in the wavenumber distribution of the N-fermion groundstate, the Fermi surface remains a meaningful notion even for interacting electrons. This result has become known as *Luttinger's theorem* [19].

Solutions to the Problems of Chapter 1

Problem 1.1: Eqs. (1.9), (1.10).

For $\mathbf{k} \neq \mathbf{K}$ one obtains by insertion of Eqs. (1.1), (1.4), (1.6), (1.8)

$$\sum_{\mathbf{R}\in V} e^{i\mathbf{k}\cdot\mathbf{R}} = \prod_{i=1}^3 \sum_{m_i=-L_i+1}^{+L_i} e^{i\pi m_i \nu_i/L_i}$$

$$= \prod_{i=1}^3 e^{-i\pi(L_i-1)\nu_i/L_i} \frac{1-e^{i2\pi m_i \nu_i}}{1-e^{i\pi\nu_i/L_i}} = 0 .$$

For $\mathbf{k} = \mathbf{K}$ this sum is L, according to (1.2), hence (1.9). For $\mathbf{R} \neq 0$ the sum in (1.10) is the same as above, except that m_i and ν_i are interchanged. For $\mathbf{R} = 0$ the sum is again L, hence (1.10).

Problem 1.2: Eq. (1.29).

With (1.24), (1.25) one obtains

$$\sum_{\mathbf{R}\in V, \nu} \frac{\mathbf{p}_\nu^2(\mathbf{R})}{M_\nu} = \sum_{qq'} \sqrt{\omega_q \omega_{q'}} \frac{1}{B} \sum_\nu \mathbf{e}_q(\nu)\cdot\mathbf{e}_{q'}^*(\nu) \frac{1}{L} \sum_{\mathbf{R}\in V} e^{i(\mathbf{q}-\mathbf{q}')\cdot\mathbf{R}} P_q^+ P_{q'}$$

$$= \sum_q \omega_q P_q^+ P_q$$

where (1.9) and (1.28) were successively used. Similarly one obtains with (1.24), (1.25)

$$\sum_{\mathbf{R},\mathbf{R}'\in V, \nu\nu'} \mathbf{u}_\nu(\mathbf{R})\mathbf{C}_{\nu\nu'}(\mathbf{R}-\mathbf{R}')\mathbf{u}_{\nu'}(\mathbf{R}') = \sum_{qq'} \frac{\hbar^2}{\sqrt{\omega_q\omega_{q'}}}\frac{1}{B}\sum_{\nu\nu'}\mathbf{e}^*(\nu)$$

$$\times \frac{1}{L}\sum_{\mathbf{R},\mathbf{R}'\in V}(M_\nu M_{\nu'})^{-1/2}\mathbf{C}_{\nu\nu'}(\mathbf{R}-\mathbf{R}')\mathbf{e}_{q'}(\nu')e^{-i\mathbf{q}\cdot\mathbf{R}}e^{i\mathbf{q}'\cdot\mathbf{R}'}Q_q^+Q_{q'}\;.$$

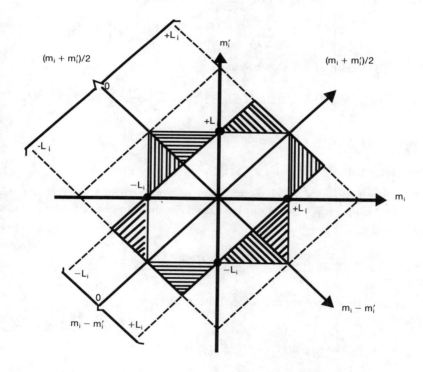

Fig. 1.4: Transformation of domains using (1.7) to shift the hatched parts.

The double sum over $V\otimes V$ is best evaluated in center-of-mass and relative coordinates $\mathbf{X} = \frac{1}{2}(\mathbf{R}+\mathbf{R}') = \sum_i \frac{1}{2}(m_i+m_i')\mathbf{a}_i$, $\mathbf{Y} = \mathbf{R}-\mathbf{R}' = \sum_i(m_i - m_i')\mathbf{a}_i$, using (1.1). The important observation now is that because of the periodicity (1.7) the domain of $\mathbf{X}\otimes\mathbf{Y}$ may again be transformed into $V\otimes V$, as is obvious from Fig. 1.4. Then application of (1.9) to the $\mathbf{X}$-sum yields a $\delta_{\mathbf{qq}'}$ ($\mathbf{K}=0$ since $\mathbf{q}$ and $\mathbf{q}'$ are in Z) so that by applying (1.26) to the

Y-sum, the above expression becomes

$$\sum_{qq'} \frac{1}{\sqrt{\omega_q \omega_{q'}}} \frac{1}{B} \sum_{\nu} \mathbf{e}_q^{\,*}(\nu) \sum_{\nu'} \mathbf{D}_{\nu\nu'}(\mathbf{q}) \mathbf{e}_{q'}(\nu') \delta_{\mathbf{q}\mathbf{q'}} Q_q^+ Q_{q'} = \sum_q \omega_q Q_q^+ Q_q$$

where in the last step (1.27) and (1.28) were used.

Problem 1.3: Eqs. (2.20), (2.21).

The first term in the expansion of (2.19) defines

$$\mathcal{H}^{\text{el-ph}} = \sum_{\mathbf{R}\in V,\nu j} \nabla\phi_\nu(\mathbf{R} + \mathbf{r}_\nu - \mathbf{r}_j) \cdot \mathbf{u}_\nu(\mathbf{R}) .$$

With (2.4), (2.9), (1.24) this becomes

$$\hbar \sum_{qq'\nu} \tilde{\phi}_\nu(\mathbf{q}') e^{i\mathbf{q}'\cdot\mathbf{r}_\nu} n(\mathbf{q}') (LBM_\nu\omega_q)^{-1/2} i\mathbf{q}' \cdot \mathbf{e}_q(\nu) \frac{1}{V} \sum_{\mathbf{R}\in V} e^{i(\mathbf{q}+\mathbf{q}')\cdot\mathbf{R}} Q_q .$$

Applying (1.9) with $V = Lv$ one finds (2.20), (2.21).

Problem 1.4: $S_\pm^2 = S_\pm$, $S_\pm^+ = S_\pm$.

The permutations of N elements form a group , hence with P and Q, $P' = QP$ is also a permutation. By definition the number of transpositions satisfies $\eta_{QP} = \eta_P + \eta_Q$, hence

$$S_\pm^2 = \frac{1}{(N!)^2} \sum_Q \sum_P (\pm)^{\eta_P + \eta_Q} QP = \frac{1}{N!} \sum_{P'} (\pm)^{\eta_{P'}} P' = S_\pm .$$

Because of this property and of definition (3.1),

$$\langle g_1 \cdots g_N | f_1 \cdots f_N \rangle^* = \langle g_1 \cdots g_N | S_\pm | f_1 \cdots f_N \rangle^* = \langle f_1 \cdots f_N | S_\pm^+ | g_1 \cdots g_N \rangle$$
$$\langle f_1 \cdots f_N | g_1 \cdots g_N \rangle = \langle f_1 \cdots f_N | S_\pm | g_1 \cdots g_N \rangle .$$

But these two expressions are the same for any states, hence $S_\pm^+ = S_\pm$.

Problem 1.5: Eq. (3.5).

Because of $S_\pm^2 = S_\pm$ and the definition (3.2),

$$\langle g_1 \cdots g_N | f_1 \cdots f_N \rangle = \langle g_1(1) \otimes \cdots \otimes g_N(N) | S_\pm | f_1(1) \otimes \cdots \otimes f_N(N) \rangle$$
$$= \frac{1}{N!} \sum_P (\pm)^{\eta_P} \langle g_1(1) \otimes \cdots \otimes g_N(N) | f_{i_1}(1) \otimes \cdots \otimes f_{i_N}(N) \rangle .$$

Here, in contrast to (3.3), we permuted the state labels instead of the particle labels, which of course is equivalent. In this way the pairing off into the one-particle scalar products of the right-hand side of (3.5) is immediate.

Problem 1.6: Eq. (3.7).

With (3.6)

$$\langle f_1 \cdots f_N | a(f_0) | g_0 g_1 \cdots g_N \rangle = \langle g_0 g_1 \cdots g_N | a^+(f_0) | f_1 \cdots f_N \rangle^*$$
$$= \sqrt{N+1} \langle f_0 f_1 \cdots f_N | g_0 g_1 \cdots g_N \rangle .$$

Now the group G_{N+1} of permutations $(0, 1 \cdots N \to k_0, k_1 \cdots k_N)$ may be decomposed into the transpositions

$$T_j \equiv (0, 1 \cdots j-1, j, j+1 \cdots N \to j, 0 \cdots j-1, j+1 \cdots N) ; \quad T_0 \equiv 1$$

with $\eta_{T_j} = j$, and the elements of G_N may be written as

$$P_j \equiv (0 \cdots j-1, j+1 \cdots N \to i_1 \cdots i_j, i_{j+1} \cdots i_N);$$
$$P_0 \equiv (1 \cdots N \to i_1 \cdots i_N) .$$

Then the $(N+1)!$ permutations $\sum_{j=0}^{N} P_j T_j$ span G_{N+1} so that, using (3.5) twice,

$$\langle f_0 f_1 \cdots f_N | g_0 g_1 \cdots g_N \rangle = \frac{1}{(N+1)!} \sum_{j=0}^{N} (\pm)^{j+\eta_{P_j}} \langle f_0 | g_j \rangle \langle f_1 | g_{i_1} \rangle \cdots \langle f_N | g_{i_N} \rangle$$

$$= \frac{1}{N+1} \sum_{j=0}^{N} (\pm)^j \langle f_0 | g_j \rangle \langle f_1 \cdots f_N | g_0 g_1 \cdots g_{j-1} g_{j+1} \cdots g_N \rangle .$$

Inserted above this yields (3.7) since $|f_1 \cdots f_N\rangle$ is an arbitrary state.

Problem 1.7: Eqs. (3.10).

With (3.6) and (3.7)

$$a(f_0) a^+(g_0) | g_1 \cdots g_N \rangle = \sqrt{N+1} a(f_0) | g_0 g_1 \cdots g_N \rangle$$

$$= \langle f_0 | g_0 \rangle | g_1 \cdots g_N \rangle + \sum_{j=1}^{N} (\pm)^j \langle f_0 | g_j \rangle | g_0 g_1 \cdots g_{j-1} g_{j+1} \cdots g_N \rangle$$

and writing for (3.9)

$$a^+(g_0) a(f_0) | g_1 \cdots g_N \rangle = \sum_{j=1}^{N} (\pm)^{j-1} \langle f_0 | g_j \rangle | g_0 g_1 \cdots g_{j-1} g_{j+1} \cdots g_N \rangle ,$$

the first relation (3.10) readily follows. Applying (3.6) twice,

$$a^+(g)a^+(f)|f_1\cdots f_N\rangle = \sqrt{(N+1)(N+2)}|gff_1\cdots f_N\rangle$$
$$= \pm a^+(f)a^+(g)|f_1\cdots f_N\rangle \, ,$$

hence the second relation (3.10).

Problem 1.8: Eq. (3.20).

Since for any permutation $P = (1\cdots N \to i_1\cdots i_N)$, $\sum_{j=1}^N O_j = \sum_{j=1}^N O_{i_j}$, the index labeling the particle on which O_j acts is arbitrary, so that

$$\sum_{j=1}^N O_j|f_1\cdots f_N\rangle$$

$$= \sum_{j=1}^N \frac{1}{N!} \sum_P (\pm)^{\eta_P} |f_1(i_1)\otimes\cdots\otimes(O_jf_j)(i_j)\otimes\cdots\otimes f_N(i_N)\rangle$$

$$= \sum_{j=1}^N |f_1\cdots f_{j-1}(O_jf_j)f_{j+1}\cdots f_N\rangle \, .$$

Inserting (3.17) and using (3.9) this becomes

$$\sum_{nm} O_{nm} \sum_{j=1}^N \langle\varphi_m|f_j\rangle|f_1\cdots f_{j-1}\varphi_n f_{j+1}\cdots f_N\rangle$$

$$= \sum_{nm} O_{nm} a_n^+ a_m |f_1\cdots f_N\rangle \, ,$$

hence (3.20).

Problem 1.9: Eqs. (3.35), (3.36).

As in Problem 1.8, $\sum_{jl}' O_{jl} = \sum_{jl}' O_{i_j i_l}$ for any permutation $P = (1\cdots N \to i_1\cdots i_N)$, the sum being from 1 to N with $j \neq l$. Therefore, the indices labeling the particles on which O_{jl} acts, are again arbitrary so that, writing $|f_1\cdots f_N\rangle = (\pm)^{l-j-1}|\cdots f_jf_lf_{j+1}\cdots f_{l-1}f_{l+1}\cdots\rangle$ one obtains

$$\sum_{jl}' O_{jl}|f_1\cdots f_N\rangle = \sum_{jl}' (\pm)^{l-j-1} \frac{1}{N!}\sum_P(\pm)^{\eta_P}$$

$$\times |f_1(i_1)\otimes\cdots\otimes(O_{jl}f_jf_l)(i_j,i_l)\otimes\cdots\otimes f_N(i_N)\rangle$$

$$= \sum_{jl}' (\pm)^{l-j-1}|f_1\cdots f_{j-1}(O^{(2)}f_jf_l)f_{j+1}\cdots f_{l-1}f_{l+1}\cdots f_N\rangle \, .$$

Inserting (3.34) and using $\langle\varphi_m\varphi_{m'}|f_jf_l\rangle = \langle\varphi_m\varphi_{m'}|f_j(1)\otimes f_l(2)\rangle$, (3.4) and the symmetry (3.33), this becomes

$$\sum_{nn'mm'} O_{nn',mm'}\sideset{}{'}\sum_{jl}(\pm)^{l-j-1}\langle\varphi_m|f_j\rangle\langle\varphi_{m'}|f_l\rangle$$

$$\times|f_1\cdots f_{j-1}\varphi_n\varphi_{n'}f_{j+1}\cdots f_{l-1}f_{l+1}\cdots f_N\rangle$$

$$= \sum_{nn'mm'} O_{nn',mm'}\sum_j\langle\varphi_m|f_j\rangle\sum_{l\neq j}\langle\varphi_{m'}|f_l\rangle$$

$$\times|f_1\cdots f_{j-1}\varphi_n f_{j+1}\cdots f_{l-1}\varphi_{n'}f_{l+1}\cdots f_N\rangle\ .$$

With (3.9) one finds for the sum over l, excluding j,

$$a_{n'}^+a_{m'}|f_1\cdots f_{j-1}\varphi_n f_{j+1}\cdots f_N\rangle - \langle\varphi_{m'}|\varphi_n\rangle|f_1\cdots f_{j-1}\varphi_{n'}f_{j+1}\cdots f_N\rangle\ .$$

Inserting back and applying (3.9) again to both terms, one arrives at

$$\sideset{}{'}\sum_{jl}O_{jl}|f_1\cdots f_N\rangle = \sum_{nn',mm'} O_{nn',mm'}\{a_{n'}^+a_{m'}a_n^+a_m - a_{n'}^+a_m\delta_{m'n}\}|f_1\cdots f_N\rangle$$

which is (3.35), (3.36) since with (3.13),

$$a_{n'}^+a_{m'}a_n^+a_m = a_{n'}^+(\pm a_n^+a_{m'} + \delta_{m'n})a_m = a_{n'}^+a_n^+a_ma_{m'} + a_{n'}^+a_m\delta_{m'n}$$

and the symmetry (3.33) establishes the correct labeling.

References to Chapter 1

1. **Ashcroft, N.W.** and **Mermin, N.D.**, *Solid State Physics* (Holt, Rinehart and Winston, New York, 1976).
2. **Ziman, J.M.**, *Electrons and Phonons* (Clarendon, Oxford, 1963).
3. **Bloch, F.**, *Z. Phys.* **52**, 555 (1928).
4. **Pauli, W.**, "Die Allgemeinen Prinzipien der Wellenmechanik", in *Hdb. Physik*, ed. Geiger, H. and Scheel, K. (Springer, Berlin, 1933), Vol. 24, Part I, Sec. 14; "General Principles of Quantum Mechanics", transl. Achuthan, P. and Ventakesan, K. (Springer, Berlin, 1980), Sec. 14.
5. **Enz, C.P.**, *Ann. Phys.* (N.Y.) **46**, 114 (1968).
6. **Enz, C.P.**, "Electron-phonon and phonon-phonon interactions", in *Theory of Condensed Matter* (IEAE, Vienna, 1968), pp. 729–782.
7. **Madelung, O.**, *Introduction to Solid-State Theory* (Springer, Berlin, 1978).
8. **Elliot, R.J.**, *Phys. Rev.* **96**, 266, 280 (1954).
9. **Dirac, P.A.M.**, *Proc. Roy. Soc.* **A114**, 243, 710 (1927).
10. **Jaggi, R.**, *J. Appl. Phys.* **69**, 816 (1991); *Bulletin Swiss Phys. Soc.* **7**, no. 2, p. 27 (1990).
11. **Enz, C.P.**, *Physica* **20**, 983 (1954). See also **Herring, C.**, in *Proc. Int. Conf. on Semiconductor Physics* (Academic, New York, 1961), p. 60.

12. **Bussmann-Holder, A., Migliori, A., Fisk, Z., Sarrao, J.L., Leisure, R.G.**, and **Cheong, S.-W.**, *Phys. Rev. Lett.* **67**, 512 (1991); **Irkhin, V.Yu., Katsnelson, M.I.** and **Trefilov, A.V.**, *Europhys. Lett.* **15**, 649 (1991).

13. **Abrikosov, A.A., Gor'kov, L.P.** and **Dzyaloshinskii, I.Ye.**, *Quantum Field Theoretical Methods in Statistical Physics*, ed. ter Haar, D. (Pergamon, Oxford, 1965).

14. **Hartree, D.R.**, *Proc. Cambridge Phil. Soc.* **24**, 89, 111 (1928).

15. **Fock, V.**, *Z. Phys.* **61**, 126 (1930); **Slater, J.C.**, *Phys. Rev.* **35**, 210 (1930).

16. See, e.g., **Schiff, L.I.**, *Quantum Mechanics, 2nd ed.* (McGraw-Hill, New York, 1955), Sec. 38.

17. **Fetter, A.L.** and **Walecka, J.D.**, *Quantum Theory of Many-Particle Systems* (McGraw-Hill, New York, 1971), Sec. 10.

18. **Thouless, D.J.**, *The Quantum Mechanics of Many-Body Systems* (Academic, New York, 1961).

19. **Luttinger, J.M.**, *Phys. Rev.* **119**, 1153 (1960); **121**, 942 (1961). See also **Migdal, A.B.**, *Soviet Phys.-JETP* **5**, 333 (1957); **Galitsky, V.M.** and **Migdal, A.B.**, *Soviet Phys.-JETP* **7**, 96 (1958).

Chapter 2

GREEN'S FUNCTION
FORMALISM IN COMPLEX TIME

Field quantization[1] and, in particular, the techniques developed by Tomonaga and Schwinger[2] and by Dyson[3] had led in the late 1940's to the spectacular success of renormalized quantum electrodynamics. However, the impact of these techniques on many-body theory and statistical mechanics took about a decade to develop. The book by Kadanoff and Baym[4] was one of the first representative texts in the West using Matsubara's technique[5] to treat Green's functions at finite temperatures; in the Soviet Union the development was parallel[6] and culminated in the book by Abrikosov, Gor'kov and Dzyaloshinskii[7]. While the number of many-body texts has grown enormously, (see, e.g. Ref. 8), there has also been a development in reverse in that, surprisingly, finite-temperature techniques have been taken over by the particle and field theorists[9]. The hinge on this two-way road was the work of Martin and Schwinger[10] which not only had been the guide in developing many-body theory but also supplied the functional techniques used today in field theory.

5. Schrödinger, Heisenberg and Interaction Representations

In general, the Hamiltonian may depend parametrically on time, e.g. through the coupling of the system to a varying external field as in Eq. (2.1). In units such that time $= \hbar t$ the time-dependent Schrödinger equation has the form

$$\frac{d}{dt}|\psi(t)\rangle = -i\mathcal{H}_t|\psi(t)\rangle \tag{5.1}$$

where $\mathcal{H}_t = \mathcal{H} + \delta\mathcal{H}_t$. It determines the time evolution of the state vector in *Schrödinger representation*,

31

$$|\psi(t)\rangle = U(t, t_o)|\psi(t_o)\rangle \tag{5.2}$$

from which the following unitarity and group properties of the time-evolution operator U follow:

$$U(t, t')U(t', t_o) = U(t, t_o) = U^+(t_o, t), \; U(t, t) = 1 \; . \tag{5.3}$$

Differentiating Eq. (5.2) and comparing with (5.1) one obtains

$$\frac{\partial}{\partial t}U(t, t_o) = -i\mathcal{H}_t U(t, t_o) \tag{5.4}$$

or equivalently

$$U(t, t_o) = 1 - i \int\limits_{t_o}^{t} dt' \mathcal{H}_{t'} U(t', t_o) \; . \tag{5.5}$$

This integral equation may be solved by iteration.

The Heisenberg representation $A_t(t)$ of a linear operator A_t (element of an operator algebra which may have a parametric time dependence) is defined by the requirement that matrix elements are representation-independent,

$$\langle\varphi(t)|A_t|\psi(t)\rangle = \langle\varphi|A_t(t)|\psi\rangle \tag{5.6}$$

where $\varphi = \varphi(0)$ and $\psi = \psi(0)$ are the state functions of the *Heisenberg representation*. From (5.6), (5.2) and (5.3) the transformation property

$$A_t(t) = U(0, t)A_t U(t, 0) \tag{5.7}$$

follows, which by differentiation yields

$$\frac{d}{dt}(A_t(t)) = i[\mathcal{H}_t(t), A_t(t)] + (\frac{d}{dt}A_t)(t) \; . \tag{5.8}$$

For a time-independent Hamiltonian $\mathcal{H}$ the solution of Eq. (5.4) is invariant under time translation $t \to t + s$. It is explicitly given by

$$U(t, t_o) = U(t + s, t_o + s) = e^{-i\mathcal{H}(t-t_o)} \; , \tag{5.9}$$

and Eq. (5.7) becomes analogous to Eq. (3.39) for spatial translations,

$$A_t(t) = e^{i\mathcal{H}t}A_t e^{-i\mathcal{H}t} \; . \tag{5.10}$$

As exemplified in Sections 1 and 2, the Hamiltonians of primary interest are of the form

$$\mathcal{H}_t = \mathcal{H}_o + \mathcal{H}_{\text{int}} \tag{5.11}$$

where $\mathcal{H}_o$ describes free excitations and $\mathcal{H}_{\text{int}}$ the interactions between them as well as their coupling to external fields, $\delta\mathcal{H}_t$. This separation is specially adapted to a perturbative treatment of $\mathcal{H}_{\text{int}}$, for which the *interaction representation* (see, e.g. Ref. 1, Section 21) of Tomonaga and Schwinger[2] is most appropriate. The reason is that the associated time-evolution operator $S(t, t_o)$ may be expressed explicitly in terms of $\mathcal{H}_{\text{int}}$. In this representation *labeled in this book by an upper index o*, the state vector follows the time evolution of the unperturbed system ,

$$|\psi^o(t)\rangle = e^{i\mathcal{H}_o t}|\psi(t)\rangle \tag{5.12}$$

and the equation defining S is analogous to (5.2),

$$|\psi^o(t)\rangle = S(t, t_o)|\psi^o(t_o)\rangle . \tag{5.13}$$

Combining Eqs. (5.12), (5.13) and (5.2) one finds

$$S(t, t_o) = e^{i\mathcal{H}_o t}U(t, t_o)e^{-i\mathcal{H}_o t_o} , \tag{5.14}$$

and from (5.13) one concludes that S has again the unitarity and group properties (5.3),

$$S(t, t')S(t', t_o) = S(t, t_o) = S^+(t_o, t), \quad S(t, t) = 1 . \tag{5.15}$$

As in Eq. (5.6), the interaction representation $A_t^o(t)$ of an operator A_t is defined by the representation-independence of matrix elements,

$$\langle\varphi^o(t)|A_t^o(t)|\psi_t^o(t)\rangle = \langle\varphi(t)|A_t|\psi(t)\rangle = \langle\varphi|A_t(t)|\psi\rangle . \tag{5.16}$$

With (5.12) and (5.7), (5.14) one then finds the transformation property

$$A_t^o(t) = e^{i\mathcal{H}_o t}A_t e^{-i\mathcal{H}_o t} = S(t, 0)A_t(t)S(0, t) . \tag{5.17}$$

Differentiation of the first equality (5.17) yields

$$\frac{d}{dt}(A_t^o(t)) = i[\mathcal{H}_o, A_t^0(t)] + (\frac{d}{dt}A_t)^o(t) \tag{5.18}$$

which shows that, in contrast to (5.8), a time-independent operator A evolves with the unperturbed Hamiltonian $\mathcal{H}_o$. This is compensated by the fact that, according to Eqs. (5.12), (5.1) and (5.17), the state vector evolves with the perturbation $\mathcal{H}_{\text{int}}$,

$$\frac{d}{dt}|\psi^o(t)\rangle = -i\mathcal{H}_{\text{int}}^o(t)|\psi^o(t)\rangle . \tag{5.19}$$

The main task of perturbation theory now is to determine the time-evolution operator S (see, e.g. Ref. 1, Section 22). Inserting (5.13) into (5.19) one finds in analogy to Eqs. (5.4) and (5.5),

$$\frac{\partial}{\partial t}S(t, t_o) = -i\mathcal{H}_{\text{int}}^o(t)S(t, t_o) \tag{5.20}$$

and

$$S(t, t_o) = 1 - i \int_{t_o}^{t} dt' \mathcal{H}_{\text{int}}^o(t') S(t', t_o) , \tag{5.21}$$

respectively. Note that in these formulas $\mathcal{H}_{\text{int}}$ may still have a parametric time dependence due to external fields. Iteration of (5.21) yields

$$S(t, t_o) = \sum_{N=0}^{\infty} S_N(t, t_o) \tag{5.22}$$

with $S_o(t, t_o) = 1$ and

$$S_N(t, t_o) = (-i)^N \int_{t_o}^{t} dt_1 \int_{t_o}^{t_1} dt_2 \cdots \int_{t_o}^{t_{N-1}} dt_N \mathcal{H}_{\text{int}}^o(t_1) \mathcal{H}_{\text{int}}^o(t_2) \cdots \mathcal{H}_{\text{int}}^o(t_N) . \tag{5.23}$$

In order to obtain a useful formula we sum the last expression over the permutations $P = (1 \ldots N \to i_1 \ldots i_N)$ of the integration variables. This yields $N!$ times the same expression so that we have the trivial generalization

$$S_N(t, t_o) = \frac{(-i)^N}{N!} \sum_{P} \int_{t_o}^{t} dt_{i_1} \int_{t_o}^{t_{i_1}} dt_{i_2} \cdots$$

$$\times \int_{t_o}^{t_{i_{N-1}}} dt_{i_N} \mathcal{H}_{\text{int}}^o(t_{i_1}) \mathcal{H}_{\text{int}}^o(t_{i_2}) \cdots \mathcal{H}_{\text{int}}^o(t_{i_N}) . \tag{5.24}$$

The complication here comes from the fact that $[\mathcal{H}_{\text{int}}^o(t), \mathcal{H}_{\text{int}}^o(t')] \neq 0$ if $t \neq t'$. Otherwise we could use the fact that the $N!$ simplexes $t > t_{i_1} > t_{i_2} > \cdots > t_{i_{N-1}} > t_{i_N} > t_o$ add up to the hypercube $t > t_i > t_o$, $i = 1, \ldots, N$, so that $S_N = \mathcal{A}^N / N!$ and $S = \exp \mathcal{A}$ where

$$\mathcal{A}(t, t_o) = -i \int_{t_o}^{t} dt' \mathcal{H}_{\text{int}}^o(t') \tag{5.25}$$

is the *action operator*.

The remarkable fact first noticed by Dyson[3] is that, due to the simplectic order of the time parameters, the operators $\mathcal{H}_{\text{int}}^o(t_i)$ in Eqs. (5.23) and (5.24) are ordered in time. Thus defining a *time-ordering operator* T for any time-dependent operator $A_i(t)$, $i = 1, \ldots, n$, by

$$T\{A_1(t_1) \cdots A_n(t_n)\} = (-)^{\eta_P} A_{i_1}(t_{i_1}) \cdots A_{i_n}(t_{i_n}),$$
$$\text{if} \quad t_{i_1} > \cdots > t_{i_n} \tag{5.26}$$

where η_P is the number of transpositions between fermionic operators in the permutation P, it is possible to write S_N as a *time-ordered product*

$$S_N(t, t_o) = \frac{1}{N!} T\{\mathcal{A}(t, t_o)\}^N \tag{5.27}$$

and S as a time-ordered exponential

$$S = T \exp \mathcal{A} . \tag{5.28}$$

This is the well-known *Dyson series* [3].

The above definition of the time-ordering operator calls for a qualification. Equation (5.26) excludes *coincident time arguments* which may be avoided in general. However, if they are unavoidable as in certain Feynman diagrams, an additional prescription has to be introduced (see Eq. (9.10) below).

6. Evolution in Complex Time and Global Equilibrium Averages

Imaginary time arguments arise naturally for a system characterized by a Hamiltonian $\mathcal{H}$ and which is in thermal equilibrium at a temperature T. Indeed, the Boltzmann factor $e^{-\beta \mathcal{H}}$ where $\beta = 1/k_B T$ (k_B is Boltzmann's constant) is nothing else than the Heisenberg time evolution $U(-i\beta, 0)$ of Eq. (5.9). As will be seen later, this formal introduction of an imaginary time argument is incompatible with the formulation of a perturbation theory at finite temperatures in terms of the real-time Dyson series (5.28), (5.25). Fortunately however, the Dyson series is readily generalized to an *arbitrary fixed oriented path C in the complex time plane*, thus allowing for both, real and imaginary time arguments.

In order to generalize the relevant formulas (5.24)–(5.28) to this case we define the path C in terms of a real parameter τ as $t = t(\tau)$ on an interval $\tau_- \leq \tau \leq \tau_+$ such that the endpoints of C are $t_\pm \equiv t(\tau_\pm)$ and its orientation corresponds to increasing values of τ. Then the action (5.25) becomes

$$\mathcal{A}(t(\tau_1), t(\tau_o)) = -i \int\limits_{\tau_o}^{\tau_1} d\tau\, t'(\tau) \mathcal{H}_{\text{int}}^o(t(\tau)) \tag{6.1}$$

where $t'(\tau) = dt/d\tau$. A set of points $t_i \equiv t(\tau_i)$, $i = 1, \cdots, n$, on C is then ordered in the sense that there exists a permutation $P = (1 \cdots n \to i_1 \cdots i_n)$ of the indices $1 \cdots n$ such that the τ_i fall into the simplex $\tau > \tau_{i_1} > \tau_{i_2} > \cdots > \tau_{i_n} > \tau_o$. Writing Eq. (5.24) in this parametrized form and defining an ordering operator T along C in analogy to (5.26),

$$T\{A_1(t_1)\cdots A_n(t_n)\} = (-)^{\eta_P} A_{i_1}(t_{i_1})\cdots A_{i_n}(t_{i_n}), \quad \text{if } \tau_{i_1} > \cdots > \tau_{i_n},$$
(6.2)

the $n!$ τ-simplexes add up to the τ-hypercube $\tau > \tau_i > \tau_o$, $i = 1, \ldots, n$. Thus Eqs. (5.27), (5.28) are recovered with the meaning (6.1) of $\mathcal{A}$ and (6.2) of T.

The objects of interest in this section are averages of observables O over a *grand-canonical ensemble* of systems of electrons, phonons or other excitations characterized by a *number operator* N. Thus second quantization introduced in Section 3 is the appropriate framework to describe the systems. Physically, this situation may be thought of as being realized by coupling each system to a reservoir of such excitations. Equilibrium then is assured by a fixed value of the associated *chemical potential* μ. Excluding for the moment time-dependent external fields, we first consider *global equilibrium* situations described by a time-independent Hamiltonian $\mathcal{H}$ of the form (5.11) which, in addition, satisfies *number conservation*,

$$[\mathcal{H}, N] = [\mathcal{H}_o, N] = 0 .$$
(6.3)

The *grand-canonical density operator* then is (see, e.g. Ref. 4)

$$\rho = e^{\beta(\Omega - \mathcal{H}^\mu)}$$
(6.4)

where $\mathcal{H}^\mu$ is obtained from $\mathcal{H}$ by a Legendre transformation between the variables N and μ,

$$\mathcal{H}^\mu = \mathcal{H} - \mu N ; \quad \mathcal{H}_o^\mu = \mathcal{H}_o - \mu N$$
(6.5)

and Ω is the *grand-canonical potential* determined by the normalization condition

$$\text{Tr}\,\rho = 1 .$$
(6.6)

The unperturbed ensemble is similarly characterized by

$$\rho_o = e^{\beta(\Omega_o - \mathcal{H}_o^\mu)} .$$
(6.7)

Since, according to (5.14), (5.9) and (6.3),

$$S(t, t_o) = e^{i\mathcal{H}_o^\mu t} e^{-i\mathcal{H}^\mu(t - t_o)} e^{-i\mathcal{H}_o^\mu t_o} ,$$
(6.8)

comparison of (6.4) and (6.7) shows that the full density operator ρ may be obtained from ρ_o with the help of the Dyson series for an imaginary time argument,

$$\rho = e^{\beta(\Omega - \Omega_o)} \rho_o S(-i\beta, 0) .$$
(6.9)

Defining full and unperturbed ensemble averages of an observable O as trace with the respective density operators,

$$\langle O \rangle = \text{Tr}\,(\rho O)\ ; \quad \langle O \rangle_o = \text{Tr}\,(\rho_o O) \tag{6.10}$$

Equation (6.9) may be used to establish a connection between them,

$$\langle O \rangle = e^{\beta(\Omega - \Omega_o)} \langle S(-i\beta, 0)O \rangle_o\ . \tag{6.11}$$

In particular, using the normalization (6.6), one obtains, for $O = 1$

$$e^{-\beta(\Omega - \Omega_o)} = \langle S(-i\beta, 0) \rangle_o\ , \tag{6.12}$$

so that Eq. (6.11) may be written in the form

$$\langle O \rangle = \langle S(-i\beta, 0)O \rangle_o / \langle S(-i\beta, 0) \rangle_o \tag{6.13}$$

in which the interaction $\mathcal{H}_{\text{int}}$ appears only in the time-evolution operator $S(-i\beta, 0)$. Insertion of the Dyson series (5.22), (5.27) then yields an expansion in terms of $\mathcal{H}_{\text{int}}$, so that Eq. (6.13) provides the basis of perturbation theory.

More explicitly, we express the operators in the chronological product (6.2) with the help of Eqs. (5.17) and (5.15) in the form

$$A(t) = S(0, t)A^o(t)S(t, 0) \tag{6.14}$$

in which the interaction is entirely localized in the S-factors. Making use of the fact that in a chronological product factors with different time arguments on the path C may be commuted freely, application of the group property (5.15) yields

$$T \prod_{j=1}^{n} A_j(t_j) = S(0, t_+)T\{S(t_+, t_-) \prod_{j=1}^{n} A_j^o(t_j)\}S(t_-, 0)\ . \tag{6.15}$$

Note that here Eq. (5.15) was also used to extend the segment of the path C covered by the times $t_i = t(\tau_i)$ in the chronological product to the complete path.

Taking the average (6.13) of the expression (6.15) we obtain the basic formula

$$\langle T \prod_{j=1}^{n} A_j(t_j) \rangle = \langle S(-i\beta, t_+)T\{S(t_+, t_-)$$

$$\times \prod_{j=1}^{n} A_j^o(t_j)\}S(t_-, 0) \rangle_o / \langle S(-i\beta, 0) \rangle_o \tag{6.16}$$

reducing a general *time-ordered n-point Green's function* to a form in which the perturbation $\mathcal{H}_{\text{int}}$ appears nowhere else than in the S-factors. A functional definition of the general Green's function (6.16) is given in Eq. (12.3) below. Now, as will be shown in Section 9, expressions of the form $\langle T \prod A_i^o(t_i)\rangle_o$ are easily evaluated by Wick's theorem. And according to Eqs. (5.27), (5.28) and (6.1), the numerator on the right of (6.16) has almost the form of a sum of such expressions, were it not for the two factors $S(-i\beta, t_+)$ and $S(t_-, 0)$ which stand outside of the chronological product. In fact, this situation is too general to be handled by the standard methods.

Thus, the condition for Eq. (6.16) to be treatable by standard perturbation theory is that the two troubling factors be trivial, or that

$$t_- = 0 ; \quad t_+ = -i\beta . \tag{6.17}$$

The obvious choice of the path C is a straight line between these two endpoints,

$$t = -i\tau ; \quad 0 \leq \tau \leq \beta \tag{6.18}$$

which leads to the *imaginary-time formalism* first introduced by Matsubara[5]. This is the usual treatment of finite-temperature perturbation theory, both in many-body problems[6-8] and in field theory[9] and will be used in the following chapters. The basic formula (6.16) now reads

$$\langle T \prod_{j=1}^{n} A_j(-i\tau_j)\rangle = \langle T\{S(-i\beta, 0) \prod_{j=1}^{n} A_j^o(-i\tau_j)\}\rangle_o / \langle S(-i\beta, 0)\rangle_o . \tag{6.19}$$

This, however, is not the only possibility since any path between the endpoints (6.17) leads to Eq. (6.19). A more general case is obtained by introducing a time shift $t \rightarrow \tilde{t} = t+s$ with an arbitrary complex value s. Since we are considering global equilibrium, i.e. a time-independent Hamiltonian $\mathcal{H}$, Eq. (5.9) holds so that, according to (5.7), we may write

$$A(t) = e^{-i\mathcal{H}s} A(\tilde{t}) e^{i\mathcal{H}s} . \tag{6.20}$$

Introducing the time-ordering $\tilde{T}$ on the shifted path $\tilde{C}$ defined by $\tilde{t}(\tau) = t(\tau) + s$ it then follows from (6.20) that

$$T \prod_{j=1}^{n} A_j(t_j) = e^{-i\mathcal{H}s} \tilde{T}\{\prod_{j=1}^{n} A_j(\tilde{t}_j)\} e^{i\mathcal{H}s} . \tag{6.21}$$

Since the density operator (6.4) commutes with $\mathcal{H}$, Eq. (6.21) leads to the same expression (6.16) but with a shifted path $\tilde{C}$. Therefore, applicability of the standard treatment now requires that the conditions (6.17) hold for the shifted endpoints $\tilde{t}_\pm$.

In Fig. 2.1 a particular path $\tilde{C}$ with these endpoints is drawn for a real positive value of s. Of particular interest is the case $s \to \infty$ since the path $\tilde{C}$ now reduces to the real axis run through in both directions. This gives rise to the *double real-time formalism* first introduced by Martin and Schwinger[10] and systematized by Keldysh[11]. For a slightly more general path see Fig. 2.4 of Ref. 9. This reference also gives the general condition, Eq. (2.1.21), a path C must satisfy.

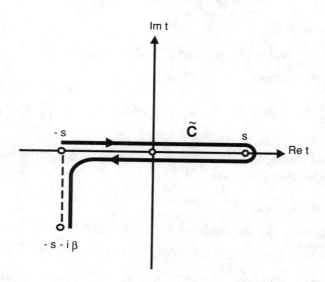

Fig. 2.1: Example of a shifted path $\tilde{C}$ in the complex t-plane.

It is obvious from the above discussion that the *standard real-time formalism* (see, e.g., Ref. 12) does not apply to finite temperatures. It is useful, however to show how it works in the limit of zero temperature , $\beta \to \infty$. Let $|\phi\rangle$ and $|\phi_o\rangle$ be, respectively, the groundstates of $\mathcal{H}$ and of $\mathcal{H}_o$ with fixed particle number N; then it follows from the closure of the respective eigenstates and from Eqs. (6.4), (6.7) that

$$\lim_{\beta \to \infty} \rho = |\phi\rangle\langle\phi| \; ; \quad \lim_{\beta \to \infty} \rho_o = |\phi_o\rangle\langle\phi_o| \; . \tag{6.22}$$

The path C now being the real axis and $t_{\pm} = \pm\infty$, the average (6.10) applied to Eq. (6.15) then yields

$$\langle\phi|T\prod_{j=1}^{n} A_j(t_j)|\phi\rangle = \langle\phi|S(0, +\infty)T\{S(+\infty, -\infty)\prod_{j=1}^{n} A_j^o(t_j)\}S(-\infty, 0)|\phi\rangle. \tag{6.23}$$

Assuming the interaction to be switched on adiabatically ,

$$\lim_{t \to -\infty} \mathcal{H}_{\text{int}} = 0 \tag{6.24}$$

it follows that

$$\lim_{t \to -\infty} \{e^{-i\mathcal{H}t}|\phi\rangle - e^{-i\mathcal{H}_o t}|\phi_o\rangle\} = 0 \tag{6.25}$$

or, in view of (6.8),

$$|\phi_o\rangle = S(-\infty, 0)|\phi\rangle \; ; \quad \langle\phi_o| = \langle\phi|S(0, -\infty) \; . \tag{6.26}$$

Hence (6.23) becomes

$$\langle\phi|T\prod_{j=1}^{n} A_j(t_j)|\phi\rangle = \langle\phi_o|S(-\infty, +\infty)T\{S(+\infty, -\infty)\prod_{j=1}^{n} A_j^o(t_j)\}|\phi_o\rangle \; . \tag{6.27}$$

The second assumption needed is that, up to a phase σ, the groundstate $|\phi\rangle$ is stable under the interaction,

$$\lim_{t \to +\infty} \{e^{-i\mathcal{H}t}|\phi\rangle - e^{i\sigma}e^{-i\mathcal{H}_o t}|\phi_o\rangle\} = 0 \; . \tag{6.28}$$

In analogy to (6.25) this condition implies that

$$e^{i\sigma}|\phi_o\rangle = S(+\infty, 0)|\phi\rangle = S(+\infty, -\infty)|\phi_o\rangle \tag{6.29}$$

which means that the S-matrix $S(+\infty, -\infty)$ acts trivially on the unperturbed groundstate. Therefore, Eq. (6.27) is equivalent to (compare Section 6 of Ref. 7)

$$\langle\phi|T\prod_{j=1}^{n} A_j(t_j)|\phi\rangle = \langle\phi_o|T\{S(+\infty, -\infty)$$

$$\times \prod_{j=1}^{n} A_j^o(t_j)\}|\phi_o\rangle/\langle\phi_o|S(+\infty, -\infty)|\phi_o\rangle \tag{6.30}$$

which has the same form as Eq. (6.19).

7. Linear Response Functions and Green's Functions

In the presence of a time-dependent external field, an additional term $\delta\mathcal{H}_t$ of the form (2.1) has to be added to the Hamiltonian $\mathcal{H}$ of the interacting system,

$$\mathcal{H}_t = \mathcal{H} + \delta\mathcal{H}_t , \qquad (7.1)$$

so that the perturbed density operator also becomes time-dependent,

$$\rho_t = \rho + \delta\rho_t . \qquad (7.2)$$

We now make the important assumption that the action of the perturbing field on the system is *nondissipative*, which means that the production of entropy, or Joule heat, is negligible. The physical interpretation of this assumption is that the external field varies so slowly that the system responds adiabatically, that is, without making transitions to other quantum states. This situation is called *local equilibrium*, in which the thermodynamic parameters T, μ, etc. may depend on the space-time coordinates $\mathbf{r}, t$ (see Section 13).

More explicitly, *adiabatic response* means that, in the statistical ensemble, the internal dynamics of the systems follow the imposed external variation. Formally, this implies that the time evolution of the Heisenberg representation (5.10) with the internal Hamiltonian $\mathcal{H}$ applied to ρ_t, compensates its parametric time evolution . In other words, $d(\rho_t(t))/dt = 0$. This is nothing else than *Liouville's theorem* which expresses constancy of phase volume during the time evolution of the systems in the ensemble[13,14]. Applying Eq. (5.8) to $\rho_t(t)$ and transforming back to the Schrödinger representation, Liouville's theorem then takes the form

$$\frac{d}{dt}\rho_t = -i[\mathcal{H}_t, \rho_t] . \qquad (7.3)$$

Dissipation may now equivalently be characterized as shrinking of phase volume. Of course, dissipative processes play an important role in nature, typical examples being the transport phenomena. Such processes will be discussed in Section 12 and in Chapter 3.

Combining (7.3) with (7.1) and (7.2) we obtain

$$\frac{d}{dt}\delta\rho_t = -i[\mathcal{H}, \delta\rho_t] - i[\delta\mathcal{H}_t, \rho] - i[\delta\mathcal{H}_t, \delta\rho_t] . \qquad (7.4)$$

The approximation of *linear response* theory now is to neglect the last term in (7.4) which is quadratic in the perturbation by the external field. Solution of Eq. (7.4) is easiest in the Heisenberg representation (5.10). Indeed, insertion of (5.8) with $A_t = \delta\rho_t$, $\mathcal{H}_t = \mathcal{H}$ leads to

$$\frac{d}{dt}(\delta\rho_t(t)) = -i[\delta\mathcal{H}_t(t), \rho] \ . \tag{7.5}$$

Assuming that the external field is switched on at time t_o, integration of (7.5) gives

$$\delta\rho_t(t) = -i \int_{t_o}^{t} dt' [\delta\mathcal{H}_{t'}(t'), \rho] \ . \tag{7.6}$$

Of particular interest is the case $t_o = -\infty$ in which the field is switched on adiabatically in analogy to Eq. (6.24). Back in Schrödinger representation, Eq. (7.6) then reads in this case

$$\delta\rho_t = -i \int_{-\infty}^{t} dt' [\delta\mathcal{H}_{t'}(t'-t), \rho] \ . \tag{7.7}$$

The change in the average value of an observable O due to this perturbation is given by the *Kubo formula* [15,16]

$$\delta\langle O\rangle_t \equiv \text{Tr}\,(\delta\rho_t O) = i \int_{-\infty}^{t} dt' \langle [\delta\mathcal{H}_{t'}, O(t-t')]\rangle \tag{7.8}$$

where in the second equality we have made use of the cyclicity of the trace, $\text{Tr}\,(AB\cdots C) = \text{Tr}\,(B\cdots CA)$, and of the fact that in the average defined by (6.10) the time dependence of the Heisenberg representation (5.10) may be shifted according to the time-translation invariance

$$\langle A(t)B(t')\rangle = \langle A(t+t_o)B(t'+t_o)\rangle \ . \tag{7.9}$$

This identity, valid for any t_o, is an immediate consequence of the commutativity of $\mathcal{H}$ and $\mathcal{H}^\mu$ and of the cyclicity of the trace. Writing the external-field Hamiltonian in terms of a density $b(\mathbf{q})$ as in Eq. (2.1),

$$\delta\mathcal{H}_t = \frac{1}{V}\sum_{\mathbf{q}} \tilde{F}(\mathbf{q}, t)b(-\mathbf{q}) \tag{7.10}$$

the Kubo formula (7.8) for a density $a(\mathbf{q})$ may now be cast into the form

$$\delta\langle a(\mathbf{q})\rangle_t = -\int_{-\infty}^{+\infty} dt' \tilde{F}(\mathbf{q}, t')\chi_{ab}(\mathbf{q}, t-t') \ . \tag{7.11}$$

Here we have introduced the *response function*

$$\chi_{ab}(\mathbf{q}; t) \equiv iV^{-1}\theta(t)\langle [a(\mathbf{q}, t), b(-\mathbf{q}, 0)]\rangle \tag{7.12}$$

where $\theta(t) \equiv (1 + \text{sgn } t)/2$ is the Heaviside function, and we made use of momentum conservation in the form (Problem 2.1)

$$\langle a(\mathbf{q})b(-\mathbf{q}')\rangle = \langle a(\mathbf{q})b(-\mathbf{q})\rangle\delta_{\mathbf{q},\mathbf{q}'} . \tag{7.13}$$

This relation readily follows from the translation invariance of the Fourier transforms according to (2.3),

$$\langle \tilde{a}(\mathbf{r})\tilde{b}(\mathbf{r}')\rangle = \langle \tilde{a}(\mathbf{r} + \mathbf{r}_o)\tilde{b}(\mathbf{r}' + \mathbf{r}_o)\rangle , \tag{7.14}$$

valid for any $\mathbf{r}_o$, which is obtained by analogy with Eq. (7.9) in terms of an operator of total momentum $\hbar\mathbf{P}$ commuting with $\mathcal{H}$ and N (Problem 2.1). For electrons and phonons $\mathbf{P}$ is explicitly given by Eqs. (3.41) and (3.46), respectively.

The response function χ_{ab} is of importance since, on the one hand, it makes direct contact with measurable quantities in that it expresses the action on a density a due to a weak external field coupled to the system through the density b. On the other hand, χ_{ab} also makes direct contact with calculable quantities in the form of *Green's functions*. Indeed, defining the *retarded two-point Green's function* as

$$G_{AB}^{\text{ret}}(t) = -i\theta(t)\langle[A(t), B(0)]_{\mp}\rangle \tag{7.15}$$

where the commutator/anticommutator applies for bosonic/fermionic operators, comparison of Eqs. (7.12) and (7.15) shows that, for densities a and b, which are observables and hence bosonic operators,

$$\chi_{ab}(\mathbf{q}; t) = -V^{-1}G_{a(\mathbf{q})b(-\mathbf{q})}^{\text{ret}}(t) . \tag{7.16}$$

In turn, the retarded Green's function is analytically connected with the *time-ordered two-point Green's function* or *propagator* $G_{AB}(t)$ defined (in accordance with Refs. 7 and 8) by

$$iG_{AB}(t) \equiv \langle T\{A(t)B(0)\}\rangle = \theta(t)\langle A(T)B(0)\rangle \pm \theta(-t)\langle B(0)A(t)\rangle , \tag{7.17}$$

which is a special case of the n-point Green's function introduced in Eq. (6.16). Here the upper/lower sign applies for bosonic/fermionic operators and the time-ordering operator T coincides with the general definition (5.26). Defining an auxiliary function

$$i\Gamma_{AB}(z) = \int_0^\infty dt \langle A(t)B(0)\rangle e^{izt} \tag{7.18}$$

which is analytic in the upper half-plane, $\text{Im } z > 0$, one finds for the Fourier-transformed propagator, making use of (7.9),

$$\tilde{G}_{AB}(\omega) \equiv \int_{-\infty}^{+\infty} dt G_{AB}(t) e^{i\omega t} = \Gamma_{AB}(\omega + i\epsilon) \pm \Gamma_{BA}(-\omega + i\epsilon) , \quad (7.19)$$

$\epsilon = 0^+$. On the other hand, the Fourier-transformed retarded function which is also analytic in $\text{Im } z > 0$, is given by

$$\tilde{G}_{AB}^{\text{ret}}(z) \equiv \int_{-\infty}^{+\infty} dt G_{AB}^{\text{ret}}(t) e^{izt} = \Gamma_{AB}(z) \pm \left(\Gamma_{A^+ B^+}(-z^*)\right)^* . \quad (7.20)$$

Therefore, knowledge of the propagator (7.17) allows to calculate in turn Γ_{AB} through (7.18), G_{AB}^{ret} through (7.20) and hence the response function through (7.16).

In view of the imaginary-time formalism to be used in subsequent chapters the following *imaginary-time propagator* defined on the path (6.18) is of particular importance (our sign convention is the same as in Refs. 7 and 8):

$$-\mathcal{G}_{AB}(\tau) \equiv \langle T\{A(-i\tau)B(0)\}\rangle$$
$$= \theta(\tau)\langle A(-i\tau)B(0)\rangle \pm \theta(-\tau)\langle B(0)A(-i\tau)\rangle . \quad (7.21)$$

Here the last term is actually zero because of (6.18). However, a continuation of $\mathcal{G}_{AB}$ beyond this interval is desirable in order to have a Fourier transform. This requires a relation between $\langle A(-i\tau)B(0)\rangle$ and $\langle B(0)A(-i\tau)\rangle$. Such a relation is easily obtained if the dynamics of the system is also defined with $\mathcal{H}^\mu = \mathcal{H} - \mu N$, as is the statistics through (6.4) and (6.8). In other words, the Heisenberg time evolution (5.9) should be redefined as

$$U(t, t_o) = e^{-i\mathcal{H}^\mu (t-t_o)} . \quad (7.22)$$

The justification runs as follows (see Section 8 of Ref. 7): The dynamics is determined by the eigenstates and eigenvalues of $\mathcal{H}$ which may be obtained from the extremum condition $\delta\langle\psi|\mathcal{H}|\psi\rangle = 0$, supplemented with orthogonality conditions and with the requirement of fixed particle number $\delta\langle\psi|N|\psi\rangle = 0$. In terms of a Lagrange parameter μ, the chemical potential, the problem reduces to $\delta\langle\psi|\mathcal{H}^\mu|\psi\rangle = 0$, plus orthogonality conditions.

Defining the Heisenberg representation (5.7) with (7.22), nothing is changed in what has been discussed so far, except for the following relation which was not true before, whose derivation is analogous to that of Eq. (7.9),

$$\langle A(t)B(t')\rangle = \langle B(t')A(t + i\beta)\rangle . \quad (7.23)$$

Applied to Eq. (7.21) one finds

$$\mathcal{G}_{AB}(\tau) = \pm\mathcal{G}_{AB}(\tau+\beta) \; ; \quad -\beta < \tau < 0 \tag{7.24}$$

which for $\tau = 0$ expresses a periodic/antiperiodic boundary condition on the interval (6.18). It is thus possible to write $\mathcal{G}_{AB}$ as a Fourier series

$$\mathcal{G}_{AB}(\tau) = \beta^{-1} \sum_{\nu} \tilde{\mathcal{G}}_{AB}(i\nu)e^{-i\nu\tau} \tag{7.25}$$

with the condition

$$e^{-i\nu\beta} = \pm 1 \; ; \quad \nu = \nu_\pm \, , \tag{7.26}$$

which determines the *Matsubara frequencies* [5]

$$\begin{matrix} \nu_+ \equiv 2n\pi/\beta & ; & \text{bosonic} \\ \nu_- \equiv (2n+1)\pi/\beta & ; & \text{fermionic} \end{matrix} \Bigg\} \text{ operators }; \quad n = 0, \pm 1, \pm 2, \dots \, . \tag{7.27}$$

Since

$$\beta^{-1} \int_0^\beta d\tau\, e^{i(\nu-\nu')\tau} = \delta_{\nu\nu'} \tag{7.28}$$

the inverse of (7.25) becomes

$$\tilde{\mathcal{G}}_{AB}(i\nu) = \int_0^\beta d\tau\, \mathcal{G}_{AB}(\tau)e^{i\nu\tau} \, . \tag{7.29}$$

In order to be able to compare with the retarded Green's function we express Eq. (7.21) for $\tau > 0$ in terms of the eigenstates $|n\rangle$ and eigenvalues E_n of $\mathcal{H}^\mu$,

$$-\mathcal{G}_{AB}(\tau) = \sum_{nn'} e^{\beta(\Omega-E_n)}\langle n|A|n'\rangle\langle n'|B|n\rangle e^{(E_n-E_{n'})\tau} \tag{7.30}$$

and insert into (7.29). Making use of (7.26) the result is

$$-\tilde{\mathcal{G}}_{AB}(i\nu) = \sum_{nn'} e^{\beta(\Omega-E_n)}\langle n|A|n'\rangle\langle n'|B|n\rangle \frac{\pm e^{\beta(E_n-E_{n'})}-1}{E_n-E_{n'}+i\nu} \tag{7.31}$$

which may also be written as

$$\tilde{\mathcal{G}}_{AB}(\pm i|\nu|) = \sum_{nn'} e^{\beta\Omega}\{e^{-\beta E_n} \mp e^{-\beta E_{n'}}\}\langle n|A|n'\rangle\langle n'|B|n\rangle$$

$$\times(-i) \int_0^{\pm\infty} dt\, e^{i(E_n-E_{n'}\pm i|\nu|)t} = -i \int_0^{\pm\infty} dt\, e^{\mp|\nu|t}\langle[A(t),B(0)]_\mp\rangle \, . \tag{7.32}$$

Thus comparison with (7.15), (7.20) yields by analytic continuation

$$\tilde{\mathcal{G}}_{AB}(z) = \tilde{G}_{AB}^{\text{ret}}(z) \; ; \quad \text{Im}\, z > 0 \, , \tag{7.33}$$

so that the imaginary-time propagators are seen also to contain all the relevant physical information.

Of particular importance in what follows are the *unperturbed electron and phonon propagators* of the form (7.21),

$$\mathcal{G}^o(k; \tau) \equiv -\langle T\{a_k^o(-i\tau)a_k^+\}\rangle_o \tag{7.34}$$

and

$$\mathcal{D}^o(q; \tau) \equiv -\langle T\{Q_q^o(-i\tau)Q_q^+\}\rangle_o \tag{7.35}$$

which will be represented by a straight line with an arrow and by a wavy line, respectively, running in time from 0 to $-i\tau$. In order to calculate them explicitly we first determine the time dependence of the operators a_k and b_q, integrating the equation of motion (5.18) with the unperturbed electron and phonon Hamiltonians (3.27) and (1.34), respectively, and making use of the relations (3.13) and (3.42). The result is

$$a_k^o(t) = e^{-i\varepsilon_k t}a_k \tag{7.36}$$

and

$$b_q^o(t) = e^{-i\omega_q t}b_q \ . \tag{7.37}$$

Noting that the time evolution of the propagators (7.34), (7.35) is given by $\mathcal{H}_o^\mu$ defined in (6.5) one finds, for $\tau > 0$,

$$\mathcal{G}^o(k; \tau) = -e^{-(\varepsilon_k - \mu)\tau}\langle a_k a_k^+\rangle_o \tag{7.38}$$

and, using Eqs. (1.31) and the fact that for phonons the chemical potential is zero,

$$\mathcal{D}^o(q; \tau) = -\frac{1}{2}\left\{e^{-\omega_q \tau}\langle b_q b_q^+\rangle_o + e^{+\omega_q \tau}\langle b_{-q}^+ b_{-q}\rangle_o\right\} \ . \tag{7.39}$$

Here the averages are easily evaluated in the number representation, (1.33) for the phonons and similarly for the electrons, to be (Problem 2.2)

$$\langle a_k^+ a_k\rangle_o = 1 - \langle a_k a_k^+\rangle_o = f_o(\varepsilon_k - \mu) \tag{7.40}$$

and

$$\langle b_q^+ b_q\rangle_o = -1 + \langle b_q b_q^+\rangle_o = n_o(\omega_q) \ , \tag{7.41}$$

where

$$f_o(\varepsilon) = 1 - f_o(-\varepsilon) = \frac{1}{e^{\beta\varepsilon} + 1} = \frac{1}{2}\left[1 - \tanh\frac{\beta\varepsilon}{2}\right] \tag{7.42}$$

and

$$n_o(\omega) = -1 - n_o(-\omega) = \frac{1}{e^{\beta\omega} - 1} = \frac{1}{2}\left[\coth\frac{\beta\omega}{2} - 1\right] \tag{7.43}$$

are the *Fermi* and *Bose distribution functions* , respectively. Taking the Fourier transform (7.29) of Eqs. (7.38) and (7.39), making use of (7.26), one finally obtains

$$\tilde{G}^o(k; i\nu_-) = \frac{1}{i\nu_- - \varepsilon_k + \mu} \qquad (7.44)$$

and

$$\tilde{D}^o(q; i\nu_+) = \frac{\omega_q}{(i\nu_+)^2 - \omega_q^2} = \frac{1}{2}\left(\frac{1}{i\nu_+ - \omega_q} - \frac{1}{i\nu_+ + \omega_q}\right) . \qquad (7.45)$$

8. Wick's Theorem for Operator Products

The aim of this and the next section is to develop the calculational scheme of *Feynman diagrams* or graphs to evaluate expressions of the form (6.19) or (6.30). In this procedure it is usual to invoke a *"Wick's theorem"*. In its original form introduced by Wick[17] this was an algebraic identity to develop chronological operator products into a sum of normal-ordered products containing c-number factors arising from certain operator pairings (see definitions (8.4) and (8.2) below). The key feature of these normal products being that they are annihilated by the unperturbed vacuum projector (6.22), the result for an expression (6.30) then simply is a sum of products of such c-number pairings, which gives rise to zero-temperature Feynman diagrams. Such diagrams were introduced for the first time in quantum electrodynamics and in meson theory by Feynman to calculate vacuum polarization and selfenergy effects[18].

Since normal products are not annihilated by the unperturbed finite-temperature density operator (6.7), the hope was that a Wick's-theorem factorization would remain valid at least for the ensemble averages (6.19). Matsubara[5] showed by a suitable redefinition of the annihilating part of the operators in the normal product that this is indeed the case (see also Ref. 19). Unfortunately, however, the physical meaning of creation and annihilation operators (see Section 3) is lost by Matsubara's trick. But this loss is actually unnecessary. Indeed, it has been shown by Enz[20] that, contrary to common belief (see Ref. 7, Section 12.2 and Ref. 8, Section 24), the ensemble average of ordinary normal products also leads to a perfectly physical Wick's-theorem type factorization which, together with the c-number pairings mentioned above, lead to the Matsubara factorization. This formalism will be developed in the next section while the present section is concerned with the operator form of Wick's theorem.

The definition of normal product is based on the division into *positive-and negative-frequency parts* $A^{(\pm)}$ of *irreducible operators* A which occur in the interactions $\mathcal{H}_{\text{int}}$. These objects, called "simple" factors in Ref. 17, are linear combinations of creation and annihilation operators (3.8),

$$A = A^{(+)} + A^{(-)} \; ; \quad A^{(-)}|\phi_o\rangle = \langle\phi_o|A^{(+)} = 0 \tag{8.1}$$

where $|\phi_o\rangle$ is the groundstate defined in (6.22). In the case of bosons/fermions the $A^{(\pm)}$ satisfy commutation/anticommutation relations (3.10),

$$
\begin{aligned}
[A^{(-)}, B^{(+)}]_{\mp} &\equiv \underline{AB} = c - \text{number} \\
[A^{(-)}, B^{(-)}]_{\mp} &= [A^{(+)}, B^{(+)}]_{\mp} = 0 \; .
\end{aligned}
\tag{8.2}
$$

From the free Hamiltonians of phonons and electrons, Eqs. (1.29) and (3.27), and from their interactions (2.22), (2.20) and (3.38) it is seen with the help of Eqs. (1.31) and (3.22) that in these cases the irreducible operators A are Q_q, P_{-q} and a_k, together with their hermitean conjugates. And the respective decomposition into positive- and negative-frequency parts $(A^{(+)}, A^{(-)})$ is $2^{-1/2}(b_{-q}^+, b_q)$, $i2^{-1/2}(b_{-q}^+, -b_q)$ and $(0, a_k)$. This shows that in second quantization phonons behave as *fields* (equal amounts of positive and negative frequency parts) and electrons as *particles* (positive or negative frequency parts only). This identification also shows that, in general, $(A^+)^{(\pm)} = (A^{(\mp)})^+$ and that $A^{(\pm)}$ satisfies an unperturbed equation of motion of the form

$$\dot{A}^{(\pm)} = i[\mathcal{H}_o, A^{(\pm)}] = \pm i\varepsilon_A A^{(\pm)} \tag{8.3}$$

which explains the meaning of positive/negative frequencies.

The *normal product* of irreducible operators, $W\{A_1 \cdots A_n\}$, called "S-product" by Wick and designated as $:A_1 \cdots A_n:$[17], is now defined by the *Wick operator* W which shifts (commutes/anticommutes) the bosonic/fermionic positive-frequency parts $A_j^{(+)}$ in the ordinary product $A_1 \cdots A_n$ to the left of all the negative-frequency parts $A_j^{(-)}$,

$$W\{A_1 \cdots A_n\} = \sum_{l=0}^{n} \sum_{P_l} (-1)^{\eta(P_l)} A_{i_1}^{(+)} \cdots A_{i_l}^{(+)} A_{i_{l+1}}^{(-)} \cdots A_{i_n}^{(-)} \; . \tag{8.4}$$

Here $P_l = (1 \cdots n \rightarrow i_1 \cdots i_n)$ are the $n!/l!(n-l)!$ permutations such that $i_1 < i_2 \cdots < i_l$ and $i_{l+1} < i_{l+2} \cdots < i_n$, and $\eta(P_l)$ is the number of transpositions between fermionic operators in P_l (fermionic operators are supposed to commute with bosonic ones). From this definition the linearity

$$W\{A_1 \cdots A_n(\lambda B + \mu C)\} = \lambda W\{A_1 \cdots A_n B\} + \mu W\{A_1 \cdots A_n C\} \tag{8.5}$$

is obvious, as is the relation

$$W\{A_1 \cdots A_n\}B^{(-)} - W\{A_1 \cdots A_n B^{(-)}\} = 0 \ . \tag{8.6}$$

Making use of the *"normal contraction* ⌣*"* defined in (8.2), which was first introduced in the Stueckelberg school by Houriet and Kind[21] and was generalized to include fermionic operators by Wick[17], one also shows that (Problem 2.3)

$$W\{A_1 \cdots A_n\}B^{(+)} - W\{A_1 \cdots A_n B^{(+)}\} = \sum_{k=1}^{n} W\{A_1 \cdots \underbrace{A_k \cdots A_n\}B}_{}{}^{(+)} \ . \tag{8.7}$$

Combining Eqs. (8.6) and (8.7) and making use of the first relation (8.1) and of the linearity (8.5) one obtains

$$W\{A_1 \cdots A_n\}B - W\{A_1 \cdots A_n B\} = \sum_{k=1}^{n} W\{A_1 \cdots \underbrace{A_k \cdots A_n B}_{}\}$$

$$\equiv W\{\sum{}'(\smile)A_1 \cdots A_n B\} \ . \tag{8.8}$$

In the first form of Eq. (8.8), which was already given by Wick (Eq. (14) of Ref. 17, see also Ref. 12, p. 165), the contraction has to be taken *before* the normal product. This rule is explicited in the second form in which the sum runs over all contractions ⌣, and the prime indicates exclusion of contractions among the $A's$. The mentioned rule is a consequence of the fact that for any A and B, $\underbrace{A^{(+)}B}_{} = \underbrace{AB}_{}{}^{(-)} = 0$ which follows from (8.1) and (8.2) and which, according to definition (8.4), implies that any ⌣-contraction taken in a normal product vanishes.

With the help of (8.8) and with the identity $W\{A\} = A$, it is possible to derive the following relation (Problem 2.4) which allows to express the interactions (2.20) and (2.22) in normal order (Eq. (3.38) is already in this form)

$$W\{A_1 \cdots A_n\} = \left[1 - \sum(\smile) + \sum(\smile)^2 - + \cdots\right]A_1 \cdots A_n \tag{8.9}$$

and its inverse (Problem 2.4)

$$A_1 \cdots A_n = W\left\{\left[1 + \sum(\smile) + \sum(\smile)^2 + \cdots\right]A_1 \cdots A_n\right\} \ . \tag{8.10}$$

Obviously, in sums over multiple contractions, $\sum(\smile)^n$, the fact has to be observed that any operator may be contracted only once. By application of Eqs. (8.9) and (8.10) one arrives at the general expansion formula for a product of normal products (Problem 2.5)

$$W\{A_1 \cdots A_n\}W\{B_1 \cdots B_m\} \cdots$$
$$= W\{[1 + \sum{}'(\smile) + \sum{}'(\smile)^2 + \cdots]A_1 \cdots A_n B_1 \cdots B_m \cdots\}$$

(8.11)

where the prime indicates exclusion of contractions within the original W-products. Equation (8.11) is mentioned by Bogoliubov and Shirkov as "Wick's theorem for normal products" (Ref. 12, Section 16.2, in particular p. 166). Wick 's original two theorems[17], however, concern chronological products.

The chronological product in Eqs. (6.19) or (6.30) contains the unperturbed time evolution (5.17) so that the existence of a normal product of operators $A^o(t)$ is not evident since irreducibility is apparently lost. Fortunately however, the equation of motion (8.3) ensures that this time dependence is trivial and has the form of Eqs. (7.36), (7.37),

$$A^{(\pm)o}(t) = e^{\pm i\varepsilon_A t} A^{(\pm)} ,$$

(8.12)

so that the contraction (8.2) also exists for irreducible operators with different arguments of their unperturbed time evolution. Defining the time ordering of this contraction along an arbitrary path C in the complex time plane such that $t = t(\tau)$ and $t' = t(\tau')$ we obtain again a c-number,

$$\overline{A^o(t)B^o(t')} \equiv T\{A^o(t)B^{\,o}(t')\} = \begin{cases} \underline{A^o(t)B^{\,o}(t')}; & \tau > \tau' \\ \pm \underline{B^o(t')A^{\,o}(t)}; & \tau < \tau' , \end{cases}$$

(8.13)

called "*spontaneous contraction*" [20] because it is independent of the ensemble average. It is indeed the only contraction occurring in Wick's theorem at zero temperature.

Since, according to (8.12) all time dependences are expressed by trivial factors, Eq. (8.10) is also valid with arbitrary time arguments $t_i = t(\tau_i)$ attached to the A_i. Applying now the T-operator to (8.10) it is easily seen with the help of the definitions (6.2) and (8.13) that

$$T\{A_1^o(t_1) \cdots A_n^o(t_n)\}$$
$$= W\{[1 + \sum(\frown) + \sum(\frown)^2 + \cdots]A_1^o(t_1) \cdots A_n^o(t_n)\} \quad (8.14)$$

which is *Wick's first theorem*[17]. In the same way we may also attach time arguments $t_A = t(\tau_A)$, $t_B = t(\tau_B)$, $\cdots$, respectively to each W-product on the left of (8.11) and apply the T-operator. The result is *Wick's second theorem* [17] (see also Ref. 12, Section 19.2)

$$T\{W\{A_1^o(t_A)\cdots A_n^o(t_A)\}W\{B_1^o(t_B)\cdots B_m^o(t_B)\}\cdots\}$$
$$= W\{[1+\sum{}'(\frown)+\sum{}'(\frown)^2$$
$$+\cdots]A_1^o(t_A)\cdots A_n^o(t_A)B_1^o(t_B)\cdots B_m^o(t_B)\cdots\} \qquad (8.15)$$

where the prime on the summation signs again means exclusion of contractions within the original W-products.

A more explicit form of Wick's first theorem (8.14) may be obtained for an even number, $2n$, of factors by introducing the sum over all permutations $P_{2n} = (1\cdots 2n \to i_1\cdots i_{2n})$, namely [20]

$$T\{\prod_{j=1}^{2n}A_j^o(t_j)\} = \sum_{P_{2n}}(-1)^{\eta(P_{2n})}\sum_{k=0}^{n}\frac{1}{2^k k!}\frac{1}{(2(n-k))!}W\{(\frown)^k\prod_{j=1}^{2n}A_{i_j}^o\}\,.$$
$$(8.16)$$

Here the factors $1/2^k k!$ and $1/(2(n-k))!$ correct for the redundancy caused by the permutations P_{2n} in the k contractions and in the remaining W-product, respectively.

It is obvious from the properties (8.1) and definition (8.4) that, as mentioned at the beginning of this section, the result (8.15) is all what is needed to reduce the zero-temperature expression (6.30) to *Feynman diagrams* in which all operators are paired off (see the next section for the general definition),

$$\langle\phi_o|T\{W\{A_1^o(t_A)\cdots A_n^o(t_A)\}W\{B_1^o(t_B)\cdots B_m^o(t_B)\}\cdots\}|\phi_o\rangle$$
$$= \sum{}'(\frown)^N A_1^o(t_A)\cdots A_n^o(t_A)B_1^o(t_B)\cdots B_m^o(t_B)\cdots\,. \qquad (8.17)$$

Here $2N$ is the total number of operators A, B, $\cdots$ (which obviously must be even in order for (8.17) to be non-zero) and the sum runs over all possible N pairings, the prime indicating exclusion of pairings between operators with the same time argument. At finite temperature only averages may be reduced in this way, as will be shown in the next section.

9. Wick's Theorem for Averages. Feynman Diagrams

The problem here is the factorization of unperturbed averages of normal products which will give rise to a third type of contraction. This is achieved with the help of the following formula for bosonic creation and annihilation operators a^+ and a (Problem 2.6), first derived independently by Thouless[19] and by Bloch and de Dominicis[22],

$$\langle (a^+)^n (a)^n \rangle_o = n! \, (\langle a^+ a \rangle_o)^n \,, \tag{9.1}$$

valid for all integers $n \geq 0$. Since this formula is trivially valid for fermionic operators and $n = 0, 1$, it is readily generalized to the form[20]

$$\langle \prod_{j=1}^n A_j^{(+)} \prod_{l=1}^m B_l^{(-)} \rangle_o = \delta_{nm} \sum_P (-1)^{\eta(P)} \prod_{j=1}^n \langle A_j^{(+)} B_{i_j}^{(-)} \rangle_o \tag{9.2}$$

where the sum runs over the $n!$ permutations $P = (1 \cdots n \to i_1 \cdots i_n)$ and $\eta(P)$ is the number of transpositions between fermionic operators.

Because of statistical independence (factorization of ρ_o, Eq. (6.7)), Eq. (9.2) may be separated into a bosonic part and a fermionic part. The latter is then seen to give rise on the right to the determinant of the matrix $\langle A_i^{(+)} B_j^{(-)} \rangle_o$ which expresses of course just the Pauli exclusion principle. Therefore, both sides of Eq. (9.2) vanish for any repetition of a fermionic operator so that it is unnecessary to exclude such repetitions explicitly. This is an expression of the *general rule* well-known in perturbation theory (see, e.g., Ref. 23, pp. 42, 51 and Ref. 24, p. 68) that *in unperturbed averages, pairings (Feynman diagrams) may be drawn irrespective of the Pauli exclusion principle*, the latter taking care automatically of the necessary cancellations.

Considering now unperturbed averages of W-products one first notices that only terms with equal numbers of positive- and negative-frequency parts survive, and hence only W-products with an even number of factors are non-zero. Applying Eq. (9.2) to the definition (8.4), one then finds

$$\langle W\{A_1 \cdots A_{2n}\} \rangle_o = \sum_{P_n} (-1)^{\eta(P_n)} \langle A_{i_1}^{(+)} \cdots A_{i_n}^{(+)} A_{i_{n+1}}^{(-)} \cdots A_{i_{2n}}^{(-)} \rangle_o$$

$$= \frac{1}{n!} \sum_P (-1)^{\eta(P)} \prod_{j=1}^n \langle A_{i_j}^{(+)} A_{i_{n+j}}^{(-)} \rangle_o \tag{9.3}$$

where the last sum runs over the $(2n)!$ permutations $P = (1 \cdots 2n \to i_1 \cdots i_{2n})$. Introducing the symmetrized average

$$\widehat{AB} \equiv \langle W\{AB\} \rangle_o = \langle A^{(+)} B^{(-)} \rangle_o \pm \langle B^{(+)} A^{(-)} \rangle_o \tag{9.4}$$

called "*induced contraction*" [20] because it exists only at finite temperature, Eq. (9.3) may finally be written in the form of a *Wick's theorem for averages of W -products* [20)]

$$\langle W\{A_1\cdots A_{2n}\}\rangle_o = \sum(\frown)^n A_1\cdots A_{2n} = \frac{1}{2^n n!}\sum_P (-1)^{\eta(P)}(\frown)^n \prod_{j=1}^{2n} A_{i_j} .$$
$$(9.5)$$

In the first expression the sum runs over all possible n pairings, giving rise to $(2n-1)!! = 2n!/2^n n!$ terms, which is manifestly the same number as that in the second expression where the sum runs over all $(2n)!$ permutations P.

Applying the second form of (9.5) to the k-th term of the averaged expression (8.16), one first notices that the sum over the permutations $P_{2(n-k)}$ is redundant and cancels against the factor $1/(2(n-k))!$. Thus

$$\langle T\{\prod_{j=1}^{2n} A_j^o(t_j)\}\rangle_o = \frac{1}{2^n n!}\sum_P (-1)^{\eta(P)}$$

$$\times \sum_{k=0}^{n} \frac{n!}{k!(n-k)!}(\frown)^k(\frown)^{n-k} \prod_{j=1}^{2n} A_{i_j}^o(t_{i_j}) . \quad (9.6)$$

But here the k-sum is just the binomial series so that, with $\underline{\;\;} \equiv \frown + \smile$,

$$\langle T\{A_1^o(t_1)\cdots A_{2n}^o(t_{2n})\}\rangle_o = \frac{1}{2^n n!}\sum_P (-1)^{\eta(P)}(\underline{\;\;})^n \prod_{j=1}^{2n} A_{i_j}^o(t_{i_j})$$

$$= \sum(\underline{\;\;})^n A_1^o(t_1)\cdots A_{2n}^o(t_{2n}) \quad (9.7)$$

which is *Wick's first theorem for averages of T-products* [20)] (see Section 12 for a functional derivation of this theorem). The c-number introduced here,

$$\overline{A^o(t)B^o(t')} \equiv A^o(t)B^o(t') + \overbrace{A^o(t)B}^{}{}^o(t') , \quad (9.8)$$

is called "*chronological contraction*" because (9.8) is just the unperturbed propagator,

$$\langle T\{A^o(t)B^o(t')\}\rangle_o = \underline{A^o(t)B^o(t')} . \quad (9.9)$$

This follows either as the special case $n = 1$ of Eq. (9.7) or by making use, on the left, of the special case $AB = W\{AB\} + \underline{AB}$ of Eq. (8.10), of the special case $T\{W\{A^o(t)B^o(t')\}\} = W\{A^o(t)B^o(t')\}$ of Eq. (8.15) and of the definitions (8.13) and (9.4).

In order to generalize Eq. (9.7) to the average of (8.15) we introduce the *rule*

$$A^o(t)B^o(t') = T\{A^o(t)B^o(t')\} - W\{A^o(t)B^o(t')\} = 0 \; ; \quad \text{if } t = t' \, , \quad (9.10)$$

which is possible because definition (8.13) does not include this case. The effect is that the sums in (8.15) are now unrestricted as in (8.14), and the result (9.7) holds again,

$$\langle T\{W\{A_1^o(t_A)\cdots A_n^o(t_A)\}W\{B_1^o(t_B)\cdots B_m^o(t_B)\}\cdots\}\rangle_o$$

$$= \sum (\underline{\quad})^N A_1^o(t_A)\cdots A_n^o(t_A)B_1^o(t_B)\cdots B_m^o(t_B)\cdots \quad (9.11)$$

where, as in Eq. (8.17), $2N$ is the total number of operators A, B, $\cdots$ and the rule (9.10) has to be used. Equation (9.11) is *Wick's second theorem for averages of T-products* .

The T-product in Eqs. (6.19) or (6.30) consists of terms

$$T_N \equiv \frac{1}{N!}\langle T\{\mathcal{A}^N \prod_j A_j^o(t_j)\}\rangle_o \quad (9.12)$$

where $\mathcal{A}$ is the action defined in (5.25), taken between the endpoints t_- and t_+ of a given time path C. A *Feynman diagram* is now defined as a *term in the Wick's theorem factorization* (9.11) *of* T_N in which the propagators (9.9) are represented by *lines*, and lines are connected by the *vertices* representing the interactions (see the examples of Figs. 1.1 and 1.2, Chapter 1). Propagators containing one (two) of the A_j are *external lines* which have one (two) open end(s). A diagram without open ends will be called *closed* .

A first classification of Feynman diagrams is obtained by the notion of connectedness. A *connected diagram* cannot be separated into two parts without cutting at least one internal line. We call *connected part* of a term T_N, and mark it by an index cp, a diagram which contains no disconnected closed parts. Thus a diagram of T_N will factorize into its connected part and l disconnected closed parts where $0 \le l \le N$. Since this factorization may be made in $N!/l!(N-l)!$ indistinguishable ways we get from (9.12)

$$T_N = \sum_{l=0}^{N} \frac{1}{l!(N-l)!}\langle T\{\mathcal{A}^{N-l} \prod_j A_j^o(t_j)\}\rangle_{o,cp}\langle T\{\mathcal{A}^l\}\rangle_o \, . \quad (9.13)$$

Summing over N from 0 to ∞ and interchanging the N- and l-summations one finds, making use of Eq. (5.28)[20],

$$\langle T\{S(t_+, t_-) \prod_j A_j^o(t_j)\}\rangle_o = \langle T\{S(t_+, t_-) \prod_j A_j^o(t_j)\}\rangle_{o,cp}\langle S(t_+, t_-)\rangle_o \, .$$

$$(9.14)$$

Since this is valid for any time path C and also in the zero-temperature limit, we find that Eqs. (6.19) and (6.30) reduce, respectively, to

$$\langle T\{\prod_j A_j(-i\tau_j)\}\rangle = \langle T\{S(-i\beta,\,0)\prod_j A_j^o(-i\tau_j)\}\rangle_{o,cp} \qquad (9.15)$$

and

$$\langle\phi|T\{\prod_j A_j(t_j)\}|\phi\rangle = \langle\phi_o|T\{S(+\infty,\,-\infty)\prod_j A_j^o(t_j)\}|\phi_o\rangle_{cp}\,. \qquad (9.16)$$

Equation (9.15) combined with Wick's theorem (9.7) or (9.11) constitutes the basis for any perturbative calculation made in this book.

As a first application we remark that, replacing the integration limit t_+ in A by an arbitrary value t on the path C gives rise to a modified definition of "connected part" in (9.13) but does not change the algebra leading to (9.14), and neither does the substitution of $\prod A_j^o(t_j)$ by $\mathcal{H}_{\text{int}}^o(t)$. Since, according to Eqs. (5.25), (5.28),

$$\frac{\partial}{\partial t}S(t,\,t_-) = -iT\{S(t,\,t_-)\mathcal{H}_{\text{int}}^o(t)\}\,, \qquad (9.17)$$

the modified Eq. (9.14) becomes

$$\langle\frac{\partial}{\partial t}S(t,\,t_-)\rangle_o = \langle\frac{\partial}{\partial t}S(t,t_-)\rangle_{o,\,\text{``}cp\text{''}}\langle S(t,t_-)\rangle_o \qquad (9.18)$$

where the index "cp" designates the modified definition of "connected part". But since t is arbitrary, the derivative may be taken outside of the average, which leads to a differential equation for $\langle S(t,\,t_-)\rangle_o$. Because of $S(t_-,\,t_-) = 1$ the solution is

$$\ln\langle S(t,\,t_-)\rangle_o = \langle S(t,\,t_-)\rangle_{o,\,\text{``}cp\text{''}} - 1\,. \qquad (9.19)$$

Replacing again t by t_+ we are back to true connected parts so that with the thermodynamic endpoints (6.17) and with the help of Eq. (6.12) we arrive at the result (see Ref. 7, Eq. (15.10) and also Ref. 24, p. 85)

$$\Omega - \Omega_o = -\beta^{-1}\langle S(-i\beta,\,0) - 1\rangle_{o,cp}\,. \qquad (9.20)$$

This formula expresses the many-body effects on the grand-canonical potential and hence on any thermodynamic quantity in terms of closed connected diagrams.

Another interesting formula is obtained in taking the zero-temperature limit of Eq. (9.20). One finds, either by direct integration or by the resolvent method (Problem 2.7),

$$E - E_o = \langle\phi_o|\mathcal{H}_{\text{int}}\sum_{N=0}^{\infty}\left(\frac{1}{E_o - \mathcal{H}_o}\mathcal{H}_{\text{int}}\right)^N|\phi_o\rangle_{cp}\,, \qquad (9.21)$$

where E and E_o are the exact and unperturbed groundstate energies, respectively. Equation (9.21) is the so-called *linked cluster expansion* (Ref. 24, p. 72, Ref. 25, p. 126) or *Goldstone theorem* (Ref. 8, p. 111) first derived by Goldstone[26] (see also Ref. 27). The approximate solution of this problem was discussed in Section 4 by means of the Hartree-Fock method of a self-consistent field.

10. Thermodynamic Feynman Diagrams

It is convenient to classify diagrams according to the number of external lines as defined in the last section. In this section we consider closed diagrams (without external lines). They will be called *thermodynamic diagrams* because they describe the many-body corrections to thermodynamic quantities contained in Eq. (9.20). They are the generalization to finite temperatures of the vacuum diagrams of field theory. However, most quantities such as the specific heat at constant volume V (often taken as volume of one mole), $c_V = T(\partial s/\partial T)_n$, are measured at constant particle number density n whereas the grand-canonical potential Ω is given as a function of the chemical potential μ. In terms of Ω, n and the entropy density s are given, respectively, by $n = -V^{-1}(\partial\Omega/\partial\mu)_T$ and $s = -V^{-1}(\partial\Omega/\partial T)_\mu$.

For free electrons, Ω_o in Eq. (9.20) has the well-known form derived for Fermi statistics[13,14]

$$\Omega_o = -\beta^{-1}\sum_k \ln(1 + e^{-\beta(\varepsilon_k - \mu)}) \tag{10.1}$$

where the label k is defined in Section 1. Introducing the Fermi distribution function f_o defined in Eq. (7.42) and making use of the identity $(1 - f_o(\varepsilon))/f_o(\varepsilon) = \exp(\beta\varepsilon)$, the entropy density may be cast into the form

$$s_o = -\frac{1}{V}\left(\frac{\partial\Omega_o}{\partial T}\right)_\mu = -k_B V^{-1}\sum_k \{f_o \ln f_o + (1 - f_o)\ln(1 - f_o)\} \tag{10.2}$$

and the specific heat at low temperatures becomes, for fixed particle number density n (Problem 2.8),

$$c_{Vo}(n, T) = T\left(\frac{\partial s_o}{\partial T}\right)_{\mu(n, T)} = \frac{2}{3}\pi^2 N(\mu)k_B^2 T + \mathcal{O}(T^3) \tag{10.3}$$

where

$$N_n(\varepsilon) = \frac{1}{2V}\sum_{k\sigma}\delta(\varepsilon - \varepsilon_{nk\sigma}) \tag{10.4}$$

is the *density of states per spin of band n.*

As an application, let us calculate the thermodynamic diagrams which give the lowest-order correction of the electron-phonon interaction (2.20) to Eq. (9.20). Making use of Eqs. (5.22), (5.27), (6.1) and (6.18) we have

$$-\beta\Delta\Omega = \langle S_2(-i\beta, 0)\rangle_{o,cp}$$
$$= \frac{1}{2}\int_0^\beta d\tau \int_0^\beta d\tau' \langle T\{\mathcal{H}^{\mathrm{el-ph}\ o}(-i\tau)\mathcal{H}^{\mathrm{el-ph}\ o}(-i\tau')\}\rangle_{o,cp} .$$

$$(10.5)$$

Neglecting Umklapp processes and using Eqs. (2.20) and (3.22) we may write

$$\mathcal{H}^{\mathrm{el-ph}} = \sum_{qk} \gamma_q a_k^+ a_{k-q} Q_q \qquad (10.6)$$

where we have put $\gamma_q \equiv \gamma_q(-\mathbf{q}) = \gamma*_{-q}$ and $k - q \equiv (\mathbf{k} - \mathbf{q}, \sigma)$.

Applying Wick's theorem (9.11) and taking into account all possibilities of pairings, the result may be written in the form

$$-\beta\Delta\Omega = \int_0^\beta d\tau \int_0^\beta d\tau' \sum_{qk} \sum_{q'k'} \gamma_q \gamma_{q'} (T^{\mathrm{H}} + T^{\mathrm{ex}}) \qquad (10.7)$$

where two types of contractions (9.9) occur which, in analogy to the Hartree-Fock problem of Section 4, we call the *"Hartree term"*

$$T^{\mathrm{H}} \equiv \langle T\{a_k^{+o}(-i\tau)a_{k-q}^o(-i\tau)Q_q^o(-i\tau)a_{k'}^{+o}(-i\tau')a_{k'-q'}^o(-i\tau')Q_{q'}^o(-i\tau')\}\rangle_{o,cp}$$

$$(10.8)$$

and the *"exchange term"*

$$T^{\mathrm{ex}} \equiv \langle T\{a_k^{+o}(-i\tau)a_{k-q}^o(-i\tau)Q_q^o(-i\tau)a_{k'}^{+o}(-i\tau')a_{k'-q'}^o(-i\tau')Q_{q'}^o(-i\tau')\}\rangle_{o,cp} .$$

$$(10.9)$$

With the aid of the translation-invariance relations (7.9) and (7.13), noting that each transposition of two fermionic operators gives a minus sign and that the equal-time propagators in (10.8) are defined according to the rule (9.10), one obtains

$$T^{\mathrm{H}} = -\mathcal{G}^o(k; 0)\delta_{\mathbf{q}, 0}\mathcal{G}^o(k'; 0)\delta_{\mathbf{q}', 0}\mathcal{D}^o(q; \tau - \tau')\delta_{q, -q'}$$
$$T^{\mathrm{ex}} = +\mathcal{G}^o(k; \tau' - \tau)\delta_{\sigma\sigma'}\delta_{\mathbf{q}', \mathbf{k}' - \mathbf{k}}$$
$$\times \mathcal{G}^o(k'; \tau - \tau')\delta_{\mathbf{q}, \mathbf{k} - \mathbf{k}'}\mathcal{D}^o(q; \tau - \tau')\delta_{q, -q'} \qquad (10.10)$$

where spin and mode conservation and the definitions (7.34) and (7.35) have been used. With these expressions Eq. (10.6) now has a simple representation in terms of Feynman diagrams as shown in Fig. 2.2 where γ_q is represented by the electron-phonon vertex of Fig. 1.1, Chapter 1, $\mathcal{G}^o$ by a straight line with arrow and $\mathcal{D}^o$ by a wavy line.

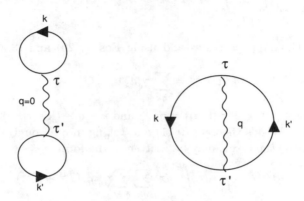

Fig. 2.2: Feynman diagrams of the "Hartree" and "exchange terms" (10.10) in the expression (10.7).

Making the simplifying assumption of a crystal without basis, $B = 1$, there is only one branch μ, namely the acoustic phonons, for which we may write $\omega_{\mathbf{q}j} = \hbar c_j |\mathbf{q}|$. Here c_j is the sound velocity of the longitudinal and transverse polarizations j. From Eq. (2.21) it then follows that only the longitudinal polarization $j = \|$ contributes,

$$\gamma_{\mathbf{q}} = -\frac{i}{v}(\hbar|\mathbf{q}|/LMc_{\|})^{1/2}\tilde{\phi}(-\mathbf{q}) \; . \tag{10.11}$$

It may be shown that the Hartree contribution (10.10) to (10.7) is just the constant (Problem 2.9)

$$\Delta\Omega^{\mathrm{H}} = -\frac{|\tilde{\phi}(0)|^2 n^2}{Mc_{\|}^2}L \tag{10.12}$$

where L is the number of unit cells in the crystal and

$$n \equiv \frac{2}{V}\sum_{\mathbf{k}} f_o(\varepsilon_k) \tag{10.13}$$

the density of conduction electrons. On the other hand, making use of the reality conditions (1.25), the exchange contribution (10.10) to (10.7) is

$$-\beta\Delta\Omega^{\text{ex}} = \int_0^\beta d\tau \int_0^\beta d\tau' \sum_{\mathbf{q}} \sum_{\mathbf{k}\sigma} |\gamma_{\mathbf{q}}|^2$$
$$\times \, \mathcal{G}^o(\mathbf{k}\sigma; \tau' - \tau)\mathcal{G}^o(\mathbf{k} - \mathbf{q}, \sigma; \tau - \tau')\mathcal{D}^o(\mathbf{q}; \tau - \tau') \quad (10.14)$$

Applying the Fourier transformation (7.25) and formula (7.28) this equation becomes

$$\Delta\Omega^{\text{ex}} = -\beta^{-2} \sum_{\mathbf{q}} \sum_{k} |\gamma_{\mathbf{q}}|^2 \sum_{\nu_-, \nu_+} \tilde{\mathcal{G}}^o(k; i\nu_-)$$
$$\times \, \tilde{\mathcal{G}}^o(k - q; i\nu_- - i\nu_+)\tilde{\mathcal{D}}^o(\mathbf{q}; i\nu_+) \,. \quad (10.15)$$

Comparison of this result with Fig. 2.2 shows the general *rule* that *each closed loop gives rise to a* $(\mathbf{q}, \nu)$*-summation* .

For the evaluation of the sums over the Matsubara frequencies we note that since, according to (7.26), the functions $(e^{\beta\omega} \mp 1)^{-1}$ have the poles $\omega = i\nu_\pm$, the following identity holds for any function $F(\omega)$ which is analytic in the vicinity of the imaginary axis (Problem 2.10):

$$\sum_{\nu_\pm} F(i\nu_\pm) = \pm\frac{\beta}{2\pi i} \int_\Gamma \frac{F(z)dz}{e^{\beta z} \mp 1} \,. \quad (10.16)$$

Here Γ is the path enclosing the imaginary axis in the positive sense shown in Fig. 2.3 below.

We first evaluate the sum over ν_- in Eq. (10.15) which according to (7.44) amounts to calculate the *"Lindhard function"* $L_-(\varepsilon_k - \mu, \varepsilon_{k-q} - \mu + i\nu_+)$ occurring, e.g. in the expression for the dielectric function of the electron gas (see Ref. 24, Chapter 4). Here we have defined, for any complex ω and ω',

$$L_\pm(\omega, \omega') \equiv \beta^{-1} \sum_{\nu_\pm} \frac{1}{i\nu_\pm - \omega} \frac{1}{i\nu_\pm - \omega'} \,, \quad (10.17)$$

to which we may apply formula (10.16),

$$L_\pm(\omega, \omega') = \pm\frac{1}{2\pi i} \int_\Gamma \frac{dz}{e^{\beta z} \mp 1} \frac{1}{z - \omega} \frac{1}{z - \omega'} \,. \quad (10.18)$$

Since the integrand vanishes faster than $|z|^{-1}$ at $|z| \to \infty$, the path Γ may be deformed to encircle the two poles at ω and ω' in the negative sense as indicated (for $\omega = \varepsilon_{\mathbf{k}}$, $\omega' = \varepsilon_{\mathbf{k}'} + i\nu_+$) in Fig. 2.3. Introducing the Bose and Fermi functions (7.43) and (7.42), respectively, the result is then easily seen to be

$$L_+(\omega, \omega') = -\frac{n_o(\omega) - n_o(\omega')}{\omega - \omega'} \; ; \quad L_-(\omega, \omega') = \frac{f_o(\omega) - f_o(\omega')}{\omega - \omega'} . \quad (10.19)$$

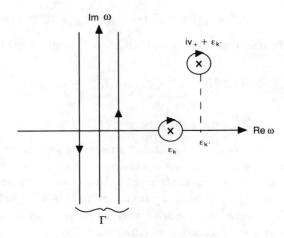

Fig. 2.3: The path Γ of Eq. (10.16) and its deformation around the poles $\omega = \varepsilon_{\mathbf{k}}$, $\omega' = \varepsilon_{\mathbf{k'}} + i\nu_+$ of Eq. (10.18).

Simplifying the notation by setting the Fermi level at zero energy, $\mu = 0$, and making use of (7.26), the Lindhard function to be calculated becomes

$$L_-(\varepsilon_k, \varepsilon_{k-q} + i\nu_+) = \frac{f_o(\varepsilon_{k-q}) - f_o(\varepsilon_k)}{i\nu_+ + \varepsilon_{k-q} - \varepsilon_k} . \quad (10.20)$$

Thus, applying definition (10.17) to the product of (10.20) and (7.45), Eq. (10.15) takes the form

$$\Delta\Omega^{\text{ex}} = -\sum_q \sum_k |\gamma_{\mathbf{q}}|^2 \frac{1}{2} [f_o(\varepsilon_{k-q}) - f_o(\varepsilon_k)]$$
$$\times \{L_+(\varepsilon_k - \varepsilon_{k-q}, \omega_q) - L_+(\varepsilon_k - \varepsilon_{k-q}, -\omega_q)\} \quad (10.21)$$

which according to (10.19) yields

$$\Delta\Omega^{\text{ex}} = \sum_q \sum_k |\gamma_{\mathbf{q}}|^2 \frac{1}{2} [f_o(\varepsilon_{k-q}) - f_o(\varepsilon_k)]$$
$$\times \left\{ \frac{n_o(\omega_q) - n_o(\varepsilon_k - \varepsilon_{k-q})}{\varepsilon_{k-q} - \varepsilon_k + \omega_q} + \frac{n_o(\varepsilon_k - \varepsilon_{k-q}) - n_o(-\omega_q)}{\varepsilon_{k-q} - \varepsilon_k - \omega_q} \right\} .$$
$$(10.22)$$

This expression may still be simplified by making, in the first term of the second bracket as well as in the first bracket, the substitutions $k - q \rightarrow k$, $k \rightarrow k - q$, $q \rightarrow -q$. Applying the identity (7.43) to the second term of the last bracket one finally obtains

$$\Delta\Omega^{\text{ex}} = \sum_{\mathbf{q}} \sum_{k} |\gamma_{\mathbf{q}}|^2 [f_o(\varepsilon_{k-q}) - f_o(\varepsilon_k)] \frac{n_o(\omega_q) - n_o(\varepsilon_{k-q} - \varepsilon_k)}{\varepsilon_{k-q} - \varepsilon_k - \omega_q} . \quad (10.23)$$

Such a formula has first been obtained by Buckingham and Schafroth[28] (see also Ref. 7, Section 21.4). Indeed, in view of the identity (see Eq. (25.7) below for a simpler identity of the same type)

$$[f_o(\varepsilon') - f_o(\varepsilon)][n_o(\omega) - n_o(\varepsilon' - \varepsilon)] \equiv [1 - f_o(\varepsilon)]f_o(\varepsilon')$$
$$\times [1 + n_o(\omega)] - f_o(\varepsilon)[1 - f_o(\varepsilon')]n_o(\omega) , \quad (10.24)$$

Equation (21) of Ref. 28 agrees with the above expression. An experimental search for such a renormalization effect in the specific heat of vanadium has been rather inconclusive[29]. For a recent detailed analysis of the contributions to the electronic specific heat due to electron-phonon interaction see Coffey and Pethick[30]. Note that the basic equation (3) of Ref. 30 is the same as our $\Delta\Omega^{\text{ex}}$, except that the ν_+-sum is transformed with the help of the identity (10.16) to an integral along the cut on the real z-axis of $L_-(\varepsilon_k, \varepsilon_{k-q} + z)$ and of $\tilde{\mathcal{D}}^o(\mathbf{q}; z)$. The temperature dependence of this contribution to the specific heat is represented in Fig. 5.19 of Ref. 31.

A much more general formula for the grand-canonical potential Ω may be obtained from Eq. (9.20) with the help of the device of a *variable coupling constant* introduced by Abrikosov , Gor'kov and Dzyaloshinskii (see Ref. 7, Section 16.2). Multiplying the interaction with a parameter λ, $0 \leq \lambda \leq 1$, and introducing the notation $\mathcal{H}_{\text{int}}(\lambda) \equiv \lambda\mathcal{H}_{\text{int}}$, we may write Eq. (9.20) as

$$\Omega(\lambda) = \Omega_o - \beta^{-1}\langle S(-i\beta, 0; \lambda) - 1\rangle_{o,cp} \quad (10.25)$$

where, according to (5.25), (5.28) and (6.18)

$$S(-i\beta, 0; \lambda) \equiv \sum_{N=0}^{\infty} \frac{\lambda^N}{N!} T\left\{ -\int_0^{\beta} d\tau \mathcal{H}_{\text{int}}^o(-i\tau) \right\}^N . \quad (10.26)$$

Now $\Omega \equiv \Omega(1)$ and $\Omega_o \equiv \Omega(0)$ so that

$$\Omega - \Omega_o = \int_0^1 d\lambda \frac{\partial\Omega}{\partial\lambda} . \quad (10.27)$$

But from (10.25), (10.26) it is straightforward to obtain the formula

$$\frac{\partial \Omega}{\partial \lambda} = -\beta^{-1} \sum_{N=1}^{\infty} \frac{\lambda^{N-1}}{(N-1)!} \left\langle T\left\{ -\int_0^\beta d\tau \mathcal{H}_{\text{int}}^o(-i\tau) \right\}^N \right\rangle_{o,cp}$$

$$= \beta^{-1} \int_0^\beta d\tau \langle T\{\mathcal{H}_{\text{int}}^o(-i\tau) S(-i\beta, 0; \lambda)\}\rangle_{o,cp} \qquad (10.28)$$

which, to lowest order, reduces to the simple expression

$$\frac{\partial \Omega}{\partial \lambda} = \langle \mathcal{H}_{\text{int}} \rangle_o = \frac{1}{\lambda} \langle \mathcal{H}_{\text{int}}(\lambda) \rangle_o . \qquad (10.29)$$

This formula, combined with (10.27), is very useful in the weak-coupling approximation of superconductivity (see Section 24).

The evaluation of Eq. (10.28) may be pushed one step further without enumerating Feynman diagrams explicitly by specifying the interaction either as the electron-electron vertex (3.38) or as the electron-phonon vertex (10.6) and by applying Wick's theorem (9.11). In both cases one has the alternative of connecting the electron operators a_k^+ and $a_{k'}$ in the explicit factor $\mathcal{H}_{\text{int}}$ of the last expression (10.28) either among themselves or to the operator S. The first possibility gives rise, according to the chronological contraction (9.9), to an unperturbed propagator $\mathcal{G}^o$ with both ends connected to the same vertex. The second possibility is to express the general structure according to Eq. (9.15) in terms of the *two-electron Green's function*

$$\mathcal{G}^2(k_1\tau_1, \cdots k_4\tau_4) \equiv -\langle T\{a_{k_1}(-i\tau_1)a_{k_2}^+(-i\tau_2)a_{k_3}(-i\tau_3)a_{k_4}^+(-i\tau_4)\}\rangle \tag{10.30}$$

in the case of the electron-electron vertex (3.38) and in terms of the *three-point vertex*

$$\Delta(k_1\tau_1, k_2\tau_2, k_3\tau_3) \equiv -\langle T\{a_{k_1}(-i\tau_1)a_{k_2}^+(-i\tau_2)Q_q(-i\tau_3)\}\rangle \tag{10.31}$$

in the case of the electron-phonon vertex (10.6). The result are general *structure equations* which are again made up of Hartree- and exchange-type contributions. We give here only the diagrammatic form of these equations.

In the case of the electron-electron vertex represented by Fig. 1.2, Chapter 1, one finds the contributions shown in Fig. 2.4 where the first two terms are of Hartree type and the last two of exchange type. $\mathcal{G}(\lambda)$ and $\mathcal{G}^2(\lambda)$ are, respectively, the electron propagator $\mathcal{G}_{a_k a_k^+}$ defined by Eq. (7.21)

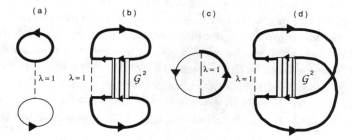

Fig. 2.4: (a), (b) Hartree-type , (c), (d) exchange-type contributions to Eq. (10.28) in the case of the electron-electron vertex of Fig. 1.2. In (a) and (c) an unperturbed electron propagator $\mathcal{G}^o$ is connected with both ends to the same vertex. Heavy lines with arrow represent the full electron propagator $\mathcal{G}(\lambda)$ and the shaded box is the two-electron Green's function $\mathcal{G}^2(\lambda)$ defined in (10.30).

and the 2-electron Green's function (10.30) generated by the interaction $\mathcal{H}^{\text{el}-\text{el}}(\lambda)$.

In the case of the electron-phonon vertex shown in Fig. 1.1, Chapter 1, Eq. (10.28) yields the Hartree- and exchange-type contributions exhibited in Fig. 2.5 where $\mathcal{G}(\lambda)$, $\mathcal{D}(\lambda)$ and $\Delta(\lambda)$ are, respectively, the full electron

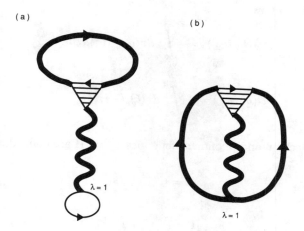

Fig. 2.5: (a) Hartree-type, (b) exchange-type contribution to Eq. (10.28) in the case of the electron-phonon vertex of Fig. 1.1. In (a) an unperturbed electron propagator $\mathcal{G}^o$ is connected with both ends to the same vertex. Heavy lines with arrow and heavy wavy lines represent, respectively, the full electron and phonon propagators $\mathcal{G}(\lambda)$ and $\mathcal{D}(\lambda)$ and the shaded triangle is the three-point vertex $\Delta(\lambda)$ defined in (10.31).

propagator, the full phonon propagator and the three-point vertex (10.31) generated by the interaction $\mathcal{H}^{el-el}(\lambda)$. Note that the diagrams of Fig. 2.2 are just the lowest-order approximation to those of Fig. 2.5. Note also the occurrence in both, Figs. 2.4 and 2.5, of the free-electron propagator $\mathcal{G}^o$. If these $\mathcal{G}^o$ were replaced by $\mathcal{G}(\lambda)$, comparison of Figs. 2.4 and 2.5, respectively with Figs. 2.9 and 2.10 of Section 11 below for the selfenergy Σ, shows that it would then be possible to express $\partial\Omega/\partial\lambda$ in both cases in terms of the product $\mathcal{G}(\lambda)\Sigma(\lambda)$ (see Ref. 7, Section 16.2). This, however, is not the case.

11. Dyson Equation, Selfenergy Diagrams and Dressed Propagators

The systematics of the diagrams introduced by Feynman[18] is due to Dyson[3] who defined interaction corrections for classes of diagrams with a given number of external lines and derived relations among them. For diagrams with no external lines, examples of such corrections were given in the last section. The next and most important class are diagrams with two external lines. They represent interaction corrections to single lines, called *selfenergies*. The total selfenergy for a given type of line is given by the *Dyson equation* [3] which, for electrons and phonons, relates the unperturbed propagators $\tilde{\mathcal{G}}^o$ and $\tilde{\mathcal{D}}^o$ of Eqs. (7.44) and (7.45) with the corrected or *dressed propagators*

$$\tilde{\mathcal{G}}(k; i\nu_-) = -\int_o^\beta d\tau \langle a_k(-i\tau)a_k^+ \rangle e^{i\nu_- \tau} \qquad (11.1)$$

and

$$\tilde{\mathcal{D}}(q; i\nu_+) = -\int_o^\beta d\tau \langle Q_q(-i\tau)Q_q^+ \rangle e^{i\nu_+ \tau} , \qquad (11.2)$$

respectively.

The electron and phonon selfenergies Σ and Π are defined, respectively by

$$\Sigma = \tilde{\mathcal{G}}^{o-1} - \tilde{\mathcal{G}}^{-1} \qquad (11.3)$$

and

$$\Pi = \tilde{\mathcal{D}}^{o-1} - \tilde{\mathcal{D}}^{-1} , \qquad (11.4)$$

or equivalently by the Dyson equations[7,8]

$$\tilde{\mathcal{G}} - \tilde{\mathcal{G}}^o = \tilde{\mathcal{G}}^o\Sigma\tilde{\mathcal{G}} = \tilde{\mathcal{G}}\Sigma\tilde{\mathcal{G}}^o \qquad (11.5)$$

and

$$\tilde{\mathcal{D}} - \tilde{\mathcal{D}}^o = \tilde{\mathcal{D}}^o \Pi \tilde{\mathcal{D}} = \tilde{\mathcal{D}} \Pi \tilde{\mathcal{D}}^o \ . \tag{11.6}$$

The last two equations have the diagrammatic representation of Fig. 2.6. Inserting (7.44) and (7.45), respectively into Eqs. (11.3) and (11.4) one finds the explicit forms

$$\tilde{\mathcal{G}}(k; i\nu_-) = \frac{1}{i\nu_- - \varepsilon_k + \mu - \Sigma(k; i\nu_-)} \tag{11.7}$$

and

$$\tilde{\mathcal{D}}(q; i\nu_+) = \frac{\omega_q}{(i\nu_+)^2 - \omega_q^2 - \omega_q \Pi(q; i\nu_+)} \ . \tag{11.8}$$

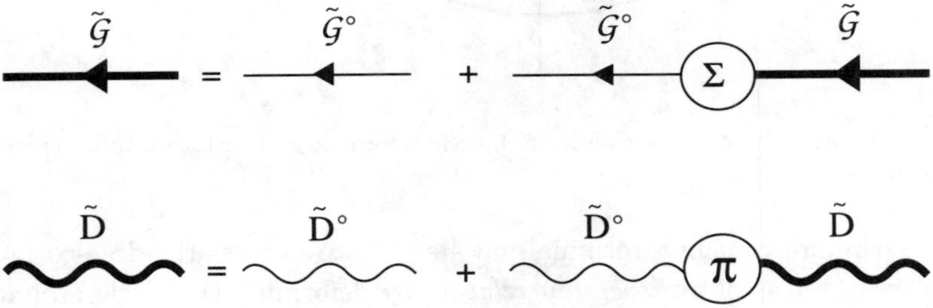

Fig. 2.6: Dyson equations (11.5) and (11.6) defining the selfenergies Σ and Π in terms of unperturbed (fine lines) and dressed (heavy lines) propagators.

From Eqs. (11.7), (11.8) follows that these dressed propagators are analytic in the neighborhood of $i\nu_\pm$ that is, along the imaginary axis of the complex frequency plane z, but have more complicated singularities than the simple poles $z = \varepsilon_k - \mu$, $z = \pm\omega_q$ of the unperturbed functions (7.44), (7.45). Indeed, Eq. (7.31) shows that $\tilde{\mathcal{G}}$ and $\tilde{\mathcal{D}}$ have cuts along the positive and negative real z-axes, $\operatorname{Re} z > 0$ and $\operatorname{Re} z < 0$, but are analytic functions in both half-planes $\operatorname{Im} z > 0$ and $\operatorname{Im} z < 0$ which are connected by complex conjugation. This follows from the fact that, according to Eqs. (7.21), (7.9) and (7.29),

$$(\tilde{\mathcal{G}}_{AA+}(z))^* = \tilde{\mathcal{G}}_{AA+}(z^*) \ . \tag{11.9}$$

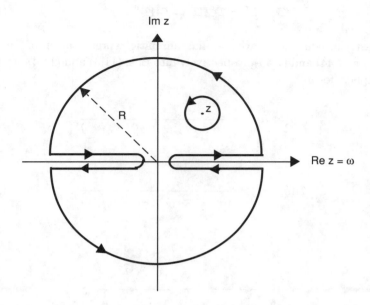

Fig. 2.7: Equivalent integration paths for Cauchy's formula and for the dispersion relation (11.10).

Therefore Cauchy's formula may be used to express the dressed propagators in terms of a *dispersion relation* by deforming the circle around z into the integration path of Fig. 2.7. Since according to (11.7), (11.8) the integrals along the large circle of radius R contain a denominator which for $R \to \infty$ grows at least as R^2, this contribution vanishes at least as R^{-1}. Hence

$$\tilde{\mathcal{G}}(k; z) = \frac{1}{\pi} \int_{-\infty}^{+\infty} \frac{d\omega}{\omega - z} \operatorname{Im} \tilde{\mathcal{G}}(k; \omega) \tag{11.10}$$

where

$$\operatorname{Im} \tilde{\mathcal{G}}(k; \omega) \equiv \frac{1}{2i} [\tilde{\mathcal{G}}(k; \omega + i\epsilon) - \tilde{\mathcal{G}}(k; \omega - i\epsilon)] \tag{11.11}$$

is the *spectral function* and $\epsilon = 0^+$.

Analogous formulas hold for the propagator $\tilde{\mathcal{D}}(q; z)$ which, however, possesses an additional symmetry. Indeed, definition (7.21) together with the reality conditions (1.25) imply that $\mathcal{D}(q; \tau)$ is an even function of τ, from which follows for the Fourier transform (7.29), making use of (7.24) and (7.26), that

$$\tilde{\mathcal{D}}(q; -i\nu_+) = \tilde{\mathcal{D}}(q; i\nu_+) \ . \tag{11.12}$$

By analytic continuation one then concludes from the analogous definition (11.11) that

$$\text{Im}\, \tilde{\mathcal{D}}(q; -\omega) = -\text{Im}\, \tilde{\mathcal{D}}(q; \omega) \ , \tag{11.13}$$

so that the corresponding dispersion relation (11.10) may by written as

$$\tilde{\mathcal{D}}(q; z) = \frac{2}{\pi} \int_0^\infty \frac{\omega d\omega}{\omega^2 - z^2} \text{Im}\, \tilde{\mathcal{D}}(q; \omega) \ . \tag{11.14}$$

Iteration of Eqs. (11.5), (11.6) or of Fig. 2.6 introduces a bare, i.e. unperturbed line between two successive selfenergies, which shows that in order to have an unambiguous definition we must require the selfenergies to be *one-particle-irreducible*. This means that by cutting one internal line a selfenergy does not fall into disconnected pieces.

Thus the electron selfenergy for a given interaction $\mathcal{H}_{\text{int}}$ is obtained by writing the propagator (11.1) in the form (9.15),

$$\tilde{\mathcal{G}}(k; i\nu_-) = - \int_0^\beta d\tau e^{i\nu_- \tau} \langle T\{S(-i\beta; 0)a_k^o(-i\tau)a_k^{+o}(0)\}\rangle_{o,cp} \tag{11.15}$$

and inserting the Dyson series

$$S(-i\beta; 0) = \sum_{N=0}^\infty \frac{(-1)^N}{N!} T\{\int_0^\beta d\tau \mathcal{H}_{\text{int}}^o(-i\tau)\}^N \ , \tag{11.16}$$

which follows from Eqs. (5.28), (6.1), (6.18). In view of Eq. (11.5) the one-particle-irreducible part of $\tilde{\mathcal{G}}(k; i\nu_-) - \tilde{\mathcal{G}}^o(k; i\nu_-)$ is recognized to be

$$\Sigma(k; i\nu_-)[\tilde{\mathcal{G}}^o(k; i\nu_-)]^2 = - \sum_{N=1}^\infty \frac{(-1)^N}{N!} \int_0^\beta d\tau \int_0^\beta d\tau_1 \cdots \int_0^\beta d\tau_N e^{i\nu_- \tau}$$
$$\times \langle T\{\mathcal{H}_{\text{int}}^o(-i\tau_1) \cdots \mathcal{H}_{\text{int}}^o(-i\tau_N)a_k^o(-i\tau)a_k^{+o}(0)\}\rangle_{o,cp,1-i} \tag{11.17}$$

where the index $1 - i$ stands for "one-particle-irreducible" and the average has to be evaluated with the help of Wick's theorem (9.11). Equation (11.17) gives a general and unambiguous prescription to calculate Σ; an analogous procedure obviously exists for Π. Note that in Section 12 an even more general expression for the selfenergy is given in Eq. (12.34).

Let us now apply Eq. (11.17) to some specific interactions and calculate the lowest-order contributions which are automatically one-particle-irreducible. For the electron-electron interaction (3.38) one obtains to order $N = 1$ (Problem 2.11)

$$\Sigma(k; i\nu_-) = V^{-1} \sum_{k'} f_o(\varepsilon_{k'})\{2\tilde{\phi}(0) - \tilde{\phi}(\mathbf{k}' - \mathbf{k})\} \tag{11.18}$$

so that, according to (11.7), the *renormalized electron energy* (see Eq. (11.34) below)

$$\varepsilon_k^r \equiv \varepsilon_k + V^{-1} \sum_{k'} f_o(\varepsilon_{k'})\{2\tilde{\phi}(0) - \tilde{\phi}(k' - k)\} \tag{11.19}$$

is just the Hartree-Fock result of Eqs. (4.26), (4.22), and the diagrams representing the two terms in (11.18) are those of Fig. 1.3, Chapter 1.

The lowest-order contribution of the electron-phonon interaction (10.6) comes for $N = 2$ so that, in analogy to the notation in Eq. (10.7),

$$\Sigma(k; i\nu_-)[\tilde{\mathcal{G}}^o(k; i\nu_-)]^2 = -\frac{1}{2} \int_0^\beta d\tau \int_0^\beta d\tau' \int_0^\beta d\tau'' e^{i\nu_- \tau}$$
$$\times \sum_{k'q} \sum_{k''q'} \gamma_q \gamma_{q'} (S^{\mathrm{H}} + S^{\mathrm{ex}}) \ . \tag{11.20}$$

Taking into account all possible pairings implied by Wick's theorem (9.11), the "Hartree" and "exchange terms" are given, respectively, by

$$S^{\mathrm{H}} = 2\langle T\{(a_{k'}^+ a_{k'-q} Q_q)^o(-i\tau')(a_{k''}^+ a_{k''-q'} Q_{q'})^o(-i\tau'')a_k^o(-i\tau)a_k^+(0)\}\rangle_{o,cp} \tag{11.21}$$

and

$$S^{\mathrm{ex}} = 2\langle T\{(a_{k'}^+ a_{k'-q} Q_q)^o(-i\tau')(a_{k''}^+ a_{k''-q'} Q_{q'})^o(-i\tau'')a_k^o(-i\tau)a_k^+(0)\}\rangle_{o,cp} \ . \tag{11.22}$$

Making use of spin and mode conservation and of the translation-invariance relations (7.9), (7.13) together with definitions (7.34), (7.35), these expressions become, noting that each transposition of fermionic operators introduces a minus sign,

$$S^{\mathrm{H}} = -2\mathcal{G}^o(k; \tau - \tau')\delta_{\sigma'\sigma}\delta_{\mathbf{k}'\mathbf{k}}\mathcal{G}^o(k; \tau')\delta_{\mathbf{k}'-\mathbf{q},\mathbf{k}}\mathcal{G}^o(k''; 0)\delta_{\mathbf{q}',0}$$
$$\times \mathcal{D}^o(q', \tau'' - \tau')\delta_{q,-q'} \ ,$$
$$S^{\mathrm{ex}} = +2\mathcal{G}^o(k; \tau - \tau'')\delta_{\sigma''\sigma}\delta_{\mathbf{k}''\mathbf{k}}\mathcal{G}^o(k; \tau')\delta_{\sigma'\sigma}\delta_{\mathbf{k}'-\mathbf{q},\mathbf{k}} \tag{11.23}$$
$$\times \mathcal{G}^o(k'; \tau'' - \tau')\delta_{\mathbf{k}',\mathbf{k}''-\mathbf{q}'}\mathcal{D}^o(q'; \tau'' - \tau')\delta_{q,-q'} \ .$$

With these expressions the second-order terms in Eq. (11.20) are represented by the Feynman diagrams of Fig. 2.8.

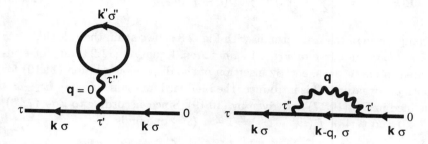

Fig. 2.8: Feynman diagrams of the "Hartree" and "exchange terms" (11.23) in the expression (11.20).

In the first Eq. (11.23) the rule (9.10) has to be invoked for the equaltime propagator which, combined with Eqs. (7.34), (7.40), states that

$$\mathcal{G}^o(k; 0) = -\langle W\{a_k a_k^+\}\rangle_o = \langle a_k^+ a_k\rangle_o = f_o(\varepsilon_k) , \qquad (11.24)$$

where we have set the Fermi energy $\mu = 0$. Inserting (11.23) into (11.20), making use of Eq. (2.21) and of the reality conditions (1.25) one then finds

$$\Sigma(k; i\nu_-)[\tilde{\mathcal{G}}^o(k; i\nu_-)]^2 = \int_0^\beta d\tau \int_0^\beta d\tau' \int_0^\beta d\tau'' e^{i\nu - \tau} \sum_q |\gamma_q|^2$$

$$\times \mathcal{D}^o(q; \tau'' - \tau') \mathcal{G}^o(k; \tau')\{\delta_{\mathbf{q},0} \sum_{\mathbf{k}''} 2f_o(\varepsilon_{k''}) \mathcal{G}^o(k; \tau - \tau')$$

$$- \mathcal{G}^o(k; \tau - \tau'') \mathcal{G}^o(k - q; \tau'' - \tau')\} . \qquad (11.25)$$

In terms of Fourier transforms (7.25), (7.28) one finds for $\Sigma = \Sigma^H + \Sigma^{ex}$

$$\Sigma^H = -nV \sum_q \frac{|\gamma_q|^2}{\omega_q} \delta_{\mathbf{q},0} , \qquad (11.26)$$

where use was made of (7.45) and (10.13), and

$$\Sigma^{ex}(k; i\nu_-) = -\sum_q |\gamma_q|^2 \beta^{-1} \sum_{\nu_+} \tilde{\mathcal{D}}^o(q; i\nu_+) \tilde{\mathcal{G}}^o(k - q; i\nu_- - i\nu_+) . \quad (11.27)$$

Note that in the case of a crystal without basis, $B = 1$, the constant (11.26) may be expressed in terms of (10.12) as $\Sigma^H = v\Delta\Omega^H/nL$ and also that there are many equivalent ways of writing Σ^{ex}.

A more rigorous result for the leading term (11.27) may be obtained by "dressing" the propagators so that, dropping the constant (11.26) and relabeling the Matsubara sum,

$$\Sigma(k;\, i\nu_-) = -\sum_q |\gamma_q|^2 \beta^{-1} \sum_{\nu'_-} \tilde{\mathcal{G}}(k-q;\, i\nu'_-)\tilde{\mathcal{D}}(q;\, i\nu_- - i\nu'_-)\,. \quad (11.28)$$

Comparison of this key formula with Fig. 2.8 shows again the *rule* that *each closed loop gives rise to a* (**q**, ν)-*summation*. Equation (11.28) may be evaluated in many ways, e.g. by insertion of the dispersion relation (11.10) for both propagators. This introduces the Lindhard function $-L_-(\omega',\, i\nu_- -\omega'')$ defined in Eq. (10.17) and given by (10.19). Since, according to Eqs. (7.26), (7.42) and (7.43), $f_o(i\nu_- - \omega'') = 1 + n_o(\omega'')$ one finds

$$\Sigma(k;\, i\nu_-) = -\sum_q |\gamma_q|^2 \frac{1}{\pi^2} \int_{-\infty}^{+\infty} d\omega' \int_{-\infty}^{+\infty} d\omega'' \mathrm{Im}\, \tilde{\mathcal{G}}(k-q;\, \omega')$$

$$\times \mathrm{Im}\, \tilde{\mathcal{D}}(q;\, \omega'')\frac{1 - f_o(\omega') + n_o(\omega'')}{\omega' + \omega'' - i\nu_-}\,. \quad (11.29)$$

Now analytic continuation $i\nu_- \to z$ is trivially possible.

Of particular interest is the imaginary part of the selfenergy since it is related with the *damping* or *transition rate*. From (11.7) one sees that the same definition (11.11) also holds for $\mathrm{Im}\,\Sigma(k;\, \omega)$. Doing first the ω''-integration we may write the denominator in the expression (11.29) for $\Sigma(k;\, \omega \pm i\epsilon)$, up to a sign, as

$$\lim_{\epsilon \to 0} \frac{1}{\omega'' + \omega' - \omega \mp i\epsilon} = \frac{P}{\omega'' + \omega' - \omega} \pm i\pi\delta(\omega'' + \omega' - \omega) \quad (11.30)$$

where P stands for *"principal part"* and the δ-function term stands for half the Cauchy integral, taken in the positive/negative sense. The result is

$$\mathrm{Im}\,\Sigma(k;\, \omega) = -\sum_q |\gamma_q|^2 \frac{1}{\pi} \int_{-\infty}^{+\infty} d\omega' \mathrm{Im}\, \tilde{\mathcal{G}}(k-q;\, \omega')$$

$$\times \mathrm{Im}\, \tilde{\mathcal{D}}(q;\, \omega - \omega')[1 - f_o(\omega') + n_o(\omega - \omega')]\,. \quad (11.31)$$

So far the manipulations were exact. A useful approximation may be obtained in the weak-coupling limit $|\gamma_q| \to 0$ to Eq. (11.31). Indeed, noting the following representation of the δ-function,

$$\lim_{\epsilon \to 0} \frac{\epsilon}{x^2 + \epsilon^2} = \pi\delta(x)\,, \quad (11.32)$$

we get from Eq. (11.7) by analytic continuation $i\nu_- \to \omega \pm i\epsilon$ and setting $\mu = 0$,

$$\mathrm{Im}\,\tilde{\mathcal{G}}(k;\, \omega) = \frac{\mathrm{Im}\,\Sigma}{(\omega - \varepsilon_k - \mathrm{Re}\,\Sigma)^2 + (\mathrm{Im}\,\Sigma)^2} \to -\pi\delta(\omega - \varepsilon_k^r) \quad (11.33)$$

where $\operatorname{Re}\Sigma \equiv [\Sigma(k; \omega + i\epsilon) + \Sigma(k; \omega - i\epsilon)]/2$; $\epsilon = 0^+$ and

$$\varepsilon_k^r \equiv \varepsilon_k + \operatorname{Re}\Sigma(k; \varepsilon_k^r) \tag{11.34}$$

is the *renormalized electron energy*. The sign has been fixed by assuming it to be the same as that of $\operatorname{Im}\tilde{\mathcal{G}}^o$ which according to Eqs. (7.44), (11.11) is minus. Inserting (11.33) into Eq. (11.31) then gives

$$\operatorname{Im}\Sigma(k; \omega) = \sum_q |\gamma_q|^2 \operatorname{Im}\tilde{\mathcal{D}}(q; \omega - \varepsilon_{k-q}^r)[1 - f_o(\varepsilon_{k-q}^r) + n_o(\omega - \varepsilon_{k-q}^r)] \; .$$
$$\tag{11.35}$$

In view of Eqs. (7.42), (7.43) this result is the same as Eq. (21.33) of Ref. 7, except for the renormalized energy value.

A further simplification is obtained in treating the phonon spectral function $\operatorname{Im}\tilde{D}$ in analogy to Eq. (11.33). Making use of Eq. (11.8) one finds in the same way as for Eq. (11.33)

$$\operatorname{Im}\tilde{\mathcal{D}}(q; \omega) = \omega_q \frac{\omega_q \operatorname{Im}\Pi}{(\omega^2 - \omega_q^2 - \omega_q \operatorname{Re}\Pi)^2 + (\omega_q \operatorname{Im}\Pi)^2}$$
$$\rightarrow -\frac{\pi}{2}\frac{\omega_q}{\omega_q^r}\{\delta(\omega - \omega_q^r) - \delta(\omega + \omega_q^r)\} \tag{11.36}$$

where

$$(\omega_q^r)^2 \equiv \omega_q^2 + \omega_q \operatorname{Re}\Pi(q; \omega_q^r) \tag{11.37}$$

defines the *renormalized phonon frequency* and the sign is determined in the same way as in Eq. (11.35). Inserting (11.36) and making use of (7.43), Eq. (11.35) finally becomes

$$\operatorname{Im}\Sigma(k; \omega) = -\frac{\pi}{2}\sum_q |\gamma_q|^2 \frac{\omega_q}{\omega_q^r}\Big\{[1 - f_o(\varepsilon_{k-q}^r) + n_o(\omega_q^r)]$$
$$\times \delta(\omega - \varepsilon_{k-q}^r - \omega_q^r) + [f_o(\varepsilon_{k-q}^r) + n_o(\omega_q^r)]$$
$$\times \delta\Big(\omega - \varepsilon_{k-q}^r + \omega_q^r\Big)\Big\}. \tag{11.38}$$

Damping or *transition rates* may now be defined by setting $\omega = \varepsilon_k^r$. Neglecting renormalization effects we define the rates for *emission* and *absorption of one phonon* respectively by[32]

$$\Gamma_+(kk') = \pi \sum_q |\gamma_q|^2 [1 - f_o(\varepsilon_{k'}) + n_o(\omega_q)]\delta(\varepsilon_k - \varepsilon_{k'} - \omega_q)\delta_{k', k-q} \tag{11.39}$$

and

$$\Gamma_-(kk') = \pi \sum_q |\gamma_q|^2 [f_o(\varepsilon_{k'}) + n_o(\omega_q)]\delta(\varepsilon_k - \varepsilon_{k'} + \omega_q)\delta_{k', k-q} \tag{11.40}$$

so that

$$-2\text{Im}\,\Sigma(k;\varepsilon_k) = \sum_{k'}[\Gamma_+(kk') + \Gamma_-(kk')] \equiv \tau_k^{-1} \tag{11.41}$$

where τ_k is the *scattering relaxation time*. At zero temperature , $n_o(\omega_q) = 0$
and the matrix element of the electron-phonon interaction (10.6) between
one-electron or one-hole states generated on the Fermi sea $|F\rangle$, $|k\rangle \equiv a_k^+|F\rangle$
and $|k'q\rangle \equiv b_q^+ a_{k'}^+|F\rangle$, is

$$\langle k'q|\mathcal{H}^{\text{el}-\text{ph}}|k\rangle = \frac{i}{\sqrt{2}}\gamma_{-q}\delta_{k',k-q} \;, \tag{11.42}$$

where use has been made of Eq. (1.31). This leads to the *Golden Rule* (see
Eqs. (5–80) of Ref. 33),

$$\Gamma_\pm(kk') = 2\pi\sum_q |\langle k'q|\mathcal{H}^{\text{el}-\text{ph}}|k\rangle|^2 f_o(\mp\varepsilon_{k'})\delta(\varepsilon_k - \varepsilon_{k'} \mp \omega_q) \;. \tag{11.43}$$

Equation (11.28) is still not the most general form of the selfenergy Σ.
Indeed, in terms of the two-electron Green's function $\mathcal{G}^2$ or the electron-
phonon vertex Δ of Figs. 2.4 and 2.5, Σ may be expressed as a general

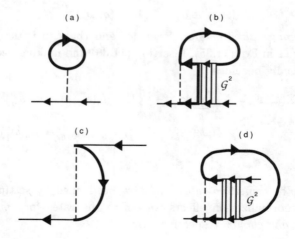

Fig. 2.9: (a), (b) Hartree-type , (c), (d) exchange-type contributions to Eq. (11.17) representing the structure equation for the selfenergy Σ in the case of the electron-electron
vertex of Fig. 1.2. (a) and (c) have the external lines connected to the same vertex while
in (b) and (d) one of them is connected to the two-electron Green's function $\mathcal{G}^2$ (shaded
box). Note that the external lines do not belong to the selfenergy.

structure equation. This is most easily obtained starting from Eq. (11.17). Making use of Wick's theorem (9.11) one has the alternative of connecting a_k and a_k^+ in the average of (11.17) to the same vertex $\mathcal{H}_{int}$ or to two different vertices. In both cases this gives rise, by application of the chronological contraction (9.9), to two unperturbed external lines $\mathcal{G}^o$ which according to Eq. (11.17) do not belong to the selfenergy Σ. In the case of the electron-electron vertex of Fig. 1.2 this alternative leads to the diagrams of Fig. 2.9, (a) and (c) being the result of connecting the external lines to the same vertex while (b) and (d) have one of them connected to the two-electron Green's function $\mathcal{G}^2$ introduced in the last section. Note that (a) and (b) are again of Hartree type while (c) and (d) are of exchange type.

In the case of the electron-phonon vertex of Fig. 1.1, Eq. (11.17) leads to the diagrams (a) and (b) of Fig. 2.10. (a) has both external lines connected to the same vertex and is of Hartree type while (b) has one of the

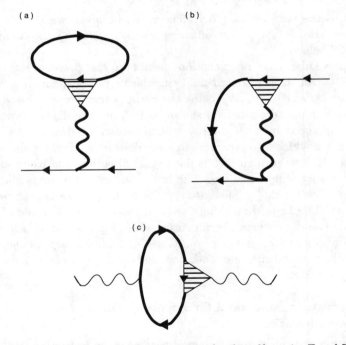

Fig. 2.10: (a), (b) and (c) are structure equations for the selfenergies Σ and Π, respectively, in the case of the electron-phonon vertex of Fig. 1.1. Except in (a) which is of Hartree type, one external line connects to the three-point vertex Δ (shaded triangle). Note that the external lines do not belong to the selfenergy.

external lines connected to the three-point vertex Δ introduced in the last section and is of exchange type. Note that the diagrams of Fig. 2.8 are the lowest-order approximation of the former. For completeness Fig. 2.10 (c) also shows the structure equation for the phonon selfenergy Π.

Due to the extremely small mass ratio $m/M \sim 10^{-4}$ between electron and lattice ions the motion of the latter is much slower than that of the former, $v_{\text{ion}}/v_{\text{el}} \sim (m/M)^{1/2}$. This gives rise to the *Born-Oppenheimer* or *adiabatic approximation*[34] in which the electronic Schrödinger equation is solved with the ion coordinates taken as fixed parameters. It is in this approximation only that the electron-phonon interaction is well defined (see, e.g. Ref. 35). But, in addition, the small mass ratio also gives rise to what is called *Migdal's theorem*[36] (see also Section 21 of Ref. 7) which states that, in the situations of Fig. 2.10 the electron-phonon vertex may be approximated as

$$\Delta(k\omega, q\nu) = \gamma_q[1 + \mathcal{O}(\sqrt{m/M})] \tag{11.44}$$

where γ_q is the bare vertex (10.6). There are, however, well-known exceptions where Eq. (11.44) is not sufficient (see, e.g. Refs. 33 and 35 and also Section 28 below).

A particular case of exception occurs in the presence of a *Kohn anomaly*[37] which is a logarithmic singularity in the phonon group velocity $\partial\omega_q/\partial\mathbf{q}$ at $q = 2k_F$, k_F being the Fermi wavenumber, caused by the electron-phonon interaction. As shown in Ref. 38, this logarithmic singularity is enhanced by a singularity in the density of states (10.4) at the Fermi surface which also gives rise to a breakdown of Migdal's theorem. In particular, if the Fermi surface is flat in one direction, the Kohn anomaly becomes a root singularity and if it is flat in both directions the singularity becomes a pole[38]. This last type of Fermi surface is referred to in Section 34. The logarithmic singularity in the density of states of a flat, i.e. locally two-dimensional, Fermi surface (see, e.g. Eq. (36.3) below) is an example of a *van Hove singularity*. These singularities are a consequence of the lattice periodicity and were first derived for the phonon density of states by van Hove[39].

12. Generating Functional for Feynman Diagrams. Ward Identities

The rules to calculate Feynman diagrams given in Section 9, although quite general, are *combinatorial* and therefore not very explicit. A much more compact *operational* formulation is obtained by coupling the system to an external source field as in Eq. (2.1) and defining the Green's functions as derivatives of an appropriate *generating functional* with respect to these source fields. This functional-integral method pioneered by Martin and Schwinger[10] leads in a first step to an elegant and compact form of Wick's theorem for averages and in a second step generates Feynman diagrams (see, e.g. Ref. 40).

In order to be able to generate fermionic Green's functions such as $\mathcal{G}^o$ in Eq. (7.34) the notion of external source field has to be formally generalized to include *anticommuting* or *Grassmann variables* which couple to fermionic (creation and annihilation) operators. Although unphysical, these fermionic source fields are such that the coupling term (2.1) still obeys Bose statistics. This trick was already used by Martin and Schwinger[10] and goes back to Schwinger[41]. In order to keep the same generality in the formulation of Wick's theorem as in Sections 8 and 9 we adapt the notation to this more general situation, designating the external source field coupled to an irreducible operator A by η_A. Then Eq. (2.1) takes the form (compare Eq. (11) of Ref. 41)

$$\delta\mathcal{H}_t = \sum_A \eta_A(t)A \qquad (12.1)$$

where η_A is a commuting/anticommuting variable according to whether A is a bosonic /fermionic operator. In particular, for electrons and phonons

$$\delta\mathcal{H}_t = \sum_k (\eta_{a_k}(t)a_k + \eta_{a_k^+}(t)a_k^+) + \sum_q (\eta_{Q_q}(t)Q_q + \eta_{\dot{Q}_q}(t)\dot{Q}_q) \qquad (12.2)$$

where $\dot{Q}_q = \omega_q P_{-q}$.

The general *time-ordered n-point Green's function* defined in Eq. (6.16) is generated by the identity

$$\langle T \prod_{j=1}^n A_j(t_j) \rangle = i^n [\delta^n/\delta\eta_{A_1}(t_1) \cdots \delta\eta_{A_n}(t_n)]$$

$$\times \langle T \exp\{-i \int_C dt \sum_B \eta_B(t)B(t)\}\rangle\Big|_{\eta=0} \qquad (12.3)$$

where all time arguments t_j are supposed to be located on the given complex time path C with the endpoints (6.17), $B(t)$ is the Heisenberg representation (5.10) and the order of the η_{A_j} must be respected for fermionic operators A_j. The average on the right-hand side of the above identity is proportional to the *partition function* of the system coupled to the external source fields as defined by the Hamiltonians (7.1), (12.1). Indeed, defining the partition function for the isolated system (all $\eta_A = 0$)

according to Eqs. (6.4), (6.6) by

$$Z[0] \equiv e^{-\beta\Omega} = Tr e^{-\beta\mathcal{H}^\mu} , \qquad (12.4)$$

application of the formulas (5.25), (5.28), (6.9) to the case where $\mathcal{H}_o$ and $\mathcal{H}_{int}$ are replaced, respectively by $\mathcal{H}^\mu$ and $\delta\mathcal{H}_t$, yields

$$Z[\eta] = Z[0]\langle T \exp\{-i \int_C dt \sum_B \eta_B(t)B(t)\}\rangle . \qquad (12.5)$$

The last formula is remarkable because it describes the thermodynamics of an *open* , i.e. *nonequilibrium system*. Indeed, *bosonic source fields* may represent external gradients (recall that fermionic source fields are unphysical) which give rise to *currents* within the system. $Z[\eta]$ then represents a *nonequilibrium partition function*, and Eq. (12.3) with bosonic $\eta_B \neq 0$ defines general *nonequilibrium Green's functions*. This interpretation, first introduced by Schwinger[10], Kadanoff and Baym[4] and by Keldysh[11], however calls for a comment. Indeed, as discussed at the beginning of Section 7, this interpretation is valid only within an approximation in which *entropy production* by the currents is negligible, that is *close to (local) equilibrium* (see Section 13) where, to first order in the currents, Liouville's theorem (7.3) is valid. Situations *far from equilibrium* where entropy production, i.e. *dissipation* (see Section 13), cannot be neglected, are far more complex because there is no unitary time evolution any more. No general formalism is available there (see, e.g. Section 4.7.1 of Ref. 16).

Having identified the generating functional, the next step now is to derive the functional form of Wick's theorem for averages. To do this we first neglect the internal interactions $\mathcal{H}_{int}$, replacing $\mathcal{H}$ by the free Hamiltonian $\mathcal{H}_o$. In this case Eq. (12.5) reads

$$Z_o[\eta] = Z_o[0]\langle \mathcal{S}\rangle_o \qquad (12.6)$$

where

$$\mathcal{S} \equiv T \exp\{-i \int_C dt \sum_B \eta_B(t)B^o(t)\} \qquad (12.7)$$

is the analog of the definition (5.25), (5.28) and $B^o(t)$ is the interaction representation (5.17). According to Wick's theorem this exponential must be entirely expressible in terms of the unperturbed propagators (9.9). Since the latter may be identified by the time dependence of the operators, $B^o(t)$, which is defined by Eqs. (8.1), (8.12), the suitable object to study this time dependence is

$$\frac{1}{Z_o[0]} \frac{i\delta Z_o[\eta]}{\delta\eta_A(t)} = \langle T\{A^o(t)\mathcal{S}\}\rangle_o . \qquad (12.8)$$

Before applying the time derivation $\partial/\partial t = \partial/t'(\tau)\partial\tau$ where $t(\tau)$ parametrizes the path C, we remultiply Eq. (12.8) by $\eta_A(t)$ in order to get rid of the complication of anticommuting variables.

In acting with $\partial/\partial t$ one encounters a serious difficulty because the time-ordering operator T on the right of Eq. (12.8) itself depends on t. We therefore expand $\mathcal{S}$ into a power series of the form (5.22), (5.24), making use of the abridged notation $H_i \equiv \eta_{B_i}(t_i)B_i^o(t_i)$ where $t_i \equiv t(\tau_i)$. Hence

$$\frac{\partial}{\partial t}\langle T\{\eta_A(t)A^o(t)\mathcal{S}\}\rangle_o = \sum_{N=1}^{\infty} \frac{(-i)^N}{N!} \sum_{B_1\cdots B_N} \int_C dt_1 \cdots \int_C dt_N$$

$$\times \sum_P \sum_{k=0}^{N} \frac{\partial}{\partial t}\{\langle H_{i_1} \cdots H_{i_k} \eta_A(t)A^o(t) H_{i_{k+1}} \cdots H_{i_N}\rangle_o$$

$$\times \theta(\tau_{i_1} - \tau_{i_2}) \cdots \theta(\tau_{i_k} - \tau)\theta(\tau - \tau_{i_{k+1}}) \cdots \theta(\tau_{i_{N-1}} - \tau_{i_N})\} \ . \ (12.9)$$

Rearranging the two terms coming from the derivatives of the θ-functions and noting that $\delta(\tau - \tau_l)d\tau = \delta(t - t_l)dt$, we obtain

$$\frac{\partial}{\partial t}\langle T\{\eta_A(t)A^o(t)\mathcal{S}\}\rangle_o = \langle T\{\frac{\partial}{\partial t}(\eta_A(t)A^o(t))\mathcal{S}\}\rangle_o$$

$$+ \sum_{N=1}^{\infty} \frac{(-i)^N}{N!} \sum_{B_1\cdots B_N} \int_C dt_1 \cdots \int_C dt_N \sum_P \sum_{k=1}^{N} \delta(t - t_{i_k})$$

$$\times \langle H_{i_1} \cdots H_{i_{k-1}} K_{i_k}(t) H_{i_{k+1}} \cdots H_{i_N}\rangle_o \theta(\tau_{i_1} - \tau_{i_2}) \cdots \theta(\tau_{i_{N-1}} - \tau_{i_N})$$

$$(12.10)$$

where

$$K_l(t) \equiv [\eta_A(t)A^o(t), \eta_{B_l}(t)B_l^o(t)] = \eta_A(t)\eta_{B_l}(t)[A^o(t), B_l^o(t)]_{\mp} \ . \ (12.11)$$

The second term on the right of Eq. (12.10) may now again be summed into a time-ordered exponential. Dropping again the field η_A, one obtains

$$\frac{\partial}{\partial t}\langle T\{A^o(t)\mathcal{S}\}\rangle_o = \langle T\{\dot{A}^o(t)\mathcal{S}\}\rangle_o - i\sum_B \eta_B(t)\langle T\{[A^o(t), B^o(t)]_{\mp}\mathcal{S}\}\rangle_o \ .$$

$$(12.12)$$

In order to gain explicit results one has to distinguish explicitly between bosonic and fermionic operators. As seen from the example (12.2), $A = Q_q$, $\dot{A} = \omega_q P_{-q}$ and $A = a_k$, $A^+ = a_k^+$ are, respectively, independent irreducible operators in the two cases. In the notation (8.1), (8.3) these operators have the following general properties:

$$\ddot{A} = -\varepsilon_A^2 A \ ; \quad \varepsilon_{\dot{A}} = \varepsilon_{A^+} = \varepsilon_A$$

$$[A, \dot{A}^+]_- = i\varepsilon_A \delta_{AB} \ ; \quad [A, B]_- = [\dot{A}, \dot{B}]_- = 0$$

$$(12.13)$$

for bosons and

$$\dot{A} = -i\varepsilon_A A \; ; \quad \dot{A}^+ = +i\varepsilon_A A^+ \; ; \quad \varepsilon_{A^+} = \varepsilon_A$$
$$[A, B^+]_+ = \delta_{AB} \; ; \quad [A, B]_+ = [A^+, B^+]_+ = 0 \tag{12.14}$$

for fermions.

With (12.13), Eq. (12.12) becomes in the bosonic case

$$\frac{\partial}{\partial t}\langle T\{A^o(t)\mathcal{S}\}\rangle_o = \langle T\{\dot{A}^o(t)\mathcal{S}\}\rangle_o + \varepsilon_A \eta_{\dot{A}^+}(t)\langle\mathcal{S}\rangle_o \, ,$$
$$\frac{\partial}{\partial t}\langle T\{\dot{A}^o(t)\mathcal{S}\}\rangle_o = -\varepsilon_A^2 \langle T\{A^o(t)\mathcal{S}\}\rangle_o - \varepsilon_A \eta_{A^+}(t)\langle\mathcal{S}\rangle_o \, . \tag{12.15}$$

Differentiating once more the first equation to eliminate $\dot{A}$ by substituting from the second equation yields, in the notation of Eqs. (12.6), (12.8),

$$\left(\frac{\partial^2}{\partial t^2} + \varepsilon_A^2\right)i\delta\log Z_o[\eta]/\delta\eta_A(t) = \varepsilon_A(\dot{\eta}_{\dot{A}^+}(t) - \eta_{A^+}(t)) \, . \tag{12.16}$$

This expression being linear in the source fields, $\log Z_o[\eta]$ cannot be more complicated than a quadratic functional of the η. Since

$$[\delta^2 \log Z_o[\eta]/\delta\eta_A(t)\delta\eta_{B^+}(t')]_{\eta=0} = -\langle T\{A^o(t)B^{+o}(t')\}\rangle_o \tag{12.17}$$

is the unperturbed propagator (see, e.g. Eq. (7.35)), integration of Eq. (12.17) yields the desired quadratic form,

$$\log Z_o[\eta] = -\frac{1}{2}\int_C dt \int_C dt' \sum_{AB} \eta_A(t)\langle T\{A^o(t)B^{+o}(t')\}\rangle_o \eta_{B^+}(t')$$
$$+ \text{const} \tag{12.18}$$

where we have left out bilinear terms with $\dot{\eta}_{\dot{A}^+}$. On the other hand, functional differentiation of Eq. (12.16) by $\eta_{B^+}(t')$ gives the correct equation of motion,

$$\left(\frac{\partial^2}{\partial t^2} + \varepsilon_A^2\right)\langle T\{A^o(t)B^{+o}(t')\}\rangle_o = -i\varepsilon_A\delta_{AB}\delta(t-t') \, . \tag{12.19}$$

In the fermionic case the result corresponding to Eq. (12.16) is obtained from Eqs. (12.12), (12.14) and reads

$$\left(\frac{\partial}{\partial t} \pm i\varepsilon_A\right)i\delta\log Z_o[\eta]/\delta\eta_X(t) = -i\eta_{X^+}(t) \; ; \quad X = \begin{cases} A \\ A^+ \end{cases} \tag{12.20}$$

which is again linear in the source fields. In determining the quadratic form of $\log Z_o[\eta]$, care has to be taken because of the anticommutativity of the η. The result, however, is formally identical with Eqs. (12.17), (12.18). By functional differentiation of Eq. (12.20) with respect to $\eta_{Y+}(t')$ one also finds the correct equation of motion,

$$\left(\frac{\partial}{\partial t} \pm i\varepsilon_A\right)\langle T\{X^o(t)Y^{+o}(t')\}\rangle_o = \delta_{XY}\delta(t-t') \; ; \quad X, Y = \begin{cases} A, B \\ A^+, B^+ \end{cases} .$$
(12.21)

Exponentiating Eq. (12.18) with the appropriate constant yields the Gaussian functional

$$Z_o[\eta] = Z_o[0] \exp\left\{ -\frac{1}{2}\int_C dt \int_C dt' \sum_{AB} \eta_A(t) \right.$$

$$\left. \times \langle T\{A^o(t)B^{+o}(t')\}\rangle_o \eta_{B+}(t') \right\} .$$
(12.22)

The importance of this result resides in the explicit factorization of any unperturbed time-ordered n-point function

$$\langle T \prod_{j=1}^n A_j^o(t_j)\rangle_o = (1/Z_o[0])[(i\delta)^n/\delta\eta_{A_1}(t_1)\cdots\delta\eta_{A_n}(t_n)]Z_o[\eta] \quad (12.23)$$

into two-point functions. This result is equivalent with *Wick's theorem for averages* given by Eq. (9.7) but is more compact due to its operational form. Still missing at this stage is the decomposition into Feynman diagrams. The latter are linked together by the vertices describing the *internal interaction* $\mathcal{H}_{int}$ which was momentarily dropped from consideration.

We are now going to show that inclusion of $\mathcal{H}_{int}$ may also be done operationally. This idea again goes back to the work of Martin and Schwinger[10]. We start by writing the partition function in the form

$$Z[\eta] = \text{Tr}\,[e^{-\beta\mathcal{H}^\mu} T \exp\{-i\int_C dt \sum_B \eta_B(t)B(t)\}] .$$
(12.24)

Summing the identity (6.15) with the endpoints (6.17) into an exponential yields the formula

$$T \exp\{-i\int_C dt \sum_B \eta_B(t)B(t)\}$$

$$= S(0, -i\beta)T\{S(-i\beta, 0) \exp(-i\int_C dt \sum_B \eta_B(t)B^o(t))\} \quad (12.25)$$

which we insert into Eq. (12.24). Using (6.8) for $S(0, -i\beta)$ and (5.25), (5.28) for $S(-i\beta, 0)$, Eq. (12.24) may be expressed in the following three forms:

$$Z[\eta] = \text{Tr}\,[e^{-\beta \mathcal{H}_o^\mu} T \exp\{-i \int_C dt (\mathcal{H}_{\text{int}}^o(t) + \sum_B \eta_B(t) B^o(t))\}]$$

$$= \text{Tr}\,[e^{-\beta \mathcal{H}_o^\mu} T \{\exp(-i \int_C dt' \mathcal{H}_{\text{int}}^o(t')) \exp(-i \int_C dt \sum_B \eta_B(t) B^o(t))\}]$$

$$= \exp\left\{-i \int_C dt' \mathcal{H}_{\text{int}}\left[\frac{\delta}{\delta \eta(t')}\right]\right\}$$

$$\times \text{Tr}\,[e^{-\beta \mathcal{H}_o^\mu} T \exp\{-i \int_C dt \sum_B \eta_B(t) B^o(t)\}]\,. \tag{12.26}$$

Since $\mathcal{H}_{\text{int}}$ is a polynomial in the operators B (see, e.g. Eqs. (2.20), (2.22), (3.38)) the differential operator in the last line of (12.26) is well-defined and therefore may be taken outside of the trace and the time-ordering dropped. Now according to (12.6), (12.7), the trace in the fourth line is just $Z_o[\eta]$, for which we substitute Eq. (12.22) to obtain the final result

$$Z[\eta] = Z_o[0] \exp\left\{-i \int_C dt'' \mathcal{H}_{\text{int}}\left[\frac{\delta}{\delta \eta(t'')}\right]\right\}$$

$$\times \exp\left\{-\frac{1}{2} \int_C dt \int_C dt' \sum_{AB} \eta_A(t)\right.$$

$$\times \left. \langle T\{A^o(t) B^{+o}(t')\}\rangle_o \eta_{B+}(t')\right\}\,. \tag{12.27}$$

This formula now generates *Feynman diagrams* for any n-point Green's function defined by (12.3), (12.5). Variations of this powerful and elegant formalism are applied in field theory[9] and in statistical physics[42].

As an example let us write down the n-point Green's functions for $n = 0$ and 2 and derive a general expression for the selfenergy. From Eqs. (12.3), (12.5) and (12.27) we obtain for $n = 0$

$$\frac{Z[0]}{Z_o[0]} = e^{-\beta(\Omega - \Omega_o)} = \langle S(-i\beta, 0)\rangle_o = \exp\left\{-i \int_C dt'' \mathcal{H}_{\text{int}}\left[\frac{\delta}{\delta \eta(t'')}\right]\right\}$$

$$\times \exp\left\{-\frac{1}{2} \int_C dt \int_C dt' \sum_{AB} \eta_A(t) \langle T\{A^o(t) B^{+o}(t')\}\rangle_o \eta_{B+}(t')\right\}\Big|_{\eta=0} \tag{12.28}$$

where Eq. (6.12) has been used, and for $n = 2$

$$\langle T\{A(t) B^+(t')\}\rangle = -\frac{Z_o[0]}{Z[0]} \exp\left\{-i \int_C dt'' \mathcal{H}_{\text{int}}\left[\frac{\delta}{\delta \eta(t'')}\right]\right\} \Phi[\eta]\Big|_{\eta=0} \tag{12.29}$$

where we have permuted the differentiation operators so that

$$\Phi[\eta] \equiv \frac{\delta^2}{\delta\eta_A(t)\delta\eta_{B+}(t')} \exp\left\{-\frac{1}{2}\int_C ds \int_C ds' \sum_{UV} \eta_U(s)\right.$$

$$\left. \times \langle T\{U^\circ(s)V^{+\circ}(s')\}\rangle_\circ \eta_{V+}(s')\right\}. \tag{12.30}$$

Equation (12.28) contains all the closed (thermodynamic) diagrams, not only the connected parts (9.20).

In evaluating the functional (12.30), care must be taken of the anti-commutativity of fermionic source fields η_A. To this end it is convenient to adopt the *rule* that functional derivatives labeled by operators A and B^+ act from the left and from the right, respectively (similar rules were used in Ref. 10). Evaluating the differentiations contained in $\Phi[\eta]$ in this way and making use of (12.28), Eq. (12.29) may be cast into the following form (Problem 2.12)

$$\langle T\{A(t)B^+(t')\}\rangle - \langle T\{A^\circ(t)B^{+\circ}(t')\}\rangle_\circ$$

$$= \int_C ds \int_C ds' \sum_{UV} \langle T\{A^\circ(t)V^{+\circ}(s')\}\rangle_\circ \Sigma_{V+U}(s', s)\langle T\{U^\circ(s)B^+(t')\}\rangle_\circ$$

$$+ \text{ one-particle-reducible terms}, \tag{12.31}$$

where

$$\Sigma_{B+A}(t', t) \equiv -\exp\left\{-i\int_C dt'' \mathcal{H}_{\text{int}}\left[\frac{\delta}{\delta\eta(t'')}\right]\right\}$$

$$\times \left[\eta_{B+}(t')\eta_A(t)\exp\left\{-\frac{1}{2}\int_C ds \int_C ds' \sum_{UV} \eta_U(s)\right.\right.$$

$$\left.\left. \times \langle T\{U^\circ(s)V^{+\circ}(s')\}\rangle_\circ \eta_{V+}(s')\right\}\right]\Bigg|_{cp,\, 1-i,\, \eta=0} \tag{12.32}$$

is a completely general expression for the selfenergy. Here the indices cp and $1-i$ designate, respectively the connected part defined in Section 9 and the one-particle-irreducible part defined in Section 11. From this construction it is also obvious that an *irreducible vertex part with n points of attachment* is defined as in Eq. (12.32) but with n source fields outside the Gaussian functional and with appropriate irreducibility prescriptions.

We now consider the application of Eqs. (12.31), (12.32) to the case of Eq. (11.25) and introduce for the source fields in Eq. (12.2) and for the time path (6.18) the abbreviated notation

$$\eta_k(\tau) \equiv \eta_{a_k}(-i\tau) \; ; \quad \bar\eta_k(\tau) \equiv \eta_{a_k^+}(-i\tau) \; ;$$
$$\xi_q(\tau) \equiv \eta_{Q_q}(-i\tau) \; ; \quad \xi_{-q}(\tau) = \eta_{Q_q^+}(-i\tau) \; ,$$

(12.33)

where the last expression is the consequence of the reality condition (1.25). Taking Eq. (10.6) for $\mathcal{H}_{\text{int}}$ we obtain, with the definitions (7.34), (7.35),

$$\Sigma_{a_k^+ a_k}(\tau', \tau) = -\exp\left\{ -\int_0^\beta d\lambda \sum_{pq} \gamma_q \frac{\delta^3}{\delta\bar\eta_p(\lambda)\delta\eta_{p-q}(\lambda)\delta\xi_q(\lambda)} \right\}$$
$$\times \left[\bar\eta_k(\tau')\eta_k(\tau)Z_o[\eta]/Z_o[0] \right]\Bigg|_{cp, 1-i, \eta=\bar\eta=\xi=0}$$

(12.34)

where Eq. (12.22) has the explicit form

$$\frac{Z_o[\eta]}{Z_o[0]} = \exp\left\{ -\frac{1}{2}\int_0^\beta d\mu \int_0^\beta d\mu' \left[\sum_{p'} \eta_{p'}(\mu)\mathcal{G}^o(p'; \mu - \mu')\bar\eta_{p'}(\mu') \right.\right.$$
$$+ \sum_{p'} \bar\eta_{p'}(\mu)\mathcal{G}^o(p'; \mu' - \mu)\eta_{p'}(\mu')$$
$$\left.\left. + \sum_{q'} \xi_{q'}(\mu)\mathcal{D}^o(q'; \mu - \mu')\xi_{-q'}(\mu') \right] \right\} \; .$$

(12.35)

Developing the exponential differentiation operator in (12.34) to second order, one then recovers the result (11.25) (Problem 2.13). It is obvious from the calculation that the *nonequilibrium selfenergy*, obtained by keeping in Eq. (12.34) $\xi \neq 0$ (but still $\eta = \bar\eta = 0$, since these variables are unphysical), contains many more terms. Since according to Eq. (1.24) the phonon term $\sum \xi_q Q_q$ in (12.2) may be written as $\sum \mathbf{F}_\nu(\mathbf{R}) \cdot \mathbf{u}_\nu(\mathbf{R})$, the source field $\mathbf{F}_\nu(\mathbf{R})$ is recognized as an *external force* acting on the basis atom at $\mathbf{R} + \mathbf{r}_\nu$. This force may, for example, be a thermal agitation acting on one side of the surface of the sample.

We close this section with a discussion of *Ward identities* . These may be defined quite generally as relations connecting irreducible vertex parts of order n and $n+2$ (usually for $n = 2$) resulting from an invariance property of the theory. The identity originally derived by Ward and by Takahashi[43] was a consequence of gauge invariance and was an important ingredient in the quantum-electrodynamic proof that charge renormalization only depends on the photon propagator but not on the propagator or the coupling vertex of the charged fields (see, e.g. Section 31 of Ref. 12). In the formulation of this section the invariance property of the theory is defined by a transformation group leaving the Hamiltonian $\mathcal{H}^\mu$ unchanged. This could be expressed directly as an invariance of Eq. (12.5) under the variation due to an infinitesimal transformation of the group (see, e.g. Ref. 44 for the group of spatial translations).

By Noether's theorem (see, e.g. Section 2.5 of Ref. 12) a transformation group always implies a local conservation law of the form

$$\dot{d}(\mathbf{q}, t) + i\mathbf{q} \cdot \mathbf{j}_d(\mathbf{q}, t) = 0 \tag{12.36}$$

$d(\mathbf{q})$ being one of the densities of Section 3. It turns out that Eq. (12.36) is a more convenient and more general starting point to derive Ward identities than the invariance of $\mathcal{H}^{\mu\,45)}$. Indeed, using definition (6.2) for $\tau_{i_1} > \cdots > \tau_{i_n}$ in the form

$$T\{A_1(t_1) \cdots A_n(t_n)\dot{d}(t)\} = (-)^{\eta_P} \sum_{l=1}^{n} \cdots \frac{\partial d(\mathbf{q}, t(\tau))}{t'(\tau)\partial\tau}$$
$$\times \theta(\tau - \tau_{i_l})A_{i_l}(t(\tau_{i_l})) \cdots \tag{12.37}$$

we may take the time derivative outside of the T-product. Since $\partial\theta(\tau - \tau_o)/\partial\tau = \delta(t(\tau) - t(\tau_o))t'(\tau)$ this leads to the identity[45]

$$T\{A_1(t_1) \cdots A_n(t_n)\dot{d}(\mathbf{q}, t)\} = \frac{\partial}{\partial t}T\{A_1(t_1) \cdots A_n(t_n)d(\mathbf{q}, t)\}$$
$$+ \sum_{l=1}^{n} \delta(t - t_l)T\{A_1(t_1) \cdots [d(\mathbf{q}, t), A_l(t_l)] \cdots A_n(t_n)\} , \tag{12.38}$$

valid for any complex time path C. Substituting here $\dot{d}$ from Eq. (12.36) and taking averages, we obtain the most general Ward identity.

To be more specific we choose electron operators but neglect spin, $A_1 = a_\mathbf{p}$, $A_2 = a_\mathbf{k}^+$, and the densities (3.22), (3.23) so that Eq. (12.36) describes local charge conservation (see Eq. (13.30) below) which is a consequence of gauge invariance (see Eq. (31.3) below). Taking real times and writing, in accordance with Eqs. (7.17) and (10.30),

$$G(\mathbf{k}, t) = -i\langle T\{a_\mathbf{k}(t)a_\mathbf{k}^+\}\rangle$$
$$G_2(\mathbf{k}_1 t_1; \mathbf{k}_2 t_2; \mathbf{k}_3 t_3; \mathbf{k}_4 t_4) = -\langle T\{a_{\mathbf{k}_1}(t_1)a_{\mathbf{k}_2}^+(t_2)a_{\mathbf{k}_3}(t_3)a_{\mathbf{k}_4}^+(t_4)\}\rangle ,$$
$$\tag{12.39}$$

substitution of Eq. (12.36) into (12.38) and averaging yields, for $\mathbf{k} - \mathbf{p} = \mathbf{q} \neq 0$,

$$\sum_{\mathbf{k}'} \left[\frac{\partial}{\partial t} + i\frac{\hbar}{m}\mathbf{q} \cdot \left(\mathbf{k}' + \frac{\mathbf{q}}{2}\right)\right] G_2(\mathbf{k} - \mathbf{q}, t'; \mathbf{k}\ 0; \mathbf{k}' + \mathbf{q}, t; \mathbf{k}'t)$$

$$= i\delta(t - t')G(\mathbf{k}, t') - i\delta(t)G(\mathbf{k} - \mathbf{q}, t') \ . \tag{12.40}$$

Taking time-Fourier transforms in t and t' according to Eq. (7.19) and making use of time-translation invariance (7.9), we obtain the Ward identity

$$\sum_{\mathbf{k}'} \int \frac{d\omega'}{2\pi}\left[\nu + \frac{\hbar}{m}\mathbf{q} \cdot \left(\mathbf{k}' + \frac{\mathbf{q}}{2}\right)\right] \tilde{G}_2(k - q, k, k' + q, k') = \tilde{G}(k) - \tilde{G}(k - q) \tag{12.41}$$

where $k \equiv (\mathbf{k}\ \omega)$, $k' \equiv (\mathbf{k}'\omega')$ and $q \equiv (\mathbf{q}\ \nu)$. Usually, the identity is given only in the limit $q \to 0$. One finds, to first order in q,

$$\nu\left\{\sum_{\mathbf{k}'} \int \frac{d\omega'}{2\pi}\tilde{G}_2(k, k, k', k') - \frac{\partial\tilde{G}(k)}{\partial\omega}\right\}$$

$$= \mathbf{q} \cdot \left\{\frac{\hbar}{m}\sum_{\mathbf{k}'}\mathbf{k}' \int \frac{d\omega'}{2\pi}\tilde{G}_2(k, k, k', k') - \frac{\partial\tilde{G}(k)}{\partial\mathbf{k}}\right\} , \tag{12.42}$$

and since ν and $\mathbf{q}$ are independent, each parenthesis must vanish.

The final step is to express $\tilde{G}_2$ in terms of the *particle-hole-irreducible vertex* Γ (see Fig. 3.8 below),

$$\tilde{G}_2(k, k, k', k') = -(\tilde{G}(k))^2\{\delta_{kk'} + \sum_{k''}\Gamma(k, k, k'', k'')\tilde{G}_2(k'', k'', k', k')\} \tag{12.43}$$

where $\delta_{kk'}$ and the sum over $k'' \equiv (\mathbf{k}''\omega'')$ include a Dirac $\delta(\omega - \omega')$ and an integral over ω'', respectively. Considering only the first relation (12.42) and approximating $\tilde{G}_2$ on the right of (12.43) by $-(\tilde{G}(k'))^2\delta_{k''k'}$, we find

$$\frac{\partial\Sigma(k)}{\partial\omega} = 1 - \frac{\partial\tilde{G}^{-1}(k)}{\partial\omega} = \sum_{k'}\Gamma(k, k, k', k')(\tilde{G}(k'))^2 \ . \tag{12.44}$$

Here use was made of the analog of Eqs. (11.7), (11.34) with $\mu = 0$,

$$\tilde{G}^{-1}(\mathbf{k}, \omega) = \omega - \varepsilon_{\mathbf{k}} - \Sigma(\mathbf{k}, \omega) \simeq z_{\mathbf{k}}^{-1}(\omega - \varepsilon_{\mathbf{k}}^r - i\mathrm{Im}\,\Sigma) \tag{12.45}$$

which near the pole defines the *residue* $z_{\mathbf{k}}$. For the free propagator this relation easily follows from definition (12.39) with the help of Eqs. (7.19), (7.36) and (7.40). Note that apart from the difference in the definition of Γ, Eq. (12.44) is the same as Eq. (19.1) of Ref. 7 which was deduced from Eq. (10.11) of that reference, this last equation playing an analogous role as our Eq. (12.40).

Solutions to the Problems of Chapter 2

Problem 2.1: Eqs. (7.13), (7.14).

Making use of Eq. (3.39), the commutativity $[\mathbf{P}, \mathcal{H}] = [\mathbf{P}, N] = 0$ and the cyclicity of the trace imply that

$$\langle \tilde{a}(\mathbf{r} + \mathbf{r}_o)\tilde{b}(\mathbf{r}' + \mathbf{r}_o)\rangle = \text{Tr}\,(e^{\beta(\Omega - \mathcal{H}^\mu)}e^{i\mathbf{P}\cdot\mathbf{r}_o}\tilde{a}(\mathbf{r})\tilde{b}(\mathbf{r}')e^{-i\mathbf{P}\cdot\mathbf{r}_o}) = \langle \tilde{a}(\mathbf{r})\tilde{b}(\mathbf{r}')\rangle$$

which is (7.14). Applying this translation invariance with $\mathbf{r}_o = -\mathbf{r}'$ and the Fourier transformation (2.3) one may write

$$\langle a(\mathbf{q})b(-\mathbf{q}')\rangle = \int_V d^3r_1 \int_V d^3r_2\, e^{-i\mathbf{q}\cdot\mathbf{r}_1} e^{i\mathbf{q}'\cdot\mathbf{r}_2} \langle \tilde{a}(\mathbf{r}_1 - \mathbf{r}_2)\tilde{b}(0)\rangle\,.$$

Here the evaluation of the double integral is exactly the same as in Eq. (3.37), it yields

$$\langle a(\mathbf{q})b(-\mathbf{q}')\rangle = \langle a(\mathbf{q})\tilde{b}(0)\rangle V\delta_{\mathbf{q}\mathbf{q}'}\,.$$

Now put $\mathbf{q}' = \mathbf{q}$, so that

$$V\langle a(\mathbf{q})\tilde{b}(0)\rangle = \langle a(\mathbf{q})b(-\mathbf{q})\rangle$$

and substitute in the former equation to obtain (7.13).

Problem 2.2: Eqs. (7.40), (7.41).

Using definition (6.10) with $M_k \equiv a_k^+ a_k$ and $N_q \equiv b_q^+ b_q$ we may write, because of (6.7), (6.6),

$$\langle a_k^+ a_k\rangle_o = \text{Tr}\,(e^{-\beta(\varepsilon_k - \mu)M_k}M_k)/\text{Tr}\,(e^{-\beta(\varepsilon_k - \mu)M_k})$$

$$= -\frac{\partial}{\partial x}\ln\text{Tr}\,(e^{-xM_k})\Big|_{x=\beta(\varepsilon_k-\mu)}$$

and

$$\langle b_q^+ b_q\rangle_o = \text{Tr}\,(e^{-\beta\omega_q N_q}N_q)/\text{Tr}\,(e^{-\beta\omega_q N_q})$$

$$= -\frac{\partial}{\partial x}\ln\text{Tr}\,(e^{-xN_q})\Big|_{x=\beta\omega_q}\,.$$

Since the fermionic/bosonic particle numbers take the values 0, 1/0, 1, 2, ... we have

$$\text{Tr}\,(e^{-xM_k}) = \sum_{M_k=0,\,1} e^{-xM_k} = 1 + e^{-x}$$

and

$$\text{Tr}\,(e^{-xN_q}) = \sum_{N_q=0}^{\infty} e^{-xN_q} = \frac{1}{1 - e^{-x}}$$

which immediately leads to Eqs. (7.40) and (7.41), respectively.

Problem 2.3: Eqs. (8.7).

Making use of the operator identity

$$B_1 \cdots B_r B - (\pm)^{\eta_r} B B_1 \cdots B_r$$
$$= \sum_{k=1}^{r} (\pm)^{\eta_{r-k}} B_1 \cdots B_{k-1} [B_k, B]_{\mp} B_{k+1} \cdots B_r \,,$$

where η_{r-k} is the number of fermionic operators among $B_{k+1} \cdots B_r$, and of the contraction (8.2), one finds that for irreducible operators

$$B_1^{(-)} \cdots B_r^{(-)} B^{(+)} - (\pm)^{\eta_r} B^{(+)} B_1^{(-)} \cdots B_r^{(-)}$$
$$= \sum_{k=1}^{r} B_1^{(-)} \cdots \underbrace{B_k^{(-)} \cdots B_r^{(-)} B^{(+)}} \,.$$

Multiplying this identity from the left with any product $C_1^{(+)} \cdots C_s^{(+)}$ we may generate on the left a term of $W(A_1 \cdots A_n) B^{(+)}$ according to (8.4) and, taking the appropriate linear combinations, we readily obtain (8.7).

Problem 2.4: Eqs. (8.9), (8.10).

In order to derive Eq. (8.9) try the recursion hypothesis

$$W\{A_1 \cdots A_n\} = W\left\{ \left[1 - {\sum}'(\smile) + {\sum}'(\smile)^2 + - \cdots \right] A_1 \cdots A_k \right\} A_{k+1} \cdots A_n$$

where the prime indicates exclusion of contractions within the W-product. By application of (8.8) in the form

$$W\{A_1 \cdots A_k\} = W\left\{ \left[1 - {\sum}'(\smile) \right] A_1 \cdots A_{k-1} \right\} A_k$$

one checks the correctness of the hypothesis for $k - 1$ terms left in the W-product. Since it is also true for $k = n - 1$, it is true for any k, in particular for $k = 1$ which is Eq. (8.9). Multiplying (8.10) to the right by A_{n+1} and applying Eq. (8.8) it is easily seen that this reproduces Eq. (8.10) for $n + 1$ factors. Hence it is true by induction.

Problem 2.5: Eq. (8.11).

Applying Eq. (8.9) to each of the W-products on the left of (8.11), and

Eq. (8.10) to the resulting expression we obtain

$$W\{A_1 \cdots A_n\}W\{B_1 \cdots B_m\}\cdots = W\{[1 + \sum(\smile) + \sum(\smile)^2 + \cdots]$$
$$\times [1 - \sum_A(\smile) + \sum_A(\smile)^2 - + \cdots]$$
$$\times [1 - \sum_B(\smile) + \sum_B(\smile)^2 - + \cdots]\cdots A_1 \cdots A_n B_1 \cdots B_m \cdots\}$$

where the indices A, B, $\cdots$ on the summation signs mean restriction to the respective W-products. Now we may factorize the first operator as follows

$$1 + \sum(\smile) + \sum(\smile)^2 + \cdots = [1 + \sum_A(\smile) + \sum_A(\smile)^2 + \cdots]$$
$$\times [1 + \sum_B(\smile) + \sum_B(\smile)^2 + \cdots]\cdots[1 + \sum{}'(\smile) + \sum{}'(\smile)^2 + \cdots]$$

where the prime means exclusion of contractions within the original W-products. But comparing Eqs. (8.9) and (8.10) we notice that the number of terms in each sum is just such that

$$[1 + \sum(\smile) + \sum(\smile)^2 + \cdots][1 - \sum(\smile) + \sum(\smile)^2 - + \cdots] = 1 .$$

Applied to the last equation this readily reduces the first equation to (8.11).

Problem 2.6: Eq. (9.1).

A simple derivation of formula (9.1) is obtained with the help of the generating function[20]

$$G(x) = \sum_{n=0}^{\infty} \frac{x^n}{n!} c_n$$

with

$$c_n \equiv \sum_{\nu=0}^{\infty} \langle\nu|(a^+)^n(a)^n|\nu\rangle w^\nu .$$

Here $|\nu\rangle$ is the state defined by Eqs. (3.14), (3.15) so that

$$\langle\nu|(a^+)^n(a)^n|\nu\rangle = \frac{\nu!}{(\nu - n)!} .$$

Interchanging the n- and ν-summations and making use of the binomial series

$$(1 + x)^\nu = \sum_{n=o}^{\infty} \frac{\nu!}{n!(\nu - n)!} x^n$$

one finds

$$G(x) = \frac{1}{1 - w(1 + x)} = \frac{1}{1 - w} \sum_{n=0}^{\infty} \left(\frac{w}{1 - w}\right)^n x^n$$

or

$$c_n = \frac{n!}{1 - w} \left(\frac{w}{1 - w}\right)^n .$$

Now the average in (9.1) is defined by Eqs. (6.7), (6.10) where $\mathcal{H}_o^\mu$ is of the form $\varepsilon a^+ a$. Hence

$$\langle (a^+)^n (a)^n \rangle_o = \sum_{\nu=0}^{\infty} e^{-\beta \varepsilon \nu} \langle \nu | (a^+)^n (a)^n | \nu \rangle \Big/ \sum_{\nu=0}^{\infty} e^{-\beta \varepsilon \nu}$$

and setting $w = e^{-\beta \varepsilon}$, it is obvious that

$$\langle (a^+)^n (a)^n \rangle_o = \frac{c_n}{c_o} = n! \left(\frac{c_1}{c_o}\right)^n = n! \left(\langle a^+ a \rangle_o\right)^n$$

which is (9.1).

Problem 2.7: Eq. (9.21).

Writing

$$E - E_o = \lim_{\beta \to \infty} \{ \langle \mathcal{H}_o + \mathcal{H}_{\text{int}} \rangle - \langle \mathcal{H}_o \rangle_o \} ,$$

the first average may be evaluated with the help of Eqs. (6.19) and (9.15):

$$\langle \mathcal{H}_o \rangle = \frac{\langle S(-i\beta, 0) \mathcal{H}_o \rangle_o}{\langle S(-i\beta, 0) \rangle_o} \xrightarrow{\beta \to \infty} \frac{\langle \phi_o | S(-i\beta, 0) \mathcal{H}_o | \phi_o \rangle}{\langle \phi_o | S(-i\beta, 0) | \phi_o \rangle} = E_o ,$$

since $\mathcal{H}_o | \phi_o \rangle = E_o | \phi_o \rangle$, so that

$$E - E_o = \lim_{\beta \to \infty} \langle \mathcal{H}_{\text{int}} \rangle = \lim_{\beta \to \infty} \langle S(-i\beta, 0) \mathcal{H}_{\text{int}} \rangle_{o,cp} .$$

The last expression may be calculated in two different ways.

The first method proceeds by application of an equation equivalent to (5.23) but with time path (6.18), so that with (5.22)

$$E - E_o = \lim_{\beta \to \infty} \sum_{N=0}^{\infty} (-1)^N \int_0^\beta d\tau_N \int_{\tau_N}^\beta d\tau_{N-1} \cdots \int_{\tau_2}^\beta d\tau_1$$
$$\times \langle \mathcal{H}_{\text{int}}^o(-i\tau_1) \cdots \mathcal{H}_{\text{int}}^o(-i\tau_{N-1}) \mathcal{H}_{\text{int}}^o(-i\tau_N) \mathcal{H}_{\text{int}} \rangle_{o,cp} .$$

Passing to the limit we have, using (6.22) and assuming that $\mathcal{H}_{\text{int}}$ commutes with the particle number so that $\mathcal{H}^{\mu}$ may be replaced by $\mathcal{H}$ in Eq. (6.8),

$$E - E_o = \sum_{N=0}^{\infty} (-1)^N \int_0^{\infty} d\tau_N \int_{\tau_N}^{\infty} d\tau_{N-1} \cdots \int_{\tau_2}^{\infty} d\tau_1$$

$$\times \langle \phi_o | \mathcal{H}_{\text{int}} e^{(E_o - \mathcal{H}_o)\tau_1} e^{\mathcal{H}_o \tau_2} \mathcal{H}_{\text{int}} e^{-\mathcal{H}_o \tau_2} \cdots e^{\mathcal{H}_o \tau_N} \mathcal{H}_{\text{int}} e^{-\mathcal{H}_o \tau_N} \mathcal{H}_{\text{int}} | \phi_o \rangle_{cp} \;.$$

Because of the connectedness condition, none of the intermediate states is $| \phi_o \rangle$ so that, assuming that $| \phi_o \rangle$ is non-degenerate,

$$\lim_{\tau \to \infty} e^{(E_o - \mathcal{H}_o)\tau} = 0 \;.$$

The successive integrations then are

$$-\int_{\tau_{i+1}}^{\infty} d\tau_i e^{(E_o - \mathcal{H}_o)\tau_i} = \frac{e^{(E_o - \mathcal{H}_o)\tau_{i+1}}}{E_o - \mathcal{H}_o} \;; \quad i = 1, 2, \ldots N \;,$$

which immediately leads to Eq. (9.21).

In the second method use is made of the resolvent

$$(z - \mathcal{H})^{-1} = (z - \mathcal{H}_o)^{-1} [1 - \mathcal{H}_{\text{int}} (z - \mathcal{H}_o)^{-1}]^{-1}$$

and of Cauchy's formula

$$e^{-\beta \mathcal{H}} = \frac{1}{2\pi i} \int_{\Lambda} \frac{e^{-\beta z}}{z - \mathcal{H}} dz$$

where the path Λ encircles the whole spectrum of $\mathcal{H}$ in the positive sense, in order to evaluate Eq. (6.8),

$$S(-i\beta, 0) = e^{\beta \mathcal{H}_o} e^{-\beta \mathcal{H}} \;.$$

Developing the expression for the resolvent and passing to the limit (6.22) we obtain

$$E - E_o = \lim_{\beta \to \infty} e^{\beta E_o} \frac{1}{2\pi i} \int_{\Lambda} dz \frac{e^{-\beta z}}{z - E_o}$$

$$\times \langle \phi_o | \sum_{N=0}^{\infty} [\mathcal{H}_{\text{int}} (z - \mathcal{H}_o)^{-1}]^N \mathcal{H}_{\text{int}} | \phi_o \rangle_{cp} \;.$$

Making use of Cauchy's formula this readily yields an expression equivalent to (9.21).

Problem 2.8: Eq. (10.3).

Insertion of (10.2) into (10.3) gives

$$c_{Vo} = -\frac{1}{\beta V} \sum_k \ln \frac{f_o}{1 - f_o} \left(\frac{\partial f_o}{\partial T}\right)_\mu$$

while from (10.1) one obtains

$$n_o = -\frac{1}{V}\left(\frac{\partial \Omega_o}{\partial \mu}\right)_T = \frac{1}{V}\sum_k f_o(\varepsilon_k - \mu) .$$

$n_o = \text{const}$ implies $\mu = \mu(T)$ so that we have from (7.42)

$$\frac{\partial f_o}{\partial T} = -\frac{1}{T}\left[\varepsilon_k - \mu - T\frac{\partial\mu}{\partial T}\right]f_o' .$$

Since $\ln f_o/(1 - f_o) = -\beta(\varepsilon_k - \mu)$ one finds, making use of (10.4),

$$c_{Vo} = -\frac{2}{T}\int_{-\infty}^{+\infty} d\varepsilon N(\varepsilon + \mu)f_o'(\varepsilon)\varepsilon\left[\varepsilon - T\frac{\partial\mu}{\partial T}\right]$$

where the density of states takes care of the precise integration limits. Taking the T-derivative of the expression for n_o, one obtains the condition

$$0 = \int_{-\infty}^{+\infty} d\varepsilon N(\varepsilon + \mu)f_o'(\varepsilon)\left[\varepsilon - T\frac{\partial\mu}{\partial T}\right] .$$

At low temperature, $f_o'(\varepsilon)$ is strongly peaked at $\varepsilon = 0$ so that we may Taylor-expand $N(\varepsilon + \mu)$ around μ. In terms of the integration variable $x = \beta\varepsilon$ this yields

$$c_{Vo} = 2k_B^2 T \sum_{n=0}^{\infty} \frac{1}{n!} N^{(n)}(\mu)(k_B T)^n\left[I_{n+2} - \frac{\partial\mu}{k_B\partial T}I_{n+1}\right]$$

and

$$0 = \sum_{n=0}^{\infty} \frac{1}{n!}N^{(n)}(\mu)(k_B T)^n\left[I_{n+1} - \frac{\partial}{k_B\partial T}I_n\right]$$

where

$$I_n \equiv \int_{-\infty}^{+\infty} \frac{x^n e^{-x}dx}{(1 + e^{-x})^2}$$

is given explicitly in Eq. (32.11) and $I_o = 1$, $I_2 = \pi^2/3$, $I_4 = 7\pi^4/15$, $I_{2n+1} = 0$ (all n). From the last equation, $\partial\mu/\partial T$ may be determined and inserted into the expression for c_{V_o}. $I_1 = 0$ implies that $\partial\mu/\partial T = \mathcal{O}(T)$ and that the first correction to the lowest-order expression of c_{V_o} is $\mathcal{O}(T^3)$. This establishes Eq. (10.3).

Problem 2.9: Eq. (10.12).

Using (7.25) and the rule (9.10) with (7.34), (7.40) (see Eq. (11.24) below) in the first line of (10.10) one finds

$$T^{\mathrm{H}} = -f_o(\varepsilon_k)f_o(\varepsilon_{k'})\beta^{-1}\sum_{\nu_+}\tilde{D}^o(q; i\nu_+)e^{-i\nu_+(\tau-\tau')}\delta_{\mathbf{q},0}\delta_{q,-q'} \ .$$

Insertion into (10.7), making use of (1.25), (7.28) and (10.13) gives

$$\Delta\Omega^{\mathrm{H}} = (nV)^2\sum_q|\gamma_q|^2\tilde{D}^o(q; 0)\delta_{\mathbf{q},0}$$

which reduces to Eq. (10.12) by taking into account Eqs. (7.45), (10.11) and $V = vL$.

Problem 2.10: Eq. (10.16).

According to (7.26) we may write

$$e^{\beta\omega}\mp 1 = \pm e^{\beta(\omega-i\nu_\pm)}\mp 1 = \pm\beta(\omega - i\nu_\pm) + \mathcal{O}(\beta^2(\omega - i\nu_\pm)^2) \ .$$

Applying Cauchy's formula we then get, for any function $F(\omega)$ which is analytic in the vicinity of $i\nu_\pm$,

$$\frac{1}{2\pi i}\oint_{\nu_\pm}\frac{F(\omega)d\omega}{e^{\beta\omega}\mp 1} = \pm\beta^{-1}F(i\nu_\pm) \ .$$

The last identity may be summed over all $\nu_\pm$ and, provided that $F(\omega)$ is analytic in the vicinity of the whole imaginary axis, the circles around each $\nu_\pm$ may be deformed into straight lines parallel to the imaginary axis which yields the identity (10.16).

Problem 2.11: Eq. (11.18).

With the expression (3.38) for $\mathcal{H}_{\mathrm{int}}$, the $N = 1$ term of (11.17) may again be written as

$$\Sigma(k; i\nu_-)[\tilde{\mathcal{G}}^o(k; i\nu_-)]^2 = \int_0^\beta d\tau\int_0^\beta d\tau' e^{i\nu_-\tau}\frac{1}{2V}\sum_{k'k''\mathbf{q}}\tilde{\phi}(\mathbf{q})(S^{\mathrm{H}} + S^{\mathrm{ex}})$$

where

$$S^{\mathrm{H}} = 2\langle T\{(a_{k'}^+a_{k''}^+a_{k''+q}a_{k'-q})^o(-i\tau')a_k^o(-i\tau)a_k^{+o}(0)\}\rangle_{o,cp}$$

and

$$S^{\text{ex}} = 2\langle T\{(a^+_{k'} a^+_{k''} a_{k''+q} a_{k'-q})^o(-i\tau') a^o_k(-i\tau) a^{+o}_k(0)\}\rangle_{o,cp} \; .$$

With (7.9), (7.13), (7.34) and spin conservation these expressions become

$$S^{\text{H}} = +2\mathcal{G}^o(k;\tau-\tau')\delta_{\sigma'\sigma}\delta_{k',k}\mathcal{G}^o(k''; 0)\delta_{q,0}\mathcal{G}^o(k;\tau')\delta_{k'-q,k}$$

and

$$S^{\text{ex}} = -2\mathcal{G}^o(k;\tau-\tau')\delta_{\sigma'\sigma}\delta_{k'k}\mathcal{G}^o(k''; 0)\delta_{\sigma''\sigma'}\delta_{k'',k'-q}\mathcal{G}^o(k;\tau')\delta_{k''+q,k} \; .$$

Using (7.25), (7.28) and (11.25) and inserting back the resulting expressions, one finds

$$\Sigma(k; i\nu_-)[\tilde{\mathcal{G}}^o(k; i\nu_-)]^2 = \frac{1}{V}\sum_{k''} f_o(\varepsilon_{k''})\{2\tilde{\phi}(0) - \tilde{\phi}(k''-k)\}[\tilde{\mathcal{G}}^o(k; i\nu_-)]^2$$

where the factor 2 comes from the spin sum. This is Eq. (11.18).

Problem 2.12: Eqs. (12.31), (12.32).

Since

$$\eta_U(s)\langle T\{U^o(s)V^{+o}(s')\}\rangle_o\eta_{V+}(s') = \eta_{V+}(s')\langle T\{V^{+o}(s')U^o(s)\}\rangle_o\eta_U(s) \; ,$$

evaluating first $\delta/\delta\eta_{B+}(t')$ from the right, one obtains two terms which, after relabeling integration and summation variables, may be combined into

$$\Phi[\eta] = \frac{\delta}{\delta\eta_A(t)}\left[-\int_C dr \sum_X \eta_X(r)\langle T\{X^o(r)B^{+o}(t')\}\rangle_o\right.$$

$$\left.\times \exp\left\{-\frac{1}{2}\int_C ds \int_C ds' \sum_{UV} \eta_U(s)\langle T\{U^o(s)V^{+o}(s')\}\rangle_o\eta_{V+}(s')\right\}\right] \; .$$

Applying now the second derivative from the left, taking great care of possible anticommutations, one finds

$$\Phi[\eta] = \left[-\langle T\{A^o(t)B^{+o}(t')\}\rangle_o\right.$$

$$\left. + \int_C dr \int_C dr' \sum_{XY}\langle T\{A^o(t)Y^{+o}(r')\}\rangle_o\eta_{Y+}(r')\eta_X(r)\langle T\{X^o(r)B^{+o}(t')\}\rangle_o\right]$$

$$\times \exp\left\{-\frac{1}{2}\int_C ds \int_C ds' \sum_{UV}\eta_U(s)\langle T\{U^o(s)V^{+o}(s')\}\rangle_o\eta_{V+}(s')\right\} \; .$$

Inserting this expression into Eq. (12.29) and making use of (12.28) one finds

$$\langle T\{A(t)B^+(t')\}\rangle = \langle T\{A^o(t)B^{+o}(t')\}\rangle_o - \frac{1}{\langle S(-i\beta, 0)\rangle_o}$$

$$\times \int_C dr \int_C dr' \sum_{XY} \langle T\{A^o(t)Y^{+o}(r')\}\rangle_o \langle T\{X^o(r)B^{+o}(t')\}\rangle_o$$

$$\times \exp\left\{-i\int_C dt'' \mathcal{H}_{\text{int}}\left[\frac{\delta}{\delta\eta(t'')}\right]\right\}\left[\eta_{Y^+}(r')\eta_X(r)\right.$$

$$\times \exp\left\{-\frac{1}{2}\int_C ds \int_C ds' \sum_{UV} \eta_U(s)\langle T\{U^o(s)V^{+o}(s')\}\rangle_o \eta_{V^+}(s')\right\}\Bigg]\Bigg|_{\eta=0}.$$

As shown in Eq. (9.14), summation of all disconnected parts just results in a factor $\langle S(-i\beta, 0)\rangle_o$ and, defining the selfenergy as one-particle-irreducible part, comparison with Eq. (11.17) readily yields Eqs. (12.31), (12.32).

Problem 2.13: Derivation of Eq. (11.25) from Eqs. (12.34), (12.35) and (12.31).

To second order in the exponential differentiation operator, the condition $\eta = \bar{\eta} = \xi = 0$ just leaves one term in Eq. (12.34) which is connected and one-particle-irreducible, namely

$$\Sigma_{a_k^+ a_k}(\tau', \tau) = -\frac{1}{2}\int_0^\beta d\lambda \int_0^\beta d\lambda' \sum_{pq}\sum_{p'q'} \gamma_q \gamma_{q'}$$

$$\times \left[\delta^6 / \delta\bar{\eta}_p(\lambda)\delta\eta_{p-q}(\lambda)\delta\xi_q(\lambda)\delta\bar{\eta}_{p'}(\lambda')\delta\eta_{p'-q'}(\lambda')\delta\xi_{q'}(\lambda')\right]$$

$$\times \left(\bar{\eta}_k(\tau')\eta_k(\tau)Z_o[\eta]/Z_o[0]\right)\Bigg|_{\eta=\bar{\eta}=\xi=0}$$

where the notation is as follows: $k = (\mathbf{k}, \sigma)$, $p = (\mathbf{p}, \tau)$, etc. and $p - q \equiv (\mathbf{p-q}, \tau)$. In this expression the prefactor $\bar{\eta}_k(\tau')\eta_k(\tau)$ must be differentiated away in order to get a non-vanishing contribution. Thereby a factor 2 comes from the alternative of differentiating $\bar{\eta}_k(\tau')$ by $\bar{\eta}_p(\lambda)$ or by $\bar{\eta}_{p'}(\lambda')$; let us choose $\bar{\eta}_{p'}(\lambda')$. The next alternative is to differentiate $\eta_k(\tau)$ by $\eta_{p-q}(\lambda)$ or by $\eta_{p'-q'}(\lambda')$. Taking into account the fact that η and $\bar{\eta}$ are anticommuting source fields, this alternative yields the factors

$$+2\delta_{kp'}\delta(\tau' - \lambda')\delta_{k,p-q}\delta(\tau - \lambda) \text{ and } -2\delta_{k,p'-q'}\delta(\tau - \lambda')\delta_{kp'}\delta(\tau' - \lambda')$$

which are recognized as leading, respectively to the "exchange" and "Hartree terms" of Fig. 2.8. Indeed, with these factors the above expression for

$\Sigma_{a_k^+ a_k}(\tau', \tau)$ simplifies as follows:

$$\Sigma_{a_k^+ a_k}(\tau', \tau) = -\sum_{qq'} \gamma_q \gamma_{q'} \left[\frac{\delta^3}{\delta\bar{\eta}_{k+q}(\tau)\delta\eta_{k-q'}(\tau')\delta\xi_q(\tau)} - \delta(\tau - \tau') \right.$$

$$\times \left. \int_0^\beta d\lambda \sum_p \delta_{\mathbf{q}',0} \frac{\delta^3}{\delta\bar{\eta}_p(\lambda)\delta\eta_{p-q}(\lambda)\delta\xi_q(\lambda)} \right] \frac{\delta}{\delta\xi_{q'}(\tau')} \frac{Z_o[\eta]}{Z_o[0]} \Bigg|_{\eta=\bar{\eta}=\xi=0} .$$

Here the derivation by $\xi_{q'}(\tau')$ brings down a factor

$$- \int_0^\beta d\mu \mathcal{D}^o(q'; \mu - \tau')\xi_{-q'}(\mu) ,$$

where use has been made of the symmetry $\mathcal{D}^o(\pm q; \tau) = \mathcal{D}^o(q; -\tau)$ which follows from (1.25), (7.45) and (7.25). The non-vanishing terms result from operating with the second ξ-differentiation on this last factor so that

$$\Sigma_{a_k^+ a_k}(\tau', \tau) = \sum_q |\gamma_q|^2 \left[\mathcal{D}^o(q; \tau - \tau') \frac{\delta^2}{\delta\bar{\eta}_{k+q}(\tau)\delta\eta_{k+q}(\tau')} \right.$$

$$\left. - \delta(\tau - \tau')\delta_{\mathbf{q},0} \int_0^\beta d\lambda \sum_p \mathcal{D}^o(q; \lambda - \tau') \frac{\delta^2}{\delta\bar{\eta}_p(\lambda)\delta\eta_p(\lambda)} \right] \frac{Z_o[\eta]}{Z_o[0]} \Bigg|_{\eta=\bar{\eta}=0} .$$

According to (12.35), the derivation by $\eta_p(\lambda)$ brings down a factor

$$- \int_0^\beta d\mu \mathcal{G}^o(p; \lambda - \mu)\bar{\eta}_p(\mu)$$

on which the $\bar{\eta}$-differentiation must act, so that

$$\Sigma_{a_k^+ a_k}(\tau', \tau) = -\sum_q |\gamma_q|^2 \left[\mathcal{D}^o(q; \tau - \tau')\mathcal{G}^o(k + q; \tau' - \tau) \right.$$

$$\left. - \delta(\tau - \tau')\delta_{\mathbf{q},0} \int_0^\beta d\lambda \sum_p \mathcal{D}^o(q; \lambda - \tau')\mathcal{G}^o(p; 0) \right] .$$

One easily checks that, inserted into the corresponding Eq. (12.31), this result yields exactly Eq. (11.25).

References to Chapter 2

1. **Pauli, W.**, *Lectures on Physics, Volume 6: Selected Topics in Field Quantization*, ed. Enz, C.P. (MIT Press, Cambridge, USA, 1973).
2. **Tomonaga, S.**, *Prog. Theor. Phys.* **1**, 27, 109 (1946); **Schwinger, J.**, *Phys. Rev.* **74**, 1439 (1948); **75**, 651 (1949).
3. **Dyson, F.J.**, *Phys. Rev.* **75**, 486, 1736 (1949).

4. **Kadanoff, L.P.** and **Baym, G.**, *Quantum Statistical Mechanics* (Benjamin, New York, 1962).
5. **Matsubara, T.**, *Prog. Theor. Phys.* **14**, 351 (1955).
6. **Alekseev, A.I.**, *Soviet Phys.-Uspekhi* **4**, 23 (1961).
7. **Abrikosov, A.A., Gor'kov, L.P.** and **Dzyaloshinskii, I.Ye.**, *Quantum Field Theoretical Methods in Statistical Physics, 2nd ed.*, ed. ter Haar, D. (Pergamon, Oxford, 1965).
8. **Fetter, A.L.** and **Walecka, J.D.**, *Quantum Theory of Many-Particle Systems* (McGraw-Hill, New York, 1971).
9. **Landsman, N.P.** and **van Weert, Ch.G.**, *Phys. Rep.* **145**, 141 (1987).
10. **Martin, P.C.** and **Schwinger, J.**, *Phys. Rev.* **115**, 1342 (1959); **Schwinger, J.**, *J. Math. Phys.* **2**, 407 (1961).
11. **Keldysh, L.V.**, *Soviet Phys.-JETP* **20**, 1018 (1965).
12. **Bogoliubov, N.N.** and **Shirkov, D.V.**, *Introduction to the Theory of Quantized Fields* (Interscience, New York, 1959).
13. **Pauli, W.**, *Lectures on Physics, Volume 4: Statistical Mechanics*, ed. Enz, C.P. (MIT Press, Cambridge, USA, 1973).
14. **Huang, K.**, *Statistical Mechanics* (Wiley, New York, 1963).
15. **Kubo, R.**, *J. Phys. Soc. Jpn.* **12**, 570 (1957).
16. **Kubo, R., Toda, M.** and **Hashitsume, N.**, *Statistical Physics II. Nonequilibrium Statistical Mechanics* (Springer, Berlin, 1985).
17. **Wick, G.C.**, *Phys. Rev.* **80**, 268 (1950).
18. **Feynman, R.P.**, *Phys. Rev.* **76**, 749, 769 (1949).
19. **Thouless, D.J.**, *Phys. Rev.* **107**, 1162 (1957); **Gaudin, M.**, *Nucl. Phys.* **15**, 89 (1960).
20. **Enz, C.P.**, *Helv. Phys. Acta* **38**, 150 (1965).
21. **Houriet, A.** and **Kind, A.**, *Helv. Phys. Acta* **22**, 319 (1949).
22. **Bloch, C.** and **de Dominicis, C.**, *Nucl. Phys.* **7**, 459 (1958).
23. **Thouless, D.J.**, *The Quantum Mechanics of Many-Body Systems* (Academic, New York, 1961).
24. **Brout, R.** and **Carruthers, P.**, *Lectures on the Many-Electron Problem* (Interscience, New York, 1963).
25. **Kittel, C.**, *Quantum Theory of Solids* (Wiley, New York, 1963).
26. **Goldstone, J.**, *Proc. Roy. Soc.* **A239**, 267 (1957).
27. **Hugenholtz, N.M.**, *Physica* **23**, 481 (1957); **Bloch, C.**, *Nucl. Phys.* **7**, 451 (1958); **Luttinger, J.M.** and **Ward, J.C.**, *Phys. Rev.* **118**, 1417 (1960).
28. **Buckingham, M.J.** and **Schafroth, M.R.**, *Proc. Phys. Soc. (London)* **A67**, 828 (1954).
29. **Krebs, K.**, *Phys. Lett.* **6**, 31 (1963).
30. **Coffey, D.** and **Pethick, C.J.**, *Phys. Rev.* **B37**, 442 (1988). See also **Herring, C.**, in *Magnetism Vol. IV*, ed. Rado, G.T. and Suhl, H. (Academic, New York, 1966), Section XI; **Kim, D.J.**, *Phys. Rep.* **171**, 129 (1988), Section 12.
31. **Grimval, G.**, *The Electron-Phonon Interaction in Metals* (North-Holland, Amsterdam, 1981).

32. Baumann, K. and Ranninger, J., *Ann. Phys. (N.Y.)* **20**, 157 (1962).
33. Schrieffer, J.R., *Theory of Superconductivity* (Benjamin, New York, 1964).
34. Born, M. and Oppenheimer, R., *Ann. Phys. (Leipzig)* **84**, 457 (1927); see also Born, M. and Huang, K., *Dynamical Theory of Crystal Lattices* (Clarendon, Oxford, 1968), Section 14.
35. Enz, C.P., "Electron-phonon and phonon-phonon interactions", in *Theory of Condensed Matter* (IAEA, Vienna, 1968), p. 729.
36. Migdal, A.B., *Soviet Phys.-JETP* **7**, 996 (1958).
37. Kohn, W., *Phys. Rev. Lett.* **2**, 393 (1959).
38. Afanas'ev, A.M. and Kagan, Yu., *Soviet Phys.-JETP* **16**, 1030 (1963).
39. Van Hove, L., *Phys. Rev.* **89**, 1189 (1953).
40. Negele, J.W. and Orland, H., *Quantum Many-Particle Systems* (Addison-Wesley, Redwood, 1988).
41. Schwinger, J., *Phys. Rev.* **91**, 713 (1953).
42. Chou, K.-C., Su, Z.-B., Hao, B.-L., and Yu, L., *Phys. Rep.* **118**, 1 (1985).
43. Ward, J.C., *Phys. Rev.* **78**, 182 (1950); **84**, 897 (1951); Takahashi, Y., *Nuovo Cimento* **6**, 371 (1957).
44. Wadati, M., *Phys. Rep.* **50**, 87 (1979).
45. Revzen, M., Toyoda, T., Takahashi, Y., and Khanna, F.C., *Phys. Rev.* **B40**, 769 (1989).

Chapter 3

TRANSPORT PHENOMENA
AND DISORDER EFFECTS

In a perfect crystal, local fluctuations of charge, matter or energy densities as described, respectively by Eqs. (2.9), (2.12), (2.14), (2.15), cannot disappear because of lack of resistance; there is no dissipation. *Dissipation is caused by disorder*. Since the measure of disorder is *entropy* and the measure of dissipation is *entropy production* (see below), transport phenomena which describe the mentioned resistances, are intimately related to the notions of entropy and entropy production. From a microscopic point of view, however, these are extremely difficult notions.

The simplest microscopic description of transport phenomena is the linear response theory presented in Section 7. However, as discussed in Section 12, this description is only valid close to equilibrium where entropy production may be neglected. In far-from-equilibrium situations which are characterized by the presence of strong currents, no general treatment is available yet. But even in the linear response regime, complications may arise due to *sample size*, as for instance in the microstructures used in modern electronic circuits[1]. In these "mesoscopic systems" quantum mechanics gives rise to nonlocal effects, the so-called "contact resistances" which are present even if the system consists of a perfect crystal[2].

Disorder in a solid, of course, has many different aspects. From the point of view of entropy, the minimum disorder presumably is a single impurity or displaced host atom or an incommensurate excitation wave in a perfect crystal. But as we know through the fundamental work of Anderson and Mott , disorder may be so important as to give rise to *wave localization* or to a *metal-insulator transition*[3-5]. We also know[6] that the effect of disorder decreases with increasing number of dimensions D, which is

97

intuitively evident since with increasing D, the number of nearest neighbors of an atom in the solid is increasing and hence also the possibilities to move. The critical dimension being $D = 2$ it is not surprising that phenomena occurring in 2-dimensional systems (inversion layers, thin metallic films) such as the quantum Hall effect[7] depend sensitively on disorder and localization of the current into filaments[8].

13. Boltzmann-equation Description of Electron Transport

The conventional treatment of transport properties is based on the Boltzmann equation for electrons in metals[9,10] and for phonons in insulators[9]. For electrons there exists, however an alternative approach due to Landauer[11] which makes use of reflection and transmission amplitudes and particle reservoirs and may be generalized to multi-terminal systems with multi-channel leads[12]. These two approaches are the subject of this and the following sections.

Let us start with the *kinetic approach* writing the *Boltzmann equation for the distribution function* $f(k, \mathbf{r}, t)$ *of the electrons* in the usual form[9,10]

$$\frac{\partial f}{\hbar \partial t} - e\left[\mathbf{E} + \left(\frac{\mathbf{v}_k}{c} \times \mathbf{H}\right)\right] \cdot \frac{\partial f}{\hbar \partial \mathbf{k}} + \mathbf{v}_k \cdot \nabla f = \left(\frac{dt}{\hbar dt}\right)_{\text{diss}} \quad (13.1)$$

where k designates the state of the electron and $\pm k = (n, \pm\mathbf{k}, \sigma)$ has the meaning defined in Section 1. Note that as in Chapter 2, time $= \hbar t$.

$$\left(\frac{df}{\hbar dt}\right)_{\text{diss}} \equiv \sum_{k'} \{\Gamma(k'k) - \Gamma(kk')\} \quad (13.2)$$

is the *dissipative time variation* due to the collisions of the electrons corresponding to Boltzmann's collision term.

$$\Gamma(kk') = W(kk')f(k)[1 - f(k')] \quad (13.3)$$

is the *transition rate* from state k to state k' (at zero temperature, $f = 1$ so that k is occupied and k' is empty). $\mathbf{E}$ and $\mathbf{H}$ are applied electric and magnetic fields, $\mathbf{v}_k = \partial \varepsilon_k / \hbar \partial \mathbf{k}$ the velocity of the electron (e is the elementary charge, $2\pi\hbar$ Planck's constant and c the light velocity). In this section the description will be phenomenological in the sense that $W(k, k')$ is assumed to be given.

Local equilibrium is defined by the Fermi distribution function f_o of Eq. (7.40) but allowing for slowly varying space-time functions $\beta^{-1} = k_B T(\mathbf{r}, t)$ and $\mu = \mu(\mathbf{r}, t)$ of temperature and chemical potential such that the collision term vanishes,

$$\left(\frac{df_o}{\hbar dt}\right)_{\text{diss}} = 0 \ . \tag{13.4}$$

Note that in a fluid, local equilibrium also allows for a *convective velocity distribution* $\mathbf{c}(\mathbf{r}, t)$ which, however is zero in a crystal lattice, except in the presence of a *supercurrent* (see Chapter 4). According to Eq. (13.2) a sufficient condition for local equilibrium is *detailed balance*[9]

$$\Gamma_o(kk') = \Gamma_o(k'k) \ . \tag{13.5}$$

Here it is important to recognize that if Γ_o is given by formula (13.3) with f replaced by f_o, Eq. (13.5) implies $\exp(\beta\varepsilon_{k'})W(kk') = \exp(\beta\varepsilon_k)W(k'k)$ (see Eq. (8.1.4) of Ref. 10); if however, *microscopic reversibility* $W(kk') = W(k'k)$ is assumed as in Ref. 9, $\Gamma_o(kk')$ cannot have this form, except for *elastic collisions*, $\varepsilon_k = \varepsilon_{k'}$. It turns out that all the microscopic examples derived in Sections 15 and 16 satisfy detailed balance.

The well-known difficulty of the Boltzmann equation (13.1) comes from the nonlinearity in Eq. (13.3). Writing[9,10]

$$f(k) = f_o(\varepsilon_k) - \phi(k)f_o'(\varepsilon_k) \tag{13.6}$$

where we have set the average chemical potential (Fermi level) $\mu = 0$, linearization in ϕ must also be extended to space-time variations of $f(k)$ which implies that

$$|\phi f_o''| \ll |f_o'| \ . \tag{13.7}$$

Since according to (7.42) $f_o' = -\beta f_o(1 - f_o)$, linearization of Eq. (13.3) gives

$$\Gamma(kk') = \Gamma_o(kk')\{1 + \beta\phi(k)[1 - f_o(\varepsilon_k)] - \beta\phi(k')f_o(\varepsilon_{k'})\} \ , \tag{13.8}$$

so that in view of detailed balance (13.5) the collision term (13.2) may be written in terms of a linear Boltzmann operator $\mathcal{B}$ as

$$\left(\frac{df(k)}{\hbar dt}\right)_{\text{diss}} = \beta\sum_{k'}\Gamma_o(kk')\{\phi(k') - \phi(k)\} \equiv -(\mathcal{B}\phi)(k) \ . \tag{13.9}$$

Note that the Boltzmann theory developed here is valid for *one-particle collisions* with an external agent; a generalization to *two-particle collisions* is given in Section 17 (see also Ref. 9, Section 7.7).

The *thermodynamic fluxes* are given by the *electric* and *thermal current densities*

$$\mathbf{j}[f] \equiv -eV^{-1}\sum_k \mathbf{v}_k f(k) \ ; \quad \mathbf{w}[f] \equiv V^{-1}\sum_k \varepsilon_k\mathbf{v}_k f(k) \ , \tag{13.10}$$

V being the crystal volume . From the symmetry $\varepsilon_{-k} = \varepsilon_k$ (see Section 1) follows immediately that $\mathbf{j}[f_o] = 0$ and $\mathbf{w}[f_o] = 0$ so that, inserting (13.6),

$$\mathbf{j} = eV^{-1} \sum_k \mathbf{v}_k \phi(k) f_o'(\varepsilon_k) \; ; \quad \mathbf{w} = -V^{-1} \sum_k \varepsilon_k \mathbf{v}_k \phi(k) f_o'(\varepsilon_k) \; . \quad (13.11)$$

Since we may write for any derivative of f_o

$$\delta f_o^{(n)}(\varepsilon_k) = -\left(\frac{\varepsilon_k}{T} \delta T + T \delta\left(\frac{\mu}{T}\right) \right) f_o^{(n+1)}(\varepsilon_k) \; ; \quad n \geq 0 \; , \quad (13.12)$$

the Boltzmann equation (13.1) becomes, making use of Eqs. (13.6), (13.7) and (13.9),

$$\tilde{\mathcal{B}}\phi \equiv \mathcal{B}\phi + e\left[\mathbf{E} + \frac{\mathbf{v}_k}{c} \times \mathbf{H} \right] \cdot \frac{\partial\phi}{\hbar\partial\mathbf{k}} f_o'$$

$$= (\mathbf{X}^{(1)} + \varepsilon_k \mathbf{X}^{(2)}) \cdot \mathbf{v}_k f_o' + \left[\frac{\varepsilon_k}{T}\dot{T} + T\left(\frac{\mu}{T}\right)^{\cdot} \right] f_o' \quad (13.13)$$

where we have defined the *thermodynamic forces* as

$$\mathbf{X}^{(1)} \equiv e\mathbf{E} + T\nabla\frac{\mu}{T} \; ; \quad \mathbf{X}^{(2)} \equiv \frac{1}{T}\nabla T \; . \quad (13.14)$$

In the following we limit the discussion to stationary situations, setting $\dot{T} = 0$ and $\dot{\mu} = 0$. Also we extend linearization to the electric field, but not to the magnetic field which is justified on physical grounds by the fact that deviations from Ohm's law (see the next section) are extremely small. This means that the field $\mathbf{E}$ may be neglected on the left-hand side of Eq. (13.13), and $\phi = \phi(k, \mathbf{H})$, independent of $\mathbf{E}$. Making the ansatz

$$\phi = \sum_{\lambda=1,2} \vec{\varphi}^{(\lambda)} \cdot \mathbf{X}^{(\lambda)} = \sum_{\lambda=1,2} \sum_{\alpha=1}^{3} \varphi_\alpha^{(\lambda)} X_\alpha^{(\lambda)} \quad (13.15)$$

and considering the thermodynamic forces as independent variables, insertion of definition (13.9) into Eq. (13.13) gives rise to the system of linear integro-differential equations

$$(\tilde{\mathcal{B}}\varphi_\alpha^{(\lambda)})(k, \mathbf{H}) = \beta \sum_{k'} \Gamma_o(kk')[\varphi_\alpha^{(\lambda)}(k, \mathbf{H}) - \varphi_\alpha^{(\lambda)}(k', \mathbf{H})]$$

$$- \frac{e}{c}\mathbf{H} \cdot \left(\frac{\partial f_o}{\hbar\partial\mathbf{k}} \times \frac{\partial\varphi_\alpha^{(\lambda)}(k, \mathbf{H})}{\hbar\partial\mathbf{k}} \right)$$

$$= \varepsilon_k^{\lambda-1} \frac{\partial f_o}{\hbar\partial k_\alpha} \; ; \quad \lambda = 1, 2 \; ; \quad \alpha = 1, 2, 3 \; . \quad (13.16)$$

In addition, the current densities (13.11) obey the *linear laws of nonequilibrium thermodynamics* [13,14]

$$\frac{1}{e}\mathbf{j} \equiv \mathbf{J}^{(1)} = L^{(11)}\mathbf{X}^{(1)} + L^{(12)}\mathbf{X}^{(2)}$$

$$-\mathbf{w} \equiv \mathbf{J}^{(2)} = L^{(21)}\mathbf{X}^{(1)} + L^{(22)}\mathbf{X}^{(2)} \tag{13.17}$$

defining the *stationary transport coefficients*

$$L_{\alpha\gamma}^{(\lambda\mu)}(\mathbf{H}) \equiv V^{-1}\sum_k (\tilde{\mathcal{B}}\varphi_\alpha^{(\lambda)})(k, \mathbf{H})\varphi_\gamma^{(\mu)}(k, \mathbf{H}) \; ; \quad \lambda = 1, 2 \; ; \quad \gamma = 1, 2, 3 \; . \tag{13.18}$$

We now show that the coefficients (13.18) satisfy the *Onsager reciprocity relations* [13-15]

$$\Delta \equiv L_{\alpha\gamma}^{(\lambda\mu)}(\mathbf{H}) - L_{\gamma\alpha}^{(\mu\lambda)}(-\mathbf{H}) = 0 \; . \tag{13.19}$$

In the proof it is useful to separate even and odd parts of the φ-functions with respect to the magnetic field by the definition

$$\left.\begin{array}{c}\mathbf{g}^{(\lambda)}(k) \\ \mathbf{u}^{(\lambda)}(k)\end{array}\right\} \equiv \frac{1}{2}[\vec{\varphi}^{(\lambda)}(k, \mathbf{H}) \pm \vec{\varphi}^{(\lambda)}(k, -\mathbf{H})] \tag{13.20}$$

and to invoke the following two identities: The first is just the odd part in $\mathbf{H}$ of Eqs. (13.16) and reads, in the notation (13.20),

$$\beta \sum_{k'} \Gamma_o(kk')[u_\alpha^{(\lambda)}(k) - u_\alpha^{(\lambda)}(k')] = \frac{e}{c}\mathbf{H} \cdot \left(\frac{\partial f_o}{\hbar \partial \mathbf{k}} \times \frac{\partial g_\alpha^{(\lambda)}}{\hbar \partial \mathbf{k}}\right) \; . \tag{13.21}$$

The second identity follows from the fact that in the k-sums the vector $\mathbf{k}$ varies within the reduced zone Z and the φ-functions are fields on the lattice , as defined in Section 1. Using the definition of Z given in Section 1 it is then possible to show by transforming to an integral over the border ∂Z that for any fields $\varphi(k)$ and $\psi(k)$

$$V^{-1}\sum_k \left(\frac{\partial \varphi}{\partial \mathbf{k}} \times \frac{\partial \psi}{\partial \mathbf{k}}\right) = 0 \tag{13.22}$$

(Problem 3.1).

In evaluating the difference (13.19) with the help of Eqs. (13.16), (13.18) and (13.20) there are many ways of writing the result, depending on where detailed balance (13.5) is used in the Γ_o-terms and where a $\mathbf{k}$-derivative is shifted with the help of the identity (13.22) between factors of φ-functions in the $\mathbf{H}$-terms. The good way turns out to be the following:

$$\Delta = 2\beta V^{-1} \sum_{kk'} \Gamma_o(kk')\{[u_\gamma^{(\mu)}(k) - u_\gamma^{(\mu)}(k')]g_\alpha^{(\lambda)}(k)$$

$$+ [u_\alpha^{(\lambda)}(k) - u_\alpha^{(\lambda)}(k')]g_\gamma^{(\mu)}(k)\}$$

$$+ 2\frac{e}{c}\mathbf{H} \cdot V^{-1} \sum_k \left(\frac{\partial f_o}{\hbar \partial \mathbf{k}} \times \left\{\frac{\partial g_\gamma^{(\mu)}}{\hbar \partial \mathbf{k}}u_\alpha^{(\lambda)} - \frac{\partial g_\alpha^{(\lambda)}}{\hbar \partial \mathbf{k}}u_\gamma^{(\mu)}\right\}\right) . \quad (13.23)$$

Using now the identity (13.21) to transform the Γ_o-sum in (13.23) into a $\mathbf{H}$-sum and vice versa, one finds zero for each of the two sums, the first by invoking again the identity (13.22) and the second because of straightforward cancellations.

With the identity (13.22) it is also immediate to demonstrate the non-negativity of the *entropy production density* which we are now going to define. To this end we first introduce the *entropy density* which for fermions is given by a slight generalisation of Eq. (10.2), namely (see, e.g. Ref. 16, Section 54)

$$s[f] = -k_B V^{-1} \sum_k \{f \ln f + (1-f)\ln(1-f)\} . \quad (13.24)$$

The entropy production density is then simply the dissipative time variation of this expression. (For the analogous case of Boltzmann statistics see Ref. 14, Section 2.8. Unfortunately, the H theorem proved there has no obvious counterpart in Fermi-Dirac statistics):

$$\left(\frac{ds}{\hbar dt}\right)_{\text{diss}} = \sum_k \frac{\delta s[f]}{\delta f(k)}\left(\frac{df(k)}{\hbar dt}\right)_{\text{diss}} . \quad (13.25)$$

With Eq. (13.24) the functional derivative of $s[f]$ is found to be

$$\frac{\delta s[f]}{\delta f(k)} = -k_B V^{-1} \ln \frac{f(k)}{1-f(k)} = -k_B V^{-1}\left\{\ln \frac{f_o}{1-f_o} + \beta\phi\right\} \quad (13.26)$$

where the second equality follows with (13.6). As an immediate consequence of Eqs. (13.4) and (13.25) we note that the local-equilibrium entropy density $s_o \equiv s[f_o]$ is non-dissipative, $(ds_o/\hbar dt)_{\text{diss}} = 0$.

Inserting Eqs. (13.9), (13.13) and (13.26) in (13.25), the two $\mathbf{H}$-terms may both be cast into the form (13.22) and therefore cancel, and the force term containing $\ln[f_o/(1-f_o)]$ cancels because of the symmetry $\varepsilon_{-k} = \varepsilon_k$. Thus[9,13]

$$\left(\frac{ds}{\hbar dt}\right)_{\text{diss}} = \beta k_B V^{-1} \sum_k \phi(k) \mathbf{v}_k \cdot f_o' \{\mathbf{X}^{(1)} + \varepsilon_k \mathbf{X}^{(2)}\}$$

$$= \frac{1}{T} \sum_{\lambda=1,2} \mathbf{J}^{(\lambda)} \cdot \mathbf{X}^{(\lambda)}$$

$$= \frac{1}{T} \sum_{\alpha\gamma} \sum_{\lambda\mu} L_{\alpha\gamma}^{(\lambda\mu)} X_\alpha^{(\lambda)} X_\gamma^{(\mu)} \tag{13.27}$$

where in the third line Eqs. (13.11) and (13.17) were used. Substituting in the last expression Eq. (13.18) with (13.16) and (13.15) one finds that the **H**-term may again be cast into the form (13.22) while detailed balance (13.5) allows the Γ_o-terms to be written in a manifestly non-negative form[9,13],

$$\left(\frac{ds}{\hbar dt}\right)_{\text{diss}} = \frac{1}{2k_B T^2} V^{-1} \sum_{kk'} \Gamma_o(kk')[\phi(k, \mathbf{H}) - \phi(k', \mathbf{H})]^2 \geq 0 . \tag{13.28}$$

This is the *second law of thermodynamics*.

As is well known, the Boltzmann equation gives rise to *local conservation* and balance laws[13]. Indeed, applying $-eV^{-1}\sum_k$ to Eq. (13.1) and noting that in view of (13.9)

$$\sum_k \left(\frac{df(k)}{\hbar dt}\right)_{\text{diss}} = 0 , \tag{13.29}$$

one finds *local charge conservation*

$$\dot{\rho} + \nabla \cdot \mathbf{j} = 0 \tag{13.30}$$

where the dot is a true time-derivation, i.e. $\dot{\rho} \equiv \partial\rho/\hbar\partial t$,

$$\rho[f] \equiv -eV^{-1} \sum_k f(k) \tag{13.31}$$

is the *charge density* and $\mathbf{j}[f]$ is defined in (13.10). Application of $-k_B V^{-1} \sum_k \ln[f/(1-f)]$ to Eq. (13.1) yields *local entropy balance*

$$\dot{s} + \nabla \cdot \mathbf{j}_s = \left(\frac{ds}{\hbar dt}\right)_{\text{diss}} \tag{13.32}$$

where again $\dot{s} \equiv \partial s/\hbar\partial t$ and

$$\mathbf{j}_s[f] \equiv -k_B V^{-1} \sum_k \mathbf{v}_k \{f \ln f + (1-f) \ln(1-f)\} \tag{13.33}$$

is the *entropy current density*. The cancellations necessary to obtain the laws (13.30) and (13.32) follow with the same arguments as used earlier in this section, particularly in Problem 3.1.

Of course, Eq. (13.32) must be equivalent with *local heat balance*

$$\dot{q} + \nabla \cdot \mathbf{w} = \left(\frac{dq}{\hbar dt} \right)_{\text{diss}} \tag{13.34}$$

where

$$q[f] \equiv V^{-1} \sum_k \varepsilon_k f(k) \tag{13.35}$$

is the *heat density* and $\mathbf{w}$ is defined in (13.10). This law is obtained by applying $V^{-1} \sum_k \varepsilon_k$ to Eq. (13.1), and its equivalence with (13.32) follows from the fact that $\ln[f_o(\varepsilon)/(1 - f_o(\varepsilon))] = -\beta\varepsilon$ so that, according to (13.6), (13.26),

$$s - s_o = -\sum_k \phi f_o' \frac{\delta s[f_o]}{\delta f_o(k)} = -\frac{1}{T} V^{-1} \sum_k \varepsilon_k \phi f_o' = \frac{q - q_o}{T} . \tag{13.36}$$

Here $q_o \equiv q[f_o]$ is the local-equilibrium heat density which is again non-dissipative, $(dq_o/\hbar dt)_{\text{diss}} = 0$; and since $\mathbf{j}_s[f_o] = 0$ we also have

$$\mathbf{j}_s = -\sum_k \phi f_o' \frac{\delta \mathbf{j}_s[f_o]}{\delta f_o(k)} = -\frac{1}{T} V^{-1} \sum_k \varepsilon_k \mathbf{v}_k \phi f_o' = \frac{1}{T} \mathbf{w} . \tag{13.37}$$

Without specifying $\Gamma_o(kk')$ further, the integro-differential equations (13.16) are still too complicated to be solved generally. Therefore, at this point the *relaxation-time appproximation* is usually introduced[10] which requires the Boltzmann term (13.9) to have the form

$$-(\mathcal{B}\phi)(k) = f_o'(\varepsilon_k - \mu)\tau_k^{-1}\phi(k) \tag{13.38}$$

where τ_k is a *k-dependent relaxation time* (remember that we have set $\mu = 0$ throughout). Equations (13.16), split into even and odd parts (13.20), may then be cast into the form

$$\begin{aligned}
\mathbf{g}^{(\lambda)}(k) &= -\tau_k \varepsilon_k^{\lambda-1} \mathbf{v}_k - \tau_k \omega_c \Omega_k \mathbf{u}^{(\lambda)}(k) ; \\
\mathbf{u}^{(\lambda)}(k) &= -\tau_k \omega_c \Omega_k \mathbf{g}^{(\lambda)}(k) .
\end{aligned} \tag{13.39}$$

Here

$$\Omega_k \equiv \hat{H} \cdot \left(m\mathbf{v}_k \times \frac{\partial}{\hbar \partial \mathbf{k}} \right) \tag{13.40}$$

is essentially the rotation (or angular momentum) operator around the axis $\hat{H} \equiv \mathbf{H}/H$, m is the mass of the electron and

$$\omega_c \equiv \frac{eH}{mc} \tag{13.41}$$

the *cyclotron frequency*. Inserting the second into the first equation (13.39) it is easily seen that, if the relaxation time has the particular form $\tau_k = \tau(\varepsilon_k)$, it may be taken to the left of Ω_k since this operator commutes with ε_k. The resulting equation may be solved by iteration and, together with the second equation (13.39), yields the formal solution (Problem 3.2)

$$\vec{\varphi}^{(\lambda)}(k, \mathbf{H}) = \varepsilon_k^{\lambda-1} \sum_{n=0}^{\infty} [-\tau(\varepsilon_k)]^{n+1} \omega_c^n \Omega_k^n \mathbf{v}_k . \tag{13.42}$$

Inserting this result into Eq. (13.18), together with Eq. (13.16), the stationary transport coefficients become

$$L_{\alpha\gamma}^{(\lambda\mu)}(\mathbf{H}) = V^{-1} \sum_k \varepsilon_k^{\lambda+\mu-2} v_{k\alpha} f_o'(\varepsilon_k) \sum_{n=0}^{\infty} [-\tau(\varepsilon_k)]^{n+1} \omega_c^n \Omega_k^n v_{k\gamma} . \tag{13.43}$$

In the case $\mathbf{H} = 0$, $\tau(\varepsilon_k)$ may be obtained by combining Eqs. (13.16) and (13.38) which yields, in the case of an electric force, $\lambda = 1$,

$$\vec{\varphi}^{(1)}(k) = -\tau(\varepsilon_k)\mathbf{v}_k . \tag{13.44}$$

Inserting this into (13.16) and multiplying by $\mathbf{v}_k$ one finds

$$\beta \sum_{k'} \Gamma_o(kk')\{\tau(\varepsilon_k)\mathbf{v}_k^2 - \tau(\varepsilon_{k'})\mathbf{v}_k \cdot \mathbf{v}_{k'}\} = -f_o'(\varepsilon_k - \mu)\mathbf{v}_k^2 . \tag{13.45}$$

If, in particular, the scattering is elastic , $\varepsilon_k = \varepsilon_{k'}$, we may use Eq. (13.3) to lowest order and obtain the well-known result (see Eq. (7.4.4) of Ref. 9)

$$\frac{1}{\tau(\varepsilon_k)} = \sum_{\sigma k'} W(kk')(1 - \hat{v}_k \cdot \hat{v}_{k'}) \tag{13.46}$$

where $\hat{v}_k \equiv \mathbf{v}_k/v_k$. Equation (13.46) shows that the main contribution comes from *backward scattering*, $\hat{v}_{k'} = -\hat{v}_k$.

14. Einstein Relation and Stationary Transport Coefficients

In order to identify the coefficients $L_{\alpha\beta}^{(\lambda\mu)}$ defined in Eq. (13.18) with the conventional transport coefficients we consider the experimental conditions to measure *Ohm's law* of charge diffusion

$$\mathbf{j} = \sigma \mathbf{E} \tag{14.1}$$

and *Fourier's law* of heat diffusion[13)]

$$\mathbf{w} = -\kappa \nabla T \tag{14.2}$$

where σ and κ are, respectively the *electric* and *thermal conductivity tensors*. The conditions for measuring Ohm's law are $\nabla \mu = 0$ and $\nabla T = 0$ so that, according to Eqs. (13.14), (13.17) and the Onsager relation (13.19),

$$\sigma(\mathbf{H}) = e^2 L^{(11)}(\mathbf{H}) = \sigma^T(-\mathbf{H}) , \tag{14.3}$$

where the upper index T designates the transposed matrix. In addition there is a heat current due to the diffusive motion of the electrons,

$$\mathbf{w} = -eL^{(21)}(\mathbf{H})\mathbf{E} . \tag{14.4}$$

To measure Fourier's law, the conditions are $\mathbf{E} = 0$, $\mathbf{j} = 0$, and $\mathbf{H} = 0$. Eliminating $\mathbf{X}^{(1)}$ from the two equations (13.17) one finds

$$\kappa = [L^{(22)} - (L^{(12)})^T (L^{(11)})^{-1} L^{(12)}]/T = \kappa^T \tag{14.5}$$

where the Onsager relation (13.19) has been used.

A third transport coefficient is defined through *Fick's law* of particle diffusion[13,17)]

$$\mathbf{j}_n = -D_n \nabla n \tag{14.6}$$

where n and $\mathbf{j}_n$ are, respectively the particle density and particle current density defined in Eqs. (2.9), (2.10) and D_n is the *tensor of diffusion constants*. For electrons, Ohm's and Fick's laws are of course not independent since, as is obvious from the first Eq. (13.10) and from (13.31), $\rho = -en$ and $\mathbf{j} = -e\mathbf{j}_n$. But more importantly, an electric current may also be produced by an external charge gradient instead of an applied electric field which, as will be described below, is Landauer's approach to electron transport[11)]. Such a connection is already evident from the fact that, at constant temperature, the thermodynamic force $\mathbf{X}^{(1)}$ in Eq. (13.14) may be produced equivalently by $e\mathbf{E}$ and by $\nabla \mu$. But, as emphasized at the beginning of Section 10, $n = n(\mu, T)$.

According to the linear response equation (7.11), Ohm's law (14.1) is local in $\mathbf{q}$-space (see Eq. (18.8) below), i.e. $j_\alpha(\mathbf{q}) = \sigma_{\alpha\gamma}(\mathbf{q})E_\gamma(\mathbf{q})$ in the stationary situation considered here. Making use of the Fourier-transformed Eq. (14.6) for fixed $\mathbf{q}$, we may write therefore

$$\sigma_{\alpha\gamma} = \left(\frac{\partial j_\alpha(\mathbf{q})}{\partial E_\gamma(\mathbf{q})}\right)_{T,\mathbf{q}\to 0,\mathbf{E}\to 0}$$

$$= -e^2(D_n)_{\alpha\lambda}(\mathbf{q})\left(\frac{q_\lambda \partial n(\mathbf{q})}{q_\gamma \partial \mu(\mathbf{q})}\right)_{T,\mathbf{q}\to 0,\mathbf{E}\to 0}$$

$$= e^2(D_n)_{\alpha\gamma}\left(\frac{\partial n}{\partial \mu}\right)_T \tag{14.7}$$

where the last equality follows by choosing $q_\lambda = q\delta_{\lambda\gamma}$. This equality may be called a *generalized Einstein relation*[14]. In the general form given here this relation was first derived by Kubo in the framework of nonequilibrium statistical mechanics (Eq. (9.4b) of Ref. 17). Note that we have only used nonequilibrium thermodynamics to derive Eq. (14.7).

At low temperatures, more familiar forms of Eq. (14.7) may be obtained by evaluating $(\partial n/\partial \mu)_T$. In taking the derivative of the particle density $n(\mu, T) = V^{-1}\sum_k f_o(\varepsilon_k - \mu)$ with respect to μ, we must distinguish between a metal and a semiconductor (see Section 16). In the case of a *metal* we may put $-f_o'(\varepsilon) = \delta(\varepsilon)$ and obtain

$$\left(\frac{\partial n}{\partial \mu}\right)_{T\sim 0} = 2N(\mu) . \tag{14.8}$$

A *semiconductor*, on the other hand, is characterized by the fact that the Fermi level μ is situated in an *energy gap* so that $\varepsilon_k - \mu > 0$ for all k. At low temperatures, $\beta(\varepsilon_k - \mu) \gg 1$, we therefore have *Boltzmann statistics* and hence $-f_o' = \beta f_o$ so that

$$\left(\frac{\partial n}{\partial \mu}\right)_{T\sim 0} = \frac{n}{k_B T} . \tag{14.9}$$

Inserting into Eq. (14.7), we obtain the classical *electron mobility*

$$\mu_e \equiv \frac{\sigma}{e^2 n} = \frac{D_n}{k_B T} . \tag{14.10}$$

This is the *classical Einstein relation*[14,17,18] contained in Einstein's famous work on Brownian motion[19].

Consider now a homogenous piece of conductor of length L, cross-sectional area A and conductivity σ. Following Landauer[11] we connect this sample to an injecting reservoir of electrons (to the left) and an extracting reservoir (to the right, see Fig. 3.1) and characterize its conduction properties by a *transmission* and a *reflection probability*, T and R respectively, such that $T + R = 1$. The current density in the sample being $\mathbf{j}_n = Tn\mathbf{v}$, integration of Eq. (14.6) gives

$$TnvAL = -D_n \oint_{\partial V} nd\mathbf{S} = -D_n(n_{\text{ex}} - n_{\text{in}})A\frac{\mathbf{v}}{|\mathbf{v}|} \, , \qquad (14.11)$$

where the particle densities on the extraction and injection sides are, respectively, $n_{\text{ex}} = Tn$ and $n_{\text{in}} = (1 + R)n$. Since $T = 1 - R$ and v is the Fermi velocity v_F, Eq. (14.11) allows to write the diffusion constant as $D_n = v_F LT/2R$. Inserting this into Eq. (14.7) one finds for the *conductance*

$$G \equiv \frac{A}{L}\sigma = e^2 N(\mu)v_F A\frac{T}{R} \, . \qquad (14.12)$$

In words, G is expressed as a *ratio between the transmission and reflection probabilities*[11].

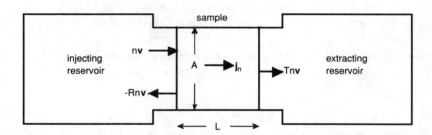

Fig. 3.1: Electric conduction as a scattering process according to Landauer .

To be more explicit, we consider a one-dimensional conductor, so that the cross-sectional area $A = 1$, and calculate the density of states $N(\mu)$ for a *single band* of *quasi-free carriers* characterized by an effective mass m and by $\mu = \varepsilon_F \equiv \hbar^2 k_F^2/2m \equiv mv_F^2/2$. In an extended-zone representation the carrier density in D dimensions is

$$n(\mu, 0) = 2 \int \frac{d^D k}{(2\pi)^D}\theta\left(\varepsilon_F - \frac{\hbar^2\mathbf{k}^2}{2m}\right) = \frac{2\Omega_D}{(2\pi)^D D}\left(\frac{2m\varepsilon_F}{\hbar^2}\right)^{D/2} \qquad (14.13)$$

where Ω_D is the surface of the unit sphere in D dimensions, $\Omega_1 = 1$, $\Omega_2 = 2\pi$, $\Omega_3 = 4\pi$. From (14.8) and (14.13) we then obtain

$$N(\mu) = \frac{nD}{4\varepsilon_F} \, , \qquad (14.14)$$

so that Eq. (14.12) becomes, for $D = 1$, the *Landauer formula*[11,12]

$$G = \frac{e^2}{h}\frac{T}{R} \qquad (14.15)$$

which expresses the resistance G^{-1} in units of the *universal resistance* h/e^2 ($h = 2\pi\hbar$ is Planck's constant). In the international system of units $e^2/4\pi\varepsilon_o\hbar c \cong 1/137$ is *Sommerfeld's finestructure constant*, $\varepsilon_o = 10^7/4\pi c^2$ sec/mΩ the permittivity of the vacuum and $c \cong 3 \times 10^8$ m/sec, so that

$$\frac{h}{e^2} \cong 137 \times 10^{-7} \times 2\pi c \ \Omega \cong 25.8 \ k\Omega \ . \tag{14.16}$$

A very important problem is the influence of a magnetic field on electric conduction. From Eqs. (13.43), (14.3) we have

$$\sigma_{\alpha\gamma}(\mathbf{H}) = e^2 V^{-1} \sum_k v_{k\alpha} f_o'(\varepsilon_k) \sum_{n=0}^{\infty} [-\tau(\varepsilon_k)]^{n+1} \omega_c^n \Omega_k^n v_{k\gamma} \ . \tag{14.17}$$

In order to be able to evaluate this expression further we resort again to a single band of quasi-free carriers, i.e. $\varepsilon_k = \hbar^2 \mathbf{k}^2/2m$ and $\mu = \varepsilon_F$. With $m\mathbf{v}_k = \hbar\mathbf{k}$ and $\hat{H} = (0, 0, 1)$ the rotation operator (13.40) becomes

$$\Omega_k = v_{k1}\frac{\partial}{\partial v_{k2}} - v_{k2}\frac{\partial}{\partial v_{k1}} \ , \tag{14.18}$$

and one finds, with some precaution, that the sum over n in Eq. (14.17) is a geometrical series. Evaluating the k-sum in the limit $T \to 0$ where $-f_o'(\varepsilon) = \delta(\varepsilon)$ one obtains, after angular averaging (Problem 3.3),

$$\sigma_{\alpha\gamma}(\mathbf{H})\Big|_{T\to 0} = \sigma_o\left[1 - \frac{\tau^2(\mu)\omega_c^2}{1 + \tau^2(\mu)\omega_c^2}(\delta_{\gamma 1} + \delta_{\gamma 2})\right]\delta_{\alpha\gamma}$$
$$- \sigma_o\frac{\tau(\mu)\omega_c}{1 + \tau^2(\mu)\omega_c^2}(\delta_{\alpha 1}\delta_{\gamma 2} - \delta_{\alpha 2}\delta_{\gamma 1}) \ . \tag{14.19}$$

Here, making use of Eq. (14.14),

$$\sigma_o \equiv \frac{2}{3}e^2 N(\mu)v_F^2\tau(\mu) = e^2\frac{n}{m}\tau(\mu) \tag{14.20}$$

is the conductivity at $\mathbf{H} = 0$ (similar results are contained in Ref. 10, Section 8.5).

As is evident from the result (14.20), there is no magnetic effect for $\mathbf{E}\|\mathbf{H}$. In the case $\mathbf{E} \perp \mathbf{H}$, we fix the current to be $\mathbf{j} = (j_1, 0, 0)$ but still keeping $\mathbf{E} = (E_1, E_2, 0)$ unspecified. With Eqs. (14.1), (14.19) the condition $j_2 = 0$ becomes

$$E_2 + \tau(\mu)\omega_c E_1 = 0 \ , \tag{14.21}$$

while

$$j_1 = \sigma_o\frac{E_1 - \tau(\mu)\omega_c E_2}{1 + \tau^2(\mu)\omega_c^2} \ . \tag{14.22}$$

In the *Hall effect*, E_2 is applied so that from Eqs. (14.21), (14.22) follows the *Hall constant* [10]

$$R \equiv \frac{E_2}{H j_1} = -\frac{\tau(\mu) \omega_c}{H \sigma_o} = -\frac{1}{ecn} \ . \tag{14.23}$$

Thus R measures the charge density $\rho = -en$ which in semiconductors may have either sign (see Section 16) [10]. In *magneto-resistance*, E_1 is applied which, however leads to $j_1 = \sigma_o E_1$. This is the well-known absence of a magneto-resistance effect in the approximation of free carriers (see Ref. 10, Section 8.523).

Going back to the general form (14.1) of Ohm's law, the condition $j_2 = 0$ gives in the two situations of the Hall effect and of the magneto-resistance,

$$j_1 = \begin{cases} -(\Delta\sigma/\sigma_{21}) E_2 \\ (\Delta\sigma/\sigma_{22}) E_1 \end{cases} \tag{14.24}$$

where $\Delta\sigma \equiv \sigma_{11}\sigma_{22} - \sigma_{12}\sigma_{21}$. Because of the Onsager relation (14.3), $\Delta\sigma$ is an even function of $\mathbf{H}$ and hence also magneto-resistance. Equation (14.24) is of particular interest in *two-dimensional conductors* which may be realized by *thin metallic films* such that in the perpendicular direction the electrons have just one quantum state available. In this case it is advantageous to introduce the *resistivity tensor* $\rho = \sigma^{-1}$ since, with the magnetic field perpendicular to the film,

$$\sigma^{-1} = \frac{1}{\Delta\sigma} \begin{pmatrix} \sigma_{22} & -\sigma_{12} \\ -\sigma_{21} & \sigma_{11} \end{pmatrix} , \tag{14.25}$$

so that Eq. (14.24) has the trivial form

$$E_1 = \rho_{11} j_1 \ ; \quad E_2 = \rho_{21} j_1 \ . \tag{14.26}$$

It turns out, however, that ρ_{11} and ρ_{21} are not sufficient to describe a two-dimensional electron gas with mesoscopic constrictions as produced by photolithography. In this case the sample may be viewed as an elastic scatterer which permits transmission and reflection of carriers between several particle reservoirs, in generalization of Fig. 3.1 and Eq. (14.15). Such a system, which may have a multiply connected topology and carry a magnetic flux ϕ, is best described by a *scattering matrix* or *S-matrix* like in field theory whose elements s_{ij} define generalized transmission probabilities $T_{ij}(\phi) = |s_{ij}|^2$ between leads i and $j \neq i$ and reflection probabilities $R_{ii}(\phi) = |s_{ii}|^2$ [12]. This description not only gives a natural explanation of the *integer quantum Hall effect* [7] but also of the *quantized conductance steps* recently discovered in ballistic point contacts by van Wees *et al.* and by Wharam *et al.*[20] As to the *fractional quantum Hall effect*[8], the strong correlations among the carriers make such a description difficult (see, however, the review by Büttiker in Ref. 20).

The idea of this S-matrix formalism is that the current flows in open trajectories or "edge states" at the Fermi energy whereas localized states correspond to closed-loop trajectories. Provided that fluctuations are negligible so that a quasi-classical description holds, each edge state carries one unit of flux and the steps in both, the Hall and the point-contact geometries, are the manifestation of the opening of a new edge state as function of the applied voltage.

Coming back to the Boltzmann-equation approach of the last section we now wish to derive an *inversion formula* for the low-temperature resistivity. We set $\mathbf{H} = 0$ and use the approximation of a single band of quasi-free carriers introduced above. The reason why such an inversion is possible is that for low temperatures, the relevant electron energies are sufficiently close to the Fermi energy, $\varepsilon_k \cong \mu$, so that most features of the bandstructure are unimportant. In particular, for quasi-free carriers, $\hat{k} \equiv \mathbf{k}/k$ is the only axis in the problem. Therefore, in the absence of a magnetic field, the Hilbert space in which the Boltzmann operator $\mathcal{B}$ acts is easily specified.

For $\mathbf{H} = 0$, $\nabla T = 0$ and $\nabla \mu = 0$, Eqs. (13.9) and (13.16) may be used to define a basis $|\psi_\alpha\rangle$ of real Hilbert vectors,

$$(\mathcal{B}\varphi_\alpha^{(1)})(k) = v_{k\alpha} f_o'(\varepsilon_k - \mu) \equiv \vartheta \psi_\alpha(k) , \qquad (14.27)$$

where ϑ is a normalization constant and

$$\langle \psi_\gamma | \psi_\alpha \rangle \equiv V^{-1} \sum_k \psi_\gamma(k) \psi_\alpha(k) = \delta_{\alpha\gamma} . \qquad (14.28)$$

Combining the last two equations and making use of the assumed isotropy one has

$$\vartheta^2 = \frac{1}{3} V^{-1} \sum_k \mathbf{v}_k^2 [f_o'(\varepsilon_k - \mu)]^2 . \qquad (14.29)$$

At *low temperatures* an explicit evaluation of ϑ^2 may be obtained with the help of the relations $[f_o'(\varepsilon)]^2 \cong -f_o'(0)\delta(\varepsilon)$ and $f_o'(0) = -\beta/4$ and using Eqs. (10.4) and (14.14) with $D = 3$, thus

$$\vartheta^2 \cong \frac{\beta}{6} N(\mu) v_F^2 = \frac{\beta n}{4m} . \qquad (14.30)$$

Since $k \cong (k_F \hat{k}, \sigma)$ and because of isotropy, integration over k or k' implies, for $\hat{k}' \neq \hat{k}$, that $\hat{k} \cdot \hat{k}'$ yields zero. Therefore the basis $|\psi_\alpha\rangle$ satisfies the approximate closure relation

$$V^{-1} \sum_{\alpha} \psi_\alpha(k)\psi_\alpha(k') = \delta_{\sigma\sigma'}\delta_{\hat{k}\hat{k}'} \cong \delta_{kk'} \tag{14.31}$$

where the value of the constant follows from Eq. (14.28). This procedure does not generalize in an obvious way to anisotropic energy surfaces.

Equation (14.27) allows to invert the Boltzmann operator,

$$\vartheta^2 \langle\psi_\gamma|\mathcal{B}^{-1}|\psi_\alpha\rangle = \vartheta\langle\varphi_\gamma^{(1)}|\psi_\alpha\rangle = \langle\varphi_\gamma^{(1)}|\mathcal{B}|\varphi_\alpha^{(1)}\rangle = e^{-2}\sigma_{\alpha\gamma}(0) \tag{14.32}$$

where definitions (13.18) and (14.3) were used. Combining now the orthonormality and closure relations (14.28) and (14.31) to write

$$\sum_{\beta} \langle\psi_\gamma|\mathcal{B}|\psi_\beta\rangle\langle\psi_\beta|\mathcal{B}^{-1}|\psi_\alpha\rangle \cong \delta_{\alpha\gamma} , \tag{14.33}$$

Eq. (14.32) yields the *inversion formula*

$$\rho_{\alpha\gamma} = (\sigma^{-1})_{\alpha\gamma} \cong e^{-2}\vartheta^{-2}\langle\psi_\gamma|\mathcal{B}|\psi_\alpha\rangle , \tag{14.34}$$

and, inserting Eq. (13.9), one arrives at the following simple semi-positive-definite quadratic form for the resistivity tensor:

$$\rho_{\alpha\gamma} = \frac{\beta}{2e^2\vartheta^2}V^{-1} \sum_{kk'} \Gamma_o(kk')\{\psi_\alpha(k) - \psi_\alpha(k')\}\{\psi_\gamma(k) - \psi_\gamma(k')\} \tag{14.35}$$

where $\psi_\alpha(k)$ is defined in Eq. (14.27) and detailed balance (13.5) has been used in the last step. Note that this result is essentially the same as Eq. (7.9.2) of Ref. 9, except that the latter still requires a variational determination of ϕ_k but, on the other hand, is not restricted to low temperatures. If we replace $f_o'(\varepsilon_k - \mu)$ in Eq. (14.27) by $f_o'(0) = -\beta/4$ and use (14.30), Eq. (14.35) finally becomes[21]

$$\rho_{\alpha\gamma} = \frac{1}{2}\left(\frac{m}{en}\right)^2 \beta V^{-1} \sum_{kk'} \Gamma_o(kk')(v_{k\alpha} - v_{k'\alpha})(v_{k\gamma} - v_{k'\gamma}) \tag{14.36}$$

where $\varepsilon_k \simeq \varepsilon_{k'} \simeq \mu + \beta^{-1}$. Note that this is a *one-band formula* . For the generalization to two-electron scattering see Section 17.

15. Residual Resistance and Dynamical Structure Factor

Before discussing specific mechanisms leading to resistivity it is of interest to establish the connection between the Boltzmann-equation approach discussed in the last two sections and a general coupling $\mathcal{H}_{\text{int}}$ between the conduction electrons and an energy density with Fourier transform $h(\mathbf{q})$. Writing this general coupling in the form (3.18), (3.19) as $\langle k'|\mathcal{H}_{\text{int}}|k\rangle \equiv (\mathcal{H})_{k'k}a_{k'}^{+}a_{k}$ with $(\mathcal{H}_{\text{int}})_{k'k} = \sum_{\mathbf{q}} \alpha_{k'k}\delta_{\mathbf{k}',\mathbf{k}-\mathbf{q}}h^{+}(\mathbf{q})$ we have

$$\langle Fk'|\mathcal{H}^{\text{int}}|Ik\rangle = \sum_{\mathbf{q}} \alpha_{k'k}\delta_{\mathbf{k}',\mathbf{k}-\mathbf{q}}\langle F|h^{+}(\mathbf{q})a_{k'}^{+}a_{k}|I\rangle . \tag{15.1}$$

Here $a_{k'}^{+}$ and a_{k} with $k \equiv (\mathbf{k}, \sigma)$ are the creation and annihilation operators for s-electrons, I and F designate initial and final states of the total system and $\alpha_{k'k}$ describes the dependence on the s-electron spin as, e.g. in Eq. (3.24).

The connection with Boltzmann theory is obtained in defining the one-band transition rate by a *Golden Rule* as in Eq. (11.43), i.e.

$$\Gamma_{o}(kk') = \frac{2\pi}{\hbar} \sum_{IF} |\langle Fk'|\mathcal{H}_{\text{int}}|Ik\rangle|^{2}\rho_{I}\delta(E_{F} - E_{I}) . \tag{15.2}$$

Here E_{I} and E_{F} are the respective energies of the total system and $\rho_{I} = \exp(\beta(\Omega - E_{I}))$ is the weight of the initial state I. Writing the δ-function in (15.2) as time integral and taking an unperturbed average over the electrons, the perturbation being taken care of by the Boltzmann equation, see (13.6), Eq. (15.2) may be cast into the form (13.3), i.e. setting the Fermi energy $\mu = 0$,

$$\Gamma_{o}(kk') = W(kk')f_{o}(\varepsilon_{s\mathbf{k}})[1 - f_{o}(\varepsilon_{s\mathbf{k}'})] \tag{15.3}$$

with (Problem 3.4)

$$W(kk') = \hbar^{-1} \sum_{\mathbf{q}} |\alpha_{k'k}|^{2}\delta_{\mathbf{k}',\mathbf{k}-\mathbf{q}} \int dt e^{i(\varepsilon_{s\mathbf{k}}-\varepsilon_{s\mathbf{k}'})t} \langle h(\mathbf{q};t)h^{+}(\mathbf{q};0)\rangle \tag{15.4}$$

where $\varepsilon_{s\mathbf{k}}$ is the s-band energy.

The time integral in Eq. (15.4) is related with the general response function defined in (7.12). In order to see this, we define the *spectral function* associated to $\chi_{ab}(\mathbf{q}; t)$ in analogy with Eq. (11.11) as

$$\chi''(\mathbf{q}; \omega) \equiv \frac{1}{2i}[\tilde{\chi}_{ab}(\mathbf{q}; \omega + i\epsilon) - \tilde{\chi}_{ab}(\mathbf{q}; \omega - i\epsilon)] \tag{15.5}$$

where

$$\tilde{\chi}_{ab}(\mathbf{q}; \omega \pm i|\nu|) = \frac{i}{V} \int_{0}^{\pm\infty} dt e^{(i\omega \mp |\nu|)t} \langle [a(\mathbf{q}; t), b(-\mathbf{q}; 0)]\rangle \tag{15.6}$$

is the *susceptibility*. With the upper sign, it is the Fourier transform of $\chi_{ab}(\mathbf{q}; t)$ while for $\omega = 0$, Eq. (7.32) relates it to $-\tilde{\mathcal{G}}_{a(\mathbf{q})b(-\mathbf{q})}(\pm i|\nu|)$.

Combining the last two equations and taking the limit $\epsilon \to 0$ one finds

$$\chi''_{ab}(\mathbf{q}; \omega) = \frac{1}{V} \int_{-\infty}^{+\infty} dt e^{i\omega t} \langle [a(\mathbf{q}; t), b(-\mathbf{q}; 0)] \rangle . \tag{15.7}$$

This, however, is not yet the form needed in Eq. (15.4). To get rid of the commutator we apply the *Kubo identity* [17] (Problem 3.5)

$$\frac{i}{\hbar}[A, \rho] = \beta \rho \dot{\hat{A}} = \beta \rho \dot{\hat{A}} \tag{15.8}$$

where ρ is the density operator (6.4), the dot is $\partial/\hbar \partial t$ and

$$\hat{A} \equiv \beta^{-1} \int_0^\beta d\tau A(-i\tau) . \tag{15.9}$$

One finds the remarkable formula (Problem 3.6)

$$\chi''_{ab}(\mathbf{q}; \omega) = \frac{1}{2V(1 + n_o(\omega))} \int_{-\infty}^{+\infty} dt e^{i\omega t} \langle a(\mathbf{q}; t) b(-\mathbf{q}; 0) \rangle . \tag{15.10}$$

Coming back to Eq. (15.4) we now may write

$$W(kk') = \frac{2}{\hbar} \sum_{\mathbf{q}} |\alpha_{k'k}|^2 \delta_{\mathbf{k}', \mathbf{k}-\mathbf{q}} V \int d\omega (1 + n_o(\omega)) \chi''_{hh}(\mathbf{q}; \omega) \delta(\varepsilon_{\mathbf{k}} - \varepsilon_{\mathbf{k}'} - \omega) . \tag{15.11}$$

As a first application of Eqs. (15.11) and (14.36) let us consider the *residual resistivity* which is due to *impurity scattering* of the conduction electrons in the limit $T \to 0$ where all excitations are frozen out. The Hamiltonian describing this scattering is of the form (2.1) where the external field is given by the static potential $U(\mathbf{r} - \mathbf{R}_j)$ of an impurity j located at $\mathbf{R}_j$. Hence, Fourier-transforming according to (2.3), (2.4) and summing over a fixed distribution $\{\mathbf{R}_j\}$ of impurities the interaction is

$$\mathcal{H}^{\text{imp}} = \int_V d^3 r \sum_j U(\mathbf{r} - \mathbf{R}_j) \tilde{n}(\mathbf{r})$$

$$= \sum_{qj} \tilde{U}(-\mathbf{q}) e^{i\mathbf{q} \cdot \mathbf{R}_j} \frac{1}{V} \sum_{kk'} a_{k'}^+ a_k \delta_{\mathbf{k}', \mathbf{k}-\mathbf{q}} \delta_{\sigma'\sigma} \tag{15.12}$$

where Eq. (3.22) has been used. This leads to a matrix element (15.1) with

$$\alpha_{k'k} = \delta_{\sigma'\sigma} ; \quad h(\mathbf{q}) = \tilde{U}(\mathbf{q}) \frac{1}{V} \sum_j e^{-i\mathbf{q} \cdot \mathbf{R}_j} = h^+(-\mathbf{q}) . \tag{15.13}$$

In general, the impurities will suffer a *recoil* which excites phonons (and/or other waves) in the lattice. Hence the position vectors $\mathbf{R}_j$ are operators which we may write as

$$\mathbf{R}_j = \mathbf{R}_j^o + \sum_{\nu=1}^{B} \sum_{\mathbf{R}} \lambda_\nu(\mathbf{R}_j^o - \mathbf{R})\mathbf{u}_\nu(\mathbf{R}) . \qquad (15.14)$$

Here $\mathbf{R}_j^o$ is the equilibrium position, $\mathbf{u}_\nu$ the displacement vector (1.24), and $\lambda_\nu(\mathbf{R}_j^o - \mathbf{R})$ is a dimensionless coupling function which describes the force between the impurity j and the atom or ion at lattice site $\mathbf{R} + \mathbf{r}_\nu$.

For the residual resistivity problem, phonon emission and absorption may be neglected because this effect may be included in the electron-phonon contribution (see Section 16). In this case $\mathbf{R}_j = \mathbf{R}_j^o$ and the interaction (15.12) may be represented by the vertex of Fig. 3.2(a) below. $h(\mathbf{q})$ in (15.13) then is a c-number and Eq. (15.11) becomes

$$W^{\mathrm{imp}}(kk') = \frac{2\pi}{\hbar}|\tilde{U}(\mathbf{k}' - \mathbf{k})|^2 \left| \frac{1}{V} \sum_j e^{i(\mathbf{k}' - \mathbf{k})\cdot\mathbf{R}_j^o} \right|^2 \delta(\varepsilon_{s\mathbf{k}'} - \varepsilon_{s\mathbf{k}})\delta_{\sigma'\sigma} . \quad (15.15)$$

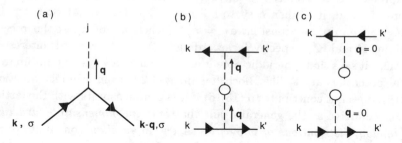

Fig. 3.2: (a) Impurity vertex (15.12) with the equilibrium position $\mathbf{R}_j^o$. (b) and (c) diagrams corresponding to the first and second term of Eq. (15.18), respectively, a circle meaning impurity-averaging, i.e. multiplication with the impurity concentration x.

This formula explicitly shows the obvious fact that recoiless scattering of the electrons is *elastic* . It has two important consequences, the first being that, not only detailed balance (13.5) is satisfied since, obviously, $\exp(\beta\varepsilon_{s\mathbf{k}'})W^{\mathrm{imp}}(kk') = \exp(\beta\varepsilon_{s\mathbf{k}})W^{\mathrm{imp}}(k'k)$, but also *microscopic reversibility* $W(kk') = W(k'k)$. The second consequence is that the Fermi functions in Eq. (15.3) have the same argument so that we may again use the relation $\beta f_o(\varepsilon)[1 - f_o(\varepsilon)] = -f_o'(\varepsilon) \cong \delta(\varepsilon)$. Inserting this and (15.15) into Eq. (14.36) one finds, with the density of states (10.4) and for s-electrons (index s),

$$\rho_{\alpha\gamma} = \hbar^{-1}\left(\frac{m_s}{en_s}\right)^2 N_s(\mu) \sum_{\mathbf{q}} |\tilde{U}(\mathbf{q})|^2 \left|\frac{1}{V}\sum_j e^{i\mathbf{q}\cdot\mathbf{R}_j^o}\right|^2 \int d\Omega_{\mathbf{k}}\delta(\varepsilon_{s,\mathbf{k}+\mathbf{q}} - \mu)$$

$$\times (v_{s\mathbf{k}\alpha} - v_{s,\mathbf{k}+\mathbf{q},\alpha})(v_{s\mathbf{k}\gamma} - v_{s,\mathbf{k}+\mathbf{q},\gamma})\Big|_{\varepsilon_{s\mathbf{k}}=\mu} \tag{15.16}$$

where the integral is over the angles of the vector $\mathbf{k}$.

So far nothing has been assumed about the impurity distribution $\{\mathbf{R}_j^o\}$. Hence it is perfectly legitimate to distribute them periodically over the lattice, e.g. by setting $\mathbf{R}_j^o = A\mathbf{R}+\mathbf{c}$ with some constant matrix A and vector $\mathbf{c}$. In this case the sum over j is *coherent*, and one finds from Eq. (1.9)

$$\left|\frac{1}{V}\sum_{\mathbf{R}\in V} e^{i\mathbf{q}\cdot(A\mathbf{R}+\mathbf{c})}\right|^2 = \left|\frac{1}{V}\sum_{\mathbf{R}\in V} e^{iA^T\mathbf{q}\cdot\mathbf{R}}\right|^2 = \frac{1}{v^2}\sum_{\mathbf{K}'}\delta_{A^T\mathbf{q},\mathbf{K}'} \tag{15.17}$$

where v is the volume of the unit cell. Since $\mathbf{v}_{\mathbf{k}+\mathbf{K}} = \mathbf{v}_{\mathbf{k}}$ for any reciprocal lattice vector $\mathbf{K}$, the last expression furnishes a sufficient condition for Eq. (15.16) to vanish, namely that, for any $\mathbf{K}$, $A^T\mathbf{K} = \mathbf{K}'$ is again a reciprocal lattice vector. Multiplying the last expression by an arbitrary vector $\mathbf{R} \neq 0$, it is then easy to see using Eqs. (1.1)–(1.4) that, writing $A\mathbf{R} = \sum_i \nu_i m_i \mathbf{a}_i$, one must have $\nu_i = n_i'/n_i$ where n_i and n_i' are the integers defining $\mathbf{K}$ and $\mathbf{K}'$, respectively, according to (1.3). This result is interesting because it says that a periodic potential, which may always be built up from local potentials at $\mathbf{R}_j^o$ like Bloch functions (1.20) from Wannier functions (1.19), does not contribute to (15.16) if it is commensurate with the lattice. Thus we arrive at the general result that only *incommensurate structures* or *random distributions of impurities* as, e.g. in an *alloy*, contribute to the residual resistance.

We now consider the latter case, assuming *statistical independence* of the random positions $\mathbf{R}_j^o$. The j-sum is then *incoherent*, and one finds with Eq. (2.6),

$$\left\langle \left|\frac{1}{V}\sum_j e^{i\mathbf{q}\cdot\mathbf{R}_j^o}\right|^2 \right\rangle_{\text{imp}} = \frac{1}{V^2}\sum_{jl}\langle e^{i\mathbf{q}\cdot(\mathbf{R}_j^o - \mathbf{R}_l^o)}\rangle_{\text{imp}} = \frac{x}{Vv} + \frac{x^2}{v^2}\delta_{\mathbf{q},\mathbf{o}} , \tag{15.18}$$

corresponding to Fig. 3.2(b) and (c). Here $\langle F(\mathbf{R})\rangle_{\text{imp}} \equiv V^{-1}\int_V d^3R\, F(\mathbf{R})$ designates the statistical average over the impurity positions $\mathbf{R}$ and x is the *impurity concentration* defined as the average number of impurities per unit cell (x/v is the average impurity density). With (15.18) and with the free-electron expression $\varepsilon_{s\mathbf{k}} = \hbar^2\mathbf{k}^2/2m_s$ which for s-electrons is quite acceptable provided that we use the *effective mass m_s*, it is straightforward to evaluate the impurity-averaged Eq. (15.16). Obviously, the second term in (15.18) does not contribute, while one finds for the first term, assuming a spherical shape of the impurity potential (Problem 3.7), $\langle \rho_{\alpha\gamma} \rangle_{\text{imp}} = \rho^{\text{imp}}(x)\delta_{\alpha\gamma}$ with

$$\rho^{\text{imp}}(x) = \frac{x}{v}\frac{N_s(\mu)}{3\pi e^2 n_s^2 v_F} \int_0^{2k_F} dq\; q^3 |\tilde{U}(q)|^2 \;. \tag{15.19}$$

In Section 18, Eq. (15.19) will be derived from the Kubo formula for the conductivity. The main feature of the above result is its *temperature-independence* which is due to the fact that impurity scattering is *elastic*. As to the x-dependence, it is valid only for $x \ll 1$. In a *binary alloy* $A_x B_{1-x}$ where the latter condition does not hold, a slight generalization of the above calculation due to Nordheim[22] is necessary since both, A and B atoms simultaneously play the role of lattice atoms and of impurity atoms. Let V_A and V_B be the respective atomic potentials. Then $V = xV_A + (1-x)V_B$ may be considered as forming the periodic potential while $U_A = V - V_A$ and $U_B = V - V_B$ are the impurity potentials which occur with concentrations x and $1-x$, respectively. Hence $x|\tilde{U}|^2$ in (15.19) should be replaced by $x|\tilde{U}_A|^2 + (1-x)|\tilde{U}_B|^2 = x(1-x)|\tilde{U}_B - \tilde{U}_A|^2$. This concentration-dependence is in fair agreement with the experimental findings[22]. Note that here the contribution $\propto x^2$ has nothing to do with the x^2-term in Eq. (15.18). The latter is in fact spurious since it represents zero average momentum exchange between electron and impurity which means no interaction in the average. A true dependence on higher powers of x comes from *multiple scatterings* (see Section 19).

It is well known since the 1930's that *magnetic impurities* may have a rather different effect at low temperatures, leading to a *resistance minimum* at a characteristic temperature, the *Kondo temperature* T_K [23],

$$\rho_K(T) = \rho_o[1 - \eta\left(\frac{T}{T_K}\right)^2 + AT^5 + \cdots] \;. \tag{15.20}$$

This behavior which depends sensitively both on the host metal and the type of impurity, has been a puzzle for 30 years (see Section 9.3 of Ref. 9 and Section 9.613 of Ref. 10). The basis for an explanation was led by Kondo[24] who in 1964 obtained a resistance minimum as a result of a calculation in second-order Born approximation of the *s-d* exchange interaction (see Eq. (17.9) below)

$$\mathcal{H}_{\text{int}} = J\mathbf{S} \cdot \tilde{\mathbf{s}}(0) \;. \tag{15.21}$$

Here $\mathbf{s}(\mathbf{r})$ is the spin density of the conduction (s-) electron whose Fourier transform is given by Eqs. (3.24), (3.25) and $\mathbf{S}$ the analogous expression for the d-electrons on the impurity which is located at $\mathbf{r} = 0$. Equation (15.21) therefore describes an *exchange interaction*[25]. $\mathbf{S}$ may be considered as a local magnetic moment provided that the ionization energy of the impurity (d-) electrons is large compared to $JN_s(\mu)$, $N_s(\mu)$ being the density of states of the s-electrons . In this case J is always antiferromagnetic ($J > 0$)[25].

Considering the denominator in Kondo's calculation as a renormalization of the coupling constant J, the latter grows with decreasing energy separation from the Fermi level or, equivalently, with decreasing temperature. This leads to a term $\propto -\log T$ which, together with the usual T^5-law of Bloch (see Eq. (16.8) below) gives rise to the minimum in (15.20). But this sensitivity of the renormalization of J with energy separation also shows that a second-order calculation is not sufficient; it is a typical case for a *renormalization group* treatment. This form of the Kondo problem was pioneered by Anderson[26] and gave rise to an enormous literature culminating in the work of Wilson[27]. But perhaps the most complete treatment is that of Wiegmann[25] who derives the scaling relations implied by the renormalization group in diagonalizing explicitly the exchange Hamiltonian with the help of the *Bethe ansatz*. To this end the Hamiltonian must first be put into a one-dimensional form which is possible since only s-waves couple to the impurity.

The transition probability (15.4) combined with (15.13) and $\sigma' = \sigma$ has yet another important interpretation, related to the scattering of a particle beam which may consist of neutrons, muons, electrons or photons. Describing the target by the *dynamical structure factor*, defined as Fourier transform of the *density-density correlation function*[28],

$$S(\mathbf{q}, \omega) = \int dt e^{i\omega t} \sum_{jl} \langle e^{-i\mathbf{q}\cdot\mathbf{R}_j(t)} e^{i\mathbf{q}\cdot\mathbf{R}_l(0)} \rangle \tag{15.22}$$

(S in Ref. 28 is our $S^*/2\pi$) one finds that

$$W^{\text{imp}}(kk') = \frac{1}{\hbar V^2} |\tilde{U}(\mathbf{k} - \mathbf{k}')|^2 S(\mathbf{k} - \mathbf{k}', \varepsilon_{\mathbf{k}} - \varepsilon_{\mathbf{k}'}) . \tag{15.23}$$

Here $\mathbf{k}$, $\varepsilon_{\mathbf{k}}$ and $\mathbf{k}'$, $\varepsilon_{\mathbf{k}'}$ are, respectively, the wavevector and energy of the incoming and scattered beam particles, $\varepsilon_{\mathbf{k}} = \hbar^2 \mathbf{k}^2/2m$ and V is the volume of the target. A *structure analysis* of the target may be carried out by measuring the *differential cross section*[28] which in the above notation is given by

$$\frac{d^2\sigma(\mathbf{kk}')}{d\Omega dE} = \frac{|\mathbf{k}'|}{|\mathbf{k}|} \frac{m^2}{(2\pi\hbar)^3} |\tilde{U}(\mathbf{k} - \mathbf{k}')|^2 S(\mathbf{k} - \mathbf{k}', \varepsilon_{\mathbf{k}} - \varepsilon_{\mathbf{k}'}) \quad (15.24)$$

where $|\mathbf{k}|$ and $|\mathbf{k}'|$ are the fluxes (multiplied by $m/\hbar$) of the incoming and outgoing beams, respectively, and m is the mass of the beam particles.

According to (15.14), the dynamical structure factor (15.22) depends in an essential way on the phonon coordinates. Writing

$$\langle e^{-i\mathbf{q}\cdot\mathbf{R}_j(t)} e^{i\mathbf{q}\cdot\mathbf{R}_l(0)} \rangle = e^{-i\mathbf{q}\cdot(\mathbf{R}_j^o - \mathbf{R}_l^o)} \langle e^{-iA_j(t)} e^{iA_l(0)} \rangle , \quad (15.25)$$

$A_j(t)$ is a linear function of the phonon emission and absorption operators,

$$iA_j(t) = \sum_{q'} \alpha_{jq'} (b_{q'}(t) + b_{-q'}^+(t)) \quad (15.26)$$

where according to Eqs. (1.24), (1.25), (1.31) and (15.14)

$$\alpha_{jq'} = -\alpha_{j,-q'}^* = i\hbar \sum_{\nu\mathbf{R}} \lambda_\nu(\mathbf{R}_j^o - \mathbf{R})(2LM_\nu\omega_{q'})^{-1/2} \mathbf{q}\cdot\mathbf{e}_{q'}(\nu) e^{i\mathbf{q}'\cdot\mathbf{R}} . \quad (15.27)$$

It is important to note that a straightforward evaluation of Eq. (15.25) is only possible if the statistical average and the time dependence are taken in the unperturbed sense with $\mathcal{H}_o = \mathcal{H}^{\mathrm{ph}}$ given in Eq. (1.29). Interaction corrections due to anharmonic or electron-phonon couplings may of course be calculated with the perturbation methods developed in the last chapter but we refrain from doing this here.

Making use of the identity (Problem 3.8)

$$e^A e^B = e^{A+B+\frac{1}{2}[A, B]} \quad (15.28)$$

which holds if the commutator $[A, B]$ commutes with both operators A and B, the following formula may be derived (Problem 3.9)

$$\langle e^C \rangle_o = \exp\langle C^2/2 \rangle_o \quad (15.29)$$

where C is an arbitrary linear combination of the phonon operators b_q, b_q^+ [28]. Taking $C = A + B$ and writing $[A, B] = \langle [A, B] \rangle_o$, Eq. (15.25) may be evaluated by combining (15.28) and (15.29), noting that according to (7.37) the unperturbed time dependence in Eq. (15.26) may be absorbed in the coefficients $\alpha_{jq'}$. Defining

$$M_{jl}(t) \equiv \langle A_j^o(t) A_l^o(0) \rangle_o = M_{lj}^*(-t) \quad (15.30)$$

one then finds

$$\langle e^{-iA_j^o(t)} e^{iA_l^o(0)} \rangle_o = \exp\{M_{jl}(t) - \frac{1}{2}M_{jj}(0) - \frac{1}{2}M_{ll}(0)\} \quad (15.31)$$

where the index o indicates unperturbed time dependence (interaction representation) and average.

The above equations are valid for arbitrary atomic positions (15.14). However, in what follows we concentrate on a regular crystal, setting $\mathbf{R}_j^o = \mathbf{R}$ and $\lambda_\nu(\mathbf{R} - \mathbf{R}') = \lambda_\nu \delta_{\mathbf{R}\mathbf{R}'}$. Then A_j in (15.26) and Eq. (15.30) may be written respectively as

$$A^o(\mathbf{R}, t) = \mathbf{q} \cdot \sum_\nu \lambda_\nu \mathbf{u}_\nu^o(\mathbf{R}, t) \equiv \mathbf{q} \cdot \mathbf{u}^o(\mathbf{R}, t) \qquad (15.32)$$

and

$$M(\mathbf{R}, t) = \langle (\mathbf{q} \cdot \mathbf{u}^o(\mathbf{R}, t))(\mathbf{q} \cdot \mathbf{u}^o(\mathbf{0}, 0)) \rangle_o . \qquad (15.33)$$

With these expressions, insertion of Eqs. (15.25), (15.31) into (15.22) yields

$$S(\mathbf{q}, \omega) = Le^{-2W} \int dt e^{i\omega t} \sum_{\mathbf{R} \in V} e^{-i\mathbf{q} \cdot \mathbf{R}} e^{M(\mathbf{R}, t)} \qquad (15.34)$$

where

$$2W = M(\mathbf{0}, 0) \qquad (15.35)$$

and $\exp(-2W)$ is the *Debye-Waller factor*[28].

Insertion into Eq. (15.33) of (15.26), (15.27) for the regular lattice, together with (15.32), remembering Eqs. (1.25) and (7.41), yields the explicit quadratic form in $\mathbf{q}$,

$$M(\mathbf{R}, t) = \frac{1}{2L} \sum_{q'} \{(1 + n_o(\omega_{q'}))e^{-i\omega_{q'}t} + n_o(\omega_{q'})e^{i\omega_{q'}t}\}$$

$$\times \frac{1}{\omega_{q'}} \left| \sum_\nu \frac{\hbar \lambda_\nu}{\sqrt{M_\nu}} \mathbf{q} \cdot \mathbf{e}_{q'}(\nu) \right|^2 e^{iq' \cdot \mathbf{R}} . \qquad (15.36)$$

Developing the last exponential in Eq. (15.34), making use of formula (1.9), the first two terms are

$$S_o(\mathbf{q}, \omega) = 2\pi L^2 e^{-2W} \delta(\omega) \sum_{\mathbf{K}} \delta_{\mathbf{q}, \mathbf{K}} \qquad (15.37)$$

and[29]

$$S_1(\mathbf{q}, \omega) = \pi L e^{-2W} \sum_{q'} \{(1 + n_o(\omega_{q'}))\delta(\omega - \omega_{q'}) + n_o(\omega_{q'})\delta(\omega + \omega_{q'})\}$$

$$\times \frac{1}{\omega_{q'}} \left| \sum_\nu \frac{\hbar \lambda_\nu}{\sqrt{M_\nu}} \mathbf{q} \cdot \mathbf{e}_{q'}(\nu) \right|^2 \sum_{\mathbf{K}} \delta_{\mathbf{q}-q', \mathbf{K}} . \qquad (15.38)$$

S_o describes the *forward* ($\mathbf{K} = 0$) and *Bragg* ($\mathbf{K} \neq 0$) *elastic scattering* and S_1 the *one-phonon emission* $\propto (1 + n_o(\omega_{q'}))$, and *absorption* $\propto n_o(\omega_{q'})$, *inelastic direct* ($\mathbf{K} = 0$) and *Umklapp* ($\mathbf{K} \neq 0$) *scattering.* Higher-order terms in M contribute to the *multi-phonon background*. The fact that the elastic and one-phonon inelastic peaks are δ-functions in ω is of course due to the unperturbed approximation (index o) used in the above calculation. Phonon-phonon and (in a metal) electron-phonon interactions give rise to line broadening and phonon-frequency renormalization.

16. Resistivity Due to Electron-phonon Interaction. Semiconductors

Historically, the first microscopic calculation of the effect of phonons on the electrical conductivity is contained in Bloch's pioneering work on the theory of metals[30] where, after considerable efforts, he derived the well-known T^5-law valid at low temperatures. Fortunately, the formalism developed in the last two sections allows us to give a much simplified and more transparent derivation of this result. Taking the electron-phonon interaction in the form (10.6), Eq. (15.1) holds with

$$\alpha_{k'k} = \delta_{\sigma'\sigma} \; ; \quad h(\mathbf{q}) = \sum_{q'} \gamma_{q'} Q_{q'} \delta_{\mathbf{q'q}} = h^+(-\mathbf{q}) \; . \tag{16.1}$$

Inserting into Eq. (15.4) and making use of (1.31), (7.37) and (7.41) one then finds for s-electrons

$$W^{\mathrm{ph}}(kk') = \frac{\pi}{\hbar} \sum_q |\gamma_q|^2 \{(1 + n_o(\omega_q))\delta(\varepsilon_{sk} - \varepsilon_{sk'} - \omega_q)$$
$$+ n_o(\omega_q)\delta(\varepsilon_{sk} - \varepsilon_{sk'} + \omega_q)\}\delta_{\mathbf{k'},\mathbf{k}-\mathbf{q}}\delta_{\sigma'\sigma} \; . \tag{16.2}$$

Note that here the correlation function in Eq. (15.4) has again been evaluated in unperturbed approximation. Note also that the emission and absorption terms in Eq. (16.2), or in the corresponding Γ_o of Eq. (15.3), agree with the expressions (11.39) and (11.40), respectively, provided that we replace $f_o(\varepsilon_\mathbf{k})$ and $1 - f_o(\varepsilon_{\mathbf{k'}})$ by 1. Equation (16.2) also agrees with the non-Umklapp, one-phonon part (15.38) in Eq. (15.23) if we identify $|\tilde{\phi}_\nu(-\mathbf{q})|^2$, defined in (2.21), with $\lambda_\nu e^{-2W}|\tilde{U}(\mathbf{q})|^2$. The most important property of the expression (16.2), however, is that it also satisfies detailed balance (13.5) in the form $\exp(\beta\varepsilon_{sk'})W^{\mathrm{ph}}(kk') = \exp(\beta\varepsilon_{sk})W^{\mathrm{ph}}(k'k) = 0$, (see Ref. 10, Section 9.33) as is easily verified with the help of Eq. (7.43).

Inserting Eqs. (16.2) and (15.3) into (14.36) we have for quasi-free s-electrons, $\varepsilon_{sk} = \hbar^2\mathbf{k}/2m_s$,

$$\rho_{\alpha\gamma} = \frac{\pi\hbar}{e^2 n_s^2}\beta \sum_q |\gamma_q|^2 V^{-1} \sum_k f_o(\varepsilon_{sk})$$

$$\times \left\{ [1 - f_o(\varepsilon_{sk} - \omega_q)](1 + n_o(\omega_q))\delta(\varepsilon_{sk} - \varepsilon_{s,k-q} - \omega_q) \right.$$

$$\left. + [1 - f_o(\varepsilon_{sk} + \omega_q)]n_o(\omega_q)\delta(\varepsilon_{sk} - \varepsilon_{s,k-q} + \omega_q) \right\} q_\alpha q_\gamma \Big|_{\varepsilon_{sk} \simeq \varepsilon_{s,k-q} \simeq \beta^{-1}}.$$

$$(16.3)$$

Here the phonon distribution function $n_o(\omega_q)$ is appreciable only for *thermal phonons*, for which $\beta\omega_q \lesssim 1$. This means that, except near the melting point where possible (if $B > 1$) optical phonons have large amplitudes, only acoustic phonons are thermally excited. We may therefore, without loss of generality, exclude optical branches by assuming that $B = 1$. In this case $\omega_{qj} = \hbar c_j |\mathbf{q}|$ where the c_j are the longitudinal ($j = \parallel$) and transverse ($j = \perp$) sound velocities and the coupling function γ_q is given by Eq. (10.11). High and low temperatures then are distinguished by the limit $k_B T = \max \omega_{qj} \equiv k_B \theta_D$ where θ_D is the *Debye temperature* which is determined by $\hbar c_\parallel q_o = k_B \theta_D$. Here $q_o^3/6\pi^2 = 1/v = L/V$ is the number of acoustic phonon modes per unit volume (in a cubic crystal $q_o = (6/\pi)^{1/3}\pi/a \cong 1.24\pi/a$). A more appropriate limiting temperature (pseudo-Debye temperature) is defined by

$$2\hbar c_\parallel k_F = k_B \theta . \tag{16.4}$$

According to Eq. (14.13) with $D = 3$, θ coincides with θ_D if there are 0.25 s-electrons per unit cell, i.e. $n_s = 1/4v$, which is fairly realistic in general.

At *low temperatures*, $T \ll \theta$, only long wavelength phonons are thermally excited so that we may approximate the coupling function (10.11) by

$$|\gamma_\mathbf{q}|^2 = V^{-1}\lambda|\mathbf{q}| ; \quad \lambda \equiv \frac{\hbar|\tilde{\phi}(0)|^2}{vMc_\parallel} . \tag{16.5}$$

On the other hand, the Fermi distribution functions $f_o(\varepsilon_{sk})$ and $1 - f_o(\varepsilon_{sk} \mp \omega_q)$ in Eq. (16.3) are appreciable only for $\varepsilon_{sk} < \mu$ and $\varepsilon_{sk} \mp \omega_q > \mu$, respectively. Since, quite generally, $k_B\theta_D \ll \mu$ (see the tables on p. 126 and p. 154 of Ref. 31), it is evident that at low temperatures ω_q is negligible as compared to μ and may therefore be dropped in the arguments of the Fermi functions in (16.3). Using then again $\beta f_o(1 - f_o) = -f_o' \cong \delta(\varepsilon_{sk} - \mu)$, the density of states (10.4) and (16.5), Eq. (16.3) becomes

$$\rho_{\alpha\gamma} = \frac{\hbar}{2e^2 n_s^2} N_s(\mu) \lambda V^{-1} \sum_{\mathbf{q}} |\mathbf{q}| \int d\Omega_{\mathbf{k}} \Big\{ (1 + n_o(\omega_q))$$

$$\times \, \delta(\mu - \varepsilon_{s,\mathbf{k}-\mathbf{q}} - \omega_q) + n_o(\omega_q)\delta(\mu - \varepsilon_{s,\mathbf{k}-\mathbf{q}} + \omega_q) \Big\} q_\alpha q_\gamma \Big|_{\varepsilon_{sk}=\mu}$$

$$(16.6)$$

where the integral is over the angles of $\mathbf{k}$ and $\omega_q = \hbar c_\parallel q$

The limit $T \to 0$ means phonon vacuum, $n_o(\omega_q) \to 0$, for all modes q. The only surviving term in (16.2) therefore is $\propto \delta(\varepsilon_{sk} - \varepsilon_{sk'} - \omega_q)$ which, according to (15.3), makes $\Gamma_o(kk') \propto f_o(\varepsilon_{sk})[1 - f_o(\varepsilon_{sk} - \omega_q)]$. But this expression vanishes in the limit $T \to 0$ for all $\mathbf{q} \neq 0$. Hence we may drop the residual term in Eq. (16.6) and obtain, using the procedure of Problem 3.7, $\rho_{\alpha\gamma} = \rho^{\mathrm{ph}}(T)\delta_{\alpha\gamma}$ with

$$\rho^{\mathrm{ph}}(T) = \frac{N_s(\mu)\lambda}{6\pi e^2 n_s^2 v_F} \int dq \, q^4 n_o(\hbar c_\parallel q)$$

$$\times \left\{ \theta\left(1 - \left(\frac{q}{2k_F} + \frac{c_\parallel}{v_F}\right)\right) + \theta\left(1 - \left|\frac{q}{2k_F} - \frac{c_\parallel}{v_F}\right|\right) \right\}. \quad (16.7)$$

Since $c_\parallel \ll v_F$ the terms $c_\parallel/v_F$ in the argument of the θ-functions may be dropped so that, with $x \equiv \beta\hbar c_\parallel q$ and with definition (16.4),

$$\rho^{\mathrm{ph}}(T) = \frac{32 m_s N_s(\mu)\lambda k_F^4}{3\pi e^2 n_s^2 \hbar} \left(\frac{T}{\theta}\right)^5 \int_0^{\theta/T} \frac{x^4}{e^x - 1}. \quad (16.8)$$

This formula is often called *Bloch-Grüneisen law* since Grüneisen had suggested such a formula for all temperatures[32]. The integral in Eq. (16.8) is related to the function[9,10]

$$J_n(z) \equiv \int_0^z \frac{x^n e^x \, dx}{(e^x - 1)^2} = n \int_0^z \frac{x^{n-1} dx}{e^x - 1} - \frac{z^n}{e^z - 1}; \quad n > 1. \quad (16.9)$$

Hence for $\theta \gg T$, the value of the integral in (16.8) may be approximated by $J_5(\infty)/5 = 4!\zeta(5) \cong 24.9$ (see Eq. (34.56) below). This leads to Bloch's famous T^5-law[30]. On the other hand, since the longitudinal acoustic spectrum may reasonably well be approximated by the linear q-dependence up to the limit $2k_F$, Eqs. (16.6)–(16.8) are still reasonably accurate at *high temperatures*, $T \gg \theta$. In this case we may develop the integrand in Eq. (16.8) to obtain the equally well-known T^1-law[9,30], see Eq. (16.28) below.

The experimental verification of these laws has a long history, dominated in the early 1930's by the work of Grüneisen (see Refs. 9 and 10). Apart from the Kondo minimum mentioned in the last section, it was also noticed quite early that *transition metals* rather follow a T^2-law at very low temperatures (see Ref. 9, Section 9.14 and Ref. 10, Section 9.612) which was explained by an interaction between s- and d-electrons[9,33]. Leaving this problem for a detailed discussion in the next section, we now wish to calculate the effect of the electron-phonon interaction on the temperature dependence of the electron and hole mobilities in semiconductors at high temperatures.

The quantitative theory of the charge distribution in semiconductors was developed by Wilson in 1931[34]. According to this theory a *semiconductor* (see, e.g. Ref. 31, Chapter 8) is characterized, at $T = 0$, by a fully occupied *valence band* and a totally empty *conduction band* so that the Fermi level lies in the *band gap* in between where there are at most some localized states. The latter are due to *imperfections* of the crystal, mostly impurity atoms, but may also be due to irregularities of the lattice such as *dislocations*. These localized states may be classified according to whether they are energetically closer to the valence or to the conduction band, in which case they are called *acceptors* and *donors*, respectively. These names express the fact that acceptors may "accept" electrons from the valence band while donors may "donate" electrons to the conduction band. A schematic energy spectrum with one acceptor and one donor level is depicted in Fig. 3.3.

At finite temperatures, holes are created at the top of the valence band and electrons at the bottom of the conduction band. These band states are well approximated by (see Fig. 3.3)

$$\varepsilon_{ek} = \frac{\hbar^2(\mathbf{k} - \mathbf{k}_e)^2}{2m_e} + E_g \; ; \quad \varepsilon_{hk} = -\frac{\hbar^2(\mathbf{k} - \mathbf{k}_h)^2}{2m_h} \tag{16.10}$$

where E_g is the gap energy and m_e^{-1} and m_h^{-1} may have tensor character so that the constant energy surfaces are ellipsoids centred at $\mathbf{k}_e$ and $\mathbf{k}_h$, respectively. If $\mathbf{k}_e \neq 0$ (which is the case in silicon and germanium, while $\mathbf{k}_h = 0$, see Ref. 34), crystal symmetry requires several equivalent positions $\mathbf{k}_e$ (or $\mathbf{k}_h$). Since $0 < \mu < E_g$ and since $\beta\mu \gg 1$, $\beta(E_g - \mu) \gg 1$ for all but very high temperatures, so that electrons and holes obey *Boltzmann statistics* and the respective densities are

$$n_e = V^{-1} \sum_{\mathbf{k}} f_o(\varepsilon_{ek} - \mu) \cong \frac{2}{V} \sum_{\mathbf{k}} e^{-\beta(\varepsilon_{sk} - \mu)}$$

$$= \frac{2}{\hbar^3} \left(\frac{m_e k_B T}{2\pi}\right)^{3/2} e^{-(E_g - \mu)/k_B T} \tag{16.11}$$

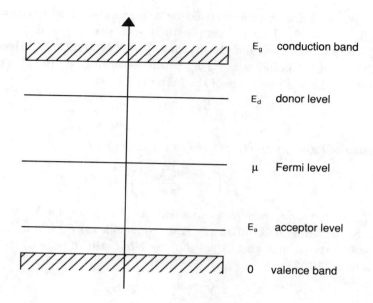

Fig. 3.3: Schematic energy spectrum of a semiconductor.

and

$$n_h = V^{-1} \sum_{\mathbf{k}} \left[1 - f_o(\varepsilon_{e\mathbf{k}} - \mu) \right] \cong \frac{2}{V} \sum_{\mathbf{k}} e^{+\beta(\varepsilon_{h\mathbf{k}} - \mu)}$$

$$= \frac{2}{\hbar^3} \left(\frac{m_h k_B T}{2\pi} \right)^{3/2} e^{-\mu/k_B T} . \tag{16.12}$$

On the other hand, "donation" of electrons to the conduction band gives rise to donor holes and "acceptance" of electrons from the valence band to acceptor electrons, the respective densities being (see Fig. 3.3)

$$n_h^d = N_d \left[1 - f_o(E_d - \mu) \right] ; \quad n_e^a = N_a f_o(E_a - \mu) . \tag{16.13}$$

Here N_d and N_a are, respectively the donor and acceptor level densities. *Charge neutrality* then implies the condition

$$n_h + n_h^d = n_e + n_e^a \tag{16.14}$$

which determines the position of the Fermi level as function of N_a and N_d.

N_a and N_d may be controlled by appropriate *doping*, giving rise to the three main categories of *intrinsic*, *p-type* or *n-type* semiconductors . The *intrinsic type* results from high purity, $N_d \cong N_a \cong 0$, or from compensation,

$N_d \cong N_a \neq 0$ and is characterized by a Fermi energy situated in mid-gap, $E_a < \mu < E_d$. Thus for temperatures such that $\beta(\mu - E_a) \gg 1$ and $\beta(E_d - \mu) \gg 1$ one sees from (16.13) that $n_h^d \cong N_d$ and $n_e^a \cong N_a$. Hence the condition (16.14) becomes $n_h \cong n_e$ which, according to (16.11), (16.12), determines the intrinsic value of the Fermi energy to be

$$\mu_i(T) = \frac{E_g}{2} + \frac{3k_B T}{4} \ln \frac{m_h}{m_e} . \tag{16.15}$$

Insertion of (16.15) into (16.11) or (16.12) then gives

$$n_e = n_h = n_n(T) \equiv \frac{2}{\hbar^3} \left(\frac{\sqrt{m_e m_h} k_B T}{2\pi} \right)^{3/2} e^{-E_g/2k_B T} . \tag{16.16}$$

An *n-type semiconductor* is obtained from the condition $N_d \gg N_a$ and is characterized by a Fermi energy situated as follows: $E_d < \mu < E_g$. Thus for temperatures such that $\beta(\mu - E_d) \gg 1$ but still $\beta(E_g - \mu) \gg 1$, one reads from (16.13) that

$$n_h^d \cong N_d e^{-\beta(\mu - E_d)} \ll N_d \tag{16.17}$$

while $n_h \cong n_e^a \cong 0$. Now the neutrality condition (16.14) is $n_h^d \cong n_e$ so that, according to (16.11) and (16.17) the n-type value of the Fermi level is

$$\mu_n(T) = \frac{E_g + E_d}{2} - \frac{k_B T}{2} \ln \left[\frac{2}{N_d \hbar^3} \left(\frac{m_e k_B T}{2\pi} \right)^{3/2} \right] , \tag{16.18}$$

and inserted into (16.11) one finds

$$n_e = n_i(T) \equiv \sqrt{N_d \frac{2}{\hbar^3} \left(\frac{m_e k_B T}{2\pi} \right)^{3/2}} \, e^{-(E_g - E_d)/k_B T} . \tag{16.19}$$

An analogous calculation applies to a *p-type semiconductor*.

Both Eqs. (16.16) and (16.19) exhibit the exponential disappearance of charge carriers as $T \to 0$ which is typical for semiconductors and, more generally for systems with an excitation threshold. It also determines their transport properties and is responsible for the fact that the resistivity formula (14.34) is not applicable. Indeed, because of Boltzmann statistics, $-f_o'(\varepsilon)$ is not the δ-function but the broad distribution $\beta f_o(\varepsilon)$ so that the closure relation (14.31) does not hold. But, fortunately, the relaxation-time approximation (13.38) works very well, provided that we assume $\tau_k = \text{const.}$ Of course, this traditional method[10] could also have been used for the derivation of the Bloch-Grüneisen law as will be seen by comparison below.

Combining Eq. (13.38) for $\tau_k = \tau = \text{const}$, respectively with (13.9) and with (13.18), (14.3), we have the linear integral equation

$$\beta \sum_{k'} \Gamma_o(kk')\{\phi(k) - \phi(k')\} = -\tau^{-1} f'_o(\varepsilon_k - \mu)\phi(k) \qquad (16.20)$$

and the traditional conductivity formula[10]

$$\sigma_{\alpha\gamma} = -\tau^{-1} e^2 V^{-1} \sum_k f'_o(\varepsilon_k - \mu)\varphi_\alpha^{(1)}(k)\varphi_\gamma^{(1)}(k) \qquad (16.21)$$

where, according to (13.16), the solution $\phi = \varphi_\alpha^{(1)}$ is now given by

$$\varphi^{(1)}(k) = -\tau \mathbf{v}_k \qquad (16.22)$$

and the energy band ε_k is still not specified.

We now determine the relaxation time τ for the electron-phonon interaction by using Eq. (16.2) without the zero-point contribution in (15.3) and inserting, together with (16.22), into (16.20). Using $-f'_o = \beta f_o(1 - f_o)$ this yields, after some simplifications,

$$\hbar^{-1}\tau\pi \sum_q |\gamma_q|^2 n_o(\omega_q)\{\delta(\varepsilon_k - \varepsilon_{k-q} - \omega_q) + \delta(\varepsilon_k - \varepsilon_{k-q} + \omega_q)\}$$

$$\times [1 - f_o(\varepsilon_{k-q} - \mu)](\mathbf{v}_k - \mathbf{v}_{k-q}) = [1 - f_o(\varepsilon_k - \mu)]\mathbf{v}_k \qquad (16.23)$$

where the explicit band structure is still left open. Applied to the conduction band of a semiconductor, Boltzmann statistics means that $f_o(\varepsilon_{ek} - \mu) \ll 1$ for all $\mathbf{k}$, so that the Fermi functions disappear from Eq. (16.23). As before, we assume the temperature to be sufficiently far away from the melting point to justify neglecting possible optical phonons. Applying (16.5) and $\omega_q = \hbar c_\parallel |\mathbf{q}|$, Eq. (16.23) becomes

$$\tau^{-1}\mathbf{v}_{ek} = \frac{\pi\lambda}{\hbar V} \sum_q |\mathbf{q}| n_o(\hbar c_\parallel |\mathbf{q}|) \sum_{e'} \left\{ \delta(\varepsilon_{ek} - \varepsilon_{e',k-q} - \hbar c_\parallel |\mathbf{q}|) \right.$$

$$\left. + \delta(\varepsilon_{ek} - \varepsilon_{e',k-q} + \hbar c_\parallel |\mathbf{q}|) \right\}(\mathbf{v}_{ek} - \mathbf{v}_{e',k-q}) \qquad (16.24)$$

where $\mathbf{v}_{ek} = \hbar(\mathbf{k} - \mathbf{k}_e)m_e$ and $\sum_{e'}$ extends over all equivalent minima $\mathbf{k}_e$ in (16.10). The terms $e' \neq e$ represent *intervalley scattering* and will be neglected since the assumption of a constant relaxation time is not really justified in this case. The reason is that $|\mathbf{k}_{e'} - \mathbf{k}_e|$ is of the order of π/a, a being the lattice constant, so that the phonons with $\mathbf{q} \cong \mathbf{k}_{e'} - \mathbf{k}_e$ have very short wavelength (see below).

Setting $e' = e$ and writing $(\mathbf{k} - \mathbf{k}_e) \cdot \mathbf{q} = \kappa q \zeta$ where $\kappa \equiv |\mathbf{k} - \mathbf{k}_e|$, the projection of (16.24) onto $\mathbf{k} - \mathbf{k}_e$ yields, after integration over the angles (compare Problem 3.7)

$$
\tau^{-1} = \frac{\lambda m_e}{4\pi\hbar^3\kappa^3} \int_0^\infty dq \; q^3 n_o(\hbar c_\| q)
$$
$$
\times \left\{ \left(\frac{q}{2} + k_o\right)\theta\left(\kappa - \left(\frac{q}{2} + k_o\right)\right) + \left(\frac{q}{2} - k_o\right)\theta\left(\kappa - \left|\frac{q}{2} - k_o\right|\right) \right\}
$$

$$(16.25)$$

where $k_o \equiv c_\| m_e/\hbar$. This expression depends explicitly on $\mathbf{k}$, contrary to the assumption $\tau = \text{const.}$ However, the role of the Fermi wavenumber is here played by the thermal wavenumber $\kappa_T \equiv \sqrt{2m_e k_B T}/\hbar$ so that we may put $\kappa \simeq \kappa_T$ in (16.25). For the same reason, the limiting temperature is the Debye temperature θ_D defined earlier as $k_B\theta_D = \max \omega_{\mathbf{q}j}$, i.e. by the relation

$$
\hbar c_\| \frac{\pi}{a} = k_B \theta_D \tag{16.26}
$$

instead of (16.4). Therefore, at *high temperatures* the dominant q-values are those near the maximum π/a. Since, on the other hand, $c_\| \ll \max|\mathbf{v}_{\mathbf{e}k}| \simeq \hbar\pi/am_e$, it follows that k_o is negligible compared to q, so that Eq. (16.25) becomes, with $n_o(\omega) \simeq 1/\beta\omega$,

$$
\tau^{-1} = \frac{\lambda m_e \kappa k_B T}{\pi\hbar^4 c_\|} \tag{16.27}
$$

where $\kappa \simeq \kappa_T$. Note that in the metallic case, $\kappa = k_F$ and, with (10.4), Eqs. (16.20), (16.21) just become (14.20). Inserting there (16.27) yields, for the s-band,

$$
\sigma_o = \frac{\pi e^2 \hbar^4 c_\| n_s}{m_s^2 \lambda k_F k_B T} \tag{16.28}
$$

which is, up to a factor 2, exactly the linear resistivity law resulting from (16.8) for $T \gg \theta$ when Eqs. (16.4) and (14.14) with $D = 3$ are used.

For the band structure (16.10) and Boltzmann statistics one finds for (16.21), (16.22), $\sigma_{\alpha\gamma} = \sigma_e \delta_{\alpha\gamma}$ with

$$
\sigma_e = \frac{e^2 \hbar^3 \beta}{3\pi^2 m_e^2} e^{-\beta(E_g - \mu)} \int_0^\infty d\kappa \; \kappa^4 e^{-\kappa^2/\kappa_T^2} \tau \; . \tag{16.29}
$$

Inserting here (16.27), the difference in taking the argument of τ to be κ or κ_T is only about a factor of ~ 0.75 which justifies the assumption of a constant relaxation time made above. With the argument κ_T one finds the exponential decrease

$$\sigma_e = \frac{e^2 \hbar^2 c_\|}{2\sqrt{\pi} m_e \lambda} e^{-\beta(E_g - \mu)} . \tag{16.30}$$

This result is actually better represented by the *electron mobility* defined in (14.10). Using (16.11) one finds the well-known $T^{-3/2}$-law (see Ref. 10, Section 9.36)

$$\mu_e = \frac{\sigma_e}{e^2 n_e} = \frac{\pi \hbar^5 c_\|}{\sqrt{2 m_e^5} \lambda} (k_B T)^{-3/2} . \tag{16.31}$$

Experimentally, deviations from the exponent $-3/2$ both, for electrons and holes in silicon and germanium have been known for a long time and were explained by contributions from optical phonons and by band structure effects[35]. In particular, for electrons in Si where optical phonons are negligible, *intervalley scattering*, $e' \neq e$ in Eq. (16.24), was proposed[35] which is reasonable since, as we have shown, for $T \geq \theta$ large wavenumbers of the order of π/a are indeed important. It should be mentioned, however, that *multi-phonon processes* as described by the Hamiltonian (2.22), as well as higher-order diagrams with one-phonon vertices, which were both invoked in the first attempt to explain the deviations from the law (16.31)[36], cannot be ruled out neither.

17. Low-temperature Resistivity Due to Electron-electron Interactions

In contradistinction to the elastic scattering of electrons by impurities and to the emission and absorption of phonons as discussed in the last two sections, electron-electron interactions give rise to *two-particle scattering processes*. As mentioned after Eq. (13.9), the Boltzmann theory developed in Section 13 must be generalized somewhat in order to be applicable to this situation. This generalization will be introduced with the example of the interaction (3.38) which here is taken as the Coulomb repulsion among s-electrons. In this case it is, however, more convenient to use instead of (3.38) the alternative expression

$$\mathcal{H}^{el-el} = \frac{1}{2V} \sum_{\mathbf{q}} \tilde{\phi}(\mathbf{q}) n(\mathbf{q}) n^+(\mathbf{q}) \tag{17.1}$$

which follows from Eq. (2.18) with (2.4) and (2.9).

Inserting Eq. (3.22) and considering for simplicity's sake the first and second pairs of operators as being statistically independent (which means excluding coinciding labels), comparison with Eq. (15.1) gives the identifications

$$\alpha_{k'k} = \delta_{\sigma'\sigma} \ ; \quad h(\mathbf{q}) = \tilde{\phi}(\mathbf{q}) \frac{1}{2V} \sum_{\mathbf{p}\tau} a^+_{\mathbf{p}-\mathbf{q},\tau} a_{\mathbf{p}\tau} = h^+(-\mathbf{q}) \ . \tag{17.2}$$

Hence the correlation function in Eq. (15.4) is, in unperturbed approximation,

$$\langle h(\mathbf{q}; t) h^+(\mathbf{q}; 0) \rangle = |\tilde{\phi}(\mathbf{q})|^2 \frac{1}{4V^2} \sum_{pp'} f_o(\varepsilon_{sp}) [1 - f_o(\varepsilon_{sp'})]$$

$$\times e^{i(\varepsilon_{sp} - \varepsilon_{sp'})t} \delta_{\tau'\tau} \delta_{\mathbf{p}',\mathbf{p}+\mathbf{q}} \ . \tag{17.3}$$

Inserting (17.2), (17.3) and (15.4) we may write Eq. (15.3) as

$$\Gamma_o(kk') = \sum_{pp'} \Gamma_o(kp; \, k'p') \tag{17.4}$$

with

$$\Gamma_o(kp; \, k'p') = \frac{\pi}{2\hbar V^2} |\tilde{\phi}(\mathbf{k} - \mathbf{k}')|^2 \delta_{\sigma'\sigma} \delta_{\tau'\tau} \delta_{\mathbf{k}+\mathbf{p},\mathbf{k}'+\mathbf{p}'}$$

$$\times \delta(\varepsilon_{sk} + \varepsilon_{sp} - \varepsilon_{sk'} - \varepsilon_{sp'}) f_o(\varepsilon_{sk}) f_o(\varepsilon_{sp})$$

$$\times [1 - f_o(\varepsilon_{sk'})][1 - f_o(\varepsilon_{sp'})] \ . \tag{17.5}$$

It is easy to verify that because of the energy-conserving δ-function, the last expression satisfies *generalized detailed balance*

$$\Gamma_o(kp; \, k'p') = \Gamma_o(k'p'; \, kp) \ . \tag{17.6}$$

The generalization of the transition rate (13.3) to the case of two-particle collisions has the form

$$\Gamma(kp; \, k'p') = W(kk') f(k)[1 - f(k')] f(p)[1 - f(p')] \tag{17.7}$$

where $W(kk')$ is determined by identifying the equilibrium version of this formula with (17.5). Since the collision term (13.2) now reads

$$\left(\frac{df(k)}{\hbar dt} \right)_{\text{diss}} = \sum_{k'pp'} \{ \Gamma(k'p'; \, kp) - \Gamma(kp; \, k'p') \} \ , \tag{17.8}$$

the generalization of (13.9) to two-particle collisions is readily found by making use of (13.6); one finds (Problem 3.10)

$$\left(\frac{df(k)}{\hbar dt} \right)_{\text{diss}} = \beta \sum_{k'pp'} \Gamma_o(kp; \, k'p') \{ \phi(k') + \phi(p') - \phi(k) - \phi(p) \}$$

$$\equiv -(\mathcal{B}\phi)(k) \ . \tag{17.9}$$

Insertion of this generalized Boltzmann operator into Eq. (14.34) and use of (14.27), (14.30) and of $\varepsilon_k \simeq \varepsilon_{k'} \simeq \varepsilon_p \simeq \varepsilon_{p'} \simeq 0$ and $f'_o(0) = -\beta/4$ finally yields the following generalized semi-positive-definite quadratic form for the resistivity tensor:

$$\rho_{\alpha\gamma} = \left(\frac{m}{2en}\right)^2 \beta V^{-1} \sum_{kpk'p'} \Gamma_o(kp; k'p')$$
$$\times (v_{k\alpha} + v_{p\alpha} - v_{k'\alpha} - v_{p'\alpha})(v_{k\gamma} + v_{p\gamma} - v_{k'\gamma} - v_{p'\gamma}) \quad (17.10)$$

where $\varepsilon_k \simeq \varepsilon_{k'} \simeq \varepsilon_p \simeq \varepsilon_{p'} \simeq \beta^{-1}$. Here we used the identity $\Gamma_o(kp; k'p') = \Gamma_o(pk; p'k')$, which easily follows from (17.5), and the generalized detailed balance (17.6), for a complete symmetrization in the velocity arguments. Note that this is still a *one-band resistivity formula*.

Now it follows immediately from the *momentum conservation* contained in Eq. (17.5) that for quasi-free *s*-electrons for which $\mathbf{v}_{sk} = \hbar k/m_s$, Eq. (17.10) gives *zero resistance* (see Ref. 9, Section 9.14). This is not the full story, however, since *Umklapp processes* which are neglected in the plane-wave expression (3.22) do contribute to (17.10)[9]. Indeed, if instead of (3.22) the rigorous form (3.28) is used for the electron density in (17.1), then the first line of Eq. (17.5) will be replaced by an expression of the form $\sum_{\mathbf{K}} M(kk'; pp'; \mathbf{K})\delta_{k+p, k'+p'+\mathbf{K}}$ where $\mathbf{K}$ is a reciprocal lattice vector so that, for quasi-free electrons, $\mathbf{v}_{sk} + \mathbf{v}_{sp} - \mathbf{v}_{sk'} - \mathbf{v}_{sp'} = \hbar\mathbf{K}/m_s \neq 0$. But of course, with rigorous matrix elements (3.29) it does not make sense to use quasi-free electrons. Using the rigorous band structure however, the velocity is periodic, $\mathbf{v}_{k+K} = \mathbf{v}_k$, so that the Umklapp vector disappears again and $\mathbf{v}_k + \mathbf{v}_p - \mathbf{v}_{k'} - \mathbf{v}_{p'} \neq 0$ for other reasons. Without going into further details we may conclude therefore that the complications introduced by intra-band Umklapp processes are not really worth the effort.

More important, as already mentioned in the last section, is the *s-d interaction* in *transition metals* whose *d*-electrons contribute to conduction and also the *s-f* interaction in lanthanides (rare earths) and actinides which gives rise to the *heavy fermion metals* to be discussed at the end of this section. The general form of the *s-d* interaction which is based on the notion of *superexchange* introduced by Anderson[37] may be written in analogy to (15.21) and (17.1) as

$$\mathcal{H}^{s-d} = \frac{4}{V} \sum_{\mathbf{q}} \sum_{i=0}^{3} J_i(\mathbf{q}) S^i(\mathbf{q}) s^i(-\mathbf{q}) \quad (17.11)$$

where

$$s^i(\mathbf{q}) = \frac{1}{2} \sum_{\mathbf{k}} \sum_{\sigma\sigma'} a_{\mathbf{k}\sigma}^+ \sigma_{\sigma\sigma'}^i a_{\mathbf{k}+\mathbf{q},\sigma'} = s^i(-\mathbf{q})^+ \; ; \quad i = 0, 1, 2, 3 \qquad (17.12)$$

are s-electron densities formed from the expressions (3.22), (3.24) and (3.25), $\sigma^o = 1$ and σ^i ($i = 1, 2, 3$) are the Pauli matrices $\sigma^x \, \sigma^y \, \sigma^z$ defined in Eq. (2.8). The S^i are the analogous d-electron densities described by creation and annihilation operators c_p^+ and c_p with $p \equiv (\mathbf{p}, \tau)$. Spin conservation requires $J_1 = J_2$ and hermiticity $J_i^*(\mathbf{q}) = J_i(-\mathbf{q})$. This interaction is of exchange type[25].

By comparison of (17.11) with Eq. (15.1) one finds that $\alpha_{k'k}^i = \sigma_{\sigma'\sigma}^i$ and

$$h_i(\mathbf{q}) = \frac{2}{V} J_i(\mathbf{q}) S^i(\mathbf{q}) = J_i(\mathbf{q}) \frac{1}{V} \sum_{pp'} c_p^+ \sigma_{\tau\tau'}^i c_{p'} \delta_{\mathbf{p}',\mathbf{p}+\mathbf{q}} \; ; \quad i = 0, 1, 2, 3 \; .$$

$$(17.13)$$

Since from (2.8) follows that

$$\sum_{\tau\tau'} \sigma_{\tau'\tau}^i \sigma_{\tau\tau'}^j = \mathrm{Tr}\,(\sigma^i \sigma^j) = 2\delta_{ij} \; , \qquad (17.14)$$

we have, in unperturbed approximation,

$$\langle h_i(\mathbf{q}; t) h_j^+(\mathbf{q}; 0)\rangle = 2\delta_{ij} |J_i(\mathbf{q})|^2 \frac{1}{V^2} \sum_{\mathbf{P}} f_o(\varepsilon_{d\mathbf{p}})$$

$$\times [1 - f_o(\varepsilon_{d,\mathbf{p}+\mathbf{q}})] e^{i(\varepsilon_{d\bar{p}} - \varepsilon_{d,\mathbf{p}+\mathbf{q}})t} \qquad (17.15)$$

where $\varepsilon_{d\mathbf{p}}$ is the d-band energy. Inserting (17.15) into (15.3), (15.4) we may sum over the spin indices and define as in (17.4)

$$\Gamma_o(\mathbf{k}\mathbf{k}') \equiv \sum_{\mathbf{P}\mathbf{P}'} \Gamma_o(\mathbf{k}p; \mathbf{k}'p') \; ; \quad \Gamma_o(\mathbf{k}p; \mathbf{k}'p') \equiv \sum_{\sigma\sigma'\tau\tau'} \Gamma_o(kp; k'p') \; . \quad (17.16)$$

Using (17.14) one then finds

$$\Gamma_o(\mathbf{k}p; \mathbf{k}'p') = \frac{8\pi}{\hbar V^2} \sum_{i=0}^{3} |J_i(\mathbf{k} - \mathbf{k}')|^2$$

$$\times \delta_{\mathbf{k}+\mathbf{p},\mathbf{k}'+\mathbf{p}'} \delta(\varepsilon_{s\mathbf{k}} + \varepsilon_{d\mathbf{p}} - \varepsilon_{s\mathbf{k}'} - \varepsilon_{d\mathbf{p}'})$$

$$\times f_o(\varepsilon_{s\mathbf{k}})[1 - f_o(\varepsilon_{s\mathbf{k}'})] f_o(\varepsilon_{d\mathbf{p}})[1 - f_o(\varepsilon_{d\mathbf{p}'})] \; . \; (17.17)$$

This expression is obviously the same for the s- and the d-band so that Γ_o does not carry a band index.

Since now we are in a *two-band situation*, the inversion formula (14.36) and hence also Eq. (17.10) are not directly applicable. However, remembering that in Section 13 the label k also included a band index n, Eq. (17.7) is immediately generalized by writing $k = (n, \mathbf{k}\sigma)$ as $k = (\mathbf{k}\sigma)$ for $n = s$ and as $p = (\mathbf{p}\tau)$ for $n = d$ so that, in obvious notation,

$$\Gamma_o(kp; k'p') = W(kk')f_s(k)[1 - f_s(k')]f_d(p)[1 - f_d(p')] \qquad (17.18)$$

and similarly with the labels skk' and dpp' permuted. Then Eq. (17.8) holds with an additional index s and d, respectively, and the generalization of Eq. (17.9) becomes (Problem 3.10)

$$\left(\frac{df_s(k)}{\hbar dt}\right)_{\text{diss}} = \beta \sum_{k'pp'} \Gamma_{so}(kp; k'p')\{\phi_s(k') + \phi_d(p') - \phi_s(k) - \phi_d(p)\}$$

$$\equiv -(\mathcal{B}\phi)_s(k) \qquad (17.19)$$

and similarly with the labels skk' and dpp' permuted.

In order to derive a generalized inversion formula we first note that Eq. (13.11) for the current density $\mathbf{j}$ includes a sum over the band index n so that Ohm's law (14.1) now has the form corresponding to *two conductors in parallel*,

$$\sigma = \sigma^{(s)} + \sigma^{(d)} . \qquad (17.20)$$

Since Eqs. (13.14)–(13.18) and (14.3) apply to each band, the contributions $\sigma^{(s)}$ and $\sigma^{(d)}$ are again given by Eq. (14.32) which now reads

$$e^{-2}\sigma_{\alpha\gamma}^{(s)} = \langle \varphi_{s\gamma}|(\mathcal{B}\varphi)_{s\alpha}\rangle = \vartheta_s^2\langle \psi_{s\alpha}|(\mathcal{B}^{-1}\psi)_{s\gamma}\rangle \qquad (17.21)$$

and similarly for $\sigma^{(d)}$.

Assuming again quasi-free carriers we write the two bands in the standard form

$$\varepsilon_{s\mathbf{k}} = \frac{\hbar^2\mathbf{k}^2}{2m_s} - E_s \; ; \quad E_s = \frac{\hbar^2 k_s^2}{2m_s}$$

$$\varepsilon_{d\mathbf{k}} = \frac{\hbar^2\mathbf{k}^2}{2m_d} - E_d \; ; \quad E_d = \frac{\hbar^2 k_d^2}{2m_d} \qquad (17.22)$$

where $\hbar k_s = m_s v_s$ and $\hbar k_d = m_d v_d$ are the Fermi momenta and v_s and v_d the Fermi velocities of the respective bands. Since we are interested in low temperatures where the interval $k_B T$ of relevant energies around the Fermi energy $\mu = 0$ is narrower than most features of the band structure, these model assumptions are not essential. Note also that the two bands in Eq. (17.22) were assumed to be centered at $\mathbf{k} = 0$. This may be generalized without change by taking any symmetry point in the Brillouin zone as origin. Assuming further approximately equal filling of the two bands, Eq. (14.13) implies that $k_s \simeq k_d$ so that v_d/v_s and E_d/E_s are both of order $m_s/m_d < 1$, the last inequality being well satisfied for the transition and heavy-fermion metals. Designating the respective band widths by W_s and W_d, the densities of states following from (17.22) are

$$N_s(\varepsilon) = N_s(0)\sqrt{1 + \frac{\varepsilon}{E_s}}\,\theta(W_s - E_s - \varepsilon)$$

$$N_d(\varepsilon) = N_d(0)\sqrt{1 + \frac{\varepsilon}{E_d}}\,\theta(W_d - E_d - \varepsilon)$$

(17.23)

where $W_s/2E_s$, $W_d/2E_d$ are both assumed to be of order 1 and $N_s(0)$, $N_d(0)$ are given by the respective expression (14.14) with $D = 3$. This band model is depicted in Fig. 3.4.

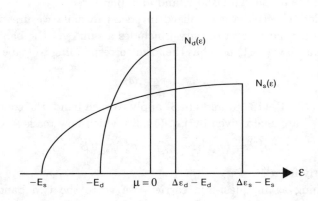

Fig. 3.4: Densities of states (17.23) for the standard two-band model (17.22).

In this *standard two-band model* the states $\psi_{s\alpha}$ and $\psi_{d\alpha}$ in Eq. (17.21) belong to the same real Hilbert space and differ only in their normalization (14.30), provided that we assume *low temperatures* in the sense that $\beta E_s > \beta E_d \gg 1$. Then the inversion formula (14.34) holds for each band,

$$\rho_{\alpha\gamma}^{(s)} \equiv (\sigma^{(s)})_{\alpha\gamma}^{-1} \cong e^{-2}\vartheta_s^{-2}\langle\psi_{s\alpha}|(\mathcal{B}\psi)_{s\gamma}\rangle$$

(17.24)

and similarly for $\rho^{(d)}$. Remembering the definitions (17.19) of $\mathcal{B}$ and (14.30) of ϑ_s and making use of the fact that, according to (17.6) and (17.17), Γ_o is invariant, respectively, under the permutations $kp \leftrightarrow k'p'$ and $skk' \leftrightarrow dpp'$, we may symmetrize the bracket in (17.19) with respect to both permutations. The result is again a semi-positive-definite quadratic form for the s-band resistivity tensor

$$\rho_{\alpha\gamma}^{(s)} \cong \left(\frac{2m_s}{\beta e n_s}\right)^2 \frac{\beta}{V} \sum_{\mathbf{kk'pp'}} \Gamma_o(\mathbf{kp}; \mathbf{k'p'})$$

$$\times \{\psi_s(\mathbf{k}) + \psi_d(\mathbf{p}) - \psi_s(\mathbf{k'}) - \psi_d(\mathbf{p'})\}_\alpha$$

$$\times \{\psi_s(\mathbf{k}) + \psi_d(\mathbf{p}) - \psi_s(\mathbf{k'}) - \psi_d(\mathbf{p'})\}_\gamma \qquad (17.25)$$

and similarly for $\rho^{(d)}$. Here the fact that the ψ's are spin-independent was used to sum over all spin indices according to (17.16). Inserting now Eqs. (17.17) and (14.27) with $\varepsilon_{sk} \simeq \varepsilon_{sk'} \simeq \varepsilon_{dp} \simeq \varepsilon_{dp'} \simeq 0$ and $f_o'(0) = -\beta/4$ one finally arrives at the formula

$$\rho_{\alpha\gamma}^{(s)} \cong \frac{2\pi\hbar}{e^2 n_s^2} \left(1 - \frac{m_s}{m_d}\right)^2 \frac{\beta}{V} \sum_{\mathbf{q}i} |J_i(\mathbf{q})|^2 q_\alpha q_\gamma$$

$$\times \frac{1}{V^2} \sum_{\mathbf{kp}} f_o(\varepsilon_{sk})[1 - f_o(\varepsilon_{s,\mathbf{k-q}})] f_o(\varepsilon_{dp})[1 - f_o(\varepsilon_{d,\mathbf{p+q}})]$$

$$\times \delta(\varepsilon_{sk} - \varepsilon_{s,\mathbf{k-q}} + \varepsilon_{dp} - \varepsilon_{d,\mathbf{p+q}}) \qquad (17.26)$$

and similarly for $\rho^{(d)}$.

Introducing the densities of state according to (10.4) and writing $\mathbf{k} \cdot \mathbf{q} = kq\zeta$, $\mathbf{p} \cdot \mathbf{q} = pq\zeta'$, Eq. (17.26) becomes for the band model (17.22)

$$\rho_{\alpha\gamma}^{(s)} \cong \frac{\pi\hbar}{2e^2 n_s^2} \left(1 - \frac{m_s}{m_d}\right)^2 N_s(0) N_d(0) \beta \int \frac{d^3q}{(2\pi)^3} \sum_i |J_i(\mathbf{q})|^2 q_\alpha q_\gamma I(q) \quad (17.27)$$

where

$$I(q) \equiv \int_{-1}^{+1} d\zeta \int_{-1}^{+1} d\zeta' \int d\varepsilon f_o(\varepsilon)[1 - f_o(\varepsilon + \Delta)] \frac{N_s(\varepsilon)}{N_s(0)}$$

$$\times \int d\varepsilon' f_o(\varepsilon')[1 - f_o(\varepsilon' + \Delta')] \frac{N_d(\varepsilon')}{N_d(0)} \delta(\Delta + \Delta') , \quad (17.28)$$

$$\Delta \equiv \varepsilon_{s,\mathbf{k-q}} - \varepsilon_{sk} = \hbar q \left(\frac{\hbar q}{2m_s} - \zeta v_s \sqrt{1 + \frac{\varepsilon}{E_s}}\right)$$

$$\Delta' \equiv \varepsilon_{d,\mathbf{p+q}} - \varepsilon_{dp} = \hbar q \left(\frac{\hbar q}{2m_d} + \zeta' v_d \sqrt{1 + \frac{\varepsilon'}{E_d}}\right) \qquad (17.29)$$

and similarly for $\rho^{(d)}$. Note that ε, ε', Δ and Δ' are all of order β^{-1}. It is interesting to note also that, had we used the resistivity formula (14.36) instead of (17.25) we would have gotten, up to a factor 2, Eq. (17.27) with $(1 - m_s/m_d)^2$ replaced by 1. This shows that formula (14.36) is only valid in the limit $m_d \to \infty$, i.e. for quasi-elastic scatterings or emission and absorption processes as in the case of impurities or phonons. The discrepancy of a factor 2 comes from the symmetrization $k \leftrightarrow p$ in Eq. (17.25) which is inconsistent with the limit $m_d \to \infty$.

Setting $\varepsilon = 0$ and $\varepsilon' = 0$ in N_s, Δ and in N_d, Δ', respectively, the ε- and ε'-integrations in Eq. (17.28) may be performed explicitly since application of the identity

$$\frac{a}{a + e^x} \frac{e^x}{b + e^x} = \frac{a}{b - a} \frac{d}{dx} \log \frac{a + e^x}{b + e^x} \qquad (17.30)$$

with $a = 1$, $b = \exp(-\beta\Delta)$ immediately gives

$$\int_{-\infty}^{+\infty} f_o(\varepsilon)[1 - f_o(\varepsilon + \Delta)]d\varepsilon = \frac{\Delta}{1 - e^{-\beta\Delta}} . \qquad (17.31)$$

Equation (17.28) then becomes

$$I(q) \cong \frac{1}{4\lambda\beta^2 E_s} \int_{-1}^{+1} d\zeta \int_{-1}^{+1} d\zeta' \delta(\mu - \lambda - \mu(\zeta' + \lambda)) \frac{x^2 e^x}{(e^x - 1)^2} \qquad (17.32)$$

where $\lambda \equiv q/2k_s \cong q/2k_d$, $\mu \equiv m_s/m_d \cong v_d/v_s < 1$ and $x \equiv \beta\Delta' = -\beta\Delta$. The δ-function in Eq. (17.32) defines lines of constant λ as shown in Fig. 3.5. It is evident from this figure that the domain of integration implies the limit $\lambda \le 1$. Neutralizing the δ-function by integrating over ζ we are left with the ζ'-integral which, for $\lambda < \lambda_o \equiv (1 - \mu)/(1 + \mu)$, has the limits $\zeta' = \pm 1$ while for $\lambda > \lambda_o$ the upper limit is $\zeta = +1$, i.e. $\zeta' = (1 - \lambda)/\mu - \lambda$. Using as integration variable $x \cong 4\lambda(\lambda + \zeta')\beta E_d$ which follows from the second Eq. (17.29) and applying definition (16.9) with $J_2(-z) = -J_2(z)$, Eq. (17.32) becomes

$$I(q) \cong \frac{1}{16\lambda^2\beta^3 E_s E_d} \begin{cases} J_2(x_-) + J_2(x_+); & \lambda < \lambda_o , \\ J_2(x_-) + J_2(x_-/\mu); & \lambda > \lambda_o \end{cases} \qquad (17.33)$$

where $x_\pm \equiv 4\lambda(1 \pm \lambda)\beta E_d$.

For simplicity we assume the s-d interaction to be of *short range* so that $J_i(\mathbf{q}) \simeq const$, insertion of Eq. (17.33) into (17.27) then yields

$$\rho_{\alpha\gamma}^{(s)} \cong \frac{\hbar(1 - \mu)^2 k_s^5}{6\pi e^2 n_s^2} N_s(0)N_d(0) \sum_i |J_i|^2 R(T)\delta_{\alpha\gamma} \qquad (17.34)$$

with

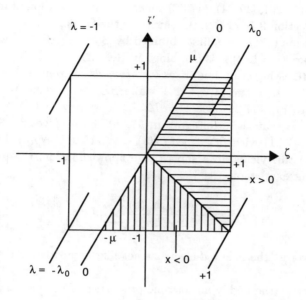

Fig. 3.5: Domain of integration of Eq. (17.32) and lines of constant λ.

$$R(T) \equiv \frac{(k_B T)^2}{E_s E_d} \int_0^1 d\lambda \lambda^2 \left\{ J_2(x_-) + \theta(\lambda_o - \lambda) J_2(x_+) + \theta(\lambda - \lambda_o) J_2(x_-) \right\}$$

(17.35)

and similarly for $\rho^{(d)}$. Since $\beta E_d \gg 1$, $R(T)$ will be dominated by the lowest power in T. Now $x_\pm \gg 1$, except in a narrow interval $\delta\lambda \simeq (\beta E_d)^{-1}$ at $\lambda = 0$ and, for x_- also at $\lambda = 1$. But $J_2(z)$ is a monotonous function between the limiting values

$$J_2(z) \cong \begin{cases} \pi^2/3; & z \gg 1 , \\ z; & z \ll 1 . \end{cases}$$

(17.36)

It is easy, therefore, to estimate the integrals in the $\delta\lambda$-intervals: The one at $\lambda = 0$ contributes a factor $\propto (\beta E_d)^{-3}$ while the one at $\lambda = 1$ contributes a factor $\propto (\beta E_d)^{-1}$. Hence the contribution from these intervals may be neglected and

$$R(T) \cong \frac{2\pi^2}{3} \frac{(k_B T)^2}{E_s E_d} \left[1 + \mathcal{O}(k_B T/E_d) \right] .$$

(17.37)

Equations (17.20), (17.34), (17.37) generalize the result found in the literature (see Section 9.14 of Ref. 9) which, for the Coulomb part ($i = 0$) of the
s-d interaction (17.9), was first obtained by Baber in 1937[33] while for the
spin-flip part ($i = 1, 2$) it is contained in Ref. 21.

An interesting situation arises for *heavy d-electrons*, $\mu \equiv m_s/m_d \ll 1$
so that $\beta E_d \ll 1$ and $\beta W_d \ll 1$ while $\beta E_s \gg 1$ remains valid. Since
according to Eq. (17.29) also $x = \beta\Delta' \ll 1$, we have $f_o(\varepsilon') \cong 1 - f_o(\varepsilon' +
\Delta') \cong 1/2$ so that the ε'-integral in (17.28) simply gives $n_d/8N_d(0)$ and,
using Eq. (17.31) with $\beta\Delta = -x$, $I(q) \cong n_d/16\lambda\beta E_s N_d(0)$. Inserted into
Eq. (17.27) this yields, in a short-range approximation, the temperature-
independent expression

$$\rho_{\alpha\gamma}^{(s)} = \frac{m_s k_s^3 n_d N_s(0)}{12\pi e^2 \hbar n_s^2} \sum_i |J_i|^2 \delta_{\alpha\gamma} . \qquad (17.38)$$

In other words, the heavy d-electrons act like impurities, the scattering is
elastic .

In view of the highly anisotropic properties of the new cuprate superconductors it is also of interest to do the above calculation in 2 dimensions.
Indeed, the surprisingly universal linear resistivity law $\rho_\parallel \propto T^1$ of the resistivity parallel to the CuO_2 planes of these materials[38] suggests that the
exponent $p = 1$ might be the result of a *dimensional cross-over* from the
Baber exponent $p = 2$ in 3 dimensions[39]. This hypothesis supposes that
a two-band model gives an appropriate description of the transport phenomena in the cuprates which, however, is not certain[38]. But in addition,
a careful evaluation of the above two-band model in 2 dimensions leads to
a resistivity $\rho \propto T^2 \ln T$ and not to the experimental linear law as will now
be shown.

To be valid in dimension $D = 2$, Eqs. (17.23), (17.27) and (17.28)
must be modified such that $N_s(\varepsilon) = N_s(0)$, $N_d(\varepsilon) = N_d(0)$, $(2\pi)^{-3}d^3q$ is
replaced by $(2\pi)^{-2}d^2q$ while $d\zeta$ and $d\zeta'$ are replaced by $2d\zeta/\pi\sqrt{1-\zeta^2}$ and
$2d\zeta'/\pi\sqrt{1-\zeta'^2}$, respectively[39]. Then Eq. (17.32) becomes

$$I_2(q) \cong \frac{1}{\pi^2\lambda\beta^2 E_s} \int_{-1}^{+1} \frac{d\zeta}{\sqrt{1-\zeta^2}} \int_{-1}^{+1} \frac{d\zeta'}{\sqrt{1-\zeta'}}$$
$$\times \delta\big(\zeta - \lambda - \mu(\zeta' + \lambda)\big)\frac{x^2 e^x}{(e^x - 1)^2} \qquad (17.39)$$

and insertion into the modified Eq. (17.27) yields, in the short-range approximation used before,

$$\rho_{\alpha\gamma}^{(s)} \cong \frac{2\hbar k_s^4}{e^2 n_s^2}(1 - \mu)^2 N_s N_d \sum_i |J_i|^2 R_2(T)\delta_{\alpha\gamma} ; \quad D = 2 \qquad (17.40)$$

with

$$R_2(T) \equiv \frac{1}{\pi^2 \beta E_s} \int_0^1 d\lambda \lambda^2 \int_{-1}^{+1} \frac{d\zeta'}{\sqrt{1 - [\lambda + \mu(\zeta' + \lambda)]^2}\sqrt{1 - \zeta'^2}}$$
$$\times \left\{ \theta(\lambda_o - \lambda) + \theta(\lambda - \lambda_o)\theta\left(\frac{1 - \lambda}{\mu} - \lambda - \zeta'\right) \right\} \frac{x^2 e^x}{(e^x - 1)^2} \qquad (17.41)$$

and similarly for $\rho^{(d)}$. Here $x \cong 4\lambda(\lambda + \zeta')\beta E_d$ and the last θ-function is equivalent to $\zeta < 1$.

Without the square-root denominators one recovers a T^2-law as in Eq. (17.37). Hence the modification can only come from the neighborhood of points where these denominators vanish simultaneously, i.e. (see Fig. 3.5) $\zeta = \pm \zeta' = 1$ or $\lambda = \lambda_o$ and $\lambda = 1$ or $x \cong 4\lambda_o(\lambda_o + 1)\beta E_d \gg 1$ and $x = 0$. Since for $x \gg 1$ the integrand in (17.41) vanishes exponentially, the only critical point is $\zeta = -\zeta' = 1$, $\lambda = 1$, $x = 0$. Parametrizing the neighborhood of this point by $\zeta' = -1 + \xi$, $\lambda = 1 - \nu$ and using the estimates (see the appendix of the Dubna paper in Ref. 39 where the error of the result $\rho \propto T$ may be traced to the fact that the fraction of phase space yielding a T^1-contribution is itself $\propto T$)

$$K(z) \equiv \int_0^z \frac{dx}{\sqrt{z - x}} \frac{x^2 e^x}{(e^x - 1)^2} \cong \begin{cases} 2; & z \ll 1 \\ \pi^2/2z; & z \gg 1 \end{cases} \qquad (17.42)$$

one finds (Problem 3.11)

$$R_2(T) \cong \frac{1 + \sqrt{\mu}}{\sqrt{1 + \mu}} \frac{(k_B T)^2}{16 E_s E_d} \left\{ \frac{4}{\pi^2} + \frac{\ln \mu}{1 + \sqrt{\mu}} + \ln \frac{8 E_d}{(1 + \mu)k_B T} \right\} . \qquad (17.43)$$

Such a $T^2 \ln T$ dependence of the resistivity in two dimensions was first derived by Hodges, Smith and Wilkins[40].

To close this section we wish to review the resistivity behavior for *heavy-fermion metals*. These are compounds with lanthanides (rare earths) and actinides which contain $4f$- and $5f$-shells, respectively. Above a characteristic temperature T^* of the order of 10 K the f-electrons are localized and give rise to an atomic magnetic moment $gJ\mu_B$ of spin J, μ_B being the Bohr magneton, and the magnetic susceptibility obeys a Curie law $\chi = n_a g^2 J(J+1)\mu_B^2/3k_B T$ (see Eqs. (29.10), (29.11) below). At $T \simeq T^*$ the exchange interaction (15.21) between the (s-, p-, d-) conduction electrons and the f-electrons gives rise to a Kondo-type quenching of the local moments accompanied by *weak hybridization* (see, e.g. Ref. 41) with the conduction-electron orbitals (note that strong hybridization may give rise to *valence fluctuations*, see e.g. Ref. 42) which delocalize the f-electrons into a narrow band of width $\sim k_B T^*$ at the Fermi energy. Thus the f-electrons become itinerant but with an effective mass m_f which is several 100 times larger than the free-electron mass[42−44].

This heavy mass is the most visible manifestation of heavy-fermion metals. It shows up at $T \ll T^*$ in the specific heat (10.3), $c_V = \gamma T$, with a *Sommerfeld constant* γ which is 2–3 orders of magnitude larger than for simple metals like Al or Pb[42−44]. The connection with the mass m_f follows, in a quasi-free electron approximation, from Eqs. (14.13), (14.14) and is given by

$$\gamma = \frac{m_f k_f k_B^2}{3\hbar^2} \tag{17.44}$$

where k_f is the Fermi wavenumber of the f-band. At the same time the susceptibility due to the quenched moments J^* goes over, in the limit $T \to 0$, into a Pauli-like susceptibility (see Eq. (31.16) below)

$$\chi(0) = g^2 J^*(J^* + 1)\mu_B^2 \frac{m_f k_f}{3\pi^2 \hbar^2} \tag{17.45}$$

which is also considerably enhanced as compared to simple metals[42−44]. However, the quotient of $\chi(0)$ in units of $g^2 J(J + 1)\mu_B^2$ and γ in units of $\pi^2 k_B^2$ called the *Wilson factor*[27] is independent of the mass m_f (and also of k_f) and therefore is expected to be of order 1 which is indeed the case (see Fig. 1 of Ref. 43).

While for an isolated Kondo impurity or a dilute alloy of f-shell atoms the resistivity is characterized by $\rho(T) - \rho(0) > 0$ for $T < T_K$, see Eq. (15.20), a *Kondo lattice* of f-shell atoms has $\rho(T) - \rho(0) < 0$ for $T < T^*$. The resistivity shows a maximum at $T \sim T^*$ which is due to the fact that the scattering at the Kondo atoms is incoherent for $T > T^*$ but becomes coherent below T^*. This coherence means nothing else than that for $T < T^*$ the f-electrons are in extended Bloch states. But this implies that the s-d scattering theory developed in this section applies so that

$$\rho = \rho_o + AT^2 \; ; \quad T \ll T^* \; . \tag{17.46}$$

Since the density of states $N_f(0) \propto m_f \gg m_s$ we conclude from Eqs. (17.34), (17.37) that $A \propto m_f^2$. Therefore the Baber term is also several orders of magnitude larger as compared to transition metals[42-44].

18. Kubo and Mori Formulas. Evaluation for Impurity Scattering

In the case of an applied electric field $\mathbf{E}(\mathbf{r}, t)$, the external Hamiltonian (2.1) for an electron (charge $-e$) is defined, in first quantization, by

$$\mathcal{H}_o + \delta\mathcal{H}_t = \sum_{i=1}^{N} \left\{ \frac{1}{2m} \left(\mathbf{p}_i + \frac{e}{c}\mathbf{A}(\mathbf{r}_i, t) \right)^2 - eU(\mathbf{r}_i, t) \right\} \tag{18.1}$$

and $\mathbf{E} = -\dot{\mathbf{A}}/c - \nabla U$ where the dot is $\partial/\hbar\partial t$, i.e. a true time derivative. Making use of definitions (2.3), (2.4), (2.9), (2.10), the freedom to choose a gauge then gives rise to the two equivalent external Hamiltonians

$$\delta\mathcal{H}_t^A = -\frac{1}{c}V^{-1}\sum_{\mathbf{q}} \tilde{\mathbf{A}}(\mathbf{q}, t) \cdot \mathbf{j}(-\mathbf{q})$$

$$- \frac{e}{2mc^2}V^{-2}\sum_{\mathbf{q}\mathbf{q}'} \tilde{\mathbf{A}}(\mathbf{q}', t) \cdot \tilde{\mathbf{A}}(\mathbf{q}-\mathbf{q}', t)\rho(-\mathbf{q}) \tag{18.2}$$

for the vector potential $\mathbf{A}$ $(U = 0)$ and

$$\delta\mathcal{H}_t^U = +V^{-1}\sum_{\mathbf{q}} \tilde{U}(\mathbf{q}, t)\rho(-\mathbf{q}) \tag{18.3}$$

for the scalar potential U $(\mathbf{A} = 0)$ where, as in (13.10), (13.31), $\rho = -en$ and $\mathbf{j} = -e\mathbf{j}_n$. Going over to the second-quantized expressions in plane-wave representation, (3.22) and (3.23), and using $\varepsilon_k = \hbar^2 k^2/2m$ for the energy, it is easy to verify the operator form of the charge conservation (13.30),

$$\dot{\rho}(\mathbf{q}) + i\mathbf{q} \cdot \mathbf{j}(\mathbf{q}) = 0 \tag{18.4}$$

where, as before, $\dot{\rho} \equiv \partial\rho/\hbar\partial t$.

We first calculate the linear response to (18.3) which is simpler[14]. To do this we write the general response function (7.12) in a more appropriate form. Using the Kubo identity (15.8) and the cyclicity of the trace it is easy to see that

$$\chi_{ab}(\mathbf{q}; t) = -\hbar\beta\theta(t)V^{-1}\langle \dot{a}(\mathbf{q}, t)b(-\mathbf{q}, 0)\rangle \tag{18.5}$$

where definition (15.9) is used. Now the current response (7.11) to the perturbation (18.3) becomes

$$\delta\langle j_\alpha(\mathbf{q})\rangle_t^U = \hbar\beta V^{-1}\int_{-\infty}^{t} dt' \langle \dot{j}_\alpha(\mathbf{q}, t-t')\rho(-\mathbf{q}, 0)\rangle\tilde{U}(\mathbf{q}, t') . \tag{18.6}$$

Since $\langle \dot{A}B \rangle = -\langle A\dot{B} \rangle$, we may shift the time derivative to ρ and apply (18.4) so that, after a change of integration variable,

$$\delta\langle j_\alpha(\mathbf{q})\rangle_t^U = \hbar\beta V^{-1} \int_0^\infty dt'' \langle \hat{j}_\alpha(\mathbf{q}, t'')j_\gamma(-\mathbf{q}, 0)\rangle \tilde{E}_\gamma(\mathbf{q}, t - t'') \qquad (18.7)$$

where $\tilde{\mathbf{E}} = -i\mathbf{q}\tilde{U}$. Equation (18.7) is represented in Fig. 3.6 below. The response to a time-periodic electric field $\tilde{\mathbf{E}}(\mathbf{q}, t) = \vec{\mathcal{E}}(\mathbf{q}, \omega)e^{-i\omega t}$ then has the general form (14.1), namely

$$i_\alpha(\mathbf{q}, \omega) = \sigma_{\alpha\gamma}(\mathbf{q}, \omega)\mathcal{E}_\gamma(\mathbf{q}, \omega) , \qquad (18.8)$$

where $\delta\langle \mathbf{j}(\mathbf{q})\rangle_t^U = \mathbf{i}(\mathbf{q}, \omega)e^{-i\omega t}$ and the conductivity tensor is expressed as a *current-current correlation function*,

$$\sigma_{\alpha\gamma}(\mathbf{q}, \omega) \equiv \hbar\beta V^{-1} \int_0^\infty dt\, e^{i\omega t} \langle \hat{j}_\alpha(\mathbf{q}, t)j_\gamma(-\mathbf{q}, 0)\rangle \qquad (18.9)$$

valid for $\text{Im}\,\omega > 0$. Eq. (18.9) is often called *Green-Kubo formula*[14] or *Kubo-Greenwood formula*[5] (see Ref. 45). The original *Kubo formula*[17] is the $\mathbf{q} = 0$ version of (18.9).

Applying a potential $U(t) = \mathcal{U}(\omega)e^{-i\omega t}$ in the direction $\mathbf{e}$ of a homogenous conductor of length L parallel to $\mathbf{e}$ and cross-sectional area A (see Fig. 3.1, Section 14), the $\mathbf{q} = 0$ components of the field and of the current response are, respectively,

$$\tilde{\mathbf{E}}(0, t) = \int_V d^3r\, \mathbf{E}(\mathbf{r}, t) = AU(t)\mathbf{e} \qquad (18.10)$$

and

$$\delta\langle \mathbf{j}(0)\rangle_t^U = \int_V d^3r\, \delta\langle \tilde{\mathbf{j}}(\mathbf{r})\rangle_t^U = LJ(t)\mathbf{e} \qquad (18.11)$$

where $J(t)$ is the total current across the sample. The $\mathbf{q} = 0$ limit of Eq. (18.8) then reads

$$J(t) = G(\omega)\mathcal{U}(\omega)e^{-i\omega t} \qquad (18.12)$$

where

$$G(\omega) \equiv \frac{A}{L}e_\alpha \sigma_{\alpha\gamma}(0, \omega)e_\gamma \qquad (18.13)$$

is the conductance (14.12).

We now consider the linear response to the vector potential Hamiltonian (18.2). In linear approximation of $\tilde{\mathbf{A}}(\mathbf{q}, t)$ we have from (7.11), (7.12) and (18.5)

$$\delta\langle j_\alpha(\mathbf{q})\rangle_t^A = -\hbar\beta V^{-1} \int_{-\infty}^{t} dt' \langle \dot{\hat{j}}_\alpha(\mathbf{q}, t-t') j_\gamma(-\mathbf{q}, 0)\rangle \frac{1}{c}\tilde{A}_\gamma(\mathbf{q}, t') \ . \quad (18.14)$$

Shifting the time-derivative to t' and making a partial integration this becomes

$$\delta\langle j_\alpha(\mathbf{q})\rangle_t^A = \beta V^{-1}\langle \hat{j}_\alpha(\mathbf{q}) j_\gamma(-\mathbf{q})\rangle \frac{1}{c}\tilde{A}_\gamma(\mathbf{q}, t)$$

$$+ \hbar\beta V^{-1} \int_0^\infty dt'' \langle \hat{j}_\alpha(\mathbf{q}, t'') j_\gamma(-\mathbf{q}, 0)\rangle \tilde{E}_\gamma(\mathbf{q}, t-t'')$$

$$(18.15)$$

where $\tilde{\mathbf{E}} = \dot{\tilde{\mathbf{A}}}/c$. This expression differs from (18.7) by the first term which may be transformed by using the following identity[14] (Problem 3.12)

$$\beta V^{-1}\langle \hat{j}_\alpha(\mathbf{q}) j_\gamma(-\mathbf{q})\rangle = \frac{e^2 n}{m}\delta_{\alpha\gamma} \ ; \quad (18.16)$$

hence

$$\delta\langle \hat{\mathbf{j}}(\mathbf{q})\rangle_t^U = \delta\langle \mathbf{j}(\mathbf{q}, t)\rangle_t^A - \frac{e^2 n}{mc}\tilde{\mathbf{A}}(\mathbf{q}, t) \ . \quad (18.17)$$

The interpretation of this result becomes evident if we define the *total current density* by substituting for the velocity $\mathbf{v}_j = \mathbf{p}_j/m$ in Eq. (2.10) the kinematical velocity $\mathbf{v}_{jt} = [\mathbf{p}_j + (e/c)\mathbf{A}(\mathbf{r}_j, t)]/m$ which occurs in (18.1). Thus, after Fourier transformation (2.4) of $\mathbf{A}(\mathbf{r}_j, t)$,

$$\mathbf{J}_t(\mathbf{q}) = \mathbf{j}(\mathbf{q}) - \frac{e^2}{mcV}\sum_{\mathbf{q}'} \tilde{\mathbf{A}}(\mathbf{q}', t) n(\mathbf{q}-\mathbf{q}') \ . \quad (18.18)$$

Taking the total A-average,

$$\langle \mathbf{J}_t(\mathbf{q})\rangle_t^A \equiv \langle \mathbf{J}_t(\mathbf{q})\rangle + \delta\langle \mathbf{J}_t(\mathbf{q})\rangle_t^A \ , \quad (18.19)$$

of (18.18), we obtain to lowest order, by comparison with (18.17) and noting that the equilibrium average $\langle \mathbf{j}(\mathbf{q})\rangle = 0$,

$$\langle \mathbf{J}_t(\mathbf{q})\rangle_t^A = \delta\langle \mathbf{j}(\mathbf{q})\rangle_t^U \ . \quad (18.20)$$

This proves the equivalence between the A- and U-gauges.

In the spirit of Landauer's approach to electric conduction discussed in Section 14, we may consider a conductor with zero applied field. In this case the thermal voltage fluctuations give rise to a current via a Langevin equation in which the resistivity ρ appears as linear response. This approach is due to Mori who finds[46]

$$\rho(\mathbf{q}, \omega) = \frac{1}{\hbar e^2 n}[\phi(\mathbf{q}, \omega) - i\omega m] \quad (18.21)$$

(note that we have set $\omega_o = 0$ in Eq. (5.4) of Ref. 46) where

$$\phi(\mathbf{q}, \omega) = \frac{\beta\hbar^2 m^2}{e^2 nV} \int_0^\infty dt e^{i\omega t} \langle (e^{iL_1 t}\hat{\mathbf{j}}(\mathbf{q}))^{\hat{}} \otimes \dot{\mathbf{j}}(-\mathbf{q}) \rangle \tag{18.22}$$

is a correlation function connecting the components of the Fourier-tranformed *force density* $(m/e)\dot{\mathbf{j}}$. In this formalism the fundamental complication is hidden in the time evolution $\exp(iL_1 t)$. Indeed, the Liouvillean L_1 applied to an operator f is defined by

$$\frac{i}{\hbar} L_1 f = (1 - P_o)\dot{f} \tag{18.23}$$

where the projector P_o acts as follows:

$$P_o f = \frac{\beta m}{e^2 nV} \langle \hat{f}\dot{\mathbf{j}}(-\mathbf{q}) \rangle \cdot \dot{\mathbf{j}}(\mathbf{q}) \ . \tag{18.24}$$

Thanks to this complication, expression (18.21) is in fact the inverse of the Kubo formula (18.9), as Mori shows by explicit construction[46].

Since "there has been unfortunate confusion in the literature regarding these [inversion] formulas" (Ref. 14, Section 4.5.4), let us have a closer look at the problem. For this purpose we introduce slightly generalized versions of the quantities $N(\epsilon)$ and $D(\epsilon)$ defined by Huberman and Chester[47]

$$N \equiv \frac{\hbar^2 m^2 \beta}{e^2 nV} \int_0^\infty dt e^{i\omega t} \langle \dot{\mathbf{j}}(\mathbf{q}, t) \otimes \dot{\mathbf{j}}(-\mathbf{q}, 0) \rangle \ , \tag{18.25}$$

and

$$D \equiv 1 + \frac{\hbar m\beta}{e^2 nV} \int_0^\infty dt e^{i\omega t} \langle \dot{\mathbf{j}}(\mathbf{q}, t) \otimes \mathbf{j}(-\mathbf{q}, 0) \rangle \ . \tag{18.26}$$

By simple partial integrations, making use of the fact that $\langle \dot{j}_\alpha(\mathbf{q}) j_\gamma(-\mathbf{q}) \rangle = 0$, one easily derives the identities

$$D = -i\frac{\omega m}{\hbar e^2 n}\sigma \ ; \quad N = -i\omega m\left(1 + i\frac{\omega m}{\hbar e^2 n}\sigma\right) \tag{18.27}$$

which, for $-i\omega \to \epsilon$ and after angular averaging, are Eqs. (2.42), (2.43) of Ref. 47. One finds with the help of the identities (18.27) that[14]

$$\rho = \frac{1}{\hbar e^2 n}\left(\frac{N}{D} - i\omega m\right) \tag{18.28}$$

and comparison with (18.21) shows that $\phi = ND^{-1}$. But ϕ and N, Eqs. (18.22) and (18.25), differ only in their time evolution, $\exp(iL_1 t)$ and $\exp(iLt)$, respectively, the correction being just D^{-1}.

In order to make the above formulas more readily accessible to many-body theoretic calculations, let us express the Kubo formula (18.9) in terms of electron Green's functions. To this end we introduce the "real part" of (18.9), defined as

$$\operatorname{Re}\sigma_{\alpha\gamma}(\mathbf{q},\,\omega) \equiv \frac{1}{2}\left(\sigma_{\alpha\gamma}(\mathbf{q},\,\omega) + \sigma_{\gamma\alpha}^{*}(\mathbf{q},\,\omega^{*})\right)$$

$$= \frac{\hbar\beta}{2V}\int_{-\infty}^{+\infty}\langle\hat{j}_{\alpha}(\mathbf{q},\,t)j_{\gamma}(-\mathbf{q},\,0)\rangle \qquad (18.29)$$

Here use was made of the identity $\langle A(\hat{B})^{+}\rangle = \langle \hat{A}B^{+}\rangle$ which follows directly from definition (15.9). Inserting for $\mathbf{j}(\mathbf{q}) = -e\mathbf{j}_n(\mathbf{q})$ from (3.23) one obtains, suppressing spin indices,

$$\operatorname{Re}\sigma_{\alpha\gamma}(\mathbf{q},\,\omega) = \frac{\hbar^{3}e^{2}}{m^{2}V}\sum_{\mathbf{k}\mathbf{k}'}k_{\alpha}k_{\gamma}'\int_{-\infty}^{+\infty}dt\,e^{i\omega t}$$

$$\times \int_{0}^{\beta}d\tau\,G_{2}(\mathbf{k}_{+},\,t-i\tau,\,\mathbf{k}_{+}',0;\,\mathbf{k}_{-}',\,0,\,\mathbf{k}_{-},\,t-i\tau)$$

$$(18.30)$$

where a factor 2 comes from the spin sum,

$$G_{2}(\mathbf{k}_{+}t_{+};\,\mathbf{k}_{+}'t_{+}';\,\mathbf{k}_{-}'t_{-}',\,\mathbf{k}_{-}t_{-}) \equiv -\langle T\{a_{\mathbf{k}_{+}}(t_{+})a_{\mathbf{k}_{+}'}^{+}(t_{+}')a_{\mathbf{k}_{-}'}(t_{-}')a_{\mathbf{k}_{-}}^{+}(t_{-})\}\rangle$$

$$(18.31)$$

is a *two-particle Green's function* of the form (10.30) and $\mathbf{k}_{\pm} \equiv \mathbf{k} \pm \mathbf{q}/2$, $\mathbf{k}_{\pm}' \equiv \mathbf{k}' \pm \mathbf{q}/2$. Expressed in terms of G_2, Eq. (18.7) has the two contributions shown in Fig. 3.6 which come from the crossed particle-hole channels.

G_2 may now be evaluated with the methods developed in Chapter 2. Here we choose a factorization approximation of the type used in Problem 3.4 whereby the contribution from Fig. 3.6 (b) is eliminated:

$$G_{2}(\mathbf{k}_{+},\,t-i\tau,\,\mathbf{k}_{+}'0;\,\mathbf{k}_{-}'0,\,\mathbf{k}_{-},\,t-i\tau) \cong G(\mathbf{k}_{+}\mathbf{k}_{+}';\,t-i\tau)G(\mathbf{k}_{-}'\mathbf{k}_{-};\,-t+i\tau)$$

$$(18.32)$$

where definition (7.17) was used with the notation $G_{a_{\mathbf{k}}a_{\mathbf{k}'}^{+}}(t) \equiv G(\mathbf{k}\mathbf{k}';\,t)$. These Green's functions may be considered to be dressed. If, in particular, the interaction is due to impurity atoms, Eq. (15.12), then translation invariance is broken and (7.13) is invalid which means that G depends indeed on both momenta. In this case an impurity-average has to be applied to (18.32). Going over to Fourier-transformed propagators (7.19), insertion of (18.32) into (18.30) and use of (7.43) leads to the formula

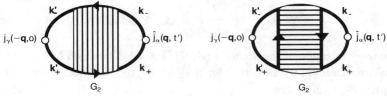

Fig. 3.6: Linear response in U-gauge, Eq. (18.7), expressed as two-electron Green's function (18.31).

$$\mathrm{Re}\,\sigma_{\alpha\gamma}(\mathbf{q},\,\omega) \cong \frac{1}{\omega(1+n_o(\omega))} \frac{\hbar^3 e^2}{m^2 V} \sum_{\mathbf{k}\mathbf{k}'} k_\alpha k'_\gamma$$

$$\times \int_{-\infty}^{+\infty} \frac{d\omega'}{2\pi} \langle \tilde{G}(\mathbf{k}_+ \mathbf{k}'_+, \,\omega_+) \tilde{G}(\mathbf{k}'_- \mathbf{k}_-, \,\omega_-) \rangle_{\mathrm{imp}}$$

$$(18.33)$$

where $\omega_\pm \equiv \omega' \pm \omega/2$. In the low-temperature limit where $n_o(\omega) \ll 1$, (18.32) is the same as Eqs. (39.10), (39.11) of Abrikosov *et al.*[48], since their $i(Q_{\alpha\gamma} - (ne^2/m)\delta_{\alpha\gamma}) = \omega\,\mathrm{Re}\sigma_{\alpha\gamma}$.

In what follows we consider the mentioned case of impurity scattering in the limit $T \to 0$. Since external electromagnetic fields are always of long wavelength as compared to the lattice spacing, we may also take the limit $\mathbf{q} \to 0$. In calculating the effect of impurities in Eq. (18.33) we first have to dress the propagators. Generalizing Eq. (11.5) to translation non-invariance we may define the selfenergy Σ by the integral equation

$$\tilde{G}(\mathbf{k}\mathbf{k}';\,\omega) = \tilde{G}^o(\mathbf{k};\omega)\{\delta_{\mathbf{k}\mathbf{k}'} + \sum_{\mathbf{q}} \Sigma(\mathbf{q})G(\mathbf{k}+\mathbf{q},\,\mathbf{k}';\,\omega)\}\,. \qquad (18.34)$$

For the free propagator $\tilde{G}^o$ which of course satisfies (7.13), one easily finds from definitions (7.17), (7.19) and (7.40)

$$\tilde{G}^o(\mathbf{k};\,\omega) = \frac{1 - f_o(\varepsilon_\mathbf{k})}{\omega - \varepsilon_\mathbf{k} + i\epsilon} + \frac{f_o(\varepsilon_\mathbf{k})}{\omega - \varepsilon_\mathbf{k} - i\epsilon} \qquad (18.35)$$

where $\epsilon = 0^+$ and we have set the Fermi energy $= 0$. Since the impurity potential U in (15.12) is a c-number, Σ is independent of frequency and hence may easily be calculated with the help of imaginary-time Green's functions. Making use of Eq. (11.17) with $N = 1$, generalized to translation non-invariance , and of Wick's theorem (9.11) one easily finds

$$\Sigma(\mathbf{q}) = \frac{1}{V} \sum_j \tilde{U}(-\mathbf{q}) e^{i\mathbf{q} \cdot \mathbf{R}_j^o} . \qquad (18.36)$$

We now consider successive iterations and impurity-averaging of Eq. (18.34). Assuming U to be small, only the first few iterations will be needed. Making use of Eq. (15.18) in the form

$$\frac{1}{V} \sum_j \langle e^{i\mathbf{q} \cdot \mathbf{R}_j^o} \rangle_{\text{imp}} = \frac{1}{V} \sum_j \delta_{\mathbf{q},0} = \frac{x}{v} \delta_{\mathbf{q},0} , \qquad (18.37)$$

the nth-order-correlated selfenergies Σ_n are as shown in Fig. 3.7; in particular

$$\sum_{\mathbf{q}} \langle \Sigma(\mathbf{q}) \tilde{G}^o(\mathbf{k} + \mathbf{q}; \omega) \Sigma(\mathbf{k}' - \mathbf{k} - \mathbf{q}) \rangle_{\text{imp}}$$

$$= \frac{x}{v} \frac{1}{V} \sum_{\mathbf{q}} |\tilde{U}(\mathbf{q})|^2 \tilde{G}^o(\mathbf{k} + \mathbf{q}; \omega) \delta_{\mathbf{k}\mathbf{k}'} \equiv \Sigma_2(\mathbf{k}; \omega) \delta_{\mathbf{k}\mathbf{k}'} . \qquad (18.38)$$

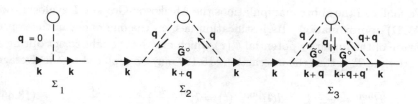

Fig. 3.7: The 1st-, 2nd- and 3rd-order-correlated selfenergies Σ_1, Σ_2 and Σ_3 due to the interaction of Fig. 3.2(a). A circle means impurity-averaging according to Eq. (18.37).

Thus, impurity-averaging restores translation invariance ,

$$\langle \tilde{G}(\mathbf{k}\mathbf{k}'; \omega) \rangle_{\text{imp}} = \bar{G}(\mathbf{k}; \omega) \delta_{\mathbf{k}\mathbf{k}'} \qquad (18.39)$$

where

$$\bar{G}(\mathbf{k}; \omega) = \tilde{G}^o(\mathbf{k}; \omega) \{ 1 + [\Sigma_1 + \Sigma_2(\mathbf{k}; \omega) + \cdots] \tilde{G}^o(\mathbf{k}; \omega) + \cdots \}$$
$$\cong \tilde{G}^o(\mathbf{k}; \omega) \{ 1 - [\Sigma_1 + \Sigma_2(\mathbf{k}; \omega)] \tilde{G}^o(\mathbf{k}; \omega) \}^{-1} . \qquad (18.40)$$

Here $\Sigma_1 = (x/v)\tilde{U}(0)$ and, in the second step, the complete iteration of the retained term has been performed.

Since at $T = 0$ we may write (18.35) as

$$\tilde{G}^o(\mathbf{k}; \omega) = \{\omega - \varepsilon_{\mathbf{k}} + i\epsilon \,\text{sgn}(\varepsilon_{\mathbf{k}})\}^{-1} \tag{18.41}$$

where $\text{sgn}(x) \equiv x/|x|$, Eq. (18.38) is easily evaluated if we assume isotropy of $\varepsilon_{\mathbf{k}}$ and of $U(\mathbf{r})$. Using the density of states (10.4) and writing (18.41) in the form (11.30) one finds, supposing that only states $\mathbf{k}$ near the Fermi energy $\mu = 0$ contribute,

$$\Sigma_2(\mathbf{k}; \omega) = \omega_o\left\{\frac{1}{\pi} \int d\varepsilon \frac{P}{\omega - \varepsilon} - i\,\text{sgn}(\omega)\right\} \tag{18.42}$$

where

$$\omega_o \equiv \frac{1}{4}\frac{x}{v}N(\mu) \int d\Omega_{\mathbf{k}'} \left|\tilde{U}(\mathbf{k} - \mathbf{k}')\right|^2 \Big|_{\varepsilon_{\mathbf{k}'} = \varepsilon_{\mathbf{k}}} \tag{18.43}$$

(see Eq. (39.6) of Ref. 48). Assuming further that $|\omega| \ll \omega_o$ so that $\text{Re}\,\Sigma_2$ is constant and considering Σ_1 and $\text{Re}\,\Sigma_2$ to be absorbed into a renormalized chemical potential $\mu^* = 0$, Eq. (18.39) finally becomes Eq. (39.8) of Ref. 48, namely

$$\bar{G}(\mathbf{k}; \omega) = \{\omega - \varepsilon_{\mathbf{k}} + i\omega_o\text{sgn}(\omega)\}^{-1} . \tag{18.44}$$

We shall call the above manipulations the Abrikosov-Gor'kov-Dzyaloshinskii (AGD) approximation . Its justification lies, on the one hand, in the static nature of the impurity potential $U(\mathbf{r})$ which may be thought to act during a finite time T so that the interaction (15.12) may be replaced by the average

$$\bar{\mathcal{H}}^{\text{imp}} = \frac{1}{2T} \int_{-T}^{+T} dt \mathcal{H}^{\text{imp}\,o}(t) = \mathcal{H}^{\text{imp}}\frac{\sin(\varepsilon_{\mathbf{k}'} - \varepsilon_{\mathbf{k}})T}{(\varepsilon_{\mathbf{k}'} - \varepsilon_{\mathbf{k}})T} . \tag{18.45}$$

On the other hand, the condition $|\omega| \ll \omega_o$ which restricts states $\mathbf{k}$ close to the Fermi surface, $|\varepsilon_{\mathbf{k}}| < \omega_o$, will be justified below.

Coming back to Eq. (18.33) we now have to take into account the correlations between the two one-particle Green's functions. A systematic way to do this is to define a *particle-hole-irreducible four-point vertex* Λ (see Section 12) by an integral equation for the impurity average,

$$L(\mathbf{k}_+\mathbf{k}'_+\omega_+; \mathbf{k}'_-\mathbf{k}_-\omega_-) \equiv \langle\tilde{G}(\mathbf{k}_+\mathbf{k}'_+; \omega_+)\tilde{G}(\mathbf{k}'_-\mathbf{k}_-; \omega_-)\rangle_{\text{imp}} \tag{18.46}$$

which reads (note the *rule* introduced in Sections 10 and 11 that there is one summation per closed loop)

$$L(\mathbf{k}_+\mathbf{k}'_+\omega_+; \mathbf{k}'_-\mathbf{k}_-\omega_-) = \bar{G}(\mathbf{k}'_+; \omega)\bar{G}(\mathbf{k}'_-; \omega_-)\{\delta_{\mathbf{k}'_+\mathbf{k}_+}\delta_{\mathbf{k}'_-\mathbf{k}_-}$$
$$+ \sum_{\mathbf{q}'} \Lambda(\mathbf{k}_+ - \mathbf{q}', \mathbf{k}'_+\omega_+; \mathbf{k}'_-, \mathbf{k}_- - \mathbf{q}', \omega_-)$$
$$\times L(\mathbf{k}_+, \mathbf{k}_+ - \mathbf{q}', \omega_+; \mathbf{k}_- - \mathbf{q}', \mathbf{k}_-\omega_-)\}$$

$$(18.47)$$

where $\mathbf{k}'_+ - \mathbf{k}'_- = \mathbf{k}_+ - \mathbf{k}_- = \mathbf{q}$ and which is shown in Fig. 3.8. The mentioned irreducibility means that Λ does not fall into disconnected pieces by dissecting, with a single cut in the plane, simultaneously a particle and a hole line.

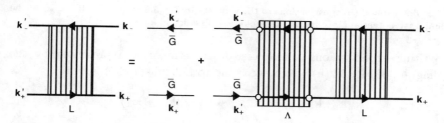

Fig. 3.8: Integral equation (18.47) defining the static particle-hole-irreducible four-point vertex Λ. $\mathbf{k}'_+ - \mathbf{k}'_- = \mathbf{k}_+ - \mathbf{k}_- = \mathbf{q}$.

Contributions to Λ are shown in Fig. 3.2(b) and in Fig. 3.9(a), (b) while Fig. 3.9(c) is particle-hole reducible. The dashed impurity lines have the particularity that they do not carry energy since the impurity potential is a c-number. This implies that the (lower) particle and the (upper) hole lines carry the same frequency labels ω_+ and ω_-, respectively, across the entire diagram which, therefore, are not indicated in Figs. 3.8 and 3.9.

Iterating Eq. (18.47) and inserting into (18.33) with $\mathbf{k}_\pm - \mathbf{k} = \mathbf{k}'_\pm - \mathbf{k}' = \pm\mathbf{q}/2 = 0$ one obtains, at $T = 0$,

$$\text{Re}\,\sigma_{\alpha\gamma}(0, \omega) \cong \frac{\hbar^3 e^2}{\omega m^2} \int_{-\infty}^{+\infty} \frac{d\omega'}{2\pi} F_{\alpha\gamma}(\omega_+, \omega_-) \qquad (18.48)$$

where

$$F_{\alpha\gamma}(\omega_+, \omega_-) \equiv \frac{1}{V} \sum_{\mathbf{k}\mathbf{k}'} k_\alpha k'_\gamma L(\mathbf{k}\mathbf{k}'\omega_+; \mathbf{k}'\mathbf{k}\omega_-)$$

$$= \frac{1}{V} \sum_{\mathbf{k}'} \bar{G}(\mathbf{k}'; \omega_+)\bar{G}(\mathbf{k}'; \omega_-)\{k'_\alpha k'_\gamma + \sum_{\mathbf{k}''} k''_\alpha k'_\gamma$$

$$\times \Lambda(\mathbf{k}''\mathbf{k}'\omega_+; \mathbf{k}'\mathbf{k}''\omega_-)\bar{G}(\mathbf{k}''; \omega_+)\bar{G}(\mathbf{k}''; \omega_-) + \cdots\}.$$

$$(18.49)$$

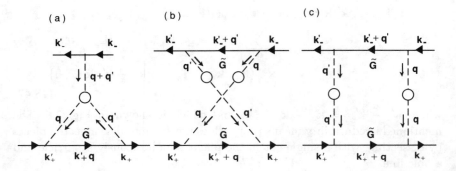

Fig. 3.9: (a) and (b) contributions to the vertex Λ in Fig. 3.8 of 3rd and 4th order in the impurity potential U; (c) is particle-hole-reducible. $\mathbf{k}'_\pm = \mathbf{k}_\pm - (\mathbf{q} + \mathbf{q}')$.

Writing explicitly only the lowest-order term of Λ, represented by the diagram of Fig. 3.2(b) which according to (18.36), (18.37) is

$$\Lambda_1(\mathbf{k}_+\mathbf{k}'_+; \mathbf{k}'_-\mathbf{k}_-) = \langle \Sigma(\mathbf{k}_+ - \mathbf{k}'_+)\Sigma(\mathbf{k}'_- - \mathbf{k}_-) \rangle_{\text{imp}}$$

$$= \frac{x}{v}\frac{1}{V}|\tilde{U}(\mathbf{k}_+ - \mathbf{k}'_+)|^2 \delta_{\mathbf{k}_+-\mathbf{k}'_+,\mathbf{k}_--\mathbf{k}'_-} \ , \quad (18.50)$$

and making use of (10.4) and (18.44), we find in AGD approximation

$$F_{\alpha\gamma}(\omega_+, \omega_-) = 2\pi i N(\mu) I(\omega_+, \omega_-)$$

$$\times \frac{1}{4\pi}\int d\Omega_{\mathbf{k}'}\, k'_\alpha k'_\gamma \{1 + 2i\omega_1 I(\omega_+, \omega_-) + \cdots\} \ . \quad (18.51)$$

Here ω_1 is defined, for an isotropic impurity potential, in analogy with (18.43) as

$$\omega_1 \mathbf{k}' \equiv \frac{1}{4}\frac{x}{v}N(\mu)\int d\Omega_{\mathbf{k}''}\mathbf{k}'' \left|\tilde{U}(\mathbf{k}'' - \mathbf{k}')\right|^2\bigg|_{\varepsilon_{\mathbf{k}''}=\varepsilon_{\mathbf{k}'}} \quad (18.52)$$

and

$$I(\omega_+, \omega_-) \equiv \int \frac{d\varepsilon}{2\pi i} \frac{1}{\omega_+ - \varepsilon + i\omega_o \text{sgn}(\omega_+)}$$

$$\times \frac{1}{\omega_- - \varepsilon + i\omega_o \text{sgn}(\omega_-)} = \frac{\theta(\omega - 2|\omega'|)}{\omega + 2i\omega_o} \ . \quad (18.53)$$

The last equality follows from the fact that the Cauchy integral I is zero unless $\text{sgn}(\omega_+) \neq \text{sgn}(\omega_-)$ which, because of $\omega_\pm = \omega' \pm \omega/2$, implies the condition $|\omega'| < \omega/2$.

Taking now the complete iteration of the term written explicitly in the bracket of (18.51), we generate the geometrical series $\{1 - 2i\omega_1 I\}^{-1}$ of all the *ladder diagrams* of which Fig. 3.9(c) is the second term while diagrams 3.9(a) and 3.9(b) are neglected (see Fig. 105 of Ref. 48). The result is

$$F_{\alpha\gamma}(\omega_+, \omega_-) = \frac{\pi k_F^2}{3} N(\mu)\delta_{\alpha\gamma} \frac{\theta(\omega - 2|\omega'|)}{\omega + 2i\omega_o} \left\{1 - \frac{2i\omega_1}{\omega + 2i\omega_o}\right\}^{-1}. \quad (18.54)$$

Defining now the *transport relaxation time* $\tau(\mu)$ in (13.46) by

$$\frac{\hbar}{\tau(\mu)} \equiv 4(\omega_o - \omega_1) = \frac{x}{v}N(\mu)$$

$$\times \int d\Omega_{\mathbf{k}'}(1 - \hat{k}\cdot\hat{k}')\left|\tilde{U}(\mathbf{k}' - \mathbf{k})\right|^2_{\varepsilon_{\mathbf{k}'}=\varepsilon_{\mathbf{k}}=\mu} \leq 4\omega_o \quad (18.55)$$

(see Eq. (39.17) of Ref. 48) while $\hbar/4\omega_o$ is the *scattering relaxation time* (11.41), the *adiabatic conductivity*

$$\lim_{\omega\to 0} \text{Re}\,\sigma_{\alpha\gamma}(0, \omega) = \sigma_o \delta_{\alpha\gamma} \quad (18.56)$$

simply yields Eq. (14.20). And with expression (18.55) one also finds that σ_o^{-1} is exactly the same, for constant $|\tilde{U}|$, as Eq. (15.19).

We now see that the AGD approximation of retaining only states $\mathbf{k}$ close to the Fermi surface, $|\varepsilon_{\mathbf{k}}| < \omega_o$, is justified by the θ-function in (18.53) and by the adiabatic limit (18.56). The final result, unfortunately, is rather deceptive. And in fact, it turns out that for a strong impurity potential U and large concentration x, meaning essentially strong disorder, the important contributions are not the ladder diagrams but the *maximally crossed ladders*, of which Fig. 3.9(b) is the second term. This situation will be investigated in the next section. The above AGD calculation is a telling example of the difficulty of the many-body theoretic evaluation of transport coefficients and correlation functions in general. An alternative method to calculate the Kubo formula (18.9) may be found in Ref. 49 which is based on a powerful Liouville-operator formalism recently developed by Loss[49]. On the other hand, a very elegant and general Green's function formalism has been put forward by Reizer[50].

19. Disorder, Localization and Length Scaling

The important notion in the phenomenon of scattering of waves is *phase coherence*. It requires that the wavelength λ stays the same before and after scattering which is the condition of *elastic scattering*. Since, quite generally, *inelastic scattering* is due to the emission or absorption by the wave of excitation modes of the scattering medium (see Section 15), phase breaking disappears at $T = 0$. In terms of a *phase breaking time* τ_φ [51] this may be expressed as [51,52]

$$\frac{1}{\tau_\varphi} \propto T^p \; ; \quad p > 0 \; . \tag{19.1}$$

The associated phase coherence length $L_\varphi = (D_n \tau_\varphi)^{1/2}$ where D_n is the diffusion constant defined in (14.6) therefore increases with decreasing temperature as $T^{-p/2}$. Applying (14.20) one expects from Eqs. (16.8), (17.34), (17.37) the power in (19.1) to be $p = 5$ and 2, respectively, for electron-phonon and for *s-d* scattering (note that in Ref. 46 $p = 3$ for electron-phonon scattering which results from defining τ_φ^{-1} by leaving out $q_\alpha q_\gamma$ in Eq. (16.6), see Ref. 18, p. 525).

Now, the residual resistivity (15.19) suggests that a different inhibiting mechanism due to *phase mixing* must exist, even at $T = 0$ as soon as there is *disorder in the scattering medium* or, more specifically, when there is a random distribution of impurities. Disorder may occur in the positions or in the amplitudes of the impurity potentials or in both. The *degree of disorder* may be defined by a *disorder parameter* Δ which measures the width of the probability distribution relative to the average distance between the impurities for *positional disorder* and relative to the band width for *amplitude disorder*. The *elastic mean free path* ℓ_{el} which at $T = 0$ is the only characterization of the scattering, then is a function of Δ and of the dimension D of the system. For weak disorder Δ, ℓ_{el} will be large compared to the wavelength $\lambda = 2\pi/k_F$ of the electrons, $\ell_{el}(\Delta, D) \gg \lambda$. In this regime of *weak localization* the electrons follow classical trajectories between collisions and disorder only gives rise to *quantum corrections to the conductivity* but is not sufficient to trap the electrons [4,51]. The latter happens when Δ is so large that $\ell_{el}(\Delta, D) \simeq \lambda$, so that the notion of wave becomes meaningless. In this regime of *strong localization* or *Anderson localization* [3,4], the wavefunction ψ (and also the transmission probability T of Section 14) is exponentially localized [53],

$$|\psi(\mathbf{r})|^2 \propto e^{-|\mathbf{r}|/\xi} \tag{19.2}$$

where $\xi = \xi(\Delta, D)$ is the *localization length*[52]. Since the notion of wave disappears in this regime, the condition $\lambda > \ell_{el}$ has no meaning which is the *Ioffe-Regel-Mott criterion*[51,53].

The passage from weak to strong localization is a *metal-insulator transition* also called *Anderson transition*[5,53] and is characterized by a critical value Δ_c of the disorder. $\Delta < \Delta_c$ means weak localization and $\xi = \infty$ that is, the wavefunction is at most algebraically localized . As mentioned, for positional disorder which is physically more important, Δ is measured in units of the average distance $x^{-1/D}$ between the impurities, x being the impurity concentration. Therefore, in this case the metal-insulator transition is also characterized by a critical concentration x_c. In addition there exists a critical dimension D_c such that $\xi < \infty$ for $D < D_c$ (see below).

Experimental verification of strong localization for electrons is difficult because condition (19.1) and also *electron-electron interaction effects*[4] forces one to go to extremely low temperatures[4]. However, as emphasized in the above discussion, localization is a wave phenomenon and therefore should also occur for classical waves, i.e. in optics and in acoustics[54]. Observation of localization of light appears to be particularly promising because, on the one hand, light quanta are so much more energetic than acoustic-type excitations (which are the only relevant modes of the medium at low temperatures) that inelastic processes are negligible. Since, on the other hand, light-light interaction is also negligible (except in nonlinear optical media), the limitation here is not determined by temperature but by *absorption*. Although in earlier experiments only weak localization had seen[55], more recently different situations of strong localization have also been realized[56]. Sound localization which appears to be much less favorable has also been studied[57], and some form of strong localization has even been seen[58].

The most striking manifestation of weak localization is a *back-scattering peak* observed in experiments of light localization[55] but which has also become an important tool in the investigation of metal surfaces and of doped bulk metal[59]. As shown by Bergmann[59], this phenomenon has a simple explanation as *multiple elastic scattering* in dimension $D \geq 2$. Indeed, compare the N-fold scattering $\mathbf{l}_+ \to \mathbf{l}'_+$ proceeding by the wavevectors

$$\mathbf{k}_{n-1} \to \mathbf{k}_n = \mathbf{k}_{n-1} + \mathbf{q}_n \ ; \quad n = 1, \cdots N \ ; \quad \mathbf{k}_o = \mathbf{l}_+, \mathbf{k}_N = \mathbf{l}'_+ \ , \quad (19.3)$$

with the scattering $\mathbf{l}'_- \to \mathbf{l}_-$ having the same recoils $\mathbf{q}_n$ but in reversed order,

$$\mathbf{k}'_{n-1} \to \mathbf{k}'_n = \mathbf{k}'_{n-1} + \mathbf{q}_{N+1-n} \ ; \quad n = 1, \cdots, N \ ; \quad \mathbf{k}'_o = \mathbf{l}'_-, \mathbf{k}'_N = \mathbf{l}_- \ ,$$
$$(19.4)$$

so that $l'_+ - l_+ = l_- - l'_- = \sum_n q_n$. Elasticity in each step,

$$|k_n| = |l_+| = |l'_+| \; ; \quad |k'_n| = |l'_-| = |l_-| \; ; \quad n = 1, \cdots, N - 1 \; , \tag{19.5}$$

obviously cannot be satisfied unless

$$l_+ + l_- = l'_+ + l'_- = 2l = 0 \; , \tag{19.6}$$

as is seen from Fig. 3.10(a).

Now the back-scattering peak is the unique result of *constructive inter-ference* of the two multiple scatterings (19.3) and (19.4) originating from the same initial wave. Indeed, in this case $l_+ = l'_- = p$ and, because of (19.6), also $l_- = l'_+ = -p$. Hence, (19.4) in reverse, namely $\{k'_n \to k'_{n-1}\}$, is the time-reversed of (19.3) or $\{k_{n-1} \to k_n\}$, as is evident from Fig. 3.10(b)[59]. But for the amplitudes this *time-reversal relation* means that[51]

$$a\{k'_n \to k'_{n-1}\} = a^*\{k_{n-1} \to k_n\} \; ; \quad \text{if } p = l_+ = -l_- = l'_- = -l'_+ = -p' \; . \tag{19.7}$$

Therefore, calling a_N and a'_N the amplitudes of the *elastic processes* (19.3) and (19.4), respectively, which satisfy condition (19.6), constructive inter-ference only happens under the back-scattering condition (19.7), in which case $a'_N = a^*\{k'_n \to k'_{n-1}\} = a\{k_{n-1} \to k_n\} = a_N$. Hence

$$\frac{|a_N + a'_N|^2}{|a_N|^2 + |a'_N|^2} = \begin{cases} 2, & \text{for back-scattering ;} \\ 1, & \text{otherwise .} \end{cases} \tag{19.8}$$

Taking into account inelastic rounding effects due to finite temperature, this gives rise to a back-scattering peak of almost double intensity[59].

(a) (b)

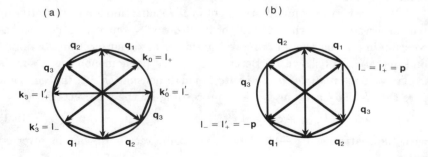

Fig. 3.10: (a) The multiple scatterings (19.3) and (19.4) for $N = 3$ under the elasticity conditions (19.6). (b) The scattering for which the time-reversal relation (19.7) holds.

Now, according to Eq. (18.37) and to Fig. 3.11(a), (b), the quantity $\langle a_N a_N'^* \rangle_{\text{imp}}$ represents the coupled scatterings (19.3) and (19.4). And since

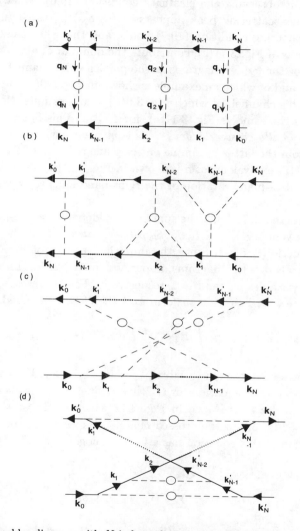

Fig. 3.11: (a) Ladder diagram with N independent vertices. (b) Example of a diagram with less than N independent vertices according to Eq. (18.37). Both diagrams, (a) and (b), are of order $2N$ in the interaction (15.12) and contribute to $\langle a_N a_N'^* \rangle_{\text{imp}}$. (c) Maximally crossed ladder obtained from (a) by an out-of-plane rotation of the lower particle line. (d) Diagram (c) after an out-of-plane rotation of the right-hand ends.

the elasticity δ-function does not appear in the interaction (15.12) but only in the transition probability (15.15), this quantity may readily be calculated without observing the elasticity condition (19.6). We wish to show that the back-scattering peak appears in $\langle a_N a_N'^* \rangle_{\text{imp}}$ as singularity when the back-scattering condition (19.7) holds, and that the so-called *Langer-Neal diagrams* of Fig. 3.11(c), contribute the dominant part[60,4]. Note that these are not the ladder diagrams in the particle-hole channel which led to Eq. (18.54) and of which an example is shown in Fig. 3.9(c) but the crossed ladders in this channel, of which Fig. 3.9(b) is an example. Note also that the latter is the same as Fig. 3.11(c), for $N = 2$. This figure which represents a *maximally crossed ladder*[60,4] is equivalent with Fig. 3.11(a) and is obtained from the latter by simple out-of-plane rotation of the lower particle line. A still equivalent form is shown in Fig. 3.11(d) which follows from (c) by an out-of-plane rotation of the right-hand ends (compare Fig. 1c of Ref. 61).

As observed in Ref. 60, the sum of all ladders in the two-particle channel may be generated by a *Bethe-Salpeter equation*[62] for the two-particle Green's function (18.31). Since the impurity vertices of Fig. 3.11 do not carry energy across the diagrams, upper and lower particle lines carry fixed frequency variables ω_+ and ω_-, respectively. For the crossed channel of Fig. 3.12 the time-Fourier transform may therefore be defined as

$$\tilde{G}_2(1'_+1_+\omega_+; 1'_-1_-\omega_-) = \int dt_+ e^{i\omega_+ t} \int dt e^{i\omega_- t} G_2(1'_+0, 1_+t_+; 1'_-0, 1_-t_-)$$

(19.9)

where, expressed in the old variables, $1'_+ = k_+$, $1_+ = k'_+$, $1'_- = k'_-$, $1_- = k_-$. The Bethe-Salpeter equation for $\tilde{G}_2$ then defines a *two-particle-irreducible four-point vertex* Γ as shown in Fig. 3.12 which in the static case is the analogue of Fig. 3.8 in the crossed channel. Suppressing spin indices, the static version of this equation has a form analogous to (18.47),

$$\tilde{G}_2(1'_+1_+\omega_+; 1'_-1_-\omega_-) = \bar{G}(1'_+; \omega_+)\bar{G}(1'_-; \omega_-)\{\delta_{1'_+1_+}\delta_{1'_-1_-}$$

$$+ \sum_{\mathbf{q}} \Gamma(1'_+, 1_+ + \mathbf{q}, omega_+; 1'_-, 1_- - \mathbf{q}, \omega_-)$$

$$\times \tilde{G}_2(1_+ + \mathbf{q}, 1_+\omega_+; 1_- - \mathbf{q}, 1_-\omega_-)\}$$

(19.10)

where $\bar{G}$ is the impurity-averaged propagator defined in (18.39). Note that, compared to the field-theoretic literature[62] we are not interested in the invariance against the rotation (c)↔(d) in Fig. 3.11 which is manifestly

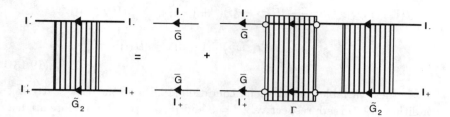

Fig. 3.12: Static Bethe-Salpeter equation for the two-particle Green's function $\tilde{G}_2$ defined by (18.31) and (19.9), $\bar{G}$ being the propagator defined in (18.39) and Γ the static two-particle-irreducible four-point vertex. $1_+ + 1_- = 1'_+ + 1'_-$.

not fulfilled. Such a symmetric Bethe-Salpeter equation would require an additional crossed zero-order term.

By iteration, Eq. (19.10) may be expressed in terms of the static two-particle-reducible *t-matrix* as follows (for a nonstatic t-matrix see Section 33):

$$
\begin{aligned}
\tilde{G}_2(1'_+1_+\omega_+; 1'_-1_-\omega_-) = \bar{G}(1'_+; \omega_+)\bar{G}(1'_-; \omega_-)\{\delta_{1'_+1_+}\delta_{1'_-1_-} \\
+ t(1'_+1_+\omega_+; 1'_-1_-\omega_-)\bar{G}(1_+; \omega_+)\bar{G}(1_-; \omega_-)\} .
\end{aligned}
$$
(19.11)

In field theory the t-matrix is defined as the nontrivial part of Heisenberg's S-matrix (see Itzykson and Zuber in Ref. 62). Thus, in our case, S is the whole parenthesis in Eq. (19.11). The iteration yields

$$
\begin{aligned}
t(1'_+1_+\omega_+; 1'_-1_-\omega_-) = \Gamma(1'_+1_+\omega_+; 1'_-1_-\omega_-) \\
+ \sum_{\mathbf{q}} \Gamma(1'_+, 1_+ + \mathbf{q}, \omega_+; 1'_-, 1_- - \mathbf{q}, \omega_-)\bar{G}(1_+ + \mathbf{q}; \omega_+) \\
\times \bar{G}(1_- - \mathbf{q}; \omega_-)\Gamma(1_+ + \mathbf{q}, 1_+\omega_+; 1_- - \mathbf{q}, 1_-\omega_-) + \cdots
\end{aligned}
$$
(19.12)

and $1_+ + 1_- = 1'_+ + 1'_-$ (note again the rule introduced in Sections 10 and 11 that there is one summation per closed loop). The ladders of Fig. 3.11(a) now result if we choose for Γ again the vertex (18.50) represented by Fig. 3.2(b). In the notation $1_\pm = 1 \pm \mathbf{p}$; $1'_\pm = 1 \pm \mathbf{p}'$, etc. where 1 is the center-of-mass wavevector and $\mathbf{p}$, $\mathbf{p}'$, etc. are the relative wavevectors of the electron pair, one finds in simplified notation

$$t(\mathbf{l}'_+\mathbf{1}_+; \omega_+\omega_-) = \frac{x}{v}\frac{1}{V}|\tilde{U}(\mathbf{p}' - \mathbf{p})|^2 + (\frac{x}{v}\frac{1}{V})^2 \sum_{\mathbf{q}} |\tilde{U}(\mathbf{p}' - \mathbf{q})|^2$$

$$\times \bar{G}(1 + \mathbf{q}; \omega_+)\bar{G}(1 - \mathbf{q}; \omega_-)|\tilde{U}(\mathbf{q} - \mathbf{p})|^2 + \cdots .$$

$$(19.13)$$

In these variables the elasticity condition (19.6) and the back-scattering condition (19.7) read, respectively, $\mathbf{l} = 0$ and $\mathbf{p}' = -\mathbf{p}$. We therefore are led to consider the sum of all ladder graphs (19.13) at $\mathbf{p}' = -\mathbf{p}$ as function of l,

$$t(\mathbf{1}_-\mathbf{1}_+; \omega_+\omega_-) = \frac{x}{vV}|\tilde{U}(2\mathbf{p})|^2\{1 - \frac{x}{v}|\tilde{U}(2\mathbf{p})|^2 J_2(l; \omega_+\omega_-)\}^{-1} \quad (19.14)$$

where[60)]

$$J_2(l; \omega_+\omega_-) \equiv \frac{1}{V} \sum_{\mathbf{q}} \left|\frac{\tilde{U}(\mathbf{q} + \mathbf{p})}{\tilde{U}(2\mathbf{p})}\right|^2 \left|\frac{\tilde{U}(\mathbf{q} - \mathbf{p})}{\tilde{U}(2\mathbf{p})}\right|^2$$

$$\times \bar{G}(1 + \mathbf{q}; \omega_+)\bar{G}(1 - \mathbf{q}; \omega_-) . \quad (19.15)$$

Assuming for simplicity point-like impurities (which means strong scattering), $\tilde{U}(\mathbf{p}) = $ const, and making use of (10.4) and (18.44), we obtain in AGD approximation $|\varepsilon_{\mathbf{l}\pm\mathbf{q}}| < \omega_o$ used in the evaluation of (18.49), writing $\varepsilon_{\mathbf{l}\pm\mathbf{q}} = \varepsilon_{\mathbf{q}} \pm \lambda\zeta + \lambda^2/4\varepsilon_F, \lambda \equiv \hbar v_F l$,

$$J_2(l; \omega_+\omega_-) = i\pi N(\mu) \int_{-1}^{+1} d\zeta \int \frac{d\varepsilon}{2\pi i}$$

$$\times \frac{1}{\omega_+ - (\varepsilon + \lambda\zeta + \lambda^2/4\varepsilon_F) + i\omega_o\text{sgn}\,\omega_+}$$

$$\times \frac{1}{\omega_- - (\varepsilon - \lambda\zeta + \lambda^2/4\varepsilon_F) + i\omega_o\text{sgn}\,\omega_-} . \quad (19.16)$$

Note that $|\varepsilon_{\mathbf{l}\pm\mathbf{q}}| < \omega_o$ also implies $\lambda < \omega_o$. Proceeding as with Eq. (18.53) one finds with $\omega_\pm = \omega' \pm \omega/2$

$$J_2(l; \omega_+\omega_-) = \frac{i\pi N(\mu)}{2\lambda}\theta(\omega - 2|\omega'|) \log\frac{\omega - 2\lambda + 2i\omega_o}{\omega + 2\lambda + 2i\omega_o} . \quad (19.17)$$

Here the back-scattering singularitiy appears in the limit $\omega \to 0$ and has the form $J_2 \propto l^{-1}$ [60)]. Note that this singularity is characteristic of the *crossed ladders* of Fig. 3.11(c) or (d) and came from the ζ-integration in (19.16) which was missing in the expression (18.53) for the straight ladders. (19.17) is of course valid only in dimensions $D > 1$.

Inserting (19.14), (19.17) into (19.11) we may now calculate the contribution $\delta\sigma_{\alpha\gamma}(0, \omega)$ of the crossed ladders to the conductivity (18.48). Since $l_+ = l'_- = k'$, $l_- = l'_+ = k$, the contribution to (18.49) is

$$\delta F_{\alpha\gamma}(\omega_+\omega_-) = \frac{1}{V} \sum_{kk'} k_\alpha k'_\gamma \bar{G}(k; \omega_+) \bar{G}(k'; \omega_-) t(kk'; \omega_+\omega_-)$$
$$\times \bar{G}(k'; \omega_+) \bar{G}(k; \omega_-) . \tag{19.18}$$

Observing that the values of interest are $l = |k + k'|/2 \ll p = |k' - k|/2 \cong k_F$, we may replace $k_\alpha k'_\gamma$ by $-p_\alpha p_\gamma$. And since for point-like impurities K is independent of p, we may again use (10.4) to write the p-sum as Cauchy integral plus integrations over the direction of p, as in (19.16). Making use of (18.43) in the form $\omega_o = \pi(x/v)N(\mu)|\bar{U}|^2$ the result is, in dimension D (Problem 3.13), $\delta F_{\alpha\gamma} = \delta F \delta_{\alpha\gamma}$ with

$$\delta F = i\omega_o \frac{\theta(\omega - 2|\omega'|)}{\omega + 2i\omega_o} \frac{k_F^2 \Omega_D}{(2\pi\hbar v_F)^D} \int_0^{\omega_o} \frac{d\lambda}{\lambda^{3-D}} \frac{\lambda + (\frac{\omega}{2} + i\omega_o)L(\lambda)}{\lambda - i\omega_o L(\lambda)} ; \quad D > 1 \tag{19.19}$$

where $L(\lambda) \equiv \frac{1}{2}\log[(\frac{\omega}{2} - \lambda + i\omega_o)/(\frac{\omega}{2} + \lambda + i\omega_o)]$. Note that for $D = 1$ the integral over ζ is replaced by the sum over $\zeta = \pm 1$ which leads to (Problem 3.13)

$$\delta F = i\omega_o \frac{\theta(\omega - 2|\omega'|)}{\omega + 2i\omega_o} \frac{k_F^2}{2\pi\hbar v_F} \int_0^{\omega_o} \frac{d\lambda}{\lambda^2 - (\frac{\omega}{2} + i\omega_o)(\frac{\omega}{2} + 2i\omega_o)} ; \quad D = 1 . \tag{19.20}$$

Evaluating (19.19), (19.20), one observes that for $D \neq 2$ the limit $\omega \to 0$ exists and insertion into (18.48) yields a correction to the adiabatic conductivity of

$$\delta\sigma \cong \frac{e^2}{h} \alpha_D \left(\frac{\hbar v_F}{\omega_o}\right)^{2-D} ; \quad \alpha_D = \begin{cases} \frac{1}{4\pi\sqrt{2}} \tan^{-1} \frac{1}{\sqrt{2}}, & D = 1; \\ -\frac{1}{(2\pi)^2} \int_0^1 dx \left(1 + 2x \frac{P}{\tan^{-1}\frac{1}{x} - x}\right), & D = 3 \end{cases} \tag{19.21}$$

where P stands for principal part. In the case $D = 2$ the integrand in (19.19) is singular in the limit $\omega \to 0$. Replacing therefore the lower limit 0 of the integral by $\epsilon = 0^+$ one finds[61]

$$\delta\sigma \cong \frac{e^2}{h} \alpha_2 \ln \frac{\omega_o}{\epsilon} ; \quad \alpha_2 = -\frac{1}{4\pi} , \quad D = 2 . \tag{19.22}$$

Inclusion of more diagrams in the calculation of Eq. (19.11) would of course modify the numerical value and even possibly the sign of the coefficients α_D. But qualitatively, the result (19.21), (19.22) is quite reasonable. For a review of more elaborate calculations of conductivity corrections see Ref. 63.

While the back-scattering singularity of Eq. (19.17) is the characteristic feature of weak localization, strong localization introduces a new length scale: For an impurity distribution such that the disorder parameter $\Delta > \Delta_c$, it is the localization length ξ which sets a limit on the motion of the electrons . We may obtain an estimate of ξ with the help of the *block-scaling argument* of Thouless[52]. Consider a long, thin wire of cross-sectional area A with given impurity concentration x such that the length ℓ of the wire is large but its transverse dimensions are small compared to ξ. Such a wire may be called one-dimensional . Now imagine the wire to be cut into blocks of length $L < \xi$. Then the one-electron wavefunction ψ is extended within such a block of dimensions L and A (see Fig. 3.1, Section 14) so that the boundary conditions uniquely determine the (discrete) energy spectrum. Hence there is a well-defined density of states per spin

$$N(\varepsilon) = \frac{1}{2LA}\frac{dZ}{d\varepsilon} \qquad (19.23)$$

where Z is the number of states with energy $< \varepsilon$. The average level spacing at the Fermi level therefore is

$$W_L = \frac{d\varepsilon}{dZ} = \frac{1}{2LAN(\mu)} \ . \qquad (19.24)$$

Electrons may pass between two adjacent blocks, provided that the energy uncertainty ΔE_L in each block is larger than W_L. An estimate of ΔE_L may be obtained via the Heisenberg uncertainty relation $\tau_d \Delta E_L > \hbar$, where τ_d is the *diffusion time* that it takes an electron to travel randomly from one end of a block to the other. τ_d is of the order L^2/D_n, as may be seen with the help of the *diffusion equation*

$$\dot{n} = D_n \nabla^2 n \qquad (19.25)$$

which follows from Eqs. (13.30) and (14.6). Since the wavefunction ψ is a superposition of standing waves with wavelength $< L$ such a wave, when inserted into (19.25), approximately yields $2\pi/\tau_d > D_n(2\pi/L)^2$, that is

$$\Delta E_L > \frac{2\pi\hbar D_n}{L^2} \ . \qquad (19.26)$$

As long as $\Delta E_L > W_L$, electrons may pass between blocks, there is conduction between them. Therefore, localization occurs when $\Delta E_L < W_L$ or, inserting (19.24), (19.26), when $L > 4\pi\hbar AD_n N(\mu)$. This determines the localization length to be[52]

$$\xi \simeq 4\pi\hbar AD_n N(\mu) = \frac{h}{e^2}\sigma A \qquad (19.27)$$

where Eqs. (14.7), (14.8) were used in the second equality. The conductance defined in (14.2) now takes the form

$$G_L = \frac{e^2}{h} \frac{\xi}{L} \tag{19.28}$$

which, like the Landauer formula (14.15), is expressed in units of the universal resistance (14.16). Another way of writing this result is to remember that L was just a tiny fraction of the length ℓ of the wire, $L = \delta\ell$, so that the inverse of (19.28) is the corresponding increase of resistance δR in units of $R \simeq h/e^2$. Hence[51)]

$$\frac{\delta R}{R} = \frac{\delta\ell}{\xi} \tag{19.29}$$

which leads to an exponential increase corresponding to (19.2).

While Thouless' block-scaling argument is limited to one dimension, the real interest of *length scaling* is to describe cross-over behaviour as function of dimension D. This is achieved in a surprisingly simple way by the *renormalization-group* scheme of Abrahams , Anderson, Licciardello and Ramakrishnan[6)]. This scheme is based on the analogy between localization and *critical phenomena* in continuous phase transitions, the localization length playing the role of the *correlation length*. Approaching the critical point T_c, the latter diverges as $(T - T_c)^{-\nu}$ where the *critical exponent* ν depends on D [64)]. The problem is to identify the driving parameter in a metal-insulator transition in analogy with $T - T_c$. According to Mott[53)], localized and extended electron states are energetically separated by an energy ε_c, the *mobility edge* [5)]. There is conduction only if the Fermi energy μ lies in the extended part of the spectrum above ε_c. This suggests that the conductivity has the Mott form[53,5,6)]

$$\sigma \propto (\mu - \varepsilon_c)^{(D-2)\nu} \tag{19.30}$$

where the value of the exponent is suggested by model calculations[65,6)] and $\mu - \varepsilon_c$ depends on the degree of disorder Δ as well as on the impurity concentration x.

In order to apply renormalization-group arguments[65)], consider an isotropic hypercube of side L in D dimensions so that the cross-sectional area is $A = L^{D-1}$ [6)]. Then Ohm's law for the total current (18.12) as expressed by the conductance (18.13) is

$$G_L = \sigma L^{D-2} . \tag{19.31}$$

This *Ohm's-law scaling* is valid for $G_L \gg G_\xi = e^2/h$. In the opposite limit $G_L \ll G_\xi$, we have *strong-localization scaling* as expressed by Eq. (19.29),

$$G_L = G_\xi e^{t_D - L/\xi} \tag{19.32}$$

where t_D defines the amplitude as function of D.

Now the idea of the *renormalization group*[66,67] is to apply successive rescalings $L \to bL$ with a dilatation parameter $b > 1$ in order to reach a *fixed point* which here is one of the two limits (19.31) or (19.32), depending on the initial length L_o from which the iteration started and on D. Introducing as iteration variable the logarithm of the "Thouless number" $g(L) \equiv G_L/G_\xi$[6],

$$t(L) \equiv \ln g(L) = \ln \frac{G_L}{G_\xi} \tag{19.33}$$

and defining the *scaling function*[67] by

$$\beta(t) \equiv \frac{d \ln g}{d \ln L} = Lt'(L) \tag{19.34}$$

the asymptotic scaling laws (19.31), (19.32) become

$$\beta(t) \cong \begin{cases} D - 2, & t \gg 1; \\ t - t_D, & t \ll -1. \end{cases} \tag{19.35}$$

Corrections to these laws may, in principle, be obtained from Eqs. (19.21), (19.22). Indeed, $\hbar v_F/\omega_o$ has the dimension of a length which, by comparison with ξ, may be identified with L/ξ. One then obtains a correction to (19.33) given by

$$\delta t(L) \cong \begin{cases} \ln[1 + \alpha_D (L/\xi)^{2-D}], & D \neq 2; \\ \ln[\alpha_2 \ln(L/\xi)] \cong t(L), & D = 2, \end{cases} \tag{19.36}$$

and hence

$$\delta\beta(t) \cong \begin{cases} (2-D)[1 + (1/\alpha_D)(L/\xi)^{D-2}]^{-1}, & D \neq 2; \\ [\ln(L/\xi)]^{-1}, & D = 2, \end{cases} \tag{19.37}$$

where L/ξ has to be substituted. For $t \gg 1$ we have according to (19.31), $t \cong (D-2)\ln(L/\xi)$ if $D \neq 2$ while, according to the second line of (19.36), $t \cong \ln[\alpha_2 \ln(L/\xi)]$. Hence

$$\delta\beta(t) \cong \begin{cases} (2-D)\alpha_D e^{-t}, & D \neq 2; \\ \alpha_2 e^{-t}, & D = 2. \end{cases} \tag{19.38}$$

In the case $t \ll -1$ Eq. (19.32) gives $t_D - t \cong L/\xi$ so that

$$\delta\beta(t) \cong \begin{cases} (2-D)[1 + (1/\alpha_D)(t_D - t)^{D-2}]^{-1}, & D \neq 2; \\ [\ln(t_D - t)]^{-1}, & D = 2. \end{cases} \tag{19.39}$$

Of course, Eqs. (19.38), (19.39) are only indicative and serve mainly pedagogical interests. See Ref. 63 for a systematic review.

If we assume that $\beta(t)$ is a smooth function, we may interpolate between the two branches of (19.35) as shown in Fig. 3.13 for $D = 1, 2, 3$. Knowing $\beta(t)$ we may obtain the *Callen-Symanzik equation*[67] by simple integration of Eq. (19.34),

$$t(bL) - t(L) = L \int_1^b db' t'(b'L) = (b-1)\beta(t) . \qquad (19.40)$$

This equation explicitly gives the scaling. However, the physics is better brought out by the flow equation (19.34). It says that when $\beta(t) > 0$, dilatation with $\Delta b = (b-1)L > 0$ results in steps $\Delta t > 0$. Thus, defining t_c by $\beta(t_c) = 0$ (see Fig. 3.13), the iteraction started at $t_o \equiv t(L_o) > t_c$ leads to the metal fixed point $t = +\infty$ while, starting at $t_o < t_c$, the insulator fixed point $t = -\infty$ is reached. t_c is an unstable fixed point from which the flow escapes to the stable ones, as indicated by the arrows in Fig. 3.13. It is seen from this figure that for $D < 2$ only the insulator fixed point exists which means that in 1 dimension there is always strong localization, at least

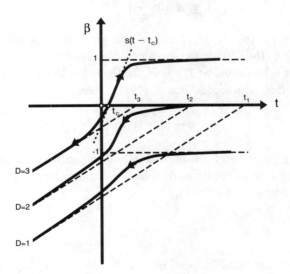

Fig. 3.13: Scaling function $\beta(t)$ interpolating between the two branches of Eq. (19.35) for dimensions $D = 1, 2, 3$ (broken lines). The arrows indicate the flow in the renormalization-group iteration. The dotted line through t_c represents Eq. (19.41).

for a homogenous system. For an impurity concentration decreasing with distance a cross-over may exist, however (see, e.g. Ref. 68).

For $D > 2$ there is a metal-insulator transition at t_c. In the neighborhood of this point we may use a linear approximation

$$\beta(t) = s(t - t_c). \tag{19.41}$$

As an indication we see that for $D = 3$ and $t \ll -1$, Eqs. (19.35) and (19.39) give $s - 1 = \delta\beta'(t_c) \simeq -\alpha_3/(t_3 - t_c)^2$ which, according to (19.21), is positive as required[6]. With (19.41), Eq. (19.34) is easily integrated and gives

$$\frac{L}{L_o} = \left(\frac{t - t_c}{t_o - t_c}\right)^{1/s} = \left(\frac{\ln(g/g_c)}{\ln(g_o/g_c)}\right)^{1/s}. \tag{19.42}$$

For $t_o > t_c$ but $t_o - t_c \ll t_c$ we may write the conductivity for a given t as

$$\sigma = \frac{e^2}{h}gL^{2-D} = \frac{e^2}{h}CL_o^{2-D}\left(\ln\frac{g_o}{g_c}\right)^{(D-2)/s} \tag{19.43}$$

where

$$C \equiv g\left(\ln\frac{g}{g_c}\right)^{(2-D)/s}. \tag{19.44}$$

Since $\ln(g_o/g_c) \cong (g_o - g_c)/g_c$, comparison with (19.30) shows that $g_o - g_c$ plays the role of the driving parameter for the Anderson transition and that the exponent is identified as $1/s = \nu$. On the other hand, in the case $D = 2$ we conclude from the second line of (19.36) that

$$g - g_o = \alpha_2 \ln\frac{L}{L_o} \tag{19.45}$$

(compare Eq. (15) of Ref. 6). If conductivity measurements are made on a sample of size L of the order of the phase coherence length $L_\varphi = (D\tau_\varphi)^{1/2}$, one obtains according to (19.1),

$$\delta\sigma \cong \frac{e^2}{h}(-\frac{p}{2}\alpha_2)\ln T. \tag{19.46}$$

This behavior has actually been observed[4]. Moreover, Eq. (19.45) shows that in two dimensions there is a cross-over from *logarithmic scaling* in weak localization to the exponential scaling (19.32)[6,61,63]. Similar results are derived with quite different methods in Section 3.8 of Ref. 51.

Solutions to the Problems of Chapter 3

Problem 3.1: Eq. (13.22).

Making use of the identity

$$\frac{\partial \varphi}{\partial \mathbf{k}} \times \frac{\partial \psi}{\partial \mathbf{k}} = \frac{\partial}{\partial \mathbf{k}} \times \left(\varphi \frac{\partial \psi}{\partial \mathbf{k}} \right) - \varphi \left(\frac{\partial}{\partial \mathbf{k}} \times \frac{\partial}{\partial \mathbf{k}} \right) \psi$$

where the second term vanishes, the left-hand side of Eq. (13.22) may be transformed with the help of Gauss' theorem into

$$V^{-1} \sum_k \left(\frac{\partial \varphi}{\partial \mathbf{k}} \times \frac{\partial \psi}{\partial \mathbf{k}} \right) = (2\pi)^{-3} \sum_{n\sigma} \oint_{\partial Z} \left(d\mathbf{S}_\mathbf{k} \times \left(\varphi \frac{\partial \psi}{\partial \mathbf{k}} \right) \right).$$

Here we inserted $k = (nk\sigma)$ and went over to continuous k-vectors according to the prescription

$$V^{-1} \sum_{k \in Z} \rightarrow \int_Z \frac{d^3 k}{(2\pi)^3}.$$

Now the definition (1.6) of the reduced zone Z implies that the border ∂Z consists of polygons in the median planes of the smallest reciprocal lattice vectors $\pm \mathbf{K} \neq 0$. Hence these polygons occur in pairs centered at $\pm \mathbf{K}/2$. On these pairs of polygons $d\mathbf{S}_\mathbf{k}$ takes opposite values, but because of the periodicity (1.11) of any field (1.5), $\varphi \partial \psi / \partial \mathbf{k}$ takes identical values. Thus the contribution of $d\mathbf{S}_\mathbf{k} \times (\varphi \partial \psi / \partial \mathbf{k})$ from any such pair cancels, which proves Eq. (13.22).

Problem 3.2: Eq. (13.42).

Substitution of the second into the first equation (13.39) yields for $\tau_k = \tau(\varepsilon_k)$

$$\mathbf{g}^\lambda = -\varepsilon_k^{\lambda-1} \tau(\varepsilon_k) \mathbf{v}_k + \tau^2(\varepsilon_k) \omega_c^2 \Omega_k^2 \mathbf{g}^\lambda.$$

Inserting here the power series

$$\mathbf{g}^{(\lambda)} = \sum_{l=0}^{\infty} \mathbf{g}_l^{(\lambda)} \omega_c^{2l} \; ; \quad \mathbf{g}_o^{(\lambda)} = -\varepsilon_k^{\lambda-1} \tau(\varepsilon_k) \mathbf{v}_k \, ,$$

ordering according to powers of ω_c^2 gives

$$\mathbf{g}_l^{(\lambda)} = \tau^2(\varepsilon_k) \Omega_k^2 \mathbf{g}_{l-1}^{(\lambda)} = \tau^{2l}(\varepsilon_k) \Omega_k^{2l} \mathbf{g}_o^{(\lambda)} \; ; \quad l \geq 1.$$

Since in $\mathbf{g}_o^{(\lambda)}$ only the factor $\mathbf{v}_k$ does not commute with Ω_k one finds

$$\mathbf{g}^{(\lambda)} = \varepsilon_k^{\lambda-1} \sum_{l=0}^{\infty} [-\tau(\varepsilon_k)]^{2l+1} \omega_c^{2l} \Omega_k^{2l} \mathbf{v}_k$$

and, from the second equation (13.39),

$$\mathbf{u}^{(\lambda)} = \varepsilon_k^{\lambda-1} \sum_{l=0}^{\infty} [-\tau(\varepsilon_k)]^{2l+2} \omega_c^{2l+1} \Omega_k^{2l+1} \mathbf{v}_k .$$

According to (13.20) the sum of the two expressions is Eq. (13.42).

Problem 3.3: Eq. (14.19).

With the form (14.18) of the rotation operator one easily verifies the relations

$$\Omega_k^{2l+2} v_{k\gamma} = -(-)^l v_{k\gamma} (\delta_{\gamma 1} + \delta_{\gamma 2}) ;$$
$$\Omega_k^{2l+1} v_{k\gamma} = (-)^l (v_{k1} \delta_{\gamma 2} - v_{k2} \delta_{\gamma 1}) ,$$

valid for $l \geq 0$. When inserting these expressions into Eq. (14.17), it is important to isolate the term $n = 0$ before writing $n = 2l + 2$ and $2l + 1$:

$$\sigma_{\alpha\gamma}(\mathbf{H}) = -e^2 V^{-1} \sum_k v_{k\alpha} f_o'(\varepsilon_k) \tau(\varepsilon_k) \{ v_{k\alpha} - \sum_{l=0}^{\infty} \tau^{2l+2}(\varepsilon_k) \omega_c^{2l+2}$$

$$\times (-)^l v_{k\gamma} (\delta_{\gamma 1} + \delta_{\gamma 2}) - \sum_{l=0}^{\infty} \tau^{2l+1}(\varepsilon_k) \omega_c^{2l+1} (-)^l (v_{k1} \delta_{\gamma 2} - v_{k2} \delta_{\gamma 1}) \} .$$

Summing the geometrical series one is left with expressions of the form

$$-e^2 V^{-1} \sum_k v_{k\alpha} v_{k\gamma} \tau^{n+1}(\varepsilon_k) f_o'(\varepsilon_k) = \sigma_o \tau^n(\mu) \delta_{\alpha\gamma}$$

where the limit $T \to 0$ was assumed and use was made of the definitions (10.4) and (14.20), noting that at the Fermi energy μ, the angular average of $v_{k\alpha} v_{k\gamma}$ is $(v_F^2/3) \delta_{\alpha\gamma}$. Putting together the pieces one readily finds Eq. (14.19).

Problem 3.4: Eqs. (15.3), (15.4).

With (15.1) and $2\pi \delta(\varepsilon) = \int dt e^{-i\varepsilon t}$, Eq. (15.2) may be written as

$$\Gamma_o(kk') = \hbar^{-1} \sum_{\mathbf{q}} |\alpha_{k'k}|^2 \delta_{\mathbf{k}',\mathbf{k}-\mathbf{q}} \int dt e^{i(\varepsilon_k - \varepsilon_{k'})t} A$$

where the auxiliary quantity

$$A \equiv \sum_{IF} \rho_I e^{i(E_I - E_F)t} \langle I | a_k^+ a_{k'} h(\mathbf{q}) | F \rangle \langle F | h^+(\mathbf{q}) a_{k'}^+ a_k | I \rangle$$

may be transformed, with the help of Eqs. (5.10), (6.4), (6.10) and using the closure relation for the states $|F\rangle$, into

$$A = \langle a_k^+(t)a_{k'}(t)h(\mathbf{q};\, t)h^+(\mathbf{q};\, 0)a_{k'}^+a_k\rangle.$$

Because of statistical independence and with an unperturbed average over the electrons, the last expression is, for $\mathbf{k'} \neq \mathbf{k}$,

$$A = \langle a_k^+a_k\rangle_o\langle a_{k'}a_{k'}^+\rangle_o\langle h(\mathbf{q};\, t)h^+(\mathbf{q};\, 0)\rangle e^{i(\varepsilon_{sk}-\varepsilon_{sk'})t}$$

so that with (7.40) one readily obtains Eqs. (15.3), (15.4).

Problem 3.5: Kubo identity (15.8).

Using definition (15.9) and Eqs. (5.10), (6.4) one may write

$$\beta\rho\dot{\hat{A}} = \rho\int_0^\beta d\tau\left(\frac{d}{-i\hbar d\tau}\right)A(-i\tau) = \frac{i}{\hbar}\rho[A(-i\beta) - A(0)] = \frac{i}{\hbar}[A,\, \rho].$$

On the other hand, one may also write

$$\beta\rho\dot{\hat{A}} = \rho\int_0^\beta d\tau\frac{d}{\hbar dt}A(t - i\beta)\Big|_{t=0} = \beta\rho\frac{d}{\hbar dt}\hat{A}\Big|_{t=0} = \beta\rho\dot{\hat{A}}$$

which establishes (15.8).

Problem 3.6: Eq. (15.10).

As in Eq. (7.8), the cyclicity of the trace allows one to write

$$\langle[a(\mathbf{q};\, t),\, b(-\mathbf{q};\, 0)]\rangle = \mathrm{Tr}\,([\rho,\, a(\mathbf{q};\, t)]b(-\mathbf{q};\, 0)).$$

Applying here the Kubo identity (15.8) and definition (15.9), the last expression becomes

$$i\hbar\beta\mathrm{Tr}\,(\rho\dot{\hat{a}}(\mathbf{q};\, t)b(-\mathbf{q};\, 0)) = i\hbar\int_0^\beta d\tau\langle\dot{a}(\mathbf{q};\, t - i\tau)b(-\mathbf{q};\, 0)\rangle$$
$$= -\langle(a(\mathbf{q};\, t - i\beta) - a(\mathbf{q};\, t))b(-\mathbf{q};\, 0)\rangle.$$

Insertion into Eq. (15.7) now gives

$$\chi_{ab}''(\mathbf{q};\, \omega) = \frac{1}{V}\int_{-\infty}^{+\infty} dt e^{i\omega t}\langle(a(\mathbf{q};\, t) - a(\mathbf{q};\, t - i\beta))b(-\mathbf{q};\, 0)\rangle.$$

Introducing in the second term of the bracket the shifted integration variable $t' = t - i\beta$ one obtains

$$\chi_{ab}''(\mathbf{q};\, \omega) = \frac{1}{V}(1 - e^{-\beta\omega})\int_{-\infty}^{+\infty} dt e^{i\omega t}\langle a(\mathbf{q};\, t)b(-\mathbf{q};\, 0)\rangle$$

which, in view of (7.43) is Eq. (15.10).

Problem 3.7: Eq. (15.19).

Substituting (15.18), $\varepsilon_{sk} = \hbar^2 k^2/2m_s$, $\mathbf{k} \cdot \mathbf{q} = kq\zeta$, $d\Omega_{\mathbf{k}} = 2\pi d\zeta$ and $V^{-1}\sum_{\mathbf{q}} = (2\pi)^{-3}\int d^3q$, the impurity-averaged Eq. (15.16) reads

$$\langle \rho_{\alpha\gamma}\rangle_{\text{imp}} = \frac{x}{v}\frac{\hbar}{e^2 n_s^2} N_s(\mu) \int \frac{d^3q}{(2\pi)^3} |\tilde{U}(\mathbf{q})|^2 q_\alpha q_\gamma 2\pi$$

$$\times \int_{-1}^{+1} d\zeta \delta\left(\hbar v_F q\left(\zeta - \frac{q}{2k_F}\right)\right).$$

Since $\tilde{U}(\mathbf{q})$ is assumed to be angle-independent, the integral over the angles of $\mathbf{q}$ is

$$\int d\Omega_{\mathbf{q}} q_\alpha q_\gamma = \frac{4\pi}{3} q^2 \delta_{\alpha\gamma},$$

and because of the identity $\delta(ax) = a^{-1}\delta(x)$, the integral over ζ gives $(1/\hbar v_F q)\theta(2k_F - q)$. Putting together the pieces one readily arrives at (15.19).

Problem 3.8: Eq. (15.28).

Equation (15.28) may be obtained by making use of the analogy with the operator $S(-i\tau, 0)$ defined by Eqs. (5.9), (5.14) and evaluated by integrating Eq. (5.20). We define

$$S(\tau) = e^{B\tau}e^{-(A+B)\tau}$$

so that

$$\dot{S}(\tau) = -e^{B\tau}Ae^{-(A+B)\tau} = -e^{B\tau}Ae^{-B\tau}S(\tau)$$

and hence

$$e^B e^{-(A+B)} = S(1) = T\exp\{-\int_o^1 d\tau e^{B\tau}Ae^{-B\tau}\}$$

where T is the "τ-ordering" operator. But due to the fact that $[AB]$ commutes with B, the exponent is easily evaluated. Indeed, in this case the identity

$$B^n A - AB^n = \sum_{k=1}^{n} B^{n-k}[B, A]B^{k-1}$$

has the value $n[B, A]B^{n-1}$ so that, taking the exponential series we find

$$e^{B\tau}A - Ae^{B\tau} = \sum_{n=1}^{\infty} \frac{\tau^n}{(n-1)!}[B, A]B^{n-1} = \tau[B, A]e^{B\tau}$$

and hence

$$e^B e^{-(A+B)} = T \exp\{-\int_o^1 d\tau(A + \tau[B, A])\}.$$

But here we may drop the T-operator and obtain

$$e^B e^{-(A+B)} = \exp\left\{-A - \frac{1}{2}[B, A]\right\}$$

which is Eq. (15.28).

Problem 3.9: Eq. (15.29).

Since in the unperturbed average, phonon operators b_q, b_q^+ belonging to different modes q are statistically independent, we may first prove Eq. (15.29) for one mode. Dropping the mode index we have, for arbitrary complex γ, γ',

$$\langle e^{\gamma b + \gamma' b^+}\rangle_o = e^{\gamma'\gamma/2}\langle e^{\gamma' b^+}e^{\gamma b}\rangle_o$$

where we used Eq. (15.28) in reverse. Making use of Eq. (9.1) we find

$$\langle e^{\gamma' b^+}e^{\gamma b}\rangle_o = \sum_{n,m=0}^{\infty} \frac{(\gamma')^n \gamma^m}{n!m!}\langle (b^+)^n b^m\rangle_o = \exp\{\gamma'\gamma\langle b^+ b\rangle_o\}$$

and

$$\langle e^{\gamma b + \gamma' b^+}\rangle_o = \exp\left\{\gamma'\gamma\left(\langle b^+ b\rangle_o + \frac{1}{2}\right)\right\} = \exp\left\langle\frac{1}{2}(\gamma b + \gamma' b^+)^2\right\rangle_o.$$

Writing now $C = \sum_q C_q$ with $C_q = \gamma_q b_q + \gamma'_q b_q^+$, statistical independence implies that

$$\langle e^C\rangle_o = \prod_q \langle e^{C_q}\rangle_o = \prod_q \exp\left\langle\frac{1}{2}C_q\right\rangle_o$$

$$= \exp\left\langle\frac{1}{2}\sum qq' C_q C_{q'}\right\rangle_o = \exp\left\langle\frac{1}{2}C^2\right\rangle_o$$

which is Eq. (15.29).

Problem 3.10: Eqs. (17.9), (17.19).

We immediately treat the general case (17.19), Eq. (17.9) being obtained simply by dropping the indices s and d. Combining Eq. (13.6) with the identity $f'_o = -\beta f_o(1 - f_o)$ we may write

$$f_s(k)n = f_o(\varepsilon_{sk})\{1 + \beta\phi_s(k)[1 - f_o(\varepsilon_{sk}]\},$$
$$1 - f_s(k) = [1 - f_o(\varepsilon_{sk})]\{1 - \beta\phi_s(k)f_o(\varepsilon_{sk})\}$$

and similarly with the labels sk and dp permuted. Hence, to first order in ϕ_s and ϕ_d,

$$f_s(k)f_d(p)[1 - f_s(k')][1 - f_d(p')] = f_o(\varepsilon_{sk})f_o(\varepsilon_{dp})$$
$$\times [1 - f_o(\varepsilon_{sk'})][1 - f_o(\varepsilon_{dp'})][1 + \beta X(kp; k'p')]$$

where

$$X(kp; k'p') \equiv \phi_s(k)[1 - f_o(\varepsilon_{sk})] + \phi_d(p)[1 - f_o(\varepsilon_{dp})]$$
$$- \phi_s(k')f_o(\varepsilon_{sk'}) - \phi_d(p')f_o(\varepsilon_{dp'}).$$

Insertion into (17.18) gives the generalization of Eq. (13.8),

$$\Gamma_s(kp; k'p') = \Gamma_{so}(kp; k'p')[1 + \beta X(kp; k'p')]$$

and similarly with index d. Using this expression in Eq. (17.8) with index s or d, together with the corresponding detailed balance (17.6), one finds

$$\left(\frac{df_s(k)}{\hbar dt}\right)_{\text{diss}} = \beta \sum_{k'pp'} \Gamma_{so}(kp; k'p')\{X(k'p'; kp) - X(kp; k'p')\}.$$

Here evaluation of the bracket immediately yields Eq. (17.19).

Problem 3.11: Eq. (17.43).

The contribution to Eq. (17.41) from the neighborhood of $\zeta' = -1$, $\lambda = 1$ is, expressed in the variables $\xi = 1 + \zeta'$, $\nu = 1 - \lambda$,

$$R_2(T) \cong \frac{1}{\pi^2 \beta E_s} \int_0^{1-\lambda_o} d\nu(1-\nu)^2 \int_0^{\nu(1+\mu)/\mu} \frac{d\xi}{S} \frac{x^2 e^x}{(e^x - 1)^2}$$

where

$$S^2 \equiv [2 - \nu(1 + \mu) + \mu\xi][\nu(1 + \mu) - \mu\xi]\xi(2 - \xi).$$

Assuming μ to be sufficiently small so that we may consider $\nu \ll 1$, we have

$$S^2 \cong 4\mu\left[\nu\frac{1+\mu}{\mu} - \xi\right]\xi.$$

Going over to the integration variable

$$x \cong 4\lambda(\lambda + \zeta')\beta E_d \cong 4\beta E_d(\xi - \nu)$$

we then have

$$R_2(T) \cong \frac{1}{2\pi^2\sqrt{\mu}\beta E_s} \int_0^{1-\lambda_o} d\nu \int_{-\mu x_o}^{x_o} \frac{dx}{\sqrt{x_o - x}\sqrt{\mu x_o + x}} \frac{x^2 e^x}{(e^x - 1)^2}$$

where $x_o \equiv 4\beta E_d \nu/\mu$. Introducing the function (17.42) we may write

$$R_2(T) \cong \frac{1}{2\pi^2\sqrt{\mu(1+\mu)}\beta E_s} \int_0^{1-\lambda_o} d\nu \left\{ \frac{K(x_o)}{\sqrt{x_o}} + \sqrt{\mu}\frac{K(\mu x_o)}{\sqrt{\mu x_o}} \right\}.$$

Splitting now the first ν-integration according to $x_o < 1$ and $x_o > 1$, i.e. $\nu < \nu_o$ and $\nu > \nu_o$ with $\nu_o \equiv \mu/4\beta E_d$ and using the estimates (17.42) one finds

$$\int_0^{1-\lambda_o} d\nu \frac{K(x_o)}{\sqrt{x_o}} \cong 2\int_0^{\nu_o} d\nu + \frac{\pi^2\nu_o}{2}\int_{\nu_o}^{1-\lambda_o}\frac{d\nu}{\nu}$$

$$= 2\nu_o\left(1 + \frac{\pi^2}{4}\ln\frac{1-\lambda_o}{\nu_o}\right)$$

and similarly for the second ν-integral,

$$\int_0^{1-\lambda_o} d\nu \frac{K(\mu x_o)}{\sqrt{\mu x_o}} \cong 2\int_0^{\nu_o/\mu} d\nu + \frac{\pi^2\nu_o}{2\mu}\int_{\nu_o/\mu}^{1-\lambda_o}\frac{d\nu}{\nu}$$

$$= 2\frac{\nu_o}{\mu}\left(1 + \frac{\pi^2}{4}\ln\frac{1-\lambda_o}{\nu_o/\mu}\right).$$

Adding the two contributions one readily finds the result (17.43).

Problem 3.12: Eq. (18.16).

Multiplying the left-hand side of (18.16) by iq_α and summing we may use Eq. (18.4). But the expression with $\dot{\rho}$ may now be transformed into a commutator by applying Eqs. (18.5) and (7.12) in reverse; hence

$$\beta V^{-1}iq_\alpha\langle\hat{j}_\alpha(\mathbf{q})j_\gamma(-\mathbf{q})\rangle = \frac{i}{\hbar V}\langle[\rho(\mathbf{q}), j_\gamma(-\mathbf{q})]\rangle.$$

Making use of (3.22), (3.23) and of the rule $[AB, C]_- = A[BC]_\mp \pm [AC]_\mp B$ one finds

$$[n(\mathbf{q}), \mathbf{j}_n(-\mathbf{q})] = \frac{\hbar}{m}\mathbf{q}n(0).$$

But

$$\langle n(0)\rangle = \sum_k f_o(\varepsilon_\mathbf{k}) = Vn,$$

so that, with $\rho(\mathbf{q}) = -en(\mathbf{q})$, $\mathbf{j}(\mathbf{q}) = -e\mathbf{j}_n(\mathbf{q})$, one finds (18.16) multiplied by iq_α and summed. The ambiguity of a part perpendicular to $\mathbf{q}$ is settled by observing that there is no other vector involved but $\mathbf{q}$ itself.

Problem 3.13: Eqs. (19.19), (19.20).

In dimensions $D > 1$ we may choose coordinates such that $\mathbf{l} = (l, \mathbf{0})$, $\mathbf{p} = k_F(\zeta, \sqrt{1 - \zeta^2}\hat{p}_\perp)$ and hence

$$\int \frac{d\Omega_{\mathbf{p}}}{\Omega_D} p_\alpha p_\gamma = k_F^2 \begin{cases} \zeta^2; & \alpha = \gamma = 1 \\ (1 - \zeta^2)/(D - 1); & \alpha = \gamma \neq 1 \end{cases}$$

where Ω_D is the solid angle in D dimensions introduced in (14.13). Averaging over the direction of $\mathbf{l}$ then leaves only $\alpha = \gamma = 1$ which, however is an arbitrary direction. Thus Eq. (19.18) may be written as $\delta F_{\alpha\gamma} = \delta F \delta_{\alpha\gamma}$ with

$$\delta F = -k_F^2 \sum_{\mathbf{l}} t(l) J_4(l) .$$

Here

$$J_4(l) = i\pi N(\mu) \int_{-1}^{+1} d\zeta \zeta^2 \int \frac{d\varepsilon}{2\pi i} \prod_{r=\pm} \prod_{s=\pm}$$

$$\times \frac{1}{\omega_r - (\varepsilon + s\lambda\zeta + \lambda^2/4\varepsilon_F) + i\omega_o \operatorname{sgn}\omega_r}$$

and $\lambda \equiv \hbar v_F l$, $\omega_\pm = \omega' \pm \omega/2$. Since, as with (18.53) and (19.16), $J_4 = 0$ unless $\operatorname{sgn}\omega_+ \neq \operatorname{sgn}\omega_-$, in which case $|\omega'| < \omega/2$, we may carry the integration path around the poles $\varepsilon = \omega_+ \pm \lambda\zeta - (\lambda^2/4\varepsilon_F) + i\omega_o$ and obtain

$$J_4(l) = i\pi N(\mu) \int_{-1}^{+1} d\zeta \frac{2\theta(\omega - 2|\omega'|)\zeta^2}{(\omega + 2i\omega_o)[(\omega + 2i\omega_o)^2 - 4\lambda^2\zeta^2]}$$

$$= -i\pi N(\mu) \frac{\theta(\omega - 2|\omega'|)}{\lambda^3(\omega + 2i\omega_o)} \left[\lambda + L\left(\frac{\omega}{2} + i\omega_o \right) L(\lambda) \right] .$$

Now insertion from (19.14) and (19.17) yields, with $\omega_o = \pi(x/\nu)N(\mu)|\tilde{U}|^2$,

$$\delta F = i\omega_o \frac{\theta(\omega - 2|\omega'|)}{\omega + 2i\omega_o} k_F^2 \Omega_D$$

$$\times \int_0^{\omega_o} \frac{l^{D-1}dl}{\lambda^3} \left\{ 1 - \frac{\omega_o}{\lambda} L(\lambda) \right\}^{-1} \left[\lambda + \left(\frac{\omega}{2} + i\omega_o \right) L(\lambda) \right] .$$

which is (19.19).

For $D = 1$ (19.16) becomes

$$J_2(l) = -i\pi N(\mu)\frac{2(\omega + 2i\omega_o)\theta(\omega - 2|\omega'|)}{(\omega + 2i\omega_o)^2 - 4\lambda^2}$$

while

$$J_4(l) = i\pi N(\mu)\frac{4\theta(\omega - 2|\omega'|)}{(\omega + 2i\omega_o)[(\omega + 2i\omega_o)^2 - 4\lambda^2]}.$$

Insertion from (19.14) and (19.17) now yields (19.20).

References to Chapter 3

1. *IBM J. Res. Develop.* **32**, no. 4 (July 1988). See also Webb, R.A. and Washburn, S., *Physics Today* **41**, no. 12 (December 1988), p. 46.
2. Stone, A.D. and Szafer, A., *IBM J. Res. Develop.* **32**, 384 (1988).
3. Anderson, P.W., *Phys. Rev.* **109**, 1492 (1958).
4. Lee, P.A. and Ramakrishnan, T.V., *Rev. Mod. Phys.* **57**, 287 (1985); Al'tshuler, B.L. and Lee, P.A., *Physics Today* **41**, no. 12 (December 1988), p. 36; Di Castro, C., in *Anderson Localization*, ed. Ando, T. and Fukuyama, H., Springer Proc. in Physics, Vol. 28 (1988), p. 96.
5. Mott, N.F., *Metal-Insulator Transitions* (Taylor and Francis, London, 1974).
6. Abrahams, E., Anderson, P.W., Licciardello, D.C., and Ramakrishnan, T.V., *Phys. Rev. Lett.* **42**, 673 (1979).
7. Von Klitzing, K., Dorda, G. and Pepper, M., *Phys. Rev. Lett.* **45**, 494 (1980).
8. *The Quantum Hall Effect*, ed. Prange, R.E. and Girvin, S.M. (Springer, New York, 1987).
9. Ziman, J.M., *Electrons and Phonons* (Clarendon, Oxford, 1962).
10. Wilson, A.H., *The Theory of Metals* (Cambridge University Press, 1953).
11. Landauer, R., *IBM J. Res. Develop.* **1**, 223 (1957); *Philos. Mag.* **21**, 863 (1970).
12. Büttiker, M., *IBM J. Res. Develop.* **32**, 317 (1988).
13. de Groot, S.R. and Mazur, P., *Non-Equilibrium Thermodynamics* (North-Holland, Amsterdam, 1962).
14. Kubo, R., Toda, M. and Hashitsume, N., *Statistical Physics II. Nonequilibrium Statistical Mechanics* (Springer, Berlin, 1985).
15. Onsager, L., *Phys. Rev.* **37**, 405 (1931); **38**, 2265 (1931); Casimir, H.B.G., *Rev. Mod. Phys.* **17**, 343 (1945).
16. Landau, L.D. and Lifshitz, E.M., *Statistical Physics* (Pergamon, Oxford, 1958).
17. Kubo, R., *J. Phys. Soc. Jpn.* **12**, 570 (1957).
18. Ashcroft, N.W. and Mermin, N.D., *Solid State Physics* (Holt, Rinehart and Winston, New York, 1976).
19. Einstein, A., *Ann. Phys. (Leipzig)* **17**, 549 (1905); *ibid.* **19**, 371 (1906).

20. Büttiker M.,*Festkörperprobleme* **30**, 41 (1990); *Phys. Rev.* **B41**, 7906 (1990); van Wees, B.J. et al., *Phys. Rev. Lett.* **60**, 848 (1988); Wharam, D.A. et al., *J. Phys.* **C21**, L209 (1988).

21. Mills, D.L. and Lederer, P., *J. Phys. Chem. Solids* **27**, 1805 (1966).

22. Nordheim, L., *Ann. Phys. (Leipzig)* (5) **9**, 641 (1931). See also Mott, N.F. and Jones, H., *The Theory of the Properties of Metals and Alloys* (Dover, New York, 1958), Chapter VII, Sec. 13.

23. Caplin, A.D. and Rizzuto, C., *Phys. Rev. Lett.* **21**, 746 (1968).

24. Kondo, J., *Prog. Theor. Phys.* **32**, 37 (1964). See also the reviews by Kondo, J. and by Heeger, A.J. in *Solid State Physics*, ed. Seitz, F., Turnbull, D. and Ehrenreich, H. (Academic, New York, 1969).

25. Wiegmann, P.B., "An exact solution of the Kondo problem", in *Quantum Theory of Solids*, ed. Lifshits, I.M. (MIR, Moscow, 1982), p. 238. See also Tsvelick, A.M. and Wiegmann, P.B., *Adv.Phys.* **32**, 453 (1983); Andrei, N., Furuya, K. and Lowenstein, J.H., *Rev. Mod. Phys.* **55**, 331 (1983).

26. Anderson, P.W., *J. Phys.* **C3**, 2346 (1970).

27. Wilson, K.G., *Rev. Mod Phys.* **47**, 773 (1975).

28. Van Hove, L., *Phys. Rev.* **95**, 249 (1954). See also Kittel, C., *Quantum Theory of Solids* (Wiley, New York, 1963), Chapter 19.

29. Baym, G., *Phys. Rev.* **135**, A1691 (1964).

30. Bloch, F., *Z. Phys.* **52**, 555 (1928); **59**, 208 (1930). See also Sommerfeld, A. and Bethe, H., "Elektronentheorie der Metalle", reprint from *Handbuch der Physik*, ed. Geiger, H. and Scheel, K., Bd. 24/2 (Springer, Berlin, 1967), Sections 35–38.

31. Kittel, C., *Introduction to Solid State Physics, 5th edition* (Wiley, New York, 1976).

32. Grüneisen, E.,*Ann. Phys. (Leipzig)* **16**, 530 (1933).

33. Baber, W.G., *Proc. Roy. Soc.* **A158**, 383 (1937).

34. Wilson, A.H., *Proc. Roy. Soc.* **A133**, 458 (1931); **134**, 277 (1931). See also Blatt, F.J., *Physics of Electronic Conduction in Solids* (McGraw-Hill, New York, 1968), Chapter 8.

35. Harrison, W.A., *Phys. Rev.* **104**, 1281 (1956).

36. Enz, C.P., *Helv. Phys. Acta* **27**, 199 (1954); *Physica* **20**, 983 (1954).

37. Anderson, P.W., *Phys. Rev.* **79**, 350 (1950).

38. Levin, K., Kim, Ju H., Lu, J.P., and Qimiao, Si, *Physica* **C175**, 449 (1991).

39. Enz, C.P., *Z. Phys.* **B80**, 317 (1990); *Proc. Int. Seminar High Temp. Supercond.*, ed. Aksenov, V.L., Plakida, N.M. and Yushankhai, V.Yu. (Dubna, 1990), p. 210.

40. Hodges, C., Smith, H. and Wilkins, J., *Phys. Rev.* **B4**, 302 (1971). See also Giuliani, G.F. and Quinn, J.J., *Phys. Rev.* **B26**, 4421 (1982).

41. Wannier, G., *Elements of Solid State Theory* (Cambridge University Press, 1960).

42. Steglich, F., *Europhysics News* **20**, 159 (1989). See also Brandt, N.B. and Moshchalkov, V.V., *Adv. Phys.* **33**, 373 (1984).

43. **Lee, P.A., Rice, T.M., Serene, J.W., Sham, L.J.,** and **Wilkins, J.W.,** *Comments on Condensed Matter Phys.* **B12**, 99 (1986). See also **Fulde, P., Keller, J.** and **Zwicknagl, G.,** *Solid State Phys.*, ed. Ehrenreich, H. and Turnbull, D. (Academic, San Diego, 1988), p. 1.

44. **Stewart, G.R.,** *Rev. Mod. Phys.* **56**, 755 (1984).

45. **Green, M.S.,** *J. Chem. Phys.* **20**, 1281 (1952); **22**, 398 (1954); **Greenwood, D.A.,** *Proc. Phys. Soc.* **71**, 585 (1958).

46. **Mori, H.,** *Prog. Theor. Phys.* **34**, 399 (1965), Section 5. See also **Suezaki, Y.** and **Mori, H.,** *Prog. Theor. Phys.* **41**, 1177 (1969); **Enz, C.P.,** *Phys. Rev.* **B25**, 6822 (1982).

47. **Huberman, M.** and **Chester, G.V.,** *Adv. Phys.* **24**, 489 (1975). See also **Argyres, P.N.** and **Resendes, D.G.,** *J. Phys. Condens. Matter* **1**, 7001 (1989).

48. **Abrikosov, A.A., Gor'kov, L.P.** and **Dzyaloshinskii, I.Ye.,** *Quantum Field Theoretical Methods in Statistical Physics, 2nd ed.* (Pergamon, Oxford, 1965).

49. **Loss, D.** and **Thellung, A.,** *Physica* **144A**, 17 (1987); **Loss, D.** and **Schoeller, H.,** *Physica* **A150**, 199 (1988) and references therein.

50. **Reizer, M.Yu.,** *Phys. Rev.* **B39**, 1602 (1989).

51. **Al'tshuler, B.L., Aronov, A.G., Khmelnitskii, D.E.,** and **Larkin, A.I.,** "Coherent effects in disordered conductors", in *Quantum Theory of Solids,* ed. Lifshits, I.M. (MIR, Moscow, 1982), p. 130.

52. **Thouless, D.J.,** *Phys. Rev. Lett.* **39**, 1167 (1977).

53. **Mott, N.F.** and **Davis, E.A.,** *Electronic Processes in Non-crystalline Materials, 2nd ed.* (Clarendon, Oxford, 1979). See also **Ioffe, A.F.** and **Regel, A.R.,** *Prog. Semicond.* **4**, 237 (1960).

54. **Anderson, P.W.,** *Philos. Mag.* **B52**, 505 (1985).

55. **Von Albada, M.P.** and **Lagendijk, A.,** *Phys. Rev. Lett.* **55**, 2692 (1985); **Wolf, P.E.** and **Maret, G.,** *Phys. Rev. Lett.* **55**, 2696 (1985); **Etemad, S., Thompson, R.** and **Andrejco, M.C.,** *Phys. Rev. Lett.* **57**, 575 (1986); **Kaveh, M., Rosenbluth M., Edrei I.** and **Freund, I.,** *Phys. Rev. Lett.* **57**, 2049 (1986).

56. **Condat, C.A.** and **Kirkpatrick, T.R.,** *Phys. Rev. Lett.* **58**, 226 (1987); **Agranovich, V.M., Kravtsov, V.E.,** and **Lerner, I.V.,** *Phys. Lett.* **A125**, 435 (1987); **De Raedt, H., Lagendijk, A.** and **de Vries, P.,** *Phys. Rev. Lett.* **62**, 47 (1989); **Genack, A.Z.** and **Garcia, N.,** *Phys. Rev. Lett.* **66**, 2064 (1991). See also **John, S.,** *Physics Today* **44**, no. 5 (May 1991), p. 32; no. 6 (June 1991), p. 17.

57. **Guazzelli, E.** and **Guyon, E.,** *J. Phys. (Paris) Lett.* **44**, L-837 (1983); **Dépollier, C., Kergomard, J.** and **Laloe, F.,** *Ann. Phys. (Paris)* **11**, 475 (1986).

58. **Courtens, E., Pelous, J., Phalippou, J., Vacher, R.,** and **Woignier, T.,** *Phys. Rev. Lett.* **58**, 128 (1987); **Aharony, A., Alexander, S., Entin-Wohlman, O.,** and **Orbach, R.,** *Phys. Rev. Lett.* **58**, 132 (1987).

59. Bergmann, G., *Phys. Rev.* **B28**, 2914 (1983); *Phys. Reports* **107**, 1 (1984). See also Khmel'nitskii, D.E., *Physica* **126B**, 235 (1984).
60. Langer, J.S. and Neal, T., *Phys. Rev. Lett.* **16**, 984 (1966).
61. Gor'kov, L.P., Larkin, A.I. and Khmel'nitskii, D.E., *Soviet Phys.-JETP Lett.* **30**, 228 (1979).
62. Salpeter, E.E. and Bethe, H.A., *Phys. Rev.* **84**, 1232 (1951). See also Itzykson, C. and Zuber, J.-B., *Quantum Field Theory* (McGraw-Hill, New York, 1980), Chapter 10.
63. Sadovskii, M.V., in *Soviet Scientific Reviews, Section A. Physics Reviews*, ed. Khalatnikov, I.M. (Harwood Academic Publ., Chur, Switzerland, 1986), Vol. 7, p. 1.
64. Stanley, H.E., *Introduction to Phase Transitions and Critical Phenomena* (Oxford University Press, New York, 1971).
65. Wegner, F.J., *Z. Phys.* **25**, 327 (1976).
66. Wilson, K.G. and Kogut, J., *Phys. Reports* **12C**, 75 (1974).
67. Itzykson, C. and Zuber, J.-B., *Quantum Field Theory* (McGraw-Hill, New York, 1980), Chapter 13.
68. Enz, C.P. and Flores, J.C., *Helv. Phys. Acta* **61**, 1079 (1988).

Chapter 4

SUPERCONDUCTIVITY

Although, historically, from Kamerlingh Onnes in 1911[1] to Bednorz and Müller in 1986[2], superconductivity was discovered by observing the disappearance of resistivity, this phenomenon is an *equilibrium state* of matter. The reason why Kamerlingh Onnes had concluded that "mercury has passed into a new state which, on account of its remarkable electrical properties, may be called the superconductive state", is that addition of impurities did not increase the resistance; mercury therefore could not be a perfect conductor (see Chapter 1 of Ref. 3 for an excellent introduction to the early history).

A *perfect conductor* is characterized by a vanishing internal electric field, $\mathbf{E} = 0$, so that Faraday's law of induction implies that $\dot{\mathbf{B}} = 0$, but allows an arbitrary value $\mathbf{B} = $ const to be frozen in. The distinguishing feature of a superconductor then is the value $\mathbf{B} = 0$ or *perfect diamagnetism*, $\chi = -1/4\pi$, that is, the *Meissner effect*[4]. The exclusion of $\mathbf{B}$ is the effect of a surface current induced by the external field H. But since current densities j are limited by a critical value j_c (see Eqs. (27.24), (27.26) below), this also implies the existence of a *critical magnetic field* H_c (see Eq. (27.27) below). This and the *critical temperature* T_c, above which superconductivity breaks down, are the most important phenomenological parameters, both from a practical and from a theoretical point of view.

The fact that theoretical understanding came only in 1957 with the fundamental work of Bardeen , Cooper and Schrieffer (BCS)[5] has several reasons. First, it had to be realized that superconductivity, particularly at $T = 0$, is a *macroscopic quantum state* analogous to superfluid ^{4}He. This analogy led Ginzburg , Feynman and Schafroth independently to show that a charged Bose gas exhibits a Meissner effect below its Bose-Einstein condensation (see Ref. 6, p. 45) and eventually motivated Schafroth[7] to postulate *electron pairs* (see Ref. 6, pp. 86–87).

Second, the *relevant interaction* between the electrons had to be identified. The most interesting among the early proposals is due to Heisenberg[8] who argued that *Coulomb repulsion* would condense a fraction ω of the electrons at the Fermi surface locally into a lattice whose motion would result in a supercurrent. These local electron lattices would form domains as in a ferromagnet but containing closed loops of supercurrents. Different fractions ω of the Fermi surface would result in different orientations of the current loops and hence to a vanishing bulk current. Although historically, Heisenberg's proposal met with severe criticism mainly by F. London[3] and was soon forgotten, it is of interest from the point of view of the history of ideas:

Wigner in 1934 had arrived at an electron lattice by calculating what he called the "correlation energy"[9]. And in the new high-T_c cuprate superconductors both, Coulomb repulsion and electron (or rather hole) correlation play a crucial role in the 2-dimensional CuO_2 sublattice perpendicular to the c-axis. This is reflected by the enormous interest in the *Hubbard model* (see Section 33) which gives a very concise formulation of the problem. In particular, at half-filling where there is one carrier per site, this model exhibits antiferromagnetic order[10] which is known[11] to be the groundstate of the undoped versions, $x = 0$, of the cuprates $La_{2-x}M_xCuO_4$ (M=Ba, Sr) of Bednorz and Müller[2] and $YBa_2Cu_3O_{6+x}$ of Wu *et al.* [12].

Historically, however it was not the Coulomb repulsion but the *electron-phonon interaction* that was recognized to be relevant for superconductivity. In 1950 Fröhlich showed that electron-phonon coupling gives rise to an effective attraction between the electrons near the Fermi surface, and this idea was supported by the independent discovery in the same year of the *isotope effect*[3]. But a third difficulty that had to be overcome was of formal nature: The perturbative methods as described in Chapter 2 turned out to be insufficient; the problem is *nonlinear* and corresponds to the summation of infinite series of diagrams. The solution of such a nonlinear problem was achieved for the first time by Fröhlich[13] in a 1-dimensional model in which the electron-phonon interaction gives rise self-consistently to a periodic lattice displacement and to a commensurate charge density wave and opens a gap at the Fermi energy. The lattice displacement and charge density waves together can move rigidly through the lattice thus resulting in a supercurrent. The solution of the general nonlinear problem of superconductivity was given by BCS in their *gap equation* (Eq. (2.39) of Ref. 5): It determines the *energy gap* Δ of the condensate formed by *Cooper pairs* [14] of loosely coupled electrons localized close to the Fermi surface, as opposed to the *Schafroth pairs* [7] which are quasi-bound states localized in physical space.

The discovery of the new cuprate superconductors has again called into question this traditional phonon-mediated pairing mechanism; the presence of electron pairs, however, is confirmed by the observation of the charge $2e$ [15]. On the one hand, the *coherence length* ξ_o which is a measure of the spatial extension of the electron pairs (see Ref. 3, p. 156), is about 100 times smaller than in conventional superconductors [3], $\xi_c \sim 1.5 - 4$ Å parallel and $\xi_{ab} \sim 14 - 30$ Å perpendicular to the c-axis [16] (see also Batlogg in Ref. 17, p. 44). This suggests that the cuprates might be closer to Schafroth pairing and strong coupling, particularly along the c-axis, and raises the question of the existence of a *pairing temperature T_p* above the *condensation temperature T_c*. But so far there is no indication of this eventuality.

On the other hand, the discussion of pairing mechanisms in the cuprate superconductors has focussed essentially around the alternative of a Fermi-liquid or a non-Fermi-liquid description. The former further subdivides into a *conventional Fermi liquid* picture dominated by a *charge-transfer mechanism* between the conduction layers of the CuO_2 planes and the out-of-plane charge-reservoir layers (see Jorgenson in Ref. 17, p. 34) and a *non-conventional Fermi liquid* picture which is dominated by strong *antiferromagnetic correlations* giving rise to "*spin bags*" (see Schrieffer in Ref. 17, p. 55).

The non-Fermi-liquid point of view has been defended with force by Anderson from the beginning of the high-T_c adventure [18]. Anderson's argument is that the strong correlations in the CuO_2 planes are best described by a single-band Hubbard model with on-site repulsions which are so strong that, within the planes, there is *separation between spin and charge* of the hole in the unit cell. The spins have the characteristics of neutral fermions, the "*spinons*" which constitute a "spin liquid" with a "pseudo Fermi surface" and, in the groundstate, form "*resonating valence bonds*" (RVB) among each other which on the average have the antiferromagnetic configuration. The charged bosons or "*holons*", on the other hand, carry the in-plane current (they cannot Bose-condense because a flux unit is attached to them) while interplane transport and pairing requires non-separated holes (see Anderson in Ref. 17, p. 55).

Finally, taking the two-dimensionality of the CuO_2 planes at face value, charged excitations with non-conventional statistics or "*anyons*" may be invoked to explain the superconductivity of the cuprates[19]. "Any-ons" may pick up "any" phase φ between 0 and 2π when a pair of them is exchanging places by an in-plane rotation of π. According to this definition, bosons and fermions are anyons with $\varphi = 0$ and π, respectively. While anyons with $\varphi \neq 0$ cannot condense, pairs $(\varphi, -\varphi)$ should be able to. However, the particular anyons with $\varphi = \pi/2$ called "*semions*" seem to be the ones of interest for the cuprates. One problem with anyons is their violation of parity and of time reversal, a feature which, in principle may be verified experimentally (see Anderson in Ref. 17, p. 60).

20. Effective Electron Attraction. Cooper Pairs and Schafroth Pairs

Starting from the electron-phonon interaction in the form (10.6), the effective coupling among the electrons may be obtained in different ways. The traditional method first used by Fröhlich (see Section 1.8 of Ref. 3) is to perform a canonical transformation[20]

$$\mathcal{H} = \mathcal{H}_o + \mathcal{H}_{\text{int}} \to \tilde{\mathcal{H}} = e^{-S}\mathcal{H}e^S = \mathcal{H} + [\mathcal{H}, S] + \frac{1}{2}[[\mathcal{H}, S]S] + \cdots , \quad (20.1)$$

$\psi \to \tilde{\psi} = \exp(-S)\psi$, and to determine S by the condition $[S\mathcal{H}_o] = \mathcal{H}_{\text{int}}$. However, since such a transformation has no immediate physical interpretation, we prefer the equivalent method of considering $\mathcal{H}_{\text{int}} = \mathcal{H}^{\text{el}-\text{ph}}$ as linear response (7.10), coupling the phonon coordinate $b(\mathbf{q}) = Q_q$ to the external field

$$\tilde{F}(\mathbf{q}, t) = V\gamma_{-q}n(\mathbf{q}, =, t)e^{\epsilon t} , \quad (20.2)$$

where $\epsilon = 0^+$, switched on at $t = -\infty$. Then, according to (7.11), (7.12), (7.16), the phonon-averaged response is

$$\delta\langle Q_q\rangle^{\text{ph}}_t = \gamma_{-q} \int_{-\infty}^{+\infty} dt' n(\mathbf{q}, t')D^{\text{ret}}(q; t - t') \quad (20.3)$$

where $D^{\text{ret}}(q; t)$ is the retarded phonon propagator (7.15) with $A = Q_q$ and $B = Q_{-q}$.

The effective electron interaction is obtained by replacing in (10.6) Q_q by $\frac{1}{2}\delta\langle Q_q\rangle^{\text{ph}}_{t=0}$ where the factor $1/2$ corrects for double counting. Using in $n(\mathbf{q}; t)$ the unperturbed time dependence (7.36) and introducing the Fourier transformed propagator (7.20), the effective interaction becomes

$$\mathcal{H}_{\text{eff}} = \frac{1}{2} \sum_q |\gamma_q|^2 \sum_{kk'} a_{k'}^+ a_{k'-q} a_k^+ a_{k+q} \tilde{D}^{\text{ret}}(q; \varepsilon_{k+q} - \varepsilon_k + i\epsilon) . \quad (20.4)$$

Apart from a trivial term of the form $\sum_k \alpha_k a_k^+ a_k$, the same result may be obtained by the canonical transformation (20.1) (Problem 4.1). Making use of (7.33), (7.35) and (7.45), the retarded phonon propagator may be approximated by the unperturbed expression

$$\tilde{D}^{\text{ret}}(q; z) \simeq \frac{\omega_q}{z^2 - \omega_q^2} ; \quad \text{Im } z > 0 \quad (20.5)$$

which, inserted into (20.4) yields the explicit form of the effective interaction used as starting point by BCS (see Eq. (2.4) of Ref. 5).

According to Eq. (20.5), the attractive part of $\mathcal{H}_{\text{eff}}$ is determined by the condition $|\varepsilon_{k+q} - \varepsilon_k| < \omega_q$ which, for quasi-free electrons $\varepsilon_{\mathbf{k}} = \hbar^2(\mathbf{k}^2 - k_F^2)/2m$ and for acoustic phonons $\omega_q = \hbar c_{\|} q$, reduces to

$$\left| \mathbf{k} \cdot \hat{q} + \frac{q}{2} \right| < k_o \quad (20.6)$$

where $\hat{q} = \mathbf{q}/q$ and $k_o = mc_{\|}/\hbar$ was introduced in Eq. (16.25). Since $k \sim k_F$ and $k_o/k_F = c_{\|}/v_F \ll 1$, condition (20.6) defines, for a given $\mathbf{k}$, a narrow domain ϕ of $\mathbf{q}$-vectors, shown in Fig. 4.1, which contribute an attractive matrix element to (20.4). The important transitions, however, involve phonons of energy of the order of $k_B T_c \ll k_B \theta_D$ [5] where θ_D is the Debye temperature defined in Eq. (16.26). This corresponds to choosing $k_o \ll mc_{\|}/\hbar$ and to $|\varepsilon_{k+q} - \varepsilon_{\mathbf{k}}| \ll \omega_q$. The contribution of this restricted domain ϕ is the interaction underlying BCS theory[5], namely

$$\mathcal{H}_{\text{eff}} = -\frac{1}{2V} \sum_{\mathbf{kk'q}} \sum_{\sigma\sigma'} \Lambda_{\mathbf{kk'q}} a_{\mathbf{k}\sigma}^+ a_{\mathbf{k'}\sigma'}^+ a_{\mathbf{k'+q},\sigma'} a_{\mathbf{k-q},\sigma} \quad (20.7)$$

where hermiticity requires that

$$\Lambda_{\mathbf{kk'q}} = \Lambda_{\mathbf{k-q,k'+q,-q}}^* \quad (20.8)$$

and the terms of (20.4) have been rearranged, resulting in an unessential contribution which may be absorbed in the free-electron Hamiltonian. Equation (20.7) is of the form (3.38) and hence may be represented by Fig. 1.2 of Chapter 1.

Equation (20.4) now is the special case of (20.7) in which the coupling function is, according to (16.5), given by

$$\Lambda_{\mathbf{kk'q}} = \begin{cases} \Lambda = V|\gamma_q|^2/\omega_q = |\tilde{\phi}(0)|^2/vMc_\parallel^2 > 0 \; ; & \mathbf{k}, \, \mathbf{k'}, \, \mathbf{q} \in \phi \\ 0 \; ; & \text{otherwise} \, . \end{cases} \tag{20.9}$$

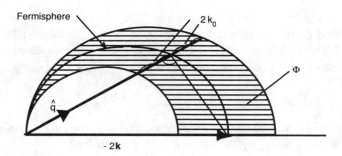

Fig. 4.1: The domain ϕ of **q**-vectors determined by condition (20.6).

This value of Λ may be shown to be independent of the atomic mass M (Problem 4.2). Instead of the complicated domain ϕ of Fig. 4.1 we define the interaction domain more generally by the ω_o-*neighborhood* of the Fermi surface,

$$\Lambda_{\mathbf{kk'q}} = \Lambda\theta(\omega_o - |\varepsilon_{\mathbf{k}}|)\theta(\omega_o - |\varepsilon_{\mathbf{k-q}}|)\theta(\omega_o - |\varepsilon_{\mathbf{k'}}|)\theta(\omega_o - |\varepsilon_{\mathbf{k'+q}}|) \tag{20.10}$$

where in the case of the phonon-mediated coupling (20.9), we may use Eq. (16.4) to set $\omega_o \simeq k_B\theta$, i.e. to define it as an average phonon energy[5]. According to Fig. 1.1 of Chapter 1, this case should be represented by a diagram with a wavy internal line instead of Fig. 1.2.

We wish to explain superconductivity as condensation of electron pairs which may be assumed to be in a singlet state (anti-parallel electron spins) since the Hamiltonian conserves angular momentum . Triplet pairing is of course also possible and is realized in the superfluid state of ${}^3\text{He}$[21]. However, there is no clear evidence for triplet superconductivity in metals. The most general singlet pair creation operator may be written as[22]

$$B_{\mathbf{q}}^+ = V^{-1/2}\sum_{\mathbf{k}} \varphi_{\mathbf{k}} a_{\mathbf{k+q}/2,\uparrow}^+ a_{-\mathbf{k+q}/2,\downarrow}^+ \, . \tag{20.11}$$

With the aid of definition (3.41) one easily shows that

$$[\mathbf{P}_{\text{el}}, B_{\mathbf{q}}^+] = \mathbf{q}B_{\mathbf{q}}^+ \tag{20.12}$$

which means that the pair has momentum $\hbar\mathbf{q}$. $\varphi_{\mathbf{k}}$ is then recognized as the Fourier transform (2.3) of the *pair wavefunction* $\tilde{\varphi}(\mathbf{r})$ in the center-of-mass system which may be assumed to be real and normalized to 1, so that $\varphi_{-\mathbf{k}} = \varphi_{\mathbf{k}}^*$ and $V^{-1} \sum_{\mathbf{k}} |\varphi_{\mathbf{k}}|^2 = 1$. The $B_{\mathbf{q}}^+$ behave somewhat like Bose operators since $[B_{\mathbf{q}}^+ B_{\mathbf{q}'}^+] = 0$. However, with the above assumptions about $\varphi_{\mathbf{k}}$ one finds that

$$[B_{\mathbf{q}} B_{\mathbf{q}'}^+] = \delta_{\mathbf{q}\mathbf{q}'} - V^{-1} \sum_{\mathbf{k}\sigma} 2\mathrm{Re}\left(\varphi_{\mathbf{k}} \varphi_{\mathbf{k}+\frac{\mathbf{q}-\mathbf{q}'}{2}}^*\right) a_{\mathbf{k}+\frac{\mathbf{q}'}{2},\sigma}^+ a_{\mathbf{k}+\mathbf{q}-\frac{\mathbf{q}'}{2},\sigma} . \quad (20.13)$$

In equilibrium, condensation must occur in the zero-momentum state. Therefore, it is interesting to see whether B_o^+ behaves as a boson operator, at least when acting on the normalized Fermi-sea state $|F\rangle$ introduced in Section 11. Since from (20.13) one easily deduces that, setting the Fermi level at $\mu = 0$,

$$\langle F|[B_{\mathbf{q}} B_{\mathbf{q}'}^+]|F\rangle = \left\{1 - \frac{4}{V} \sum_{\mathbf{k}} |\varphi_{\mathbf{k}}|^2 \theta(-\varepsilon_{\mathbf{k}+\frac{\mathbf{q}}{2}})\right\} \delta_{\mathbf{q}\mathbf{q}'} , \quad (20.14)$$

the Bose property of B_o^+ is obtained, provided that $\varphi_{\mathbf{k}} = 0$ for $\varepsilon_{\mathbf{k}} < 0$, i.e. for states which are occupied in $|F\rangle$. This in turn gives the correct normalization of the one-pair state $B_o^+|F\rangle$ since

$$\langle F|B_{\mathbf{q}} B_{\mathbf{q}}^+|F\rangle = V^{-1} \sum_{\mathbf{k}} |\varphi_{\mathbf{k}}|^2 \theta(\varepsilon_{\mathbf{k}+\frac{\mathbf{q}}{2}}) \theta(\varepsilon_{\mathbf{k}-\frac{\mathbf{q}}{2}}) . \quad (20.15)$$

Defining the system by the Hamiltonian $\mathcal{H} = \mathcal{H}_o + \mathcal{H}_{\mathrm{eff}}$ with $\mathcal{H}_o = \sum \varepsilon_{\mathbf{k}} a_k^+ a_k$ and $\mathcal{H}_{\mathrm{eff}}$ given by (20.7), (20.10), the energy in this single-pair state as compared to the energy in the Fermi-sea state is given by (Problem 4.3)

$$E \equiv \langle F|B_o \mathcal{H} B_o^+|F\rangle = \langle F|\mathcal{H}|F\rangle = \frac{2}{V} \sum_{\mathbf{k}} \varepsilon_{\mathbf{k}} |\varphi_{\mathbf{k}}|^2 \theta(\varepsilon_{\mathbf{k}})$$
$$- \Lambda \left| V^{-1} \sum_{\mathbf{k}} \varphi_{\mathbf{k}} \theta(\varepsilon_{\mathbf{k}}) \theta(\omega_o - |\varepsilon_{\mathbf{k}}|) \right|^2 . \quad (20.16)$$

Setting here $\varphi_{\mathbf{k}} = 0$ for $\varepsilon_{\mathbf{k}} < 0$ makes $\theta(\varepsilon_{\mathbf{k}}) = 1$ and gives the correct normalization so that E is the true binding energy.

The pair wavefunction may be determined by minimizing the functional $E[\varphi_{\mathbf{k}}]$, subject to the normalization condition on $\varphi_{\mathbf{k}}$,

$$0 = \frac{V}{2} \frac{\partial}{\partial \varphi_{\mathbf{k}}^*} \left(E - \frac{\lambda}{V} \sum_{\mathbf{k}} |\varphi_{\mathbf{k}}|^2 \right) = (2\varepsilon_{\mathbf{k}} - \lambda)\varphi_{\mathbf{k}}$$

$$- \frac{\Lambda}{V} \sum_{\mathbf{k}'} \varphi_{\mathbf{k}'} \theta(\omega_o - \varepsilon_{\mathbf{k}'}) \theta(\omega_o - \varepsilon_{\mathbf{k}}) \ . \tag{20.17}$$

Multiplication with $V^{-1}\varphi_{\mathbf{k}}^*$ and summation determines the Lagrange multiplier: $\lambda = E$. The solution of (20.17) then is $\varphi_{\mathbf{k}} = 0$ if $\varepsilon_{\mathbf{k}} > \omega_o$ and hence

$$\varphi_{\mathbf{k}} = \frac{\Lambda}{V} \gamma \frac{\theta(\omega_o - \varepsilon_{\mathbf{k}})}{2\varepsilon_{\mathbf{k}} - E} \ . \tag{20.18}$$

Here $\gamma = \sum \varphi_{\mathbf{k}}$ is determined by the normalization condition. Inserted into (20.11) this solution defines a *Cooper pair* . By summing over $\mathbf{k}$ and introducing the density of states (10.4) one obtains

$$1 = N(0)\Lambda \int_0^{\omega_o} \frac{d\varepsilon}{2\varepsilon - E} \ , \tag{20.19}$$

and integrating one finds the binding energy first calculated by Cooper[14] (see Ref. 3, Section 4.2),

$$E = -2\omega_o \exp\left(-\frac{2}{N(0)\Lambda} \right) \ . \tag{20.20}$$

This binding is extremely weak and vanishes rapidly when Λ goes to zero. The remarkable feature of this solution is that it is non-analytic at $\Lambda = 0$ and hence cannot be obtained by perturbation theory.

It was the solution of this problem of Cooper's that led BCS to the correct form of their groundstate . Obviously, Cooper's single-pair state $B_o^+|F\rangle$ has to be generalized to a state in which a large number of pairs from the energy interval $|\varepsilon_{\mathbf{k}}| < \omega_o$ is excited. It turns out that this is easier to formulate by operating directly on the vacuum state $|0\rangle$ and not on $|F\rangle$. To see this, we first turn to the opposite limit of *Schafroth pairs*, defined here as strongly bound electron pairs (see, however, Ref. 6, Section III.9). The condition is that the spatial extension of the pair wavefunction, measured by the coherence length ξ_o (see Eq. (22.7) below), is small compared to the average distance between pairs, $(n/2)^{-1/3}$, or $n\xi_o^3 \ll 1$, n being the electron density. In this case all electrons of the band are paired and the $N = Vn/2$ pairs form a dilute Bose gas[22]. And since there is no Fermi sea, the pair operators act on the vacuum state. From (20.13) then follows that

$$\langle 0|[B_{\mathbf{q}}B_{\mathbf{q}'}^+]|0\rangle = \delta_{\mathbf{q}\mathbf{q}'} \ , \tag{20.21}$$

i.e. the $B_{\mathbf{q}}^+$ behave as true Bose operators. Note that this situation corresponds to the "no overlap" approximation of Schafroth , Butler and Blatt (see Ref. 6, Section III.8).

A dilute gas of bosons has a *Bose condensation* such that, at zero temperature, all N bosons are in the (normalized) condensate state $|C\rangle$ which must have the form $\Gamma(B_o^+)|0\rangle$. Here the function Γ is determined by the condition that

$$\langle C|N_B|C\rangle = \langle C|B_o^+ B_o|C\rangle = N \qquad (20.22)$$

where $N_B \equiv \sum B_{\mathbf{q}}^+ B_{\mathbf{q}}$ is the boson-number operator. This condition has two solutions, namely the *Fock state* with $\Gamma(B_o^+) = (N!)^{-1/2}(B_o^+)^N$ and the *coherent state* defined below which has $\Gamma(B_o^+) = \exp(-(N/2) + \sqrt{N}B_o^+)$. The corresponding states $|C\rangle$ are, respectively the *canonical* and the *grand-canonical groundstates*, which are distinguished by the respective value 0 and $\sqrt{N}$ of the *number fluctuation*

$$\Delta N_B = (\langle C|(N_B - N)^2|C\rangle)^{1/2} . \qquad (20.23)$$

It is interesting to note at this point that Schafroth failed to find the correct form of the groundstate because he considered only Fock states and not coherent states (see Eqs. (1.9), (1.11) of Ref. 6).

Quite generally, a *coherent state* $|\gamma\rangle$ of a system of bosons is defined as eigenstate of the annihilation operator (we drop momentarily the index o),

$$B|\gamma\rangle = \gamma|\gamma\rangle \qquad (20.24)$$

where γ is any complex number[23]. One then shows (Problem 4.4) that this (normalized) state is given by

$$|\gamma\rangle = \exp\left\{ -\frac{1}{2}|\gamma|^2 + \gamma B^+ \right\}|0\rangle \qquad (20.25)$$

and that

$$\langle \gamma|B^+ B|\gamma\rangle = |\gamma|^2 ; \quad \langle \gamma|(B^+ B)^2|\gamma\rangle = |\gamma|^2 + |\gamma|^4 . \qquad (20.26)$$

Here we are of course interested in a grand-canonical description. Therefore, setting $\gamma = \sqrt{N}$ in Eq. (20.25), the condensate state takes the form mentioned above[22], namely

$$|C\rangle = \exp\left\{ -\frac{1}{2}N + \sqrt{N}B_o^+ \right\}|0\rangle , \qquad (20.27)$$

and $\Delta N_B = \sqrt{N}$ follows from Eqs. (20.22), (20.23) and (20.26). Schrieffer was the first to write the pair state in the form (20.27), see Section 2–4 of Ref. 24.

Inserting now the pair operator (20.11) into (20.27) one finds, by application of the Pauli principle $(a_k^+)^2 = 0$ [22],

$$|C\rangle = e^{-N/2} \prod_{\mathbf{k}} \left(1 + \sqrt{\frac{N}{V}} \varphi_{\mathbf{k}} a_{\mathbf{k}\uparrow}^+ a_{-\mathbf{k}\downarrow}^+ \right) |0\rangle \qquad (20.28)$$

and the normalization becomes

$$\langle C|C \rangle = e^{-N} \prod_{\mathbf{k}} \left(1 + \frac{N}{V} |\varphi_{\mathbf{k}}|^2 \right) . \qquad (20.29)$$

Because of the assumption of strong localization, $\varphi_{\mathbf{k}}$ is nearly constant up to $|\mathbf{k}| \sim \pi/\xi_o$ so that by normalizing $\varphi_{\mathbf{k}}$ one finds

$$\frac{N}{V} |\varphi_{\mathbf{k}}|^2 \simeq \frac{3}{\pi} n\xi_o^3 \theta\left(\frac{\pi}{\xi_o} - |\mathbf{k}| \right) ; \quad n\xi_o^3 \ll 1 , \qquad (20.30)$$

and one easily verifies that, to first order in $n\xi_o^3$, $\langle C|C \rangle = 1$. Fourier transformation of the pair wavefunction defined by (20.30) gives, for any value of ξ_o, the expression

$$\tilde{\varphi}(r) \simeq \sqrt{\frac{3\pi}{2\xi_o^3}} \left(\sin\frac{\pi r}{\xi_o} - \frac{\pi r}{\xi_o} \cos\frac{\pi r}{\xi_o} \right) \left(\frac{\xi_o}{\pi r} \right)^3 \qquad (20.31)$$

which has the correct extension to $r \simeq \xi_o$.

The remarkable feature of the state (20.28) is that it makes sense also outside of the domain $n\xi_o^3 \ll 1$ of strong localization . There $|C\rangle$ is not any more normalized to unity but this deficiency is easily corrected by using (20.29). Calling the general normalized state $|v\rangle$, one finds from (20.28), (20.29) that [22]

$$|v\rangle = \prod_{\mathbf{k}} (u_{\mathbf{k}} + v_{\mathbf{k}} a_{\mathbf{k}\uparrow}^+ a_{-\mathbf{k}\downarrow}^+)|0\rangle \qquad (20.32)$$

where

$$u_{\mathbf{k}} = \left(1 + \frac{N}{V} |\varphi_{\mathbf{k}}|^2 \right)^{-1/2} ; \quad \frac{v_{\mathbf{k}}}{u_{\mathbf{k}}} = \sqrt{\frac{N}{V}} \varphi_{\mathbf{k}} , \qquad (20.33)$$

automatically satisfies the normalization condition $|u_{\mathbf{k}}|^2 + |v_{\mathbf{k}}|^2 = 1$.

Formally, the Fermi-sea state $|F\rangle$ may be obtained from (20.32) as the limiting case $u_{\mathbf{k}} = 0$, $v_{\mathbf{k}} = 1$ if $|\mathbf{k}| < k_F$ and $v_{\mathbf{k}} = 0$ otherwise. However, according to (20.33) the pair wavefunction $\varphi_{\mathbf{k}}$ does not exist in this limit. On the other hand, taking the limit $\pi/\xi_o = k_F$ in the expression (20.30) for $\varphi_{\mathbf{k}}$, one finds with Eq. (14.13) for $D = 3$ that $n\xi_o^3 = \pi/3$ and hence $|u_{\mathbf{k}}|^2 = |v_{\mathbf{k}}|^2 = 1/2$ if $|\mathbf{k}| < k_f$ and $v_{\mathbf{k}} = 0$ otherwise. The crucial difference between these two limits is that, while the Fermi-sea is a *Fock state* with sharp electron number nV, the electron-number operator $N_{\mathrm{el}} = \sum_k a_k^+ a_k$ fluctuates maximally in the pair state $|v\rangle$ which, therefore may be called a *coherent pair state*. An explicit calculation yields (Problem 4.5)

$$\bar{N}_{\mathrm{el}} \equiv \langle v|N_{\mathrm{el}}|v\rangle = 2 \sum_k |v_{\mathbf{k}}|^2 \tag{20.34}$$

and

$$(\Delta N_{\mathrm{el}})^2 \equiv \langle v|(N_{\mathrm{el}} - \bar{N}_{\mathrm{el}})^2|v\rangle = 2 \sum_k |u_{\mathbf{k}}|^2 |v_{\mathbf{k}}|^2 \tag{20.35}$$

This result confirms the values $\bar{N}_{\mathrm{el}} = nV$ and $\Delta N_{\mathrm{el}} = 0$ for the Fermi-sea state, but gives a maximum fluctuation for $|u_{\mathbf{k}}|^2 = |v_{\mathbf{k}}|^2 = 1/2$.

In the case (20.7), (20.10), the effective attraction causes the ω_o-neighbourhood of the Fermi surface to become unstable while $v_{\mathbf{k}} \simeq 1$ for $\varepsilon_{\mathbf{k}} < -\omega_o$ and $v_{\mathbf{k}} \simeq 0$ for $\varepsilon_{\mathbf{k}} > \omega_o$. Assuming $\omega_o \simeq k_B \theta_D \ll \varepsilon_F$, (20.32) is indeed the groundstate postulated by Bardeen, Cooper and Schrieffer in their fundamental work[5]. In this case the state $|v\rangle$ differs only little from the Fermi-sea state $|F\rangle$. The pair wavefunction does not exist which is the reason for Schafroth's intuition to speak of "resonant states of electron pairs" already in 1954[7] and which forces one to define ξ_o as *coherence length*; all pairs overlap and $n\xi_o^3 \gg 1$. This limit $\omega_o \ll \varepsilon_F$, $n\xi_o^3 \gg 1$ defines *Cooper pairs*. In the opposite limit $\omega_o \simeq \varepsilon_F$, $n\xi_o^3 \ll 1$ which was taken as definition of *Schafroth pairs*, we have the Bose condensation in the dilute gas limit discussed earlier.

Both parameters ω_o and ξ_o of course depend on the physics underlying the effective electron attraction which may be quite complex. In the case of genuinely bound pairs it is of interest to compare the spatial extension ξ_o with the lattice constant a. For a band-filling of approximately one electron per unit cell, $n \simeq 1/v$ (in a cubic system $v = a^3$, see Section 1), strong localization means that $\xi_o \ll a$. However, in order to reduce the strong Coulomb repulsion, this situation would require unreasonably strong correlation effects giving rise to *polarons* which would localize[25]. If, on the other hand, $n \ll 1/v$ which is the case in the cuprate superconductors[26], it is possible to have $\xi_o \simeq a$. Indeed, as mentioned in the introduction to this chapter, ξ_o in this case has the values $\xi_c \sim 1.5 - 4$ Å and $\xi_{ab} \sim 14 - 30$ Å which are to be compared with the respective lattice constants $c \sim 12$ Å and $a \sim b \sim 3.8$ Å[27].

In order to gain some idea about the strength of the binding, let us estimate the effective electron attraction by retaining in Eq. (20.7) just the most important terms describing singlet pairs. Making use of (20.11) with $\varphi_{\mathbf{k}}$ given by Eq. (20.30) which at the same time replaces the definition (20.10) of the interaction domain ϕ, these terms are contained in the reduced Hamiltonian

$$\mathcal{H}_{\text{red}} = -\frac{\Lambda}{V} \sum_{\mathbf{k},\mathbf{k}',\mathbf{q}} \theta\left(\frac{\pi}{\xi_o} - |\mathbf{k}|\right)\theta\left(\frac{\pi}{\xi_o} - |\mathbf{k}'|\right)$$

$$\times \, a^+_{\mathbf{k}+\frac{\mathbf{q}}{2},\uparrow} a^+_{-\mathbf{k}+\frac{\mathbf{q}}{2},\downarrow} a_{-\mathbf{k}'+\frac{\mathbf{q}}{2},\downarrow} a_{\mathbf{k}'+\frac{\mathbf{q}}{2},\uparrow} = -\frac{\pi\Lambda}{6\xi_o^3} N_B \ . \qquad (20.36)$$

In the strong-coupling limit where $|v\rangle = |C\rangle$ we find with (20.22) that the binding energy per electron pair is

$$\varepsilon_B \simeq \frac{\pi\Lambda}{6\xi_o^3} \ . \qquad (20.37)$$

This energy has to match the effective Coulomb repulsion $\varepsilon_C = e_{\text{eff}}^2/\xi_o$ where the effective charge $e_{\text{eff}} \leq e$ because of correlation effects (screening). Then $\varepsilon_B = \varepsilon_C$ determines the pair size ξ_o. Taking the latter to be $\xi_o \sim 5$ Å and $e_{\text{eff}} = e$ one finds from (20.37) $\varepsilon_B \sim 3$ eV. This value is of the same order as the width of the oxygen-hole band in the cuprates[28] which is thought to be responsible for superconductivity. Such an enormous binding energy for the cuprates is of course totally unrealistic as shows the weak-coupling calculation of ε_B in Eq. (22.8) below. This means, first that strong correlations among the holes in the cuprates drastically reduce the Coulomb energy, i.e. $e_{\text{eff}} \ll e$, and second that the coupling cannot be too strong, at least not in the ab-planes (see Ref. 29, in particular Fig. 1).

21. BCS Groundstate, Gap Equation and Bogoljubov Transformation

As discussed in the last section, the BCS groundstate may be defined as a coherent pair state (20.32) in which $v_{\mathbf{k}} \simeq 1$ for $\varepsilon_{\mathbf{k}} < -\omega_o$ and $v_{\mathbf{k}} \simeq 0$ for $\varepsilon_{\mathbf{k}} > \omega_o$. In analogy with the Cooper problem, $v_{\mathbf{k}}$ will be determined by minimizing the groundstate energy $E = \langle v|\mathcal{H}|v\rangle$ where, as before, $\mathcal{H} = \mathcal{H}_o + \mathcal{H}_{\text{eff}}$. Making use of the first part of Problem 4.5 one immediately finds

$$E_o = \langle v| \sum_k \varepsilon_{\mathbf{k}} a^+_k a_k |v\rangle = 2 \sum_k \varepsilon_{\mathbf{k}} |v_{\mathbf{k}}|^2 \ . \qquad (21.1)$$

The interaction (20.7) gives rise to the three contributions shown in Fig. 4.2 where the first two may again be obtained with the help of the first part of Problem 4.5,

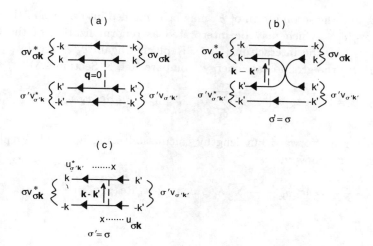

Fig. 4.2: The contributions (21.2), (21.3) and (21.4) to the groundstate energy of interaction $\langle\varphi|\mathcal{H}_{\text{eff}}|\varphi\rangle$.

$$E_a \equiv -\frac{1}{2V}\sum_{kk'}\Lambda_{\mathbf{kk'}0}\langle v|a_k^+ a_{k'}^+ a_{k'} a_k|v\rangle$$

$$= -\frac{2}{V}\sum_{\mathbf{kk'}}(1-\frac{1}{2}\delta_{\mathbf{kk'}})\Lambda_{\mathbf{kk'}0}|v_\mathbf{k}|^2|v_{\mathbf{k'}}|^2 \tag{21.2}$$

and

$$E_b \equiv -\frac{1}{2V}\sum_{kk'}\Lambda_{\mathbf{kk'},\mathbf{k-k'}}\langle v|a_k^+ a_{k'}^+ a_k a_{k'}|v\rangle\delta_{\sigma\sigma'}$$

$$= +\frac{1}{V}\sum_{\mathbf{kk'}}(1-\delta_{\mathbf{kk'}})\Lambda_{\mathbf{kk'},\mathbf{k-k'}}|v_\mathbf{k}|^2|v_{\mathbf{k'}}|^2 \tag{21.3}$$

where the $\delta_{\mathbf{kk'}}$-terms take care of the exclusion principle $(a_k^+)^2 = 0$. Similarly one finds for the third term (second part of Problem 4.5)

$$E_c \equiv -\frac{1}{2V}\sum_{kk'}\lambda_{\mathbf{k},-\mathbf{k},\mathbf{k-k'}}\langle v|a_k^+ a_{-k}^+ a_{-k'} a_{k'}|v\rangle\delta_{\sigma\sigma'}$$

$$= -\frac{1}{V}\sum_{\mathbf{kk'}}\lambda_{\mathbf{k},-\mathbf{k},\mathbf{k-k'}}u_{\mathbf{k'}}u_\mathbf{k}\text{Re}\,(v_{\mathbf{k'}}v_\mathbf{k}^*)$$

$$+ \frac{1}{V}\sum_{\mathbf{k}}\Lambda_{\mathbf{k},-\mathbf{k},0}|v_\mathbf{k}|^2(|u_\mathbf{k}|^2 - |v_\mathbf{k}|^2)\,. \tag{21.4}$$

E_a, E_b and the second term of E_c may be combined into a term of the form $2\sum_{\mathbf{k}} \delta\varepsilon_{\mathbf{k}}|v_{\mathbf{k}}|^2$ which may be interpreted as renormalization of the band energy $\varepsilon_{\mathbf{k}}$ by the interaction (20.7). Replacing again $\varepsilon_{\mathbf{k}} + \delta\varepsilon_{\mathbf{k}}$ by $\varepsilon_{\mathbf{k}}$ one obtains for the groundstate energy (compare Ref. 3, Section 4.3)

$$E = 2\sum_{\mathbf{k}} \varepsilon_{\mathbf{k}}|v_{\mathbf{k}}|^2 - \frac{1}{V}\sum_{\mathbf{k}\mathbf{k}'} \Lambda_{\mathbf{k},-\mathbf{k},\mathbf{k}-\mathbf{k}'} u_{\mathbf{k}'} u_{\mathbf{k}} \mathrm{Re}\,(v_{\mathbf{k}'} v_{\mathbf{k}}^*) \,. \tag{21.5}$$

A straightforward but lengthy calculation using $u_{\mathbf{k}} = (1 - |v_{\mathbf{k}}|^2)^{1/2}$ and $v_{\mathbf{k}}^* = v_{-\mathbf{k}}$ now yields

$$\frac{\partial E}{\partial v_{\mathbf{k}}^*} = 4\varepsilon_{\mathbf{k}} v_{\mathbf{k}} - \frac{2}{V}\sum_{\mathbf{k}'} \lambda_{\mathbf{k}\mathbf{k}'}[v_{\mathbf{k}'} u_{\mathbf{k}}^2 - v_{\mathbf{k}}\mathrm{Re}\,(v_{\mathbf{k}} v_{\mathbf{k}'}^*)]\frac{u_{\mathbf{k}'}}{u_{\mathbf{k}}} \tag{21.6}$$

where

$$\lambda_{\mathbf{k}\mathbf{k}'} \equiv \frac{1}{4}(\Lambda_{\mathbf{k},-\mathbf{k},\mathbf{k}-\mathbf{k}'} + \Lambda_{-\mathbf{k},\mathbf{k},-\mathbf{k}+\mathbf{k}'} + \Lambda_{\mathbf{k}',-\mathbf{k}',\mathbf{k}'-\mathbf{k}}$$
$$+ \Lambda_{-\mathbf{k}',\mathbf{k}',-\mathbf{k}'+\mathbf{k}}) = \lambda_{\mathbf{k}'\mathbf{k}} = \lambda_{-\mathbf{k},-\mathbf{k}'} \,. \tag{21.7}$$

Since according to (20.8) $\lambda_{\mathbf{k}\mathbf{k}'}$ is real, it is natural to fix the phase of $v_{\mathbf{k}}$ by setting $v_{\mathbf{k}} = |v_{\mathbf{k}}|$. Equating then (21.6) to zero and defining the *gap function* as

$$\Delta_{\mathbf{k}} \equiv \frac{1}{v}\sum_{\mathbf{k}'} \lambda_{\mathbf{k}\mathbf{k}'} u_{\mathbf{k}'} v_{\mathbf{k}'} \tag{21.8}$$

one finds

$$2\varepsilon_{\mathbf{k}} u_{\mathbf{k}} v_{\mathbf{k}} = \Delta_{\mathbf{k}}(u_{\mathbf{k}}^2 - v_{\mathbf{k}}^2) \,. \tag{21.9}$$

Taking the square of this equation, the terms may be rearranged as $4E_{\mathbf{k}}^2 u_{\mathbf{k}}^2 v_{\mathbf{k}}^2 = \Delta_{\mathbf{k}}^2(u_{\mathbf{k}}^2 + v_{\mathbf{k}}^2) = \Delta_{\mathbf{k}}^2$ where

$$E_{\mathbf{k}}^2 \equiv \varepsilon_{\mathbf{k}}^2 + \Delta_{\mathbf{k}}^2 \,. \tag{21.10}$$

Assuming that $\Delta_{\mathbf{k}} \geq 0$ and $E_{\mathbf{k}} > 0$, the square root of the last expression leads to the relation

$$u_{\mathbf{k}} v_{\mathbf{k}} = \frac{\Delta_{\mathbf{k}}}{2E_{\mathbf{k}}} \tag{21.11}$$

which shows that for $\Delta_{\mathbf{k}} = 0$, either $u_{\mathbf{k}} = 1$, $v_{\mathbf{k}} = 0$ or $u_{\mathbf{k}} = 0$, $v_{\mathbf{k}} = 1$. Inserting (21.11) into (21.8) leads to the *gap equation*

$$\Delta_{\mathbf{k}} = \frac{1}{V}\sum_{\mathbf{k}'} \lambda_{\mathbf{k}\mathbf{k}'} \frac{\Delta_{\mathbf{k}'}}{2E_{\mathbf{k}'}} \tag{21.12}$$

which, because of (21.10), is a nonlinear, non-homogeneous integral equation. Elimination of $\Delta_{\mathbf{k}}$ from (21.9), (21.11) finally yields

$$u_{\mathbf{k}}^2 = \frac{1}{2}\left(1 + \frac{\varepsilon_{\mathbf{k}}}{E_{\mathbf{k}}}\right) \; ; \quad v_{\mathbf{k}}^2 = \frac{1}{2}\left(1 - \frac{\varepsilon_{\mathbf{k}}}{E_{\mathbf{k}}}\right) . \tag{21.13}$$

These relations give for the trivial solution $\Delta_{\mathbf{k}} = 0$ of the gap equation (21.12), $u_{\mathbf{k}} = 0$, $v_{\mathbf{k}} = 1$ if $\varepsilon_{\mathbf{k}} < 0$ and $u_{\mathbf{k}} = 1$, $v_{\mathbf{k}} = 0$ if $\varepsilon_{\mathbf{k}} > 0$, for which the BCS groundstate (20.32) becomes the Fermi sea $|F\rangle$. On the other hand, in the limit of strongly bound electron pairs, Eq. (20.30) holds which, together with (20.33), implies that $v_{\mathbf{k}}^2 \ll 1$, $u_{\mathbf{k}}^2 \simeq 1$. In this limit (21.11) determines the pair wavefunction, $\Delta_{\mathbf{k}}/2E_{\mathbf{k}} \simeq v_{\mathbf{k}} \simeq \sqrt{N/V}\varphi_{\mathbf{k}}$ and, combined with (21.9), $\Delta_{\mathbf{k}} \simeq 2\varepsilon_{\mathbf{k}}\sqrt{N/V}\varphi_{\mathbf{k}}$. Furthermore, since there is no Fermi sea, $\varepsilon_{\mathbf{k}} = (\hbar^2 k^2/2m) - \mu$, so that the gap equation (21.12) goes over, in this limit, into the Schrödinger equation with eigenfunction $\varphi_{\mathbf{k}}$ and eigenvalue 2μ. But this eigenvalue must be $-\varepsilon_B$ where $\varepsilon_B > 0$ is the binding energy, hence $\mu = -\varepsilon_B/2$ (compare Eq. (8) of Ref. 22).

With the particular choice (20.10), the coupling function (21.7) factorizes,

$$\lambda_{\mathbf{k}\mathbf{k}'} = \Lambda\theta(\omega_o - |\varepsilon_{\mathbf{k}}|)\theta(\omega_o - |\varepsilon_{\mathbf{k}'}|) \tag{21.14}$$

so that the gap equation (21.12) is explicitly solved by

$$\Delta_{\mathbf{k}} = \Delta_o\theta(\omega_o - |\varepsilon_{\mathbf{k}}|) . \tag{21.15}$$

In this approximation $E_{\mathbf{k}}$ has the form shown in Fig. 4.3. In terms of the density of states (10.4) and with (21.10) and $E_o^2 \equiv \varepsilon^2 + \Delta_o^2$, integration of Eq. (21.12) then gives, for $\omega_o \ll \varepsilon_F$,

$$\frac{1}{\Lambda N(0)} = \int_0^{+\omega_o} \frac{d\varepsilon}{E_o} = \ln\left(\frac{\omega_o}{\Delta_o} + \sqrt{\frac{\omega_o^2}{\Delta_o^2} + 1}\right) \tag{21.16}$$

or, since in general $\Lambda N(0) \ll 1$,

$$\Delta_o = \omega_o\left(\sinh\frac{1}{\Lambda N(0)}\right)^{-1} \cong 2\omega_o e^{-1/\Lambda N(0)} \ll \omega_o , \tag{21.17}$$

which is the famous *weak-coupling solution of BCS*[5].

As pointed out by Bogoljubov and by Valatin[30], the generation of the state $|v\rangle$ from the vacuum $|0\rangle$ by Eq. (20.32) may be interpreted as a canonical transformation of the operators $a_{\mathbf{k}\uparrow}$, $a_{-\mathbf{k}\downarrow}$ into new operators $\alpha_{\mathbf{k}}$, $\beta_{\mathbf{k}}$ such that the anticommutation relations are left invariant. This *Bogoljubov transformation* is defined by

$$\alpha_{\mathbf{k}} = u_{\mathbf{k}}a_{\mathbf{k}\uparrow} - v_{\mathbf{k}}a_{-\mathbf{k}\downarrow}^+ \; ; \quad \beta_{\mathbf{k}} = v_{\mathbf{k}}a_{\mathbf{k}\uparrow}^+ + u_{\mathbf{k}}a_{-\mathbf{k}\downarrow} \tag{21.18}$$

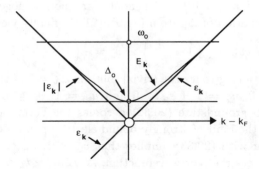

Fig. 4.3: $E_{\mathbf{k}}$ defined by (21.10) in the approximation (21.15).

and by the hermitean conjugate equations. This transformation is not generated, however, by the operator $(u_{\mathbf{k}} + v_{\mathbf{k}} a_{\mathbf{k}\uparrow}^{+} a_{-\mathbf{k}\downarrow}^{+})$ of Eq. (20.32) which, in fact, is not unitary. The correct unitary operator will be derived later (see Eq. (21.39) below). At this point two observations are of relevance. First, it is straightforward to verify that the new operators $\alpha_{\mathbf{k}}$, $\beta_{\mathbf{k}}$ satisfy the same anticommutation relations (3.13) as the old ones provided that $u_{\mathbf{k}}^{2} + v_{\mathbf{k}}^{2} = 1$. Second, Eqs. (21.18) imply that both, $\alpha_{\mathbf{k}}$ and $\beta_{\mathbf{k}}$ annihilate the state $(u_{\mathbf{k}} + v_{\mathbf{k}} a_{\mathbf{k}\uparrow}^{+} a_{\mathbf{k}\downarrow}^{+})|0\rangle$; hence it follows from (20.32) that for any $\mathbf{k}$,

$$\alpha_{\mathbf{k}}|v\rangle = \beta_{\mathbf{k}}|v\rangle = 0 \ . \tag{21.19}$$

Therefore, $|v\rangle$ is the vacuum for the excitations created by $\alpha_{\mathbf{k}}^{+}$ and $\beta_{\mathbf{k}}^{+}$ which may be called *bogolons*[6].

In order to determine the energy of free bogolons we introduce the new operators into the Hamiltonian, using the reduced BCS-form of the interaction, Eq. (20.7) namely[5]

$$\mathcal{H}_{\mathrm{BCS}} = -\frac{1}{V} \sum_{\mathbf{k}\mathbf{k}'} \lambda_{\mathbf{k}\mathbf{k}'} b_{\mathbf{k}}^{+} b_{\mathbf{k}'} \tag{21.20}$$

where we made use of (20.8) and (21.7) and put

$$b_{\mathbf{k}}^{+} \equiv a_{\mathbf{k}\uparrow}^{+} a_{-\mathbf{k}\downarrow}^{+} \ . \tag{21.21}$$

Note that in the special case (20.10) this is just the $\mathbf{q} = 0$ version of Eq. (20.36). Since making the substitution (21.18) in $\mathcal{H} = \mathcal{H}_{o} + \mathcal{H}_{\mathrm{BCS}}$ gives rise to complicated algebraic manipulations, it is highly recommended to introduce the compact *Nambu representation*[31] of two-component operators

$\Psi_{\mathbf{k}}$, $\Phi_{\mathbf{k}}$ and of real matrices $R_{\mathbf{k}}$, defined by

$$\Psi_{\mathbf{k}} = \begin{pmatrix} a_{\mathbf{k}\uparrow} \\ a^+_{-\mathbf{k}\downarrow} \end{pmatrix} ; \quad \Phi_{\mathbf{k}} = \begin{pmatrix} \alpha_{\mathbf{k}} \\ \beta^+_{\mathbf{k}} \end{pmatrix} ; \quad R_{\mathbf{k}} = \begin{pmatrix} u_{\mathbf{k}} & -v_{\mathbf{k}} \\ v_{\mathbf{k}} & u_{\mathbf{k}} \end{pmatrix} . \tag{21.22}$$

Then the transformation (21.18) is recognized as the rotation

$$\Phi_{\mathbf{k}} = R_{\mathbf{k}}\Psi_{\mathbf{k}} ; \quad R^T_{\mathbf{k}} R_{\mathbf{k}} = 1 \tag{21.23}$$

where the superscript T designates the transposed matrix.

Introducing the Pauli matrices (2.8) and defining $\delta_{\pm} \equiv (1 \pm \sigma^z)/2$ one has

$$a^+_{\mathbf{k}\uparrow} a_{\mathbf{k}'\uparrow} = \Psi^+_{\mathbf{k}} \delta_+ \Psi_{\mathbf{k}'} ; \quad a_{-\mathbf{k}\downarrow} a^+_{-\mathbf{k}'\downarrow} = \Psi^+_{\mathbf{k}} \delta_- \Psi_{\mathbf{k}'} \tag{21.24}$$

and

$$a^+_{\mathbf{k}\uparrow} a^+_{-\mathbf{k}'\downarrow} = \Psi^+_{\mathbf{k}} \sigma^+ \Psi_{\mathbf{k}'} ; \quad a_{-\mathbf{k}\downarrow} a_{\mathbf{k}'\uparrow} = \Psi^+_{\mathbf{k}} \sigma^- \Psi_{\mathbf{k}'} . \tag{21.25}$$

Applying the inverse of the transformation (21.23) to these expressions, the Hamiltonians $\mathcal{H}_o = \sum \varepsilon_{\mathbf{k}} a^+_k a_k$ and (21.20) become, respectively,

$$\mathcal{H}_o = \sum_{\mathbf{k}} \varepsilon_{\mathbf{k}} (\Phi^+_{\mathbf{k}} \sigma^{(z)}_{\mathbf{kk}} \Phi_{\mathbf{k}} + 1) \tag{21.26}$$

and

$$\mathcal{H}_{\text{BCS}} = -\frac{1}{V} \sum_{\mathbf{kk}'} \lambda_{\mathbf{kk}'} (\Phi^+_{\mathbf{k}} \sigma^{(+)}_{\mathbf{kk}} \Phi_{\mathbf{k}})(\Phi^+_{\mathbf{k}'} \sigma^{(-)}_{\mathbf{k}'\mathbf{k}'} \Phi_{\mathbf{k}'}) \tag{21.27}$$

where we define for later use $\sigma^{(\pm)}_{\mathbf{kk}'} \equiv R_{\mathbf{k}} \sigma^\pm R^T_{\mathbf{k}'}$ and $\delta^{(\pm)}_{\mathbf{kk}'} \equiv R_{\mathbf{k}} \delta_\pm R^T_{\mathbf{k}'}$ so that

$$\delta^{(+)}_{\mathbf{kk}'} = \begin{pmatrix} u_{\mathbf{k}} u_{\mathbf{k}'}, & u_{\mathbf{k}} v_{\mathbf{k}'} \\ v_{\mathbf{k}} u_{\mathbf{k}'}, & v_{\mathbf{k}} v_{\mathbf{k}'} \end{pmatrix} ; \quad \delta^{(-)}_{\mathbf{kk}'} = \begin{pmatrix} v_{\mathbf{k}} v_{\mathbf{k}'}, & -v_{\mathbf{k}} u_{\mathbf{k}'} \\ -u_{\mathbf{k}} v_{\mathbf{k}'}, & u_{\mathbf{k}} u_{\mathbf{k}'} \end{pmatrix} ;$$

$$\sigma^{(+)}_{\mathbf{kk}'} = \sigma^{(-)T}_{\mathbf{k}'\mathbf{k}} = \begin{pmatrix} -u_{\mathbf{k}} v_{\mathbf{k}'}, & u_{\mathbf{k}} u_{\mathbf{k}'} \\ -v_{\mathbf{k}} v_{\mathbf{k}'}, & v_{\mathbf{k}} u_{\mathbf{k}'} \end{pmatrix} . \tag{21.28}$$

For the determination of the energy of free bogolons we only need the diagonal parts of (21.26), (21.27), obtained by replacing the $\sigma^{()}_{\mathbf{kk}}$ by their diagonal parts. With (21.22), (21.28) we find

$$\text{diag}\{\mathcal{H}_o\} = \sum_{\mathbf{k}} \varepsilon_{\mathbf{k}} \{(u^2_{\mathbf{k}} - v^2_{\mathbf{k}})(\alpha^+_{\mathbf{k}} \alpha_{\mathbf{k}} - \beta_{\mathbf{k}} \beta^+_{\mathbf{k}}) + 1\} \tag{21.29}$$

and, making use of definition (21.7),

$$\text{diag}\{\mathcal{H}_{\text{BCS}}\} = -\frac{1}{V} \sum_{\mathbf{kk}'} \lambda_{\mathbf{kk}'} u_{\mathbf{k}} v_{\mathbf{k}} u_{\mathbf{k}'} v_{\mathbf{k}'}$$
$$\times (-\alpha^+_{\mathbf{k}} \alpha_{\mathbf{k}} + \beta_{\mathbf{k}} \beta^+_{\mathbf{k}})(-\alpha^+_{\mathbf{k}'} \alpha_{\mathbf{k}'} + \beta_{\mathbf{k}'} \beta^+_{\mathbf{k}'}) . \tag{21.30}$$

Using the anticommutation relations and keeping in (21.30) only the bi-linear terms, leaving out the constant terms arising from commutations, one finds by applying Eqs. (21.8), (21.10), (21.11) and (21.23) that the free bogolon Hamiltonian is given by

$$\mathcal{H}_B = \sum_{\mathbf{k}} E_{\mathbf{k}}(\alpha_{\mathbf{k}}^+ \alpha_{\mathbf{k}} + \beta_{\mathbf{k}}^+ \beta_{\mathbf{k}}) \; . \qquad (21.31)$$

According to Fig. 4.3, the minimum energy for breaking a Cooper pair into two bogolons therefore is $2\Delta_o$.

Finally, we wish to derive the unitary operator which generates the canonical transformation $(a_{\mathbf{k}\uparrow}, a_{-\mathbf{k}\downarrow}) \rightarrow (\alpha_{\mathbf{k}}, \beta_{\mathbf{k}})$ and $|0\rangle \rightarrow |v\rangle$. Again, the analogous problem for the coherent states (20.25) is of great help since the unitary operator is well known there[23]. The idea is to multiply the operator $\exp(\gamma B^+)$ which is not unitary, from the right by $\exp(-\gamma^* B)$. Since the latter gives 1 when acting on $|0\rangle$, this does not affect the state $|\gamma\rangle$. Making use now of the identity (15.28) this modified operator becomes

$$U(\gamma) = e^{-\frac{1}{2}|\gamma|^2} e^{\gamma B^+} e^{-\gamma^* B} = e^{\gamma B^+ - \gamma^* B} = U^+(-\gamma) \qquad (21.32)$$

and unitarity follows as a special case of the relation

$$U(\gamma_1)U(\gamma_2) = e^{i\mathrm{Im}(\gamma_1 \gamma_2^*)} U(\gamma_1 + \gamma_2) \qquad (21.33)$$

which is easily proven by repeated application of (15.28).

Replacing now the boson operator B^+ by the electron pair operator $b_{\mathbf{k}}^+$, the operator corresponding to $\exp(\gamma B^+)$ is $\exp(\gamma b_{\mathbf{k}}^+) = 1 + \gamma b_{\mathbf{k}}^+$ where $\gamma = v_{\mathbf{k}}/u_{\mathbf{k}}$ may be chosen real. The analogue of Eq. (21.32) then is[32]

$$U(\gamma) = e^{\gamma(b_{\mathbf{k}}^+ - b_{\mathbf{k}})} \; . \qquad (21.34)$$

Here, however, the identity (15.28) does not apply since $b_{\mathbf{k}}$ and $b_{\mathbf{k}}^+$ do not commute; indeed from (21.21) it follows that

$$[b_{\mathbf{k}}, b_{\mathbf{k'}}^+] = (a_{-\mathbf{k}\downarrow} a_{-\mathbf{k}\downarrow}^+ - a_{\mathbf{k}\uparrow}^+ a_{\mathbf{k}\uparrow}) \delta_{\mathbf{k}\mathbf{k'}} \; . \qquad (21.35)$$

But we may take advantage of the exclusion principle in the form $(a_k^+)^2 = (b_{\mathbf{k}}^+)^2 = 0$. In fact, it turns out that (21.34) is easily expanded in powers of the antihermitean operator

$$Q \equiv b_{\mathbf{k}}^+ - b_{\mathbf{k}} = -Q^+ \; . \qquad (21.36)$$

With (21.35) one finds

$$Q^2 = -b_{\mathbf{k}}^+ b_{\mathbf{k}} - b_{\mathbf{k}} b_{\mathbf{k}}^+ \; ; \quad Q^3 = -Q \qquad (21.37)$$

and hence $Q^4 = Q^3Q = -Q^2$ so that $-Q^2$ is a projection operator,

$$P \equiv b_{\mathbf{k}}^+ b_{\mathbf{k}} + b_{\mathbf{k}} b_{\mathbf{k}}^+ = P^2 = P^+ . \tag{21.38}$$

Therefore, $Q^{2n} = (-1)^n P^n = (-1)^n P$, $Q^{2n+1} = (-1)^n QP = (-1)^n PQ$, $n \geq 1$ so that

$$U(\gamma) = 1 + \gamma Q + \sum_{n=1}^{\infty} \left(\frac{\gamma^{2n}}{(2n)!} Q^{2n} + \frac{\gamma^{2n+1}}{(2n+1)!} Q^{2n+1} \right)$$

$$= (1 + \gamma Q)(1 - P) + (\cos\gamma + Q\sin\gamma)P = U_{\mathbf{k}}^+(-\gamma) . \tag{21.39}$$

With this expression it is easy to show that for any real γ

$$U(\gamma)U^+(\gamma) = U^+(\gamma)U(\gamma) = \cos^2\gamma + \sin^2\gamma = 1 . \tag{21.40}$$

It remains to show that for an appropriate value of γ, $U_{\mathbf{k}}(\gamma)$ indeed generates the Bogoljubov transformation . To start with, application of (21.39) and (21.35) yields

$$U(\gamma)|0\rangle = (\cos\gamma + b_{\mathbf{k}}^+ \sin\gamma)|0\rangle , \tag{21.41}$$

so that, by comparing with (20.32), one has $\cos\gamma = u_{\mathbf{k}}$, $\sin\gamma = v_{\mathbf{k}}$ and

$$|v\rangle = \prod_{\mathbf{k}} U_{\mathbf{k}}|0\rangle ; \quad U_{\mathbf{k}} \equiv U\left(\tan^{-1} \frac{v_{\mathbf{k}}}{u_{\mathbf{k}}} \right) . \tag{21.42}$$

Making use of Eq. (21.39) and of the identities

$$Q\Psi_{\mathbf{k}}Q = 0 ; \quad (1 - P)Q\Psi_{\mathbf{k}} = \Psi_{\mathbf{k}}Q(1 - P) = 0 \tag{21.43}$$

and

$$\{\Psi_{\mathbf{k}}P\} = \Psi_{\mathbf{k}} ; \quad [\Psi_{\mathbf{k}}Q] = i\sigma^y \Psi_{\mathbf{k}} , \tag{21.44}$$

which readily follow from Eqs. (21.21), (21.22) and (21.35)–(21.38), it is straightforward to show that[32]

$$U_{\mathbf{k}}\Psi_{\mathbf{k}}U_{\mathbf{k}}^+ = R_{\mathbf{k}}\Psi_{\mathbf{k}} = \Phi_{\mathbf{k}} . \tag{21.45}$$

22. Binding Energy, Correlation Amplitudes and Coherence Length

The constant terms mentioned after Eq. (21.30) are of importance because they give rise to the binding energy per electron pair, ε_B which in the limit of strong localization was given by Eq. (20.37). The general definition of ε_B is

$$-\frac{nV}{2}\varepsilon_B \equiv \langle v|\mathcal{H}|v\rangle - \langle F|\mathcal{H}|F\rangle \tag{22.1}$$

where $\mathcal{H} = \mathcal{H}_o + \mathcal{H}_{\text{BCS}}$ and $nV/2$ is the number of electron pairs. Note that this definition coincides with the W_o of BCS , Ref. 5, Eq. (2.28). Since $\langle F|a_k^+ a_k|F\rangle = \theta(-\varepsilon_{\mathbf{k}})$ and $\langle F|b_{\mathbf{k}}^+ b_{\mathbf{k}'}|F\rangle = \theta(-\varepsilon_{\mathbf{k}})\delta_{\mathbf{k}\mathbf{k}'}$ one finds, making use of the anticommutation relations and of Eqs. (21.7), (21.19), (21.22) and (21.26)–(21.28),

$$\varepsilon_B = -\frac{4}{nV}\sum_{\mathbf{k}}\varepsilon_{\mathbf{k}}\{v_{\mathbf{k}}^2 - \theta(-\varepsilon_{\mathbf{k}})\}$$

$$+ \frac{2}{nV^2}\sum_{\mathbf{k}\mathbf{k}'}\lambda_{\mathbf{k}\mathbf{k}'}\{u_{\mathbf{k}}v_{\mathbf{k}}u_{\mathbf{k}'}v_{\mathbf{k}'} + (v_{\mathbf{k}}^4 - \theta(-\varepsilon_{\mathbf{k}}))\delta_{\mathbf{k}\mathbf{k}'}\} . \qquad (22.2)$$

Here the term $\propto v_{\mathbf{k}}^4\delta_{\mathbf{k}\mathbf{k}'}$ is the contribution from the non-diagonal part of $\mathcal{H}_{\text{BCS}}$. Evaluation of this expression in the approximation (21.14), (21.15) yields (Problem 4.6)

$$\varepsilon_B = \frac{N(0)}{n}\Delta_o^2\left(1 - \frac{\pi}{2}\frac{\Lambda}{\Delta_o V}\right) \qquad (22.3)$$

where the term $\propto V^{-1}$ coming from the $\delta_{\mathbf{k}\mathbf{k}'}$ is negligible as $V \to \infty$. In this approximation the result (22.3) is related with Eq. (2.42) of Ref. 5 by $W_o = -(n/2)\varepsilon_B$.

The absolute value of the function $v_{\mathbf{k}}^2 - \theta(-\varepsilon_{\mathbf{k}})$ occurring in Eq. (22.2) may be viewed as squared Fourier component of the BCS pair wavefunction. Making use of Eqs. (2.4) and (21.13) this wavefunction then reads

$$\tilde{\varphi}(\mathbf{r}) = \frac{C}{V}\sum_{\mathbf{k}}\sqrt{1 - \frac{|\varepsilon_{\mathbf{k}}|}{E_{\mathbf{k}}}}\,e^{i\mathbf{k}\cdot\mathbf{r}} \qquad (22.4)$$

where C is a normalization constant. In the approximation (21.15) and $\omega_o \ll \varepsilon_F$ we may write $\varepsilon_{\mathbf{k}} \simeq \hbar v_F(k - k_F)$ (see Fig. 4.3) so that with (10.4)

$$\tilde{\varphi}(r) = CN(0)\int_{-\omega_o}^{\omega_o}d\varepsilon\sqrt{1 - \frac{\varepsilon}{E_o}}\frac{\sin(k_F r + \varepsilon r/\hbar v_F)}{k_F r + \varepsilon r/\hbar v_F}$$

$$= CN(0)\frac{\sin k_F r}{k_F r}\int_0^{\omega_o}d\varepsilon\sqrt{1 - \frac{\varepsilon}{E_o}}\cos\frac{\varepsilon r}{\hbar v_F} \qquad (22.5)$$

where, as before, $E_o^2 \equiv \varepsilon^2 + \Delta_o^2$ and the ε-term was neglected in the denominator. Similarly one finds

$$C^{-2} = \frac{1}{V}\sum_{\mathbf{k}}\left(1 - \frac{|\varepsilon_{\mathbf{k}}|}{E_{\mathbf{k}}}\right) = 2N(0)\int_0^{\omega_o}d\varepsilon(1 - \frac{\varepsilon}{E_o}) \cong 2N(0)\Delta_o . \qquad (22.6)$$

The radial probability distribution being $r^2\tilde{\varphi}^2(r)$, it oscillates with wavelength $2\pi/k_F$ which is much shorter than the range of $\tilde{\varphi}(r)$. The latter is determined by the fact that the contribution to the integral in (22.5) from $\varepsilon \gg \Delta_o$ is negligible. Hence the range is of the order of the *coherence length* defined as[5)]

$$\xi_o = \frac{\hbar v_F}{\pi \Delta_o} . \tag{22.7}$$

Using this expression to eliminate Δ_o from Eq. (22.3) and substituting $N(0)$ with the help of (14.14) with $D = 3$ one finds a formula analogous to (20.37),

$$\varepsilon_B = \frac{3\Delta_o^2}{\varepsilon_F} = \frac{6\hbar^2}{\pi^2 m \xi_o^2} . \tag{22.8}$$

The first of these expressions shows that in the approximation (21.17), ε_B is exceedingly small since $\varepsilon_B \ll \Delta_o \ll \omega_o \ll \varepsilon_F$. It is therefore of no surprise that ε_B is also much smaller than the Coulomb repulsion e^2/ξ_o which in the strong-coupling case was of order unity. Indeed

$$\frac{\varepsilon_B}{e^2/\xi_o} = \frac{6m}{\pi^2 m} \frac{a_B}{\xi_o} \ll 1 \tag{22.9}$$

where $a_B \equiv \hbar^2/me^2$ is the Bohr radius. This shows that the Cooper pairs are not really bound, a fact that had been anticipated by Schafroth[7)] (see Section 20).

Rather than evaluating the complicated integral in (22.5) we turn to the pair-correlation function which gives another estimate of the pair size. We start with the static density-density correlation function in the BCS groundstate (see Section 4.8 of Ref. 3) which is the analog of Eq. (15.22),

$$S(\mathbf{q}) = \langle v|n(\mathbf{q})n(-\mathbf{q})|v\rangle = S_{\uparrow\uparrow}(\mathbf{q}) + S_{\uparrow\downarrow}(\mathbf{q}) \tag{22.10}$$

where $n(\mathbf{q})$ is given by (3.22). Here we are interested only in the antiparallel-spin part which describes the *correlation function of Cooper pairs*,

$$S_{\uparrow\downarrow}(\mathbf{q}) = S_{\uparrow\downarrow}(-\mathbf{q}) = 2\sum_{\mathbf{kk'}}\langle v|a_{\mathbf{k}\uparrow}^+ a_{-\mathbf{k'}\downarrow}^+ a_{-\mathbf{k'}-\mathbf{q}\downarrow} a_{\mathbf{k}+\mathbf{q}\uparrow}|v\rangle . \tag{22.11}$$

In the notation (21.25) and with (21.23) this may be written as

$$S_{\uparrow\downarrow}(\mathbf{q}) = 2\sum_{\mathbf{kk'}}\langle v|(\Phi_{\mathbf{k}}^+ \sigma_{\mathbf{kk'}}^{(+)} \Phi_{\mathbf{k'}})(\Phi_{\mathbf{k'}+\mathbf{q}}^+ \sigma_{\mathbf{k'}+\mathbf{q},\mathbf{k}+\mathbf{q}}^{(-)} \Phi_{\mathbf{k}+\mathbf{q}})|v\rangle . \tag{22.12}$$

Applying Eqs. (21.19), (21.22), (21.28) and the anticommutation relations one then finds

$$S_{\uparrow\downarrow}(\mathbf{q}) = 2\left(\sum_{\mathbf{k}} v_{\mathbf{k}}^2\right)^2 \delta_{\mathbf{q},0} + 2\sum_{\mathbf{k}} u_{\mathbf{k}} v_{\mathbf{k}} u_{\mathbf{k}+\mathbf{q}} v_{\mathbf{k}+\mathbf{q}} . \tag{22.13}$$

Fourier-transforming this expression according to Eq. (2.4) it may be written in the form

$$\tilde{S}_{\uparrow\downarrow}(\mathbf{r}) = 2V\{(\tilde{G}(0))^2 + (\tilde{F}(\mathbf{r}))^2\} \tag{22.14}$$

where

$$\tilde{G}(\mathbf{r} - \mathbf{r}') = \langle v|\psi_\sigma^+(\mathbf{r})\psi_\sigma(\mathbf{r}')|v\rangle \; ; \quad \sigma = \uparrow, \downarrow \tag{22.15}$$

(for the spin-independence see Eq. (22.18) below) and

$$\tilde{F}(\mathbf{r} - \mathbf{r}') = \langle v|\psi_\uparrow^+(\mathbf{r})\psi_\downarrow^+(\mathbf{r}')|v\rangle \tag{22.16}$$

are, respectively, the particle-particle and particle-hole *correlation amplitudes*
in the groundstate and

$$\psi_\sigma^+(\mathbf{r}) = V^{-1/2} \sum_{\mathbf{k}} a_{\mathbf{k}\sigma}^+ e^{i\mathbf{k}\cdot\mathbf{r}} \tag{22.17}$$

is the *field operator* in Schrödinger representation (it is the analogue of Eq. (1.24) for phonons). That Eqs. (22.15) and (22.16) give indeed the Fourier transform of (22.13) follows from the fact that they are the Fourier transforms of

$$G(\mathbf{k}) = \langle v|a_{\mathbf{k}\sigma}^+ a_{\mathbf{k}\sigma}|v\rangle = v_{\mathbf{k}}^2 \; ; \quad \sigma = \uparrow, \downarrow \tag{22.18}$$

and

$$F(\mathbf{k}) = \langle v|b_{\mathbf{k}}^+|v\rangle = u_{\mathbf{k}} v_{\mathbf{k}} \, , \tag{22.19}$$

respectively, where the second equalities follow from (21.19), combined with the inverse of (21.18) or with Eqs. (21.24), (21.25) and (21.28).

The $\mathbf{r}$-dependence of the correlation function (22.14) is given by $\tilde{F}(\mathbf{r})$ which may be evaluated explicitly with the help of (22.19) and (21.13). In the approximation (21.15) one may write $\varepsilon_{\mathbf{k}} \cong \hbar v_F(k - k_F)$ so that, introducing the density of states (10.4), one obtains in view of (21.10)

$$\tilde{F}(r) = \frac{1}{V} \sum_{\mathbf{k}} u_{\mathbf{k}} v_{\mathbf{k}} e^{i\mathbf{k}\cdot\mathbf{r}} = \frac{1}{2} N(0)\Delta_o \int_{-\omega_o}^{+\omega_o} \frac{d\varepsilon}{E_o} \frac{\sin((\varepsilon/\hbar v_F) + k_F)r}{((\varepsilon/\hbar v_F) + k_F)r} \cdot \tag{22.20}$$

This expression is quite similar to Eq. (22.5) so that, making use of definition (22.7) one finds, using as integration variable $t = \varepsilon/\Delta_o$ and setting $\omega_o/\Delta_o \simeq \infty$,

$$\tilde{F}(r) = N(0)\Delta_o \frac{\sin k_F r}{k_F r} K_o\left(\frac{r}{\pi \xi_o}\right) \tag{22.21}$$

where

$$K_o(y) \equiv \int_0^\infty \frac{ds}{\sqrt{1+s^2}} \cos ys = \sqrt{\frac{\pi}{2y}} e^{-y}[1 + \mathcal{O}(y^{-1})] \tag{22.22}$$

is a Bessel function of order 0 (see Eqs. (8.432.5) and (8.451.6) of Ref. 33). Unlike the case of the wavefunction (22.4) where we had to rely on qualitative arguments to see that $\tilde{\varphi}(r)$ has a range $r \simeq \xi_o$, Eqs. (22.21), (22.22) explicitly show that the radial correlation function $r^2 \tilde{F}^2(r)$ falls off at $r \simeq \xi_o$ and oscillates with wavelength $2\pi/k_F \ll \xi_o$. Obviously, this oscillation is due to the restriction of the superconducting electrons to the vicinity of the Fermi level, $|\varepsilon_{\mathbf{k}}| < \omega_o \ll \varepsilon_F$.

The correlation amplitudes (22.18), (22.19) may also serve to identify the Hamiltonian of free bogolons (21.31) as *random phase approximation* (RPA) of the Hamiltonian $\mathcal{H} = \mathcal{H}_o + \mathcal{H}_{\text{BCS}}$. RPA is the approximation which linearizes the equations of motion by averaging over the phases of the dynamical variables[34]. For a Hamiltonian with quartic terms this is equivalent with taking pairwise averages which is nothing else than the Hartree-Fock prescription of replacing the interaction (4.7) by the Hartree-Fock operator (4.25). Thus RPA is just the *mean-field approximation*

$$\text{RPA}\{b_{\mathbf{k}}^+ b_{\mathbf{k}'}\} = \{G(-\mathbf{k})a_{\mathbf{k}\uparrow}^+ a_{\mathbf{k}\uparrow} + G(\mathbf{k})a_{-\mathbf{k}\downarrow}^+ a_{-\mathbf{k}\downarrow}\}\delta_{\mathbf{k}\mathbf{k}'} \\ + F(\mathbf{k})b_{\mathbf{k}'} + F^*(\mathbf{k}')b_{\mathbf{k}}^+ . \tag{22.23}$$

Applying this prescription to $\mathcal{H}$ and absorbing the G-terms of Eq. (22.23) as renormalizations into $\varepsilon_{\mathbf{k}}$, we find using (21.8),

$$\text{RPA}\{\mathcal{H}\} = \sum_{\mathbf{k}\sigma} \varepsilon_{\mathbf{k}} a_{\mathbf{k}\sigma}^+ a_{\mathbf{k}\sigma} - \sum_{\mathbf{k}} \Delta_{\mathbf{k}}(b_{\mathbf{k}}^+ + b_{\mathbf{k}})$$
$$= \mathcal{H}_B - 2\sum_{\mathbf{k}}(E_{\mathbf{k}} - \varepsilon_{\mathbf{k}}) \tag{22.24}$$

where in the second step Eqs. (21.10), (21.11), (21.13) and (21.22)–(21.25), (21.28) were used. The meaning of this result is that the Bogoljubov transformation diagonalizes the hermitean quadratic form RPA$\{\mathcal{H}\}$ and, according to (21.8) and (22.19), the gap function is given by

$$\Delta_{\mathbf{k}} = \frac{1}{V} \sum_{\mathbf{k}'} \lambda_{\mathbf{k}\mathbf{k}'} \langle v|b_{\mathbf{k}'}^+|v\rangle . \tag{22.25}$$

23. Statistical Mechanics of Bogolons. The Transition Temperature

The Bogoljubov Hamiltonian (21.31) may be used in conjunction with Eq. (6.7) to define the ensemble of free bogolons,

$$\rho_B = e^{\beta(\Omega_B - \mathcal{H}_B)} = \prod_{\mathbf{k}} Z_{\mathbf{k}}^{-2} e^{-\beta E_{\mathbf{k}} \alpha_{\mathbf{k}}^+ \alpha_{\mathbf{k}}} e^{-\beta E_{\mathbf{k}} \beta_{\mathbf{k}}^+ \beta_{\mathbf{k}}} \qquad (23.1)$$

where

$$Z_{\mathbf{k}} = \sum_{n=0,1} e^{-\beta E_{\mathbf{k}} n} = 1 + e^{-\beta E_{\mathbf{k}}} \qquad (23.2)$$

is the partition function and we have put the chemical potential $\mu = 0$ (see the remark after Eq. (23.8) below). This immediately results in relations analogous to (7.40), namely

$$\begin{aligned}
\langle \alpha_{\mathbf{k}}^+ \alpha_{\mathbf{k}} \rangle_B &= \langle \beta_{\mathbf{k}}^+ \beta_{\mathbf{k}} \rangle_B = f_o(E_{\mathbf{k}}) , \\
\langle \alpha_{\mathbf{k}} \alpha_{\mathbf{k}}^+ \rangle_B &= \langle \beta_{\mathbf{k}} \beta_{\mathbf{k}}^+ \rangle_B = 1 - f_o(E_{\mathbf{k}}) ,
\end{aligned} \qquad (23.3)$$

all other bilinear averages being zero.

The correlation amplitudes (22.18), (22.19) are then easily generalized to finite temperatures by writing

$$G(\mathbf{k}) \equiv \langle a_{\mathbf{k}\sigma}^+ a_{\mathbf{k}\sigma} \rangle_B \; ; \quad F(\mathbf{k}) \equiv \langle b_{\mathbf{k}}^+ \rangle_B . \qquad (23.4)$$

Applying Eqs. (21.22)–(21.25), (21.28) and (23.3) one deduces

$$G(\mathbf{k}) = v_{\mathbf{k}}^2 + (u_{\mathbf{k}}^2 - v_{\mathbf{k}}^2) f_o(E_{\mathbf{k}}) , \qquad (23.5)$$

which is still independent of spin and

$$F(\mathbf{k}) = u_{\mathbf{k}} v_{\mathbf{k}} \tanh \frac{\beta E_{\mathbf{k}}}{2} , \qquad (23.6)$$

where in the last relation Eq. (7.42) was used. Expressing now the definition of the groundstate gap function (21.8) in terms of $F(\mathbf{k})$ given by (22.19) and switching to the finite-temperature form (23.6) of $F(\mathbf{k})$ but maintaining relation (21.11), one arrives at the *finite-temperature gap equation*

$$\Delta_{\mathbf{k}} = \frac{1}{V} \sum_{\mathbf{k}} \lambda_{\mathbf{k}\mathbf{k}'} \frac{\Delta_{\mathbf{k}'}}{2 E_{\mathbf{k}'}} \tanh \frac{\beta E_{\mathbf{k}'}}{2} . \qquad (23.7)$$

This last manipulation, of course, is quite ambiguous since it is not evident *a priori* that Eqs. (21.10) and (21.11) stay the same when going over to finite temperatures. A more careful derivation of the gap equation (23.7) is therefore necessary. In analogy with the minimization of the groundstate energy which led to the gap equation (21.12), we now have to minimize the free energy or, equivalently, the grand-canonical potential Ω determined by the Hamiltonian. Since the interaction enters the gap equation only in first order (compare Fig. 4.2), it is sufficient to choose as Hamiltonian Eqs. (21.29) and (21.30) writing

$$\text{diag}\{\mathcal{H}\} = \mathcal{H}_B + \mathcal{H}' \, . \tag{23.8}$$

Since, in addition, the finite-temperature form of $E_\mathbf{k}$ is not known *a priori*, we must consider $v_\mathbf{k}$ and $E_\mathbf{k}$ (or equivalently $f_o(E_\mathbf{k})$, see Ref. 5) as independent variation parameters. Therefore, we may also include an undetermined value of μ in $E_\mathbf{k}$ which is formally the same as having set $\mu = 0$ in Eqs. (23.1), (23.2).

We now have to determine the thermodynamic potential Ω. From (23.1), (23.2) we deduce the equivalent of Eq. (10.1),

$$\Omega_B = -2\beta^{-1} \sum_\mathbf{k} \ln Z_\mathbf{k} = 2\beta^{-1} \sum_\mathbf{k} \ln[1 - f_o(E_\mathbf{k})] \tag{23.9}$$

where use was made of Eq. (7.42). For the interaction part $\Omega - \Omega_B$ we apply Eq. (9.20) to first order in $\mathcal{H}'$ which according to (5.28), (6.1) and (7.9) is simply given by

$$\Omega - \Omega_B = \langle \mathcal{H}' \rangle_B = 2 \sum_\mathbf{k} \{ \varepsilon_\mathbf{k} v_\mathbf{k}^2 + [\varepsilon_\mathbf{k}(u_\mathbf{k}^2 - v_\mathbf{k}^2) - E_\mathbf{k}] f_o(E_\mathbf{k}) \}$$
$$- \frac{1}{V} \sum_{\mathbf{kk'}} \lambda_{\mathbf{kk'}} u_\mathbf{k} v_\mathbf{k} u_{\mathbf{k'}} v_{\mathbf{k'}} [1 - 2f_o(E_\mathbf{k})][1 - 2f_o(E_{\mathbf{k'}})]$$
$$\tag{23.10}$$

where in the second step Eqs. (23.8), (21.29)–(21.31) and (23.3) were used and a term $\propto \delta_{\mathbf{kk'}}$ in the double sum has been neglected since it contributes an extra factor V^{-1} as in Eq. (22.3). From (23.9), (23.10) one then deduces (compare Section 5.2 of Ref. 3)

$$\frac{1}{2}\frac{\partial \Omega}{\partial E_\mathbf{k}} = \{ \varepsilon_\mathbf{k}(u_\mathbf{k}^2 - v_\mathbf{k}^2) - E_\mathbf{k} + 2u_\mathbf{k} v_\mathbf{k} \frac{1}{V} \sum_{\mathbf{k'}} \lambda_{\mathbf{kk'}} u_{\mathbf{k'}} u_{\mathbf{k'}}$$
$$\times [1 - 2f_o(E_{\mathbf{k'}})] \} f_o'(E_\mathbf{k}) \tag{23.11}$$

and

$$\frac{1}{2}\frac{\partial \Omega}{\partial v_\mathbf{k}} = \{ 2\varepsilon_\mathbf{k} v_\mathbf{k} - \frac{u_\mathbf{k}^2 - v_\mathbf{k}^2}{u_\mathbf{k}} \frac{1}{V} \sum_{\mathbf{k'}} \lambda_{\mathbf{kk'}} u_{\mathbf{k'}} v_{\mathbf{k'}}$$
$$\times [1 - 2f_o(E_{\mathbf{k'}})] \} [1 - 2f_o(E_\mathbf{k})] \, . \tag{23.12}$$

Defining now the *finite-temperature gap function* or *order parameter* as

$$\Delta_{\mathbf{k}} \equiv \frac{1}{V} \sum_{\mathbf{k}} \lambda_{\mathbf{k}\mathbf{k}'} u_{\mathbf{k}'} v_{\mathbf{k}'} [1 - 2f_o(E_{\mathbf{k}'})] \tag{23.13}$$

the extremum conditions following from (23.11) and (23.12) read, respectively,

$$E_{\mathbf{k}} = \varepsilon_{\mathbf{k}}(u_{\mathbf{k}}^2 - v_{\mathbf{k}}^2) + 2u_{\mathbf{k}} v_{\mathbf{k}} \Delta_{\mathbf{k}} \tag{23.14}$$

and

$$2\varepsilon_{\mathbf{k}} u_{\mathbf{k}} v_{\mathbf{k}} = (u_{\mathbf{k}}^2 - v_{\mathbf{k}}^2)\Delta_{\mathbf{k}} , \tag{23.15}$$

the latter being formally identical with Eq. (21.9). Eliminating $\Delta_{\mathbf{k}}$ from these two equations one easily derives the expressions (21.13) and, substituting these in Eq. (23.15) squared, one also confirms (21.10). This, finally justifies the gap equation (23.7) since, of course, relation (21.11) also holds.

As in the case of the groundstate, the gap equation (23.7) is explicitly solved, in the approximation (21.14), by Eq. (21.15) but with a temperature-dependent gap $\Delta(T)$,

$$\Delta_{\mathbf{k}} = \Delta(T)\theta(\omega_o - |\varepsilon_{\mathbf{k}}|) . \tag{23.16}$$

One finds in exactly the same way

$$\frac{1}{\Lambda N(0)} = \int_0^{\omega_o} \frac{d\varepsilon}{E} \tanh\frac{\beta E}{2} \tag{23.17}$$

where $E^2 \equiv \varepsilon^2 + \Delta^2(T)$. In the limit $\beta \to \infty$ where the hyperbolic tangent may be replaced by 1, we recover Eq. (21.16) and $\Delta(0) = \Delta_o$.

The other limit of interest is the transition temperature T_c where by definition the gap vanishes, $\Delta(T_c) = 0$. In this case Eq. (23.17) becomes

$$\frac{1}{\Lambda N(0)} = \int_0^{\omega_o} \frac{d\varepsilon}{\varepsilon} \tanh\frac{\beta_c \varepsilon}{2} \tag{23.18}$$

where $\beta_c^{-1} = k_B T_c$ and partial integration gives

$$\int_0^{\omega_o} \frac{d\varepsilon}{\varepsilon} \tanh\frac{\beta_c \varepsilon}{2} = \left\{ \log x \tanh x - \int dx \frac{\ln x}{\cosh^2 x} \right\}_0^{\beta_c \omega_o/2} . \tag{23.19}$$

In the weak-coupling limit $\beta_c \omega_o \gg 1$, the integral here may be approximated by (see Eq. (4.371.3) of Ref. 33)

$$-\int_0^{\infty} dx \frac{\ln x}{\cosh^2 x} = \ln\frac{4e^\gamma}{\pi} ; \quad \frac{2e^\gamma}{\pi} \cong 1.134 \tag{23.20}$$

where $\gamma = -\psi(1) \cong 0.5772$ is Euler's constant and $\psi(x)$ the *Digamma function* (see Ref. 33, Sections 8.36 and 9.73) which plays an important role in the theory of superconductivity. From Eqs. (23.18)–(23.20) follows the famous result[5]

$$k_B T_c \cong \frac{2e^\gamma}{\pi} \omega_o e^{-1/\Lambda N(0)} \ . \tag{23.21}$$

Comparison with the zero-temperature gap (21.17) shows that

$$\frac{2\Delta_o}{k_B T_c} \cong 2\pi e^{-\gamma} \cong 3.528 \tag{23.22}$$

is independent of the limiting energy ω_o.

In the case of a phonon-mediated coupling (20.9), $\omega_o \simeq k_B \theta$ which, according to (16.4), is proportional to the sound velocity $c_\parallel$. Now it was shown in Problem 4.2 that $Mc_\parallel^2$ and hence Λ are independent of the isotopic mass M. This leads to an *isotope effect* of the transition temperature[5,3]

$$T_c \propto M^{-1/2} \ . \tag{23.23}$$

As mentioned at the beginning of Section 10, the grand-canonical potential is not well adapted to the usual experimental situation in which the total electron density n (in distinction to the densities n_n and n_s defined in Eqs. (25.15) and (25.16) below) rather than the chemical potential μ is fixed. Fortunately, n is the same in the *superconducting state* S as in the *normal state* N. Indeed, the respective total densities n_S^{tot} and n_N^{tot} may be expressed, respectively, in terms of Eqs. (23.4), (23.5) and (21.13), i.e. of

$$G_S(\mathbf{k}) = \frac{1}{2}\left(1 - \frac{\varepsilon_{\mathbf{k}}}{E_{\mathbf{k}}}\right) + \frac{\varepsilon_{\mathbf{k}}}{E_{\mathbf{k}}} f_o(E_{\mathbf{k}}) \tag{23.24}$$

and of

$$G_N(\mathbf{k}) = f_o(\varepsilon_{\mathbf{k}}) = \frac{1}{2}\left(1 - \frac{\varepsilon_{\mathbf{k}}}{|\varepsilon_{\mathbf{k}}|}\right) + \frac{\varepsilon_{\mathbf{k}}}{|\varepsilon_{\mathbf{k}}|} f_o(|\varepsilon_{\mathbf{k}}|) \ . \tag{23.25}$$

With (10.4) and (23.16) these expressions give

$$n_S^{\text{tot}} - n_N^{\text{tot}} = \frac{2}{V} \sum_{\mathbf{k}} \{G_S(\mathbf{k}) - G_N(\mathbf{k})\}$$

$$\cong N(0) \int_{-\omega_o}^{\omega_o} d\varepsilon \ \varepsilon \left\{\frac{1}{E} - \frac{1}{|\varepsilon|} + \frac{f_o(E)}{E} - \frac{f_o(|\varepsilon|)}{|\varepsilon|}\right\} = 0 \ , \tag{23.26}$$

since the integrand is an odd function (see Eq. (37.51) of Ref. 35). This result has the important consequence that for differences between the states S and N, the grand-canonical potential Ω and the *free energy*

$$F(T, V, n) = \Omega(T, V, \mu) + V\mu n \tag{23.27}$$

are equivalent,

$$F_S - F_N \cong \Omega_S - \Omega_N \; . \tag{23.28}$$

24. Thermodynamic Properties Near $T = T_c$ and Near $T = 0$

For arbitrary temperatures $T < T_c$ the solution of the gap equation (23.17) cannot be given in analytic form except in the vicinity of $T = 0$ and $T = T_c$. A form which is particularly useful near T_c is obtained by writing the hyperbolic tangent in (23.17) as $f_o(-E) - f_o(E)$ and expressing it as half a Matsubara sum (Problem 4.7),

$$\tanh \frac{\beta E}{2} = \frac{4E}{\beta} \sum_{\nu > 0} \frac{1}{\nu^2 + E^2} \; . \tag{24.1}$$

Note that this series representation may be obtained by taking the difference between Eqs. (1.217.1) and (1.217.2) of Ref. 33. It was used by Mühlschlegel to calculate numerically the thermodynamic quantities at all temperatures $T < T_c$ [36]. Inserting (24.1), the gap equation (23.17) becomes

$$\frac{1}{\Lambda N(0)} = \frac{4}{\beta} \int_0^{\omega_o} d\varepsilon \sum_{\nu > 0} \frac{1}{\nu^2 + \varepsilon^2 + \Delta^2} \; . \tag{24.2}$$

Near T_c we may develop the denominator in (24.2) in powers of Δ since, according to (7.27), the fermionic Matsubara frequencies satisfy $|\nu| \geq \pi/\beta$ and hence

$$\frac{\Delta^2}{\nu^2 + \varepsilon^2} \leq \left(\frac{\Delta}{\pi k_B T} \right)^2 \cong \left(\frac{\Delta}{\Delta_o} \right)^2 \left(\frac{2\Delta_o}{2\pi k_B T_c} \right)^2 < \frac{\Delta^2}{\Delta_o^2} \tag{24.3}$$

where in the last step we used (23.22). Equation (24.2) then becomes

$$\frac{1}{\Lambda N(0)} = \frac{4}{\beta} \int_0^{\omega_o} d\varepsilon \sum_{\nu > 0} \frac{1}{\nu^2 + \varepsilon^2} \left\{ 1 - \frac{\Delta^2}{\nu^2 + \varepsilon^2} + \mathcal{O}\left(\frac{\Delta^4}{\Delta_o^4} \right) \right\} \; . \tag{24.4}$$

Applying Eq. (24.1) for $\Delta = 0$ and evaluating according to (23.19), (23.20) one has

$$\frac{4}{\beta} \int_0^{\omega_o} d\varepsilon \sum_{\nu > 0} \frac{1}{\nu^2 + \varepsilon^2} = \ln \left(\frac{2e^\gamma}{\pi} \beta \omega_o \right) \tag{24.5}$$

and subtracting the same expression for $\beta = \beta_c$, Eq. (24.4) yields

$$\ln \frac{\beta}{\beta_c} = \frac{e\Delta^2}{\beta} \int_0^{\omega_o} d\varepsilon \sum_{\nu > 0} \frac{1}{(\nu^2 + \varepsilon^2)^2} + \mathcal{O}\left(\frac{\Delta^4}{\Delta_o^4} \right) \; . \tag{24.6}$$

With (7.27) this may be written as

$$\ln \frac{T}{T_c} \cong -4\Delta^2 \beta^2 \sum_{n=0}^{+\infty} I_n \tag{24.7}$$

where, in the weak-coupling limit $\beta\omega_o \gg 1$,

$$I_n \equiv \int_0^{\beta\omega_o} \frac{dx}{(x^2 + x_n^2)^2} \cong \int_0^{\infty} \frac{dx}{(x^2 + x_n^2)^2} = \frac{\pi}{4|x_n|^3} \tag{24.8}$$

and $x_n \equiv \beta\nu = (2n + 1)\pi$.

Insertion of (24.8) into (24.7) introduces another special function of importance in superconductivity theory, the *Riemann zeta-function* defined as (see Ref. 33, Section 9.5)

$$\zeta(z) \equiv \sum_{n=1}^{\infty} \frac{1}{n^z} . \tag{24.9}$$

Indeed, one finds (Problem 4.8)

$$\sum_{n=0}^{+\infty} I_n \cong \frac{7\zeta(3)}{32\pi^2} . \tag{24.10}$$

Therefore, defining an energy Δ_1 such that

$$\left(\frac{\Delta_1}{k_B T_c}\right)^2 = \frac{8\pi^2}{7\zeta(3)} = 9.384 ; \quad \frac{\Delta_1}{\Delta_o} = 1.737 , \tag{24.11}$$

one obtains[5]

$$\Delta(T) \cong \Delta_1 t^{1/2} ; \quad t \equiv 1 - \frac{T}{T_c} \ll 1 . \tag{24.12}$$

Near $T = 0$ where $y \equiv \beta\Delta \gg 1$ we may approximate $\tanh(\beta E/2)$ by $1 - 2\exp(-\beta E)$ and write Eq. (23.17) as

$$\frac{1}{\Lambda N(0)} \cong \int_\theta^{\omega_o/\Delta} \frac{ds}{\sqrt{1 + s^2}} \{1 - 2e^{-y\sqrt{1+s^2}}\} . \tag{24.13}$$

Since according to (21.17) $\omega_o/\Delta \gg 1$, the second term on the right may again be expressed in terms of the Bessel function K_o defined in Eq. (22.22) but now in the representation (see Eq. (8.432.9) of Ref. 33)

$$K_o(y) \equiv \int_0^{\infty} \frac{ds}{\sqrt{1 + s^2}} e^{-y\sqrt{1+s^2}} \tag{24.14}$$

so that Eq. (24.13), combined with (21.16), becomes

$$\ln \frac{2\omega_o}{\Delta_o} \cong \ln \frac{2\omega_o}{\Delta} - 2K_o(y) . \tag{24.15}$$

Because of $y \gg 1$, the approximation (22.22) may again be used for $K_o(y)$ which implies that $0 < 1 - \Delta/\Delta_o \ll 1$, and (24.15) then yields

$$\frac{\Delta(T)}{\Delta_o} \cong 1 - \sqrt{\frac{2\pi k_B T}{\Delta_o}} e^{-\Delta_o/k_B T} \; ; \quad T \ll T_c \; . \tag{24.16}$$

Equations (24.12) and (24.16) give the behavior of $\Delta(T)$ shown in Fig. 4.4.

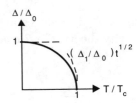

Fig. 4.4: Temperature dependence of the gap function according to Eqs. (24.12), (24.16).

The thermodynamic quantities may be obtained following an idea of a variable coupling constant due to Gor'kov[37] (see also Ref. 38, Sections 16.2 and 36.2) which is to calculate the contribution of the pairing interaction to the grand-canonical potential by multiplying the interaction (21.20) with a parameter λ and letting it increase from 0 (normal state N) to 1 (superconducting state S),

$$\Delta\Omega(T) \equiv \Omega_S(T) - \Omega_N(T) = \int_0^1 \frac{\partial\Omega}{\partial\lambda} d\lambda \; . \tag{24.17}$$

But this is Eq. (10.27) and, since in weak-coupling approximation the interaction is treated in lowest order, we may use Eq. (10.29).

To calculate $\langle \mathcal{H}_{\text{BCS}} \rangle_B$ we use RPA in the form (22.23) together with (23.4) to get

$$\langle \text{RPA}\{b_{\mathbf{k}}^+ b_{\mathbf{k}'}\} \rangle_B = \langle b_{\mathbf{k}}^+ \rangle_B \langle b_{\mathbf{k}'} \rangle_B + \cdots \tag{24.18}$$

where the left-out contributions are unessential diagonal renormalization terms. In order to obtain manageable expressions we use the constant coupling function (21.14). Using (23.4) and (23.6) the gap function (23.13) then takes the form (23.16) with

$$\Delta = \frac{\Lambda}{V} \sum_{\mathbf{k}} \theta(\omega_o - |\varepsilon_{\mathbf{k}}|) \langle b_{\mathbf{k}}^+ \rangle_B \; . \tag{24.19}$$

Since, on the other hand, the interaction (21.20) becomes, after multiplication with λ,

$$\mathcal{H}_{BCS}(\lambda) = -\lambda \frac{\Lambda}{V} \left| \sum_{\mathbf{k}} \theta(\omega_o - |\varepsilon_{\mathbf{k}}|) b_{\mathbf{k}}^+ \right|^2 , \tag{24.20}$$

we find from (24.18)–(24.20)

$$\langle \mathrm{RPA}\{\mathcal{H}_{BCS}(\lambda)\}\rangle_B = -\frac{V}{\lambda\Lambda}\Delta^2(\lambda) . \tag{24.21}$$

Inserting this result into (10.29), Eq. (24.17) takes the form

$$\Delta\Omega = -\frac{V}{\Lambda} \int_0^1 \frac{d\lambda}{\lambda^2}\Delta^2(\lambda) \tag{24.22}$$

where the functional dependence $\Delta(\lambda)$ is given by Eq. (23.17) as

$$\lambda^{-1} = \Lambda N(0) \int_0^{\omega_o} \frac{d\varepsilon}{\sqrt{\varepsilon^2 + \Delta^2(\lambda)}} \tanh \frac{\beta}{2}\sqrt{\varepsilon^2 + \Delta^2(\lambda)} . \tag{24.23}$$

However, since the explicitly given function is $\lambda^{-1}(\Delta^2)$, we transform the integration variable in Eq. (24.22) to $x = \Delta^2$ and obtain after a partial integration

$$\Delta\Omega = \frac{V}{\Lambda}\left\{ \Delta^2\lambda^{-1}(\Delta^2) - \int_0^{\Delta^2} dx\,\lambda^{-1}(x) \right\} . \tag{24.24}$$

Near T_c the last expression may be evaluated by substituting in (24.23) again the Matsubara sum (24.1) which, formally, may be interchanged with the x-integration to yield

$$\Delta\Omega = \frac{4VN(0)}{\beta} \int_0^{\omega_o} d\varepsilon \sum_{\nu>0}\left\{ \frac{\Delta^2}{\nu^2 + \varepsilon^2 + \Delta^2} - \ln\frac{\nu^2 + \varepsilon^2 + \Delta^2}{\nu^2 + \varepsilon^2} \right\} . \tag{24.25}$$

Developing in powers of $(\Delta/\Delta_o)^2$ as in Eq. (24.4) one finds, with (24.8) and (24.10)–(24.12), the result

$$\Delta\Omega(T) = -\frac{1}{2}VN(0)\Delta_1^2 t^2(1 + \mathcal{O}(t)) ; \quad t \equiv 1 - \frac{T}{T_c} \ll 1 \tag{24.26}$$

which shows that the superconducting state is more stable than the normal one. This formula also shows that the entropy is continuous at T_c,

$$\Delta S = -\left(\frac{\partial\Delta\Omega}{\partial T}\right)_{\mu=0} = VN(0)\Delta_1^2\frac{t}{T_c} . \tag{24.27}$$

On the other hand, the specific heat has a discontinuity[5]

$$\Delta c_V = \frac{T_c}{V}\left(\frac{\partial\Delta S}{\partial T}\right)_n = \frac{N(0)\Delta_1^2}{T_c} \tag{24.28}$$

so that the transition is of *second order*. Note that in passing from (24.27) to (24.28) we changed thermodynamic variables from μ to n which, according to Eq. (23.28), is perfectly legitimate. This may also be seen directly from the fact that, since the μ-dependence is contained solely in $N(\mu)$, we may consider the latter as variable or, according to Eq. (14.14), equivalently n. In terms of the normal-state specific heat (10.3),

$$c_{VN}(T) = \frac{2\pi^2}{3} N(0) k_B^2 T , \qquad (24.29)$$

this discontinuity has the value

$$\frac{\Delta c_V}{c_{VN}(T_c)} = \frac{12}{7\zeta(3)} = 1.426 . \qquad (24.30)$$

Formula (24.24) together with (24.23) may also be applied near $T = 0$ (see Section 51 of Ref. 35). Changing the integration variable from x to $E' = \sqrt{\varepsilon^2 + x}$ one has to evaluate

$$\int_0^{\Delta^2} \frac{dx}{E'} \tanh\frac{\beta E'}{2} = \frac{4}{\beta} \ln\frac{\cosh(\beta E/2)}{\cosh(\beta\varepsilon/2)} = 2(E - \varepsilon)$$
$$+ \frac{4}{\beta} \ln(1 + e^{-\beta E}) - \frac{4}{\beta} \ln(1 + e^{-\beta\varepsilon}) \quad (24.31)$$

and to insert into (24.24),

$$\Delta\Omega = VN(0) \int_0^{\omega_o} d\varepsilon \left\{ \frac{\Delta^2}{E} \tanh\frac{\beta E}{2} - 2(E - \varepsilon) \right.$$
$$\left. - \frac{4}{\beta} \ln(1 + e^{-\beta E}) + \frac{4}{\beta} \ln(1 + e^{-\beta\varepsilon}) \right\} . \qquad (24.32)$$

Now according to Eqs. (10.1), (10.4) we may write

$$\Omega_N = -\frac{2V}{\beta} \int_{-\varepsilon_F}^{\infty} d\varepsilon N(\varepsilon) \ln(1 + e^{-\beta\varepsilon}) + \text{const} \qquad (24.33)$$

and, separating the contributions from $\varepsilon < 0$ and $\varepsilon > 0$,

$$\Omega_N = -2V \int_0^{\varepsilon_F} \varepsilon N(-\varepsilon) d\varepsilon + \Omega_{oN} + \text{const} \qquad (24.34)$$

where

$$\Omega_{oN} \equiv -\frac{4V}{\beta} \int_0^{\infty} d\varepsilon \frac{1}{2}(N(\varepsilon) + N(-\varepsilon)) \ln(1 + e^{-\beta\varepsilon}) \qquad (24.35)$$

and we have replaced the upper limit ε_F by ∞ in the second term. Since the first term in (24.34) is temperature independent we may use it to compensate the constant. Hence (compare Eq. (3.33) of BCS, Ref. 5)

$$\Omega_N = \Omega_{oN} \cong -\frac{4V}{\beta^2}N(0)\int_0^1 \frac{dt}{t}\ln(1+t) = -\frac{\pi^2}{3}VN(0)(k_BT)^2 \quad (24.36)$$

(see Eq. (4.291.1) of Ref. 33 for the last step).

Equation (24.32) may now be evaluated by using (23.17) for the first term of the bracket and (24.36) for the last while, according to (23.9), the third term may be identified with Ω_{oS}, and the second is easily integrated. The result is

$$\Delta\Omega = \frac{V}{\Lambda}\Delta^2 - VN(0)\left(\omega_o\sqrt{\omega_o^2 + \Delta^2}\right.$$
$$\left. + \Delta^2\ln\frac{\omega_o + \sqrt{\omega_o^2 + \Delta^2}}{\Delta} - \omega_o^2\right) + \Omega_{oS} - \Omega_N . \quad (24.37)$$

Using Eq. (21.16) for the logarithm and neglecting terms of order Δ^4/ω_o^4 and Δ_o^4/ω_o^4 this becomes

$$\Omega_S \equiv \Omega_N + \Delta\Omega = \Omega_{oS} - VN(0)\Delta^2$$
$$\times \left\{\frac{1}{2} + \ln\left[\frac{\Delta_o}{\Delta}\left(1 + \frac{\Delta^2 - \Delta_o^2}{4\omega_o^2}\right)\right]\right\} \quad (24.38)$$

and, replacing ω_o/Δ_o by ∞,

$$\Omega_S = \Omega_{oS} - \frac{1}{2}VN(0)\Delta^2\left(1 - \ln\frac{\Delta^2}{\Delta_o^2}\right) . \quad (24.39)$$

On the other hand,

$$\Omega_{oS} \equiv -4VN(0)\beta^{-1}\int_0^{\omega_o} d\varepsilon\ln(1 + e^{-\beta E})$$
$$\cong -4VN(0)\Delta\beta^{-1}\int_0^\infty dt e^{-\beta\Delta\sqrt{1+t^2}} . \quad (24.40)$$

Comparing with the representation (24.14) of the Bessel function K_o, the last integral is recognized to be $-K_o'(\beta\Delta)$ which, upon using the asymptotic form (22.22) for $\beta\Delta \gg 1$ is just $+K_o(\beta\Delta)$. Since, according to (24.16), $(\Delta/\Delta_o) - 1$ is exponentially small, we may develop in powers of this quantity. In (24.40) this introduces higher powers of $\exp(-\Delta_o/k_BT)$ which are manifestly negligible, hence

$$\Omega_{oS} \cong -\frac{V}{\pi}N(0)\Delta_o^2\left(\frac{2\pi k_BT}{\Delta_o}\right)^{3/2}e^{-\Delta_o/k_BT} . \quad (24.41)$$

This result now has to be compared with the analogous development of the second term in (24.39); one finds

$$\Delta^2\left(1 - \ln\frac{\Delta^2}{\Delta_o^2}\right) = \Delta_o^2\left[1 + \mathcal{O}\left(\left(\frac{\Delta}{\Delta_o} - 1\right)^2\right)\right] \qquad (24.42)$$

so that, finally,

$$\Omega_S \cong \Omega_{oS} = -\frac{1}{2}VN(0)\Delta_o^2 . \qquad (24.43)$$

In view of the remark made after Eq. (24.28) the specific heat immediately follows from this result. In leading power in T one finds (see Eq. (51.61) of Ref. 35)

$$c_{VS}(T) = -\frac{T}{V}\left(\frac{\partial^2\Omega_S}{\partial T^2}\right)_{\mu=0} \cong 2\sqrt{2\pi}k_B N(0)\Delta_o\left(\frac{\Delta_o}{k_B T}\right)^{3/2}e^{-\Delta_o/k_B T} \qquad (24.44)$$

or with (24.29) (see Eq. (36.8) of Ref. 38)

$$\frac{c_{VS}(T)}{c_{VN}(T_c)} \cong \frac{3}{2}\left(\frac{2}{\pi}\right)^{3/2}\left(\frac{\Delta_o}{k_B T_c}\right)^{5/2}\left(\frac{T_c}{T}\right)^{3/2}e^{-\Delta_o/k_B T} \; ; \quad T \ll T_c . \quad (24.45)$$

The behavior of $c_V(T)$ given by Eqs. (24.29), (24.30) and (24.45) is shown in Fig. 4.5.

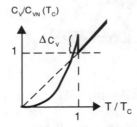

Fig. 4.5: Temperature dependence of the specific heat according to Eqs. (24.29), (24.30), (24.45).

25. Weak Static Magnetic Field. London Equation and Critical Field

As observed in the introduction to this chapter, the crucial feature of superconductivity is the current response to an external magnetic field giving rise to the Meissner effect. This was recognized quite early by Schafroth who in 1951 noticed that, in order to satisfy the London equation (25.18) below, the current response to the Fourier component $\tilde{\mathbf{A}}(\mathbf{q})$ of a vector potential must have a constant term $\propto q^0$ as compared to the dependence $\propto q^2$ in the normal state[39]. But of course, there is a question of gauge invariance to which we shall return later in this section.

In order to derive Schafroth's criterion we calculate the total current (18.20). It may be expressed with the aid of Eq. (18.17) as

$$\langle \mathbf{J}_t(\mathbf{q}) \rangle_t^A = \delta \langle \mathbf{j}(\mathbf{q}) \rangle_t^A - \frac{e^2 n}{mc} \tilde{\mathbf{A}}(\mathbf{q}, t) \tag{25.1}$$

where the first term on the right-hand side is the response (7.10), (7.11), to the linear part of the external Hamiltonian (18.2). Since the vector potential couples to $-\mathbf{j}/c$, this response is

$$\delta \langle j_\alpha(\mathbf{q}) \rangle_t^A = \frac{1}{c} \int_{-\infty}^{+\infty} dt' \chi_{\alpha\gamma}(\mathbf{q}; t - t') \tilde{A}_\gamma(\mathbf{q}, t') \tag{25.2}$$

where the current-current response function is defined by Eq. (7.12). The response to a time-periodic vector potential $\tilde{\mathbf{A}}(\mathbf{q}, t) = \tilde{A}(\mathbf{q}, \omega) e^{-i\omega t}$ may then be written in the form[39]

$$\tilde{I}_\alpha(\mathbf{q}, \omega) = K_{\alpha\gamma}(\mathbf{q}, \omega) \tilde{A}_\gamma(\mathbf{q}, \omega) \tag{25.3}$$

where $\langle \mathbf{J}_t(\mathbf{q}) \rangle_t^A = \tilde{\mathbf{I}}(\mathbf{q}, \omega) e^{-i\omega t}$ and

$$K_{\alpha\gamma}(\mathbf{q}, \omega) \equiv \frac{1}{c} \{ \tilde{\chi}_{\alpha\gamma}(\mathbf{q}, \omega) - \frac{e^2 n}{m} \delta_{\alpha\gamma} \} \; ; \quad \mathrm{Im}\,\omega > 0 \; . \tag{25.4}$$

The first term on the right-hand side of Eq. (25.4) is the *current-current susceptibility* defined in Eq. (15.6). As mentioned thereafter, one may use relation (7.32) to evaluate this expression. Indeed, making use of definitions (7.21) and (7.29) one obtains

$$\tilde{\chi}_{ab}(\mathbf{q}; i\nu) = \frac{1}{V} \int_0^\beta d\tau e^{i\nu\tau} \langle a(\mathbf{q}, -i\tau) b(-\mathbf{q}, 0) \rangle \tag{25.5}$$

which is the key formula for the calculation of susceptibilities . Note that ν in Eq. (25.5) is a bosonic Matsubara frequency. Making use of expression (3.23) for the current density $\mathbf{j} = -e\mathbf{j}_n$ we obtain in the normal state

$$\tilde{\chi}_{\alpha\gamma}^N(\mathbf{q}; i\nu) = \frac{e^2 \hbar^2}{m^2 V} \int_0^\beta d\tau e^{i\nu\tau} \sum_{\mathbf{k}\sigma} \sum_{\mathbf{k}'\sigma'} k_\alpha k_\gamma'$$
$$\times \langle (a_{\mathbf{k}-\mathbf{q}/2,\sigma}^+ a_{\mathbf{k}+\mathbf{q}/2,\sigma})(-i\tau)(a_{\mathbf{k}'+\mathbf{q}/2,\sigma'}^+ a_{\mathbf{k}'-\mathbf{q}/2,\sigma'}) \rangle \; . \tag{25.6}$$

Evaluating the average we may assume $\mathbf{q} \neq 0$ and an unperturbed average, so that the bracket factorizes as in the case of Eq. (15.3). Making use of the identity

$$f_o(\varepsilon)[1 - f_o(\varepsilon')] \equiv [f_o(\varepsilon') - f_o(\varepsilon)]n_o(\varepsilon - \varepsilon') \qquad (25.7)$$

and of Eqs. (7.26), (7.36), (7.43) one then finds

$$\tilde{\chi}^N_{\alpha\gamma}(\mathbf{q}; \omega) = \frac{2e^2\hbar^2}{m^2 V} \sum_{\mathbf{k}} k_\alpha k_\gamma \frac{f_o(\varepsilon_+) - f_o(\varepsilon_-)}{\varepsilon_- - \varepsilon_+ + \omega} \qquad (25.8)$$

where $\varepsilon_\pm \equiv \varepsilon_{\mathbf{k}\pm\mathbf{q}/2}$ and the analytic continuation $i\nu \to \omega$ was made.

We are interested in the *isothermal limit* $\omega = 0$, $\mathbf{q} \to 0$, in which case Eq. (25.8) becomes

$$\tilde{\chi}^N_{\alpha\gamma}(\mathbf{q}; 0) = -\frac{2e^2\hbar^2}{3m^2 V} \sum_{\mathbf{k}} \mathbf{k}^2 f'_o(\varepsilon_{\mathbf{k}})\delta_{\alpha\gamma} + \mathcal{O}(q^2) . \qquad (25.9)$$

Taking the limit $T \to 0$ and introducing the density of states (10.4) this becomes $(2e^2\hbar^2 k_F^2 N(0)/3m^2)\delta_{\alpha\gamma}$ which is $-4c^2 k_F^2$ times the *orbital magnetic susceptibility* χ^{orb} (see Section 31). Here the minus sign is a reminder that the response (25.2) to the vector potential $\tilde{\mathbf{A}}$ is $-(1/c)\chi_{\alpha\gamma}$, i.e. it is diamagnetic , and the remaining factors reflect the fact that in the absence of transport or polarization currents, $\mathbf{j} = ic\mathbf{q} \times \mathbf{M}^{\mathrm{orb}}$ where $\mathbf{M}^{\mathrm{orb}}$ is the orbital magnetization. This shows that the current response is related to orbital diamagnetism. However, $|\chi^{\mathrm{orb}}|$ is orders of magnitude smaller than the value $1/4\pi$ of perfect diamagnetism resulting from the Meissner effect. Indeed, we may use Eqs. (14.13) and (14.14) to write $-\chi^{\mathrm{orb}} = e^2 k_F/12\pi^2 mc^2$ which is of the order of 10^{-7}. This shows that the current response must change drastically at the transition to superconductivity.

We now turn to the calculation of the current-current susceptibility in the superconducting state. To this end we express the current density (3.23) in terms of the Nambu representation (21.22). Assuming $\mathbf{q} \neq 0$ we may anticommute the operators $a^+_{\mathbf{k}-\mathbf{q}/2,\downarrow}$ and $a_{\mathbf{k}+\mathbf{q}/2,\downarrow}$ in the current density (3.23) and change the summation variable from $\mathbf{k}$ to $-\mathbf{k}$. Then

$$\mathbf{j}_S(\mathbf{q}) = -\frac{e\hbar}{m} \sum_{\mathbf{k}} \mathbf{k} \Psi^+_{\mathbf{k}-\mathbf{q}/2} \Psi_{\mathbf{k}+\mathbf{q}/2} ; \quad \mathbf{q} \neq 0 . \qquad (25.10)$$

Inserting this expression into Eq. (25.5) and applying the Bogoljubov transformation (21.23), factorization as in (25.6) and use of Eqs. (23.3) yields, after some calculation (Problem 4.9) and after analytic continuation $i\nu \to \omega$

(compare Eq. (52.29) of Ref. 35),

$$\tilde{\chi}_{\alpha\gamma}^S(\mathbf{q};\omega) = \frac{e^2\hbar^2}{m^2V} \sum_{\mathbf{k}} k_\alpha k_\gamma \left\{ (f_+ - f_-)\left(1 + \frac{\varepsilon_-\varepsilon_+ + \Delta^2}{E_-E_+}\right)\right.$$
$$\times \frac{E_- - E_+}{(E_- - E_+)^2 - \omega^2} + (1 - f_+ - f_-)$$
$$\left. \times \left(1 - \frac{\varepsilon_-\varepsilon_+ + \Delta^2}{E_-E_+}\right)\frac{E_- + E_+}{(E_- + E_+)^2 - \omega^2} \right\} . \tag{25.11}$$

Here we have introduced the abbreviations

$$\varepsilon_\pm \equiv \varepsilon_{\mathbf{k}\pm\mathbf{q}/2} ; \quad E_\pm \equiv E_{\mathbf{k}\pm\mathbf{q}/2} ; \quad f_\pm \equiv f_o(E_\pm) . \tag{25.12}$$

Taking again the isothermal limit $\omega = 0$, $\mathbf{q} \to 0$, Eq. (25.11) takes a form similar to (25.9),

$$\tilde{\chi}_{\alpha\gamma}^S(\mathbf{q};0) = -\frac{2e^2\hbar^2}{3m^2V}\sum_{\mathbf{k}} \mathbf{k}^2 f_o'(E_{\mathbf{k}})\delta_{\alpha\gamma} + \mathcal{O}(q^2) . \tag{25.13}$$

In order to see the physics behind this result we first note that for free electrons the electron density may be written as (Problem 4.10)

$$n = -\frac{2\hbar^2}{3mV}\sum_{\mathbf{k}} \mathbf{k}^2 f_o'(\varepsilon_{\mathbf{k}}) . \tag{25.14}$$

Defining similarly the *density of normal electrons* as

$$n_n \equiv -\frac{2\hbar^2}{3mV}\sum_{\mathbf{k}} \mathbf{k}^2 f_o'(E_{\mathbf{k}}) \tag{25.15}$$

and the *density of super-electrons* as

$$n_s \equiv n - n_n , \tag{25.16}$$

we have a *two-fluid description*[40,41] as for superfluid helium, and insertion into the isothermal limit of Eq. (25.4) gives

$$K_{\alpha\gamma}(\mathbf{q};0) = -\frac{e^2 n_s}{mc}\delta_{\alpha\gamma} + \mathcal{O}(q^2) . \tag{25.17}$$

This is *Schafroth's criterion*[39] since in the normal state $n_s = 0$.

Substituting Eq. (25.17) in (25.3) and taking the vector product with $i\mathbf{q}$, we obtain in $\mathbf{r}$-space the *London equation*[41]

$$\nabla \times \mathbf{j}_S(\mathbf{r}) = -\frac{e^2 n_s}{mc}\mathbf{H}(\mathbf{r}) \tag{25.18}$$

where $\mathbf{H} = \nabla \times \mathbf{A}$ is the applied magnetic field. Combination with the Maxwell equation

$$\nabla \times \mathbf{H} = \frac{4\pi}{c}\mathbf{j} \tag{25.19}$$

then yields, in the stationary limit $\nabla \cdot \mathbf{j} = 0$ of Eq. (18.4),

$$\nabla^2 \mathbf{j}_S = \lambda_L^{-2} \mathbf{j}_S \tag{25.20}$$

where

$$\lambda_L \equiv \sqrt{\frac{mc^2}{4\pi e^2 n_s}} \tag{25.21}$$

is the *London penetration depth*[41]. For a superconductor situated in the half-space $z > 0$, Eq. (25.20) has the solution

$$\mathbf{j}_S(z) = \mathbf{j}_o e^{-z/\lambda_L} \tag{25.22}$$

with $\mathbf{j}_o = (j_{ox}, j_{oy}, 0)$. This is the *Meissner effect*[4]. Equation (25.20) may be viewed as the stationary case of the *Klein-Gordon equation* with mass $\hbar/c\lambda_L$. Since the field $\mathbf{H}$ satisfies the same equation, this mass term signals the fact that *gauge invariance is broken because of pairing* .

The temperature dependence of λ_L is determined by that of n_s. Evaluating Eqs. (25.15), (25.16) one finds by applying the techniques which led to Eqs. (24.12) and (24.16) that (Problem 4.10)

$$\frac{n_s}{n} = \begin{cases} 2t \ ; & t \equiv 1 - T/T_c \ll 1 \\ 1 - (2\pi\Delta_0/k_B T)^{1/2} \exp(-\Delta_0/k_B T) \ ; & T \ll T_c \ . \end{cases} \tag{25.23}$$

Note the difference with the phenomenological law $1 - (T/T_c)^4$ of F. London, Ref. 41, Section 8.

Long before BCS , Pippard[42] argued that the superconductive state must be characterized by a finite coherence length ξ_o and that this fact must render the London equation (25.18) nonlocal . His argument was that since only electrons with energy ε limited by $|\varepsilon - \varepsilon_F| < k_B T_c$ contribute to superconductivity, the latter have a momentum spread of $\Delta p \simeq k_B T_c/v_F$. Application of the Heisenberg uncertainty relation then yields $\xi_o \simeq \hbar v_F/k_B T_c$ which in view of (23.22) agrees well with the BCS value (22.7). ξ_o is therefore often called the *Pippard coherence length*. Assuming now that the variation of the vector potential $\mathbf{A}$ over distances of the order of ξ_o is negligible, Eq. (25.18) is equivalent with the *nonlocal relation*

$$\mathbf{j}_S(\mathbf{r}) = -\frac{3}{4\pi\xi_o\lambda_L^2} \int d^3x \frac{\mathbf{x}(\mathbf{x} \cdot \mathbf{A}(\mathbf{r} + \mathbf{x}))}{|\mathbf{x}|^4} e^{-|\mathbf{x}|/\xi_o} \ . \tag{25.24}$$

Pippard[42] also introduced a generalized coherence length ξ which charac-terizes *dirty superconductors* by taking into account the mean free path ℓ for impurity scattering ,

$$\frac{1}{\xi} = \frac{1}{\xi_o} + \frac{1}{\ell} . \tag{25.25}$$

In the above equations for n_s and n_n we had put the chemical poten-tial $\mu = 0$. Now an interesting situation arises at an interface between a superconductor and a normal metal since, by applying a voltage, we may create a value $\mu \neq 0$. In this case $n_n(\mu)$ represents the *tunneling electron density*. In particular, near $T = 0$ and for $0 < \mu \ll \varepsilon_F$ we then obtain from Eq. (25.15), using (14.14) with $D = 3$,

$$\frac{n_n(\mu)}{n} \cong \frac{1}{N(0)V} \sum_{\mathbf{k}} \delta(E_{\mathbf{k}} - \mu) \equiv 2N_T(\mu)/N(0) . \tag{25.26}$$

This relation defines the *tunneling density of states* $N_T(\mu)$ [24] which is easily calculated by using the density of states (10.4) and changing the integration variable from ε to E (compare Eq. (3.26) of BCS, Ref. 5),

$$\frac{N_T(\mu)}{N(0)} = \frac{\mu}{\sqrt{\mu^2 - \Delta^2}} ; \quad 0 < \Delta < \mu \ll \varepsilon_F . \tag{25.27}$$

This expression where, in general, Δ also depends on μ, may be compared with the *current-voltage characteristics* obtained from tunneling experi-ments at low temperatures which, therefore provide a direct determination of the gap . This type of experiments was pioneered by Giaever in 1960[43] (see Ref. 3, Chapter 10 and in particular Fig. 10.3).

Turning now to the question of gauge invariance raised at the beginning of this section, the discussion is much simplified by going over to a 4-vector notation. Thus introducing $\tilde{A}(q) \equiv (\vec{\tilde{A}}, \tilde{A}_o)$, $\tilde{I}(q) \equiv (\tilde{\mathbf{I}}, \tilde{I}_o)$ and $q \equiv (\mathbf{q}, \omega)$, Eq. (25.3) may be generalized to

$$\tilde{I} = K\tilde{A} \tag{25.28}$$

where K now is a 4-tensor. Since gauge transformations are defined by

$$\tilde{A} \rightarrow \tilde{A}' + iq\tilde{\phi} \tag{25.29}$$

with an arbitrary scalar function $\tilde{\phi}(q)$ and since $\tilde{I}$ is gauge-invariant, Eqs. (25.28), (25.29) imply the condition[39]

$$Kq = 0 . \tag{25.30}$$

On the other hand, charge conservation, Eq. (18.4), reads in this notation

$$q\tilde{I} = 0 , \tag{25.31}$$

so that Eq. (25.28) also implies[39)]

$$qK = 0 \qquad (25.32)$$

and K must be $\propto q^2 \mathbf{1} - q \otimes q$[39)].

In the isothermal limit $\omega = 0$, Eqs. (25.28), (25.30) imply in particular that, in the previous notation,

$$K_{\alpha\gamma}(\mathbf{q}; 0)q_\gamma = q_\gamma K_{\gamma\alpha}(\mathbf{q}; 0) = 0 . \qquad (25.33)$$

But according to Eq. (25.17) these conditions are manifestly violated. This means that the Bogoljubov Hamiltonian (21.31) underlying the whole calculation gives an incomplete description of the system which, of course, is no surprise. Indeed, the Bogoljubov transformation (21.42) leads from a state with charge $\langle 0| - eN_{el}|0\rangle = 0$ to a state with charge $-e\bar{N}_{el}$ given by Eq. (20.34) where, according to (21.13), $\bar{N}_{el}/V \simeq n$. Because of the symmetry of the tensor K, violation of charge conservation then carries along violation of gauge covariance. The remedy is to introduce the collective charge-carrying degrees of freedom through their equations of motion. This approach was pioneered by Anderson[34)] and gave rise to a rather technical but not very illuminating literature (see Ref. 3, Sections 6.4–6.6; Ref. 24, Sections 8.5, 8.6).

We now turn to the problem of the critical magnetic field. As observed above, the orbital Landau susceptibility of free electrons χ_L is orders of magnitude smaller than $1/4\pi$, and this is also true for the Pauli spin susceptibility $\chi_P = 2\mu_B^2 N(0)$ since $\chi_P = -3\chi_L$ (see Eqs. (31.16), (31.48) below). Therefore, the magnetic energy in the normal state is (in Gaussian cgs units)

$$\Omega_N(H) - \Omega_N(0) = -\frac{V}{8\pi}(1 + 4\pi\chi_N)H^2 \cong -\frac{V}{8\pi}H^2 . \qquad (25.34)$$

On the other hand, the Meissner effect, i.e. perfect diamagnetism, means that

$$\Omega_S(H) - \Omega_S(0) = -\frac{V}{8\pi}(1 + 4\pi\chi_S)H^2 = 0 . \qquad (25.35)$$

Since the *critical magnetic field* H_c is defined by the condition

$$\Omega_S(H_c) = \Omega_N(H_c) , \qquad (25.36)$$

we obtain by combining the last three relations

$$\Omega_N(0) - \Omega_S(0) = \frac{V}{2}(\chi_N - \chi_S)H_c^2 \cong \frac{V}{8\pi}H_c^2 \qquad (25.37)$$

where the left-hand side is $-\Delta\Omega(T)$ as defined in Eq. (24.17). Making use of Eqs. (24.36), (24.41), (24.43) in the limit $T \to 0$ and of Eq. (24.26) in the limit $T \to T_c$, one obtains (see Eqs. (3.43), (3.53) of BCS, Ref. 5)

$$H_c(T) = H_{co} \times \begin{cases} [1 - (\pi k_B T/\Delta_o)^2/3] \; ; \quad T \ll T_c \\ (\Delta_1/\Delta_o)t \; ; \quad t \equiv 1 - T/T_c \ll 1 \end{cases} \tag{25.38}$$

where

$$H_{co} \equiv \sqrt{4\pi N(0)}\Delta_o \; . \tag{25.39}$$

Making use of the numerical values (23.22), (24.11) one obtains for $(H/H_{co})^2$ as function of $(T/T_c)^2$ the curve shown in Fig. 4.6 which clearly deviates from the linear law of the two-fluid model of Gorter and Casimir[40] but also from most experimental curves as shown, e.g. in Fig. 5.4 of Ref. 3.

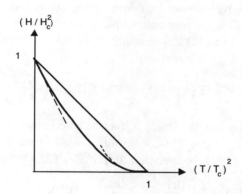

Fig. 4.6: The temperature dependence of the critical magnetic field interpolated between the limits of Eqs. (25.38) (broken lines). The straight line corresponds to the two-fluid model of Gorter and Casimir.

In the case of an *incomplete Meissner effect* we still have $\chi_S < 0$. On the other hand, if the normal state is paramagnetic, e.g. due to *magnetic impurities*, $\chi_N \geq \chi_P$ so that Eq. (25.37) leads to the inequality

$$\Omega_N(0) - \Omega_S(0) > \frac{V}{2}\chi_P H_c^2 \tag{25.40}$$

and, using Eqs. (25.37)–(25.39) at $T = 0$, to

$$H_c < H_{CC} \equiv \sqrt{\frac{N(0)}{\chi_P}}\Delta_o = \frac{1.764}{\sqrt{2}\mu_B}k_B T_c = 18400G \times T_c \; . \tag{25.41}$$

This is the *Clogston-Chandrasekhar limit*[44] which is of some importance for thin films in a parallel magnetic field, see Ref. 45, Section 8.2. Magnetic impurities give rise to a *molecular magnetic field* which cannot be expulsed and therefore prevents a complete Meissner effect. This undercritical magnetic field breaks some, but not all, Cooper pairs giving rise to a normal current $\mathbf{j}_n$. In the groundstate ($T = 0$), $\mathbf{j}_n$ is compensated by a supercurrent which causes the remaining Cooper pairs to have momentum $\hbar\mathbf{q}$. This gives rise to a gap function Δ oscillating in space as $e^{i\mathbf{q}\cdot\mathbf{r}}$, in close analogy with a *spin density wave* (see Section 36).

26. Weak Electromagnetic Field. Ginzburg-Landau Equation

As first shown by Gor'kov[37], the phenomenological equation of Ginzburg and Landau[46] describing the order parameter of an inhomogeneous superconductor may be derived from the equations of motion for the microscopic Green's functions. This is most conveniently done in the Nambu representation (21.22) in which the unperturbed Green's functions are defined by the 2×2 matrix

$$\mathcal{G}^B(\mathbf{k}; \tau) \equiv -\langle T\{\Psi_\mathbf{k}^o(-i\tau) \otimes \Psi_\mathbf{k}^+(0)\}\rangle_B = \begin{pmatrix} \mathcal{G}_\uparrow^B & \mathcal{F}^B \\ \mathcal{F}^{B*} & \mathcal{G}_\downarrow^B \end{pmatrix} \qquad (26.1)$$

where $\mathcal{F}^B$ is the extraordinary Green's function of Gor'kov[47] in the Bogoljubov groundstate. This definition is easily recognized as the natural generalization of the correlation amplitudes (23.4) since

$$\mathcal{G}^B(\mathbf{k}; 0^+) = \begin{pmatrix} G(\mathbf{k}) - 1, F(\mathbf{k}) \\ F(\mathbf{k}), -G(\mathbf{k}) \end{pmatrix} = \mathcal{G}^B(\mathbf{k}; 0^-) - \mathbf{1} . \qquad (26.2)$$

Taking the Fourier transform (7.29) of Eq. (26.1) we obtain with (21.22), (21.23) and (23.3)

$$R_\mathbf{k}\tilde{\mathcal{G}}^B(\mathbf{k}; i\nu)R_\mathbf{k}^T = \int_0^\beta d\tau e^{i\nu\tau} \begin{pmatrix} e^{-E_\mathbf{k}\tau}[1 - f_o(E_\mathbf{k})], 0 \\ 0, e^{E_\mathbf{k}\tau} f_o(E_\mathbf{k}) \end{pmatrix}$$
$$= \begin{pmatrix} (i\nu - E_\mathbf{k})^{-1}, 0 \\ 0, (i\nu + E_\mathbf{k})^{-1} \end{pmatrix} \qquad (26.3)$$

where in the last step use was made of Eqs. (7.26) and (7.42). Allowing for a complex gap function $\Delta_\mathbf{k}$, this gives, with (21.11), (21.13),

$$(\tilde{\mathcal{G}}^B)^{-1}(\mathbf{k}; i\nu) = \begin{pmatrix} i\nu - \varepsilon_\mathbf{k}, \Delta_\mathbf{k} \\ \Delta_\mathbf{k}^*, -(i\nu - \varepsilon_\mathbf{k})^* \end{pmatrix} \qquad (26.4)$$

or

$$\tilde{\mathcal{G}}_\uparrow^B = -\tilde{\mathcal{G}}_\downarrow^{B*} = -\frac{i\nu + \varepsilon_{\mathbf{k}}}{\nu^2 + \varepsilon_{\mathbf{k}}^2} \; ; \quad \tilde{\mathcal{F}}^B = \frac{\Delta_{\mathbf{k}}}{\nu^2 + \varepsilon_{\mathbf{k}}^2} \; . \tag{26.5}$$

Applying to the last expression Eqs. (24.1) and (23.7), one finds the inverse relation

$$\Delta_{\mathbf{k}} = \frac{1}{\beta V} \sum_{\mathbf{k}'} \lambda_{\mathbf{k}\mathbf{k}'} \sum_\nu \tilde{\mathcal{F}}^B(\mathbf{k}'; i\nu) \; . \tag{26.6}$$

According to Eq. (18.1) an electromagnetic field may be introduced by the gauge-invariant substitutions $\mathbf{k} \rightarrow -i\nabla + (e/\hbar c)\mathbf{A}(\mathbf{r}, t)$ and $\nu \rightarrow \partial/\partial t - ieU(\mathbf{r}, t)$. Thus, multiplying Eq. (26.4) from the right by $\tilde{\mathcal{G}}^B(\mathbf{k}; i\nu)$ and taking the Fourier transform in all 4 variables, these substitutions yield the general equations of motion

$$\begin{pmatrix} i\frac{\partial}{\partial t} + eU + \frac{\hbar^2}{2m}(\nabla + \frac{ie}{\hbar c}\mathbf{A})^2 + \mu, & \Delta(\mathbf{r}) \\ \Delta^*(\mathbf{r}), & -\{i\frac{\partial}{\partial t} + eU + \frac{\hbar^2}{2m}(\nabla + \frac{ie}{\hbar c}\mathbf{A})^2 + \mu\}^* \end{pmatrix} \mathcal{G}(\mathbf{r}\mathbf{r}'; tt')$$
$$= \delta(\mathbf{r} - \mathbf{r}')\delta(t - t') \tag{26.7}$$

where we have replaced the Fermi energy ε_F by a general value μ of the chemical potential and, as in (26.1),

$$\mathcal{G} \equiv \begin{pmatrix} \mathcal{G}_\uparrow & \mathcal{F} \\ \mathcal{F}^* & \mathcal{G}_\downarrow \end{pmatrix} \; . \tag{26.8}$$

Equations (26.7) imply that, in analogy to (26.5),

$$\mathcal{G}_\uparrow = -\mathcal{G}_\downarrow^* \; , \tag{26.9}$$

and they reduce to the two equations[37]

$$\left\{ i\frac{\partial}{\partial t} + eU + \frac{\hbar^2}{2m}\left(\nabla + \frac{ie}{\hbar c}\mathbf{A}\right)^2 + \mu \right\}\mathcal{G}_\uparrow + \Delta\mathcal{F}^* = \delta(\mathbf{r} - \mathbf{r}')\delta(t - t') \; ,$$
$$\Delta^*\mathcal{G}_\uparrow - \left\{ -i\frac{\partial}{\partial t} + eU + \frac{\hbar^2}{2m}\left(\nabla - \frac{ie}{\hbar c}\mathbf{A}\right)^2 + \mu \right\}\mathcal{F}^* = 0 \; .$$
$$\tag{26.10}$$

While Eqs. (26.10) may also be applied to the case of a superconductor interacting with radiation, here we are interested only in the situation of the Ginzburg-Landau equation which involves stationary fields $\mathbf{A} = \mathbf{A}(\mathbf{r})$, $U = 0$ only. This implies that the time arguments occur in $\mathcal{G}$ only in the form $t - t' = -i\tau$ and we may go back to the variable $i\nu$ by applying the Fourier transform (7.29). Since ν is a constant throughout we will not indicate it explicitly in $\mathcal{G}$. Introducing the "Hamiltonian"

$$\mathcal{H}(\mathbf{r}) \equiv -\frac{\hbar^2}{2m}\left(\nabla + \frac{ie}{\hbar c}\mathbf{A}(\mathbf{r})\right)^2 - \mu \tag{26.11}$$

and taking the complex conjugate of the second equation (26.10), these may be written in the form

$$\{i\nu - \mathcal{H}(\mathbf{r})\}\mathcal{G}_\uparrow(\mathbf{rr}') + \Delta(\mathbf{r})\mathcal{F}^*(\mathbf{rr}') = \delta(\mathbf{r} - \mathbf{r}') \ ,$$
$$\Delta(\mathbf{r})\mathcal{G}_\uparrow^*(\mathbf{rr}') = \{i\nu - \mathcal{H}(\mathbf{r})\}\mathcal{F}(\mathbf{rr}') \ . \tag{26.12}$$

In the following derivation which is due to Werthamer[48] we shall make use of *two assumptions*, the *first* being that $\Delta(\mathbf{r})$ is small, i.e. that T is close to T_c. This means that we shall express $\mathcal{G}$ in terms of the Green's function $\mathcal{G}_N$ of the normal state. From (26.12) we have for $\Delta = 0$ that $\mathcal{G}_\uparrow = \mathcal{G}_N$, $\mathcal{F} = 0$ and

$$\mathcal{G}_N(\mathbf{rr}') = \{i\nu - \mathcal{H}(\mathbf{r})\}^{-1}\delta(\mathbf{r} - \mathbf{r}') \ . \tag{26.13}$$

With this expression, Eqs. (26.12) may be written in the iterative form

$$\mathcal{G}_\uparrow(\mathbf{rr}') = \mathcal{G}_N(\mathbf{rr}') - \int_V d^3s\, \mathcal{G}_N(\mathbf{rs})\Delta(\mathbf{s})\mathcal{F}^*(\mathbf{sr}') \ ,$$
$$\mathcal{F}(\mathbf{rr}') = \int_V d^3s\, \mathcal{G}_N(\mathbf{rs})\Delta(\mathbf{s})\mathcal{G}_\uparrow^*(\mathbf{sr}') \ . \tag{26.14}$$

Introducing center-of-mass and relative coordinates $\mathbf{R} \equiv (\mathbf{r} + \mathbf{r}')/2$ and $\mathbf{x} \equiv \mathbf{r} - \mathbf{r}'$, we make the *second assumption* that the $\mathbf{R}$-dependence of physical quantities, but not necessarily of the phase of $\Delta(\mathbf{R})$, is slow (macroscopic inhomogeneities) so that we may develop physical quantities in powers of $\partial_\mathbf{R} \equiv \partial/\partial\mathbf{R}$. Performing the partial Fourier transformation

$$\tilde{\mathcal{G}}(\mathbf{k}; \mathbf{R}) \equiv \int_V d^3x\, \mathcal{G}\left(\mathbf{R} + \frac{\mathbf{x}}{2}, \mathbf{R} - \frac{\mathbf{x}}{2}\right)e^{-i\mathbf{k}\cdot\mathbf{x}} \tag{26.15}$$

in Eqs. (26.14) it is convenient to define new variables $\mathbf{u}$ and $\mathbf{v}$ such that, for $\mathbf{R}$ fixed, $\mathbf{r} = \mathbf{s} - \mathbf{v}$ and $\mathbf{r}' = \mathbf{s} - \mathbf{u}$ so that the transformation $(\mathbf{x}, \mathbf{s}) \to (\mathbf{u}, \mathbf{v})$ is given by

$$\mathbf{s} = \mathbf{R} + \frac{\mathbf{u} + \mathbf{v}}{2} \ ; \quad \mathbf{x} = \mathbf{u} - \mathbf{v} \ ; \quad \left|\frac{\partial(\mathbf{x}, \mathbf{s})}{\partial(\mathbf{u}, \mathbf{v})}\right| = 1 \ . \tag{26.16}$$

Then introducing the inverse of the transformation (26.15) on the right-hand side of Eqs. (26.14), the Fourier transform of the latter become

$$\tilde{\mathcal{G}}_\uparrow(\mathbf{k}; \mathbf{R}) - \tilde{\mathcal{G}}_N(\mathbf{k}; \mathbf{R}) = -\frac{1}{V^2} \sum_{\mathbf{pp'}} \int_V d^3u \int_V d^3v e^{-i(\mathbf{p'}+\mathbf{k})\cdot\mathbf{u}} e^{-i(\mathbf{p}-\mathbf{k})\cdot\mathbf{v}}$$

$$\times \tilde{\mathcal{G}}_N\left(\mathbf{p}; \mathbf{R}+\frac{\mathbf{u}}{2}\right)\Delta\left(\mathbf{R}+\frac{\mathbf{u}+\mathbf{v}}{2}\right)$$

$$\times \tilde{\mathcal{F}}^*\left(\mathbf{p'}; \mathbf{R}+\frac{\mathbf{v}}{2}\right),$$

$$\tilde{\mathcal{F}}(\mathbf{k}; \mathbf{R}) = \frac{1}{V^2} \sum_{\mathbf{pp'}} \int_V d^3u \int_V d^3v e^{-i(\mathbf{p'}+\mathbf{k})\cdot\mathbf{u}} e^{-i(\mathbf{p}-\mathbf{k})\cdot\mathbf{v}}$$

$$\times \tilde{\mathcal{G}}_N\left(\mathbf{p}; \mathbf{R}+\frac{\mathbf{u}}{2}\right)\Delta\left(\mathbf{R}+\frac{\mathbf{u}+\mathbf{v}}{2}\right)$$

$$\times \tilde{\mathcal{G}}_\uparrow^*\left(\mathbf{p'}; \mathbf{R}+\frac{\mathbf{v}}{2}\right).$$

$$(26.17)$$

Formally, these equations may be simplified by first introducing the translation operators which generate the Taylor series

$$f(\mathbf{R}+\mathbf{a}) = \exp(\mathbf{a}\cdot\partial_\mathbf{R})f(\mathbf{R}) \qquad (26.18)$$

and then use a similar identity with $\partial_\mathbf{q} \equiv \partial/\partial\mathbf{q}$, namely

$$e^{i\mathbf{q}\cdot\mathbf{a}}e^{\mathbf{a}\cdot\mathbf{b}} = \exp(-i\mathbf{b}\cdot\partial_\mathbf{q})e^{i\mathbf{q}\cdot\mathbf{a}}, \qquad (26.19)$$

with $\mathbf{b} = \partial_\mathbf{R}$. After partial integration (summation) in $\mathbf{p}$ and $\mathbf{p'}$, Eqs. (26.17) then become[48] (Problem 4.11)

$$\tilde{\mathcal{G}}_\uparrow(\mathbf{k};\mathbf{R}) - \tilde{\mathcal{G}}_N(\mathbf{k};\mathbf{R}) = -\exp\left\{-\frac{i}{2}\partial_{\mathbf{p'}}\cdot(\partial_\mathbf{R}+\partial_{\mathbf{R''}}) - \frac{i}{2}\partial_\mathbf{p}\cdot(\partial_{\mathbf{R'}}+\partial_{\mathbf{R''}})\right\}$$

$$\times \tilde{\mathcal{G}}_N(\mathbf{p}; \mathbf{R})\Delta(\mathbf{R''})\tilde{\mathcal{F}}^*(\mathbf{p'}; \mathbf{R'})\Big|_{\mathbf{R}=\mathbf{R'}=\mathbf{R''}, \mathbf{p}=-\mathbf{p'}=\mathbf{k}},$$

$$(26.20)$$

$$\tilde{\mathcal{F}}(\mathbf{k}; \mathbf{R}) = \exp\left\{-\frac{i}{2}\partial_{\mathbf{p'}}\cdot(\partial_\mathbf{R}+\partial_{\mathbf{R''}}) - \frac{i}{2}\partial_\mathbf{p}\cdot(\partial_{\mathbf{R'}}+\partial_{\mathbf{R''}})\right\}$$

$$\times \tilde{\mathcal{G}}_N(\mathbf{p}; \mathbf{R})\Delta(\mathbf{R''})\tilde{\mathcal{G}}_\uparrow^*(\mathbf{p'}; \mathbf{R'})\Big|_{\mathbf{R}=\mathbf{R'}=\mathbf{R''}, \mathbf{p}=-\mathbf{p'}=\mathbf{k}}.$$

Applying the analogue of the identity (26.18) to functions of $\mathbf{p}$ and $\mathbf{p'}$, Eqs. (26.20) become

$$\tilde{\mathcal{G}}_\uparrow(\mathbf{k}; \mathbf{R}) - \tilde{\mathcal{G}}_N\left(\mathbf{k}; \mathbf{R}\right) = -\tilde{\mathcal{G}}_N(\mathbf{k} - \frac{i}{2}(\partial_{\mathbf{R}'} + \partial_{\mathbf{R}''}); \mathbf{R})$$

$$\times \tilde{\mathcal{F}}^*\left(-\mathbf{k} - \frac{i}{2}(\partial_{\mathbf{R}} + \partial_{\mathbf{R}''}); \mathbf{R}'\right)\Delta(\mathbf{R}'')\Big|_{\mathbf{R}=\mathbf{R}'=\mathbf{R}''},$$

$$\tilde{\mathcal{F}}(\mathbf{k}; \mathbf{R}) = \tilde{\mathcal{G}}_N\left(\mathbf{k} - \frac{i}{2}(\partial_{\mathbf{R}'} + \partial_{\mathbf{R}''}); \mathbf{R}\right)$$

$$\times \tilde{\mathcal{G}}_\uparrow^*\left(-\mathbf{k} - \frac{i}{2}(\partial_{\mathbf{R}} + \partial_{\mathbf{R}''}); \mathbf{R}'\right)\Delta(\mathbf{R}'')\Big|_{\mathbf{R}=\mathbf{R}'=\mathbf{R}''} \tag{26.21}$$

where $\partial_{\mathbf{R}}$ is understood to act backward. However, according to the second assumption we may drop both derivatives, $\partial_{\mathbf{R}}$ and $\partial_{\mathbf{R}'}$. Alternatively, according to the first assumption, Eqs. (26.21) may be solved iteratively thus generating a power series in $\Delta(\mathbf{R})$. Starting with $\tilde{\mathcal{G}}_\uparrow^{(0)} = \tilde{\mathcal{G}}_N$, it is easily seen that $\tilde{\mathcal{G}}_\uparrow$ and $\tilde{\mathcal{F}}$ are, respectively, even and odd functions of Δ,

$$\tilde{\mathcal{G}}_\uparrow = \sum_{n=0}^\infty \tilde{\mathcal{G}}_\uparrow^{(2n)} \; ; \quad \tilde{\mathcal{F}} = \sum_{n=0}^\infty \tilde{\mathcal{F}}^{(2n+1)} , \tag{26.22}$$

and Eqs. (26.21) imply the iteration

$$\tilde{\mathcal{G}}_\uparrow^{(2n+2)}(\mathbf{k}; \mathbf{R}) = -\tilde{\mathcal{G}}_N\left(\mathbf{k} - \frac{i}{2}\partial_{\mathbf{R}}; \mathbf{R}\right)\tilde{\mathcal{F}}^{(2n+1)*}\left(-\mathbf{k} - \frac{i}{2}\partial_{\mathbf{R}}; \mathbf{R}\right)\Delta(\mathbf{R}) , \tag{26.23}$$

$$\tilde{\mathcal{F}}^{(2n+1)}(\mathbf{k}; \mathbf{R}) = \tilde{\mathcal{G}}_N\left(\mathbf{k} - \frac{i}{2}\partial_{\mathbf{R}}; \mathbf{R}\right)\tilde{\mathcal{G}}_\uparrow^{(2n)*}\left(-\mathbf{k} - \frac{i}{2}\partial_{\mathbf{R}}; \mathbf{R}\right)\Delta(\mathbf{R})$$

where $\partial_{\mathbf{R}}$ acts only on $\Delta(\mathbf{R})$.

Everything is now determined by $\tilde{\mathcal{G}}_N$ which may be obtained from (26.11), (26.13) and (26.15). It is commendable although, because of the second assumption, not essential for this calculation to adopt the *London gauge* (see F. London , Ref. 41, Section 10)

$$\nabla \cdot \mathbf{A} = 0 , \tag{26.24}$$

in order to avoid introducing derivatives of $\mathbf{A}$. Making use of the second assumption the result then is (Problem 4.11)

$$\tilde{\mathcal{G}}_N(\mathbf{k}; \mathbf{R}) = \{i\nu - \varepsilon(\mathbf{k} + \frac{e}{\hbar c}\mathbf{A}(\mathbf{R}))\}^{-1} \tag{26.25}$$

where

$$\varepsilon(\mathbf{k}) \equiv \frac{\hbar^2 \mathbf{k}^2}{2m} - \mu \tag{26.26}$$

is the band energy $\varepsilon_{\mathbf{k}}$ but with an arbitrary value of the chemical potential μ which in *local equilibrium* also depends on $\mathbf{R}$, as does the temperature T. Introducing the gauge-covariant kinematic momentum operator for the *pair charge* $e^* = 2e$,

$$\vec{\Pi} \equiv -i\hbar\partial_{\mathbf{R}} + \frac{e^*}{c}\mathbf{A}(\mathbf{R}) \tag{26.27}$$

and applying again the analog of identity (26.18) to the $\mathbf{k}$-dependence, we may write

$$\tilde{\mathcal{G}}_N\left(\mathbf{k} - \frac{i}{2}\partial_{\mathbf{R}};\ \mathbf{R}\right) = \exp\left(\frac{1}{2\hbar}\vec{\Pi}\cdot\partial_{\mathbf{k}}\right)\tilde{\mathcal{G}}^o(\mathbf{k};i\nu)$$

$$= \tilde{\mathcal{G}}^o + \frac{1}{2\hbar}\tilde{\mathcal{G}}^o_\lambda\Pi_\lambda + \frac{1}{8\hbar^2}\tilde{\mathcal{G}}^o_{\lambda\mu}\Pi_\lambda\Pi_\mu + \cdots \tag{26.28}$$

where

$$\tilde{\mathcal{G}}^o(\mathbf{k};i\nu) = \{i\nu - \varepsilon(\mathbf{k})\}^{-1} \tag{26.29}$$

is the unperturbed Green's function (7.44) or (26.5) and, with (26.26),

$$\tilde{\mathcal{G}}^o_\lambda \equiv \frac{\partial}{\partial k_\lambda}\tilde{\mathcal{G}}^o = (\tilde{\mathcal{G}}^o)^2\frac{\hbar^2 k_\lambda}{m}\ . \tag{26.30}$$

The main content of the pehnomenological Ginzburg-Landau theory is a generalized gap equation which may be obtained from (26.6) using the model (21.14), (21.15). Replacing Δ_o by the local gap function $\Delta(\mathbf{R})$ one has, setting again $\mu = \varepsilon_F$, i.e. $\varepsilon(\mathbf{k}) = \varepsilon_{\mathbf{k}}$,

$$\Delta(\mathbf{R}) = \frac{\Lambda}{\beta V}\sum_\nu\sum_{\mathbf{k}}\tilde{\mathcal{F}}(\mathbf{k};\mathbf{R})\theta(\omega_o - |\varepsilon_{\mathbf{k}}|) \tag{26.31}$$

where it suffices to calculate $\tilde{\mathcal{F}}$ to third order in Δ and to second order in $\vec{\Pi}$. From (26.23), (26.28) one finds

$$\tilde{\mathcal{F}}^{(1)}(\mathbf{k};\mathbf{R}) = \left\{\tilde{\mathcal{G}}^o + \frac{1}{2\hbar}\tilde{\mathcal{G}}^o_\lambda\Pi_\lambda + \frac{1}{8\hbar^2}\tilde{\mathcal{G}}^o_{\lambda\mu}\Pi_\lambda\Pi_\mu + \cdots\right\}$$

$$\times\left\{\tilde{\mathcal{G}}^{o*} - \frac{1}{2\hbar}\tilde{\mathcal{G}}^{o*}_\lambda\Pi_\lambda + \frac{1}{8\hbar^2}\tilde{\mathcal{G}}^{o*}_{\lambda\mu}\Pi_\lambda\Pi_\mu + \cdots\right\}\Delta(\mathbf{R}) \tag{26.32}$$

and

$$\tilde{\mathcal{F}}^{(3)}(\mathbf{k};\mathbf{R}) = -\{|\tilde{\mathcal{G}}^o|^4 + \cdots\}|\Delta(\mathbf{R})|^2\Delta(\mathbf{R})\ . \tag{26.33}$$

Taking first the average over the direction $\hat{k} = \mathbf{k}/k$ we obtain, with (26.30),

$$\langle \tilde{\mathcal{F}}^{(1)} \rangle_{\hat{k}} \simeq |\tilde{\mathcal{G}}^o|^2 \left\{ 1 + \frac{1}{4m} \mathrm{Re} \left[\mathcal{G}^o + \frac{2}{3}(\varepsilon_{\mathbf{k}} + \varepsilon_F)(2(\tilde{\mathcal{G}}^o)^2 \right. \right.$$

$$\left. \left. - |\tilde{\mathcal{G}}^o|^2) \right] \vec{\Pi}^2 \right\} \Delta(\mathbf{R}) \ . \tag{26.34}$$

Introducing the density of states (10.4), Eq. (26.31) becomes

$$\Delta(\mathbf{R}) = \frac{\Lambda N(0)}{\beta} \int_{-\omega_o}^{\omega_o} d\varepsilon \sum_{\nu} \{ \langle \tilde{\mathcal{F}}^{(1)} \rangle_{\hat{k}} + \tilde{\mathcal{F}}^{(3)} \} \tag{26.35}$$

where (26.33), (26.34) have to be inserted. Making use of (23.21), (24.5), (24.8), (24.10), (24.11) and (26.29) one finally arrives, to lowest order in $t \equiv 1 - T/T_c$, at the result (Problem 4.12)[37,48]

$$\left\{ -at + \frac{a}{\Delta_1^2} |\Delta|^2 + \frac{1}{2m^*} \vec{\Pi}^2 \right\} \Delta = 0 \tag{26.36}$$

where $m^* = 2m$ is the *effective pair mass* and

$$a \equiv \frac{3\Delta_1^2}{4\varepsilon_F} = \frac{6\pi^2}{7\zeta(3)} \frac{(k_B T_c)^2}{\varepsilon_F} \ . \tag{26.37}$$

Gauge invariance guarantees that the relation $e^* = 2e$ is exact, i.e. that there are *no renormalization effects* (see, e.g. Ref. 49). More explicitly, this follows directly from the *Ward identity* (12.44) since the propagator $\tilde{G}$ *defines* the charge of the electron while the vertex Γ serves to *measure* this charge as electromagnetic response (see Figs. 3.6 and 5.11). Any interaction the electron may experience (e.g. by being confined to a metal) leaves intact the propagator line which runs through a self-energy diagram while extra lines occur in closed loops (see Figs. 2.9 and 2.10). Therefore, the propagator always carries exactly one unit of charge while any polarization cloud (the closed loops) has charge zero, irrespective of the residue $z_{\mathbf{k}}$, defined in Eq. (12.45), which the pole in $G(k)$ may acquire through interactions. The Ward identity then says that the charge measured by a susceptibility described by the vertex Γ also is exactly one unit. The same rigour does not apply to $m^* = 2m$ (e and m are the free-electron values). Since m^* is defined by Eqs. (26.27) and (26.36), deviations from the last relation may either be of kinematic origin or result from the definition of $\mathbf{A}$. A recent determination of m^* for niobium indicates that there are indeed small deviations[50].

For constant Δ and $\mathbf{A} = 0$, Eq. (26.36) is identical with (24.4) and has the solution (24.12). This result could have been obtained in a much simpler way by iteration of the equation $\tilde{\mathcal{G}}^{o-1} \tilde{\mathcal{G}} = 1$ defined by Eqs. (26.4), (26.8) and (26.9), writing the solution in the form (26.22). This is actually the equation originally used by Gor'kov in his Green's function derivation of the BCS theory[47]. In the case considered here, the general solution of (26.36) has the typical form of an *order parameter* describing the neighborhood of a *second-order phase transition*[51],

$$|\Delta| = \begin{cases} 0; & T > T_c \\ \Delta_1 t^{1/2}; & T < T_c \, , \end{cases} \tag{26.38}$$

and according to (22.23), (22.24) this is a *mean-field approximation*.

In fact, it is this aspect that led Ginzburg and Landau to their phenomenological theory[46]. Defining, for $\Delta = $ const and $\mathbf{A} = 0$, a free energy by

$$F_S = \nu V \left\{ -at|\Delta|^2 + \frac{a}{2\Delta_1^2}|\Delta|^4 \right\} + F_N \tag{26.39}$$

where ν is a normalization constant and F_N is independent of Δ, Eq. (26.36) then simply is the equilibrium condition $\partial F_S/\partial\Delta^* = 0$. The normalization constant ν is determined by identifying $F_S - F_N$ with the magnetic energy (25.37) through (23.28). Making use of (25.38), (25.39) and of (26.38), (26.39) one finds for $t \ll 1$

$$\nu = \frac{N(0)}{a} = \frac{n}{\Delta_1^2} \tag{26.40}$$

where in the last step Eq. (14.14) with $D = 3$ was used. Note that Eqs. (26.38) and (26.39) are analogous, respectively, to the ferromagnetic relations (29.25) and (29.27) below which shows that, historically, the mean-field idea is a generalization of Weiss' molecular field.

This identification of the gap function as order parameter is also possible in the general case $\Delta(\mathbf{R}, T)$ of Eq. (26.36). Including also the magnetic energy, Eq. (26.39) is then generalized to a *free energy functional*

$$F_S\Big[\Delta(\mathbf{R})\Big] = \int_V d^3R \left\{ -at|\Psi|^2 + \frac{1}{2m^*}|\vec{\Pi}\Psi|^2 + \frac{b}{2}|\Psi|^4 + \frac{1}{8\pi}\mathbf{H}^2 \right\} + F_N \tag{26.41}$$

where we have defined a *renormalized order parameter* Ψ and a constant b by

$$\Psi \equiv \nu^{1/2}\Delta \; ; \quad b \equiv \frac{a}{\nu\Delta_1^2} = \frac{a}{n} \, , \tag{26.42}$$

in agreement with Eqs. (53.23)–(53.25) of Ref. 35. The *equilibrium condition*

$$\delta F_S/\delta\Psi^*(\mathbf{R}) = 0 \tag{26.43}$$

then gives the full equation (26.36) where the "kinetic energy" term is obtained by a partial integration using definition (26.27). But Eq. (26.41) contains still more information. Indeed, comparison with (18.2) shows that

$$\delta F_S/\delta \mathbf{A}(\mathbf{R}) = -\frac{1}{c}\mathbf{j}_S(\mathbf{R}, T) \tag{26.44}$$

which yields the *current density in the superconducting state S*,

$$\mathbf{j}_S = -\frac{e^*}{2m^*}\{\Psi^*\vec{\Pi}\Psi + \Psi(\vec{\Pi}\Psi)^*\}$$

$$= \frac{i\hbar e^*}{2m^*}\{\Psi^*\partial_{\mathbf{R}}\Psi - \Psi\partial_{\mathbf{R}}\Psi^*\} - \frac{e^{*2}}{m^*c}|\Psi|^2\mathbf{A} . \tag{26.45}$$

Note that this result could also have been obtained similarly to (26.36) by starting from a Green's function expression for $\mathbf{j}$ analogous to (26.31)[37]. Using expression (26.41) for different geometries and boundary conditions, various physical situations may be studied[45,52]. Equation (26.41) may also easily be generalized to anisotropic superconductors, in particular to the layer structures of the cuprates. To this end it suffices to introduce an *effective mass tensor* and write (26.41) in the principle-axes coordinates of this tensor which implies replacing m^* by m_l^*, $l = 1, 2, 3$[53].

27. Type-I and -II Superconductors. Flux Quantum, Critical Current

The simplest and most important application of the Ginzburg-Landau theory is to a superconductor with $\mathbf{A} = 0$ in the half-space $z > 0$ as in Eq. (25.22). At $z = \infty$ the gap function assumes the bulk value (26.38) or, with Eqs. (26.40), (26.42), (25.23),

$$\Psi_\infty^2 = nt = \frac{n_s}{2} . \tag{27.1}$$

Writing the solution of the equilibrium condition

$$\left\{-at + b|\Psi|^2 + \frac{1}{2m^*}\vec{\Pi}^2\right\}\Psi = 0 \tag{27.2}$$

as $\Psi_\infty f(z)$ and using (26.27), (26.42) the equation for $f(z)$ is found to be

$$\xi_{\mathrm{GL}}^2 f'' = f^3 - f . \tag{27.3}$$

Here

$$\xi_{\mathrm{GL}}(T) \equiv \frac{\hbar}{\sqrt{2m^*at}} \tag{27.4}$$

is the *Ginzburg-Landau coherence length* which may be compared with the BCS value (22.7). Making use of (26.37) and (24.11) one obtains

$$\frac{\xi_{\mathrm{GL}}(T)}{\xi_o} = \frac{\pi}{\sqrt{6t}}\frac{\Delta_o}{\Delta_1} = 0.738\, t^{-1/2} . \tag{27.5}$$

Equation (27.3) may be integrated in analogy to Newton's equation. Indeed, identifying z, f and ξ_{GL}^2 as "time", "position" and "mass", respectively, the right-hand side of (27.3) may be written as $-dV/df$ with the "potential" $V(f) = (2f^2 - f^4)/4$. Then the "energy" $E = \xi_{GL}^2 f'^2/2 + V(f)$ is a first integral whose value is $1/4$ since $f = 1$ at $z \to \infty$. Taking the square root with the appropriate sign then leads to the equation

$$f' = \frac{1 - f^2}{\sqrt{2}\xi_{GL}} \tag{27.6}$$

which has the solution (see Eq. (50.20) of Ref. 35)

$$\frac{\Psi(z)}{\Psi_\infty} \equiv f(z) = \tanh\frac{z}{\sqrt{2}\xi_{GL}} . \tag{27.7}$$

This result shows that ξ_{GL} measures the distance over which the order parameter falls off at the boundary of the superconductor. It therefore plays an analogous role as the London penetration depth λ_L for the magnetic field, see Eq. (25.22). It turns out that the *Ginzburg-Landau parameter*

$$\kappa \equiv \frac{\lambda_L}{\xi_{GL}} = \frac{m^* c}{e^* \hbar}\sqrt{\frac{b}{2\pi}} \tag{27.8}$$

is of crucial importance for the magnetic penetration properties of a superconductor (we used (25.21), (25.23), (26.42) and (27.4) in the second step).

It was essentially the problem of the surface energy σ_{ns} separating the normal and superconducting phases of a metal that motivated the work of Ginzburg and Landau[46]. They found that $\sigma_{ns} = 0$ for $\kappa = 1/\sqrt{2}$ and that penetration of the magnetic field deep into the superconducting state, $\kappa \gg 1$, leads to an unstable situation characterized by $\sigma_{ns} < 0$. The sign of σ_{ns} gives rise to a classification due to Abrikosov[54] of superconductors into *type-I* for which $\sigma_{ns} > 0$ and *type-II* characterized by $\sigma_{ns} < 0$:

$$\text{Type-I} : \kappa < \frac{1}{\sqrt{2}} ; \quad \sigma_{ns} > 0 ,$$
$$\text{Type-II} : \kappa > \frac{1}{\sqrt{2}} ; \quad \sigma_{ns} < 0 . \tag{27.9}$$

As expressed by Eq. (22.25), BCS theory is a *mean-field approximation* which should cease to be valid close to T_c. According to the Ginzburg criterion that the fluctuation of the order parameter Ψ becomes comparable to $|\Psi|^{55)}$, this temperature interval is limited by

$$|t| < t_G \equiv \frac{(k_B T_c)^2 b^2}{32\pi^2 a} \left(\frac{2m^*}{\hbar^2}\right)^3 = \frac{e^{*4}(k_B T_c)^2}{m^* c^4 \hbar^2 a} \kappa^4 . \qquad (27.10)$$

While *critical fluctuations* have never been seen in any superconductor, this formula shows that the systems of interest are strongly type-II. Now the new high-T_c superconductors satisfy this criterion since they have κ's of the order of $100^{56)}$ (more along the c-axis and less in the ab-plane, see Table 1 of Ref. 53). One finds $t_G \sim 10^{-4} - 10^{-3}$ which should be accessible to observation since $T_c \sim 100$ K$^{53,56)}$.

For magnetic fields in the neighborhood of H_c a superconductor of *type-I* develops a laminar *intermediate state* of alternating superconducting and normal regions parallel to the field with macroscopic (or mesoscopic) period (see Ref. 45, Chapter 3). On the other hand, Abrikosov showed[54] that for a *type-II superconductor* the negative surface energy gives rise to a microscopic distribution of normal regions such that in each one, $\int \mathbf{H} \cdot d\mathbf{s}$ is just one *flux quantum*

$$\phi_o = \frac{hc}{2e} \qquad (27.11)$$

(see Ref. 45, Chapter 5). This *mixed state* forms a regular array, the *Abrikosov lattice* for fields in a range $H_{c1} < H_c < H_{c2}$ (see Fig. 5.2 of Ref. 45).

That magnetic flux is quantized with the value (27.11) may be seen with the aid of the current density (26.45). Considering first a pure supercurrent $\mathbf{j}_S$ we may again put $\mathbf{A} = 0$. Writing the order parameter as

$$\Psi = |\Psi|e^{i\varphi} \qquad (27.12)$$

and taking for $|\Psi|$ the bulk value (27.1), insertion into (26.45) yields the two-fluid expression

$$\mathbf{j}_S = -\frac{\hbar e^*}{m^*}\Psi_\infty^2 \nabla\varphi = en_s \mathbf{v}_s \qquad (27.13)$$

where the *superfluid velocity* is defined by

$$\mathbf{v}_s \equiv -\frac{\hbar}{m^*}\nabla\varphi . \qquad (27.14)$$

Adding now the vector potential in (26.45) and integrating over the border ∂a of an area a within the superconductor we obtain

$$\phi \equiv \oint_{\partial a} \left(\mathbf{A} + \frac{2m^* c}{e^{*2} n_s}\mathbf{j}_S\right) \cdot d\mathbf{l} = -\frac{\hbar c}{e^*} \oint_{\partial a} \nabla\varphi \cdot d\mathbf{l} . \qquad (27.15)$$

Here the first term on the left may be transformed by Stokes' theorem into $\int_a \mathbf{H} \cdot d\mathbf{s}$ while the integral on the right is $\Delta\varphi = 2\pi N$, $|N|$ being an integer. Thus London's "*fluxoid*" $\phi^{41)}$ is quantized in units of ϕ_o, Eq. (27.11).

The last result shows that *the phase difference is a quantum-mechanical observable*. This fact is even more striking if we consider a *tunnel junction* which consists of two superconductors S_1 and S_2 between which a thin layer of a normal metal N is sandwiched. In this case the phase difference $\varphi = \varphi_2 - \varphi_1$ is an observable which was shown by Josephson[57] to give rise to two new laws, a current-phase relation

$$J_S = J_o \sin \varphi \qquad (27.16)$$

where J_S is the supercurrent across the junction, and a time-dependent voltage-phase relation

$$\dot\varphi = \frac{2e}{\hbar} V \qquad (27.17)$$

where V is the voltage across the junction (see Ref. 45, Chapter 6). Note that Eq. (27.17) is intrinsically quantum mechanical since the limit $\hbar \to 0$ does not exist.

Introducing in addition the capacity C of the junction and forcing the system by an external current J_{ex}, we have the situation of Fig. 4.7 described by $J_{\text{ex}} = J_S + CV$. But this is nothing else than the equation of a pendulum[58]

$$\ddot\varphi + \omega_o^2 \sin \varphi = \nu^2 \qquad (27.18)$$

where

$$\omega_o^2 \equiv \frac{2e}{\hbar C} J_o, \ \nu^2 = \frac{2e}{\hbar C} J_{\text{ex}} \ . \qquad (27.19)$$

This equation may be derived from the Hamiltonian

$$H(\pi, \varphi) = \frac{1}{2}\pi^2 + U(\varphi) \qquad (27.20)$$

where $\pi = \dot\varphi = 2eQ/\hbar C$ is the *canonical momentum*, $Q = CV$ being the charge on the capacitor, and

$$U(\varphi) = -\omega_o^2 \cos \varphi - \nu^2 \varphi \qquad (27.21)$$

the potential. The truly remarkable fact is that, not only are π and φ observables but they may be quantized by imposing the commutation relation $[\pi, \varphi] = -i$. In fact, quantum tunneling out of the local minima of $U(\varphi)$ has been observed (see Ref. 58). Another situation of interest is obtained with the equilibrium condition $J_{\text{ex}} = 0$. In this case $U(\varphi)$ is periodic modulo

2π so that the energy spectrum separates into bands and the wavefunction $\psi(\varphi)$ has the Bloch form (1.14)[58].

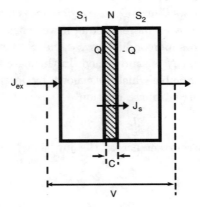

Fig. 4.7: Tunnel junction consisting of the superconductors S_1, S_2 and the thin layer of normal metal N having voltage V, external current J_{ex}, supercurrent J_S and charge $Q = CV$.

Equations (27.12) and (27.14) may also be used to obtain an estimate of the critical current. Indeed, inserted into (26.41) gives, with $\mathbf{A} = 0$ in (26.27),

$$F_S = V\left\{ \left(-at + \frac{m^*}{2}\mathbf{v}_s^2 \right)|\Psi|^2 + \frac{b}{2}|\Psi|^4 + \frac{1}{8\pi}\mathbf{H}^2 \right\} . \tag{27.22}$$

Using now as equilibrium condition $\partial F_S/\partial|\Psi| = 0$ instead of (26.43) we obtain with (26.42) and (27.1)

$$|\Psi|^2 = \frac{n_s}{2}\left(1 - \frac{m^*}{2}\frac{\mathbf{v}_s^2}{at} \right) . \tag{27.23}$$

Using this value in (26.45) one finds the expression

$$|\mathbf{j}_s| = en_s\left(1 - \frac{m^*}{2}\frac{\mathbf{v}_s^2}{at} \right)|\mathbf{v}_s| , \tag{27.24}$$

the maximum of which is the critical current density j_c. Thus, j_c is obtained from the condition $\partial j_s/\partial v_s = 0$ which, in turn, determines the *critical velocity* v_c as maximum of v_s. We find, using (26.37) and (22.8),

$$\frac{m^*}{2}v_c^2 = \frac{1}{3}at = \frac{\Delta_1^2 t}{4\varepsilon_F} = \frac{\varepsilon_B}{12}\frac{\Delta_1^2}{\Delta_o^2}t . \tag{27.25}$$

In view of (24.11) this result shows that the critical velocity is attained when the kinetic energy of the Cooper pairs becomes comparable with their binding energy ε_B. This is analogous to the Landau criterion for the critical velocity in superfluid ^{4}He if ε_B is interpreted as the roton gap energy and m^* as the effective mass of the roton[59].

Inserting (27.25) back into (27.24) the *critical current density* is found to be (see Eq. (4.36) of Ref. 45)

$$j_c = \frac{2}{3} e n_s v_c \ . \tag{27.26}$$

Using this value in (25.18), (25.22) we may determine a critical field H_c^* from the breakdown of the Meissner effect at the critical current j_c. With the aid of (25.21), (25.23), (25.38), (25.39), (27.25) and (14.14) one finds

$$H_c^* = \frac{4\pi}{c} \lambda_L j_c = \left(\frac{2}{3} \right)^{3/2} H_c \tag{27.27}$$

which shows the consistency of the definitions used.

28. Strong-coupling Theory and Coulomb Repulsion

Quite apart from being a mean-field approximation, BCS theory gives an oversimplified description of the electrons, phonons and their interactions. This can be seen, e.g., from realistic electron and phonon densities of state (see Ref. 60 for the case of the transition metals). In particular, the Coulomb repulsion between the electrons is completely left out. Therefore, a many-body theoretic formulation in terms of Gor'kov's Green's functions[37,47] must be developed, introducing dressed electron and phonon lines and vertices .

Now "*Migdal's theorem*" [61] (see Section 11) states that the corrections to the electron-phonon vertex are of order $\omega_D / \varepsilon_F \sim \sqrt{m/M}$ (ω_D is the Debye frequency and M the ionic mass) and hence may be neglected. Thus Migdal's theorem is recognized as the many-body theoretic form of the Born-Oppenheimer approximation (see Section 2 and Ref. 62). As shown by Eliashberg[63], Migdal's theorem is still true in the superconducting state . In calculating the electron self-energy due to the electron-phonon interaction, first for $T = 0$ then also at $T > 0$, Eliashberg derived a set of nonlinear equations for the frequency-dependent gap function $C(\omega)$ and frequency renormalization factor $f_o(\omega)/\omega$[63]. In modern treatments the self-energy due to the Coulomb interaction is also included and the notation is $C(\omega) \sim \Delta(\omega)$ and $f_o(\omega)/\omega \sim 1 - Z(\omega)$. In this form the set of equations is called the *Eliashberg equations* (see Refs. 24, 60, 62, 64). We shall adopt this generalized point of view and notation.

Writing the Hamiltonian

$$\mathcal{H} = \mathcal{H}_o + \mathcal{H}_{ph} + \mathcal{H}_C \qquad (28.1)$$

in the Nambu representation (21.22), (21.24) we have from (21.26), (1.29) and (10.6), (3.38), respectively

$$\mathcal{H}_o = \sum_{\mathbf{k}} \varepsilon_{\mathbf{k}}\{\Psi_{\mathbf{k}}^+ \sigma^z \Psi_{\mathbf{k}} + 1\} + \frac{1}{2}\sum_q \omega_q(P_q^+ P_q + Q_q^+ Q_q) \qquad (28.2)$$

where the constant $\sum \varepsilon_{\mathbf{k}}$ may be neglected and, with $\mathbf{q} \neq 0$, the *phonon-and Coulomb parts of the interaction* are

$$\mathcal{H}_{ph} = \sum_{q\mathbf{k}} \gamma_q(\Psi_{\mathbf{k}}^+ \sigma^z \Psi_{\mathbf{k}-q})Q_q \,,$$

$$\mathcal{H}_C = \frac{1}{2V}\sum_{q\mathbf{k}\mathbf{k}'} V_C(\mathbf{q})(\Psi_{\mathbf{k}}^+ \sigma^z \Psi_{\mathbf{k}-q})(\Psi_{\mathbf{k}'}^+ \sigma^z \Psi_{\mathbf{k}'+q}) \,. \qquad (28.3)$$

The full propagator matrix to be calculated is that of Eq. (26.8),

$$\mathcal{G}(\mathbf{k}; \tau) \equiv -\langle T\{\Psi_{\mathbf{k}}(-i\tau) \otimes \Psi_{\mathbf{k}}^+(0)\}\rangle = \begin{pmatrix} \mathcal{G}_\uparrow & \mathcal{F} \\ \mathcal{F}^* & \mathcal{G}_\downarrow \end{pmatrix} \,. \qquad (28.4)$$

But in distinction to the Bogoljubov groundstate used in Eq. (26.1), perturbation theory now starts from the Fermi level so that the inverse of the free propagator matrix is

$$(\tilde{\mathcal{G}}^o)^{-1}(\mathbf{k}; i\nu) = \begin{pmatrix} i\nu - \varepsilon_{\mathbf{k}}, & 0 \\ 0, & i\nu + \varepsilon_{\mathbf{k}} \end{pmatrix} = i\nu\mathbf{1} - \varepsilon_{\mathbf{k}}\sigma^z \qquad (28.5)$$

instead of (26.4). $\tilde{\mathcal{G}}(\mathbf{k}; i\nu)$ is obtained by calculating the self-energy defined by Eq. (11.3) in the form (11.17), that is

$$(\tilde{\mathcal{G}}^o \Sigma \tilde{\mathcal{G}}^o)(\mathbf{k}; i\nu) = -\sum_{N=1}^{\infty} \frac{(-1)^N}{N!} \int_0^\beta d\tau \int_0^\beta d\tau_1 \cdots \int_0^\beta d\tau_N e^{i\nu\tau}$$
$$\times \langle T\{\mathcal{H}_{int}(-i\tau_1)\cdots\mathcal{H}_{int}(-i\tau_N)\Psi_{\mathbf{k}}^o(-i\tau) \otimes \Psi_{\mathbf{k}}^+(0)\}\rangle_{o,1-i} \qquad (28.6)$$

where $\mathcal{H}_{int} = \mathcal{H}_{ph} + \mathcal{H}_C$ and the index $1 - i$ means one-particle-irreducible parts.

According to Migdal's theorem it suffices to retain diagrams with bare electron-phonon vertex but keeping the lines dressed. Doing the same for the Coulomb vertex, the only terms in (28.6) are $N = 2$ for $\mathcal{H}_{ph}$ and $N = 1$

for $\mathcal{H}_C$. Introducing the phonon propagator through Eqs. (11.2) and (7.25), the self-energy generated by $\mathcal{H}_{\mathrm{ph}}$ is found from (28.3), (28.6) to be given by

$$(\tilde{\mathcal{G}}^o \Sigma_{\mathrm{ph}} \tilde{\mathcal{G}}^o)(\mathbf{k}; i\nu) = \frac{1}{2} \int_0^\beta d\tau \int_0^\beta d\tau_1 \int_0^\beta d\tau_2 e^{i\nu\tau} \sum_{\mathbf{q}\mathbf{k}'\mathbf{k}''} |\gamma_q|^2$$
$$\times \beta^{-1} \sum_{\nu'} \tilde{D}(q; i\nu') e^{-i\nu'(\tau_1 - \tau_2)} M(\tau_1 \tau_2) \quad (28.7)$$

where

$$M(\tau_1 \tau_2) \equiv \langle T\{(\Psi_{\mathbf{k}'}^+ \sigma^z \Psi_{\mathbf{k}'-\mathbf{q}})^o(-i\tau_1)(\Psi_{\mathbf{k}''}^+ \sigma^z \Psi_{\mathbf{k}''+\mathbf{q}})^o(-i\tau_2)$$
$$\times \Psi_{\mathbf{k}}^o(-i\tau) \otimes \Psi_{\mathbf{k}}^+(0)\}\rangle_o . \quad (28.8)$$

Similarly one finds with $\mathcal{H}_C$

$$(\tilde{\mathcal{G}}^o \Sigma_C \tilde{\mathcal{G}}^o)(\mathbf{k}; i\nu) = \frac{1}{2} \int_0^\beta d\tau \int_0^\beta d\tau_1 e^{i\nu\tau} \frac{1}{V} \sum_{\mathbf{q}\mathbf{k}'\mathbf{k}''} V_C(\mathbf{q}) M(\tau_1 \tau_1) . \quad (28.9)$$

The diagrams corresponding to (28.7) and (28.9) are given, respectively, by Figs. 2.8 and 1.3.

Considering first the Hartree contributions of Figs. 2.8 and 1.3, they are characterized by a Wick's-theorem factorization of (28.8) containing a simultaneous T-product $C_{\mathbf{k}''}$,

$$M^H(\tau_1) = 2\mathcal{G}^o(\mathbf{k}; \tau - \tau_1)\sigma^z \mathcal{G}^o(\mathbf{k}; \tau_1) C_{\mathbf{k}''}\delta_{\mathbf{q}0}\delta_{\mathbf{k}\mathbf{k}'} . \quad (28.10)$$

According to rule (9.10) the simultaneous T-product is Wick-ordered so that one deduces from Eqs. (21.22), (8.4) and (7.40)

$$C_{\mathbf{k}''} = \langle W(\Psi_{\mathbf{k}''}^+ \sigma^z \Psi_{\mathbf{k}''})\rangle_o = 2f_o(\varepsilon_{\mathbf{k}''}) . \quad (28.11)$$

Inserting this with (28.10) into (28.7) and (28.9) one finds

$$\Sigma_{\mathrm{ph},C}^H = \mu_{\mathrm{ph},C}\sigma^z \quad (28.12)$$

where μ_{ph} and μ_C are constants which according to (28.5) and (11.3) may simply be absorbed into the chemical potential (as before we put the Fermi energy at $\varepsilon_{\mathbf{k}} = 0$).

Hence only the exchange parts of Figs. 2.8 and 1.3 are relevant. In this case Wick's theorem applied to (28.8) yields

$$M^{\mathrm{ex}}(\tau_1 \tau_2) = -2\mathcal{G}^o(\mathbf{k}; \tau - \tau_1)\sigma^z \mathcal{G}^o(\mathbf{k} - \mathbf{q}; \tau_1 - \tau_2)\sigma^z$$
$$\times \mathcal{G}^o(\mathbf{k}; \tau_2)\delta_{\mathbf{k}\mathbf{k}'}\delta_{\mathbf{k}''+\mathbf{q},\mathbf{k}} . \quad (28.13)$$

Inserting this expression into (28.7) and making use of (7.25) and (7.28) one easily finds (compare Eq. (53) of Ref. 62)

$$\Sigma_{\text{ph}}^{\text{ex}}(\mathbf{k}; i\nu) = -\sum_{q} |\gamma_q|^2 \beta^{-1} \sum_{\nu'} \tilde{D}(q; i\nu - i\nu')\sigma^z \tilde{\mathcal{G}}(\mathbf{k} - \mathbf{q}; i\nu')\sigma^z \quad (28.14)$$

where we have replaced the bare propagator $\tilde{\mathcal{G}}^o$ by a dressed one. In exactly the same way one deduces (compare Eq. (52) of Ref. 62)

$$\Sigma_C^{\text{ex}}(\mathbf{k}; i\nu) = -\frac{1}{V} \sum_{\mathbf{q}} V_C(\mathbf{q})\beta^{-1} \sum_{\nu'} \sigma^z \tilde{\mathcal{G}}(\mathbf{k} - \mathbf{q}; i\nu')\sigma^z . \quad (28.15)$$

With the Fourier transform of (28.4) we have

$$\sigma^z \tilde{\mathcal{G}} \sigma^z = \begin{pmatrix} \tilde{\mathcal{G}}_\uparrow, & -\tilde{\mathcal{F}} \\ -\tilde{\mathcal{F}}^*, & \tilde{\mathcal{G}}_\downarrow \end{pmatrix} . \quad (28.16)$$

$\tilde{\mathcal{G}}^{-1}$ then is determined by (11.3) with Eqs. (28.5) and (28.14)–(28.16),

$$\tilde{\mathcal{G}}^{-1} = (\tilde{\mathcal{G}}^o)^{-1} - \Sigma_{\text{ph}} - \Sigma_C = \begin{pmatrix} \Omega_+, & -\Phi_+ \\ -\Phi_-, & \Omega_- \end{pmatrix} , \quad (28.17)$$

and the connection of the last expression with the conventional notation[60,62,64] is (see Eqs. (50) and (51) of Ref. 62)

$$\Omega_\pm(\mathbf{k}; i\nu) = i\nu Z \mp (\varepsilon_{\mathbf{k}} + \chi) ; \quad \Phi_\pm(\mathbf{k}; i\nu) = \phi \mp i\phi' . \quad (28.18)$$

$\Omega_\pm$ and $\Phi_\pm$ are renormalized energies and gap functions , respectively. Now the inverse of a 2×2 matrix is easily found, indeed,

$$\left.\begin{matrix} \tilde{\mathcal{G}}_\uparrow \\ \tilde{\mathcal{G}}_\downarrow \end{matrix}\right\} = \frac{\Omega_\mp}{D} ; \quad \left.\begin{matrix} \tilde{\mathcal{F}} \\ \tilde{\mathcal{F}}^* \end{matrix}\right\} = \frac{\Phi_\pm}{D} \quad (28.19)$$

where

$$D \equiv \Omega_+ \Omega_- - \Phi_+ \Phi_- \quad (28.20)$$

is the determinant of $\tilde{\mathcal{G}}^{-1}$. This leads to the 4 Eliashberg equations

$$\Omega_\pm(\mathbf{k}; i\nu) = i\nu \mp \varepsilon_{\mathbf{k}} + \beta^{-1} \sum_{\nu'} \frac{1}{V} \sum_{\mathbf{k}'} K(\mathbf{k} - \mathbf{k}'; i\nu - i\nu') \frac{\Omega_\mp(\mathbf{k}'; i\nu')}{D(\mathbf{k}'; i\nu')} \quad (28.21)$$

$$\Phi_\pm(\mathbf{k}; i\nu) = \beta^{-1} \sum_{\nu'} \frac{1}{V} \sum_{\mathbf{k}'} K(\mathbf{k} - \mathbf{k}'; i\nu - i\nu') \frac{\Phi_\pm(\mathbf{k}'; i\nu')}{D(\mathbf{k}'; i\nu')} \quad (28.22)$$

with the kernel $K = K_{\text{ph}} + V_C$ where

$$K_{\text{ph}}(\mathbf{q}; z) \equiv V \sum_{q'} |\gamma_{q'}|^2 \tilde{D}(q'; z)\delta_{\mathbf{q}'\mathbf{q}} \quad (28.23)$$

and the sum over q' in fact runs only over the phonon branch and polarization indices as defined in Section 1.

The above derivation is formally quite simple. However, Eqs. (28.19)–(28.23) are extremely complicated coupled nonlinear integral equations which call for a number of simplifications in order to arrive at the standard form of the Eliashberg equations[60,62,64]. First, the structure of (28.17) suggests that the band renormalization χ in (28.18) is not much affected by the superconducting state[62]. But considering it as a normal-state effect we may absorb it into $\varepsilon_{\mathbf{k}}$ and hence put $\chi = 0$. Second, the reality of the BCS gap function in the absence of external fields suggests that we may put $\phi' = 0$ and choose Z and ϕ to be real even analytic functions of frequency, i.e.,

$$Z^*(z^*) = Z(z) = Z(-z) ; \quad \phi^*(z^*) = \phi(z) = \phi(-z) . \tag{28.24}$$

Then Eqs. (28.18) and (28.20) become

$$\begin{aligned} \Omega_\pm &= i\nu Z(\mathbf{k}; i\nu) \mp \varepsilon_{\mathbf{k}} ; \quad \Phi_\pm = \phi(\mathbf{k}; i\nu) \\ D &= -\varepsilon_{\mathbf{k}}^2 - \phi^2(\mathbf{k}; i\nu) - \nu^2 Z^2(\mathbf{k}; i\nu) . \end{aligned} \tag{28.25}$$

Neglecting $\varepsilon_{\mathbf{k}'}$ in $\Omega_{\mp}(\mathbf{k}; i\nu')$ and using the symmetry (28.24), the Coulomb part of the kernel drops out of Eq. (28.21) which becomes

$$\begin{aligned} i\nu[Z(\mathbf{k}; i\nu) - 1] &= -\beta^{-1} \sum_{\nu'} \frac{1}{V} \sum_{\mathbf{k}'} K_{\mathrm{ph}}(\mathbf{k} - \mathbf{k}'; i\nu - i\nu') \\ &\times \frac{i\nu' Z(\mathbf{k}'; i\nu')}{\varepsilon_{\mathbf{k}'}^2 + \phi^2(\mathbf{k}'; i\nu') + \nu'^2 Z^2(\mathbf{k}'; i\nu')} \end{aligned} \tag{28.26}$$

while Eq. (28.22) may be separated into phonon and Coulomb parts , $\phi = \phi_{\mathrm{ph}} + \phi_C$ with

$$\begin{aligned} \phi_{\mathrm{ph}}(\mathbf{k}; i\nu) &= -\beta^{-1} \sum_{\nu'} \frac{1}{V} \sum_{\mathbf{k}'} K_{\mathrm{ph}}(\mathbf{k} - \mathbf{k}'; i\nu - i\nu') \\ &\times \frac{\phi(\mathbf{k}'; i\nu')}{\varepsilon_{\mathbf{k}'}^2 + \phi^2(\mathbf{k}'; i\nu') + \nu'^2 Z^2(\mathbf{k}'; i\nu')} \end{aligned} \tag{28.27}$$

and

$$\begin{aligned} \phi_C(\mathbf{k}) &= -\beta^{-1} \sum_{\nu'} \frac{1}{V} \sum_{\mathbf{k}'} V_C(\mathbf{k} - \mathbf{k}') \\ &\times \frac{\phi(\mathbf{k}'; i\nu')}{\varepsilon_{\mathbf{k}'}^2 + \phi^2(\mathbf{k}'; i\nu') + \nu'^2 Z^2(\mathbf{k}'; i\nu')} . \end{aligned} \tag{28.28}$$

Since according to (20.10) the effective interaction (20.7) resulting from the electron-phonon coupling is restricted to a narrow interval $|\varepsilon_{\mathbf{k}}| < \omega_o \ll \varepsilon_F$ around the Fermi level, we may limit the wavevectors in (28.26) and (28.27) by $|\mathbf{k}| \sim |\mathbf{k}'| \sim k_F$ and average over the angles of $\mathbf{k}$ and $\mathbf{k}'$[62]. This amounts to neglecting the wavevector dependence in these equations. Introducing the electron density of states (10.4) this leads to

$$i\nu[Z(i\nu) - 1] = -\beta^{-1} \sum_{\nu'} N(0)\langle K_{\mathrm{ph}}(i\nu - i\nu')\rangle$$

$$\times \int_{-\omega_o}^{+\omega_o} \frac{i\nu' Z(i\nu')d\varepsilon}{\varepsilon^2 + \phi^2(i\nu') + \nu'^2 Z^2(i\nu')} \quad (28.29)$$

and

$$\phi_{\mathrm{ph}}(i\nu) = -\beta^{-1} \sum_{\nu'} N(0)\langle K_{\mathrm{ph}}(i\nu - i\nu')\rangle$$

$$\times \int_{-\omega_o}^{+\omega_o} \frac{\phi(i\nu')d\varepsilon}{\varepsilon^2 + \phi^2(i\nu') + \nu'^2 Z^2(i\nu')} \quad (28.30)$$

where

$$\langle K_{\mathrm{ph}}(z)\rangle \equiv \frac{1}{4\pi} \oint d\Omega_{\mathbf{k}'} K_{\mathrm{ph}}(\mathbf{k} - \mathbf{k}'; z)\Big|_{|\mathbf{k}|=|\mathbf{k}'|=k_F} \quad (28.31)$$

and isotropy in $\mathbf{k}$ is assumed.

Unfortunately, Eq. (28.28) cannot be treated in the same way because the slow energy dependence of the Coulomb potential, $V_C(\mathbf{k}) \sim \mathrm{const}/(\varepsilon_F + \varepsilon_{\mathbf{k}})$. The idea, originally due to Bogoljubov *et al.*[65], is to replace V_C by a *pseudo-potential* U which is self-consistently determined such that the $\mathbf{k}'$-sum in (28.28) is also restricted to an interval $|\varepsilon_{\mathbf{k}'}| < \omega_C \ll \varepsilon_F$ with an appropriately chosen cutoff ω_C[62]. Thus, dividing the $\mathbf{k}'$-sum into two terms with $|\varepsilon_{\mathbf{k}'}| < \omega_C$ and $|\varepsilon_{\mathbf{k}'}| > \omega_C$, respectively, and assuming $\omega_C > \omega_0$, the $\mathbf{k}'$-states of the second term lie outside of the domain of importance for superconductivity. But since V_C assumes its dominant values for $\mathbf{k} - \mathbf{k}' \sim 0$, we may put $\phi(\mathbf{k}'; i\nu') \sim \phi_C(\mathbf{k}')$ and $Z(\mathbf{k}', i\nu') \sim Z_C = \mathrm{const}$ in the second term and make use of (24.1) for the ν'-sum. With these manipulations Eq. (28.28) becomes

$$\phi_C(\mathbf{k}) = -\frac{1}{V} \sum_{\mathbf{k}'} V_C(\mathbf{k} - \mathbf{k}') \left\{ \theta(\omega_C - |\varepsilon_{\mathbf{k}'}|) \frac{\phi(\mathbf{k}'; i\nu')}{\varepsilon_{\mathbf{k}'}^2 + \phi^2(\mathbf{k}'; i\nu') + \nu'^2 Z^2(\mathbf{k}'; i\nu')} \right.$$

$$\left. + \theta(|\varepsilon_{\mathbf{k}'}| - \omega_C) \frac{\phi_C(\mathbf{k}')}{2|\varepsilon_{\mathbf{k}'}| Z_C} \tanh \frac{\beta |\varepsilon_{\mathbf{k}'}|}{2 Z_C} \right\} \quad (28.32)$$

where we may approximate $\tanh \sim 1$.

Defining now

$$P(\mathbf{k}') \equiv -\theta(\omega_C - |\varepsilon_{\mathbf{k}'}|)\beta^{-1} \sum_{\nu'} \frac{\phi(\mathbf{k}'; i\nu')}{\varepsilon_{\mathbf{k}'}^2 + \phi^2(\mathbf{k}'; i\nu') + \nu'^2 Z^2(\mathbf{k}'; i\nu')}$$

$$\Omega(\mathbf{k}, \mathbf{k}') \equiv \frac{\theta(|\varepsilon_{\mathbf{k}'}| - \omega_C)}{2|\varepsilon_{\mathbf{k}'}|Z_C} V_C(\mathbf{k} - \mathbf{k}') , \qquad (28.33)$$

Eq. (28.32) may be written formally as[62)]

$$\phi_C = (1 + \Omega)^{-1} V_C P \equiv UP \qquad (28.34)$$

and the pseudo-potential U is the solution of the integral equation $(1 + \Omega)U = V_C$. Since the original equation (28.28) is the limit $\omega_C \to \infty$, $\Omega \to 0$ of (28.34) and may be written as $\phi_C = V_C P_\infty$, comparison with (28.34) shows that in replacing V_C by U we may limit the $\mathbf{k}'$-sum in (28.28) to $|\varepsilon_{\mathbf{k}'}| < \omega_C$ around the Fermi level. Hence, we may now also approximate the wavevectors by $|\mathbf{k}| \sim |\mathbf{k}'| \sim k_F$ and average over the angles of $\mathbf{k}$ and $\mathbf{k}'$ so that Eq. (28.28) takes a form analogous to (28.30),

$$\phi_C = -\beta^{-1} \sum_{\nu'} \mu^* \int_{-\omega_C}^{+\omega_C} \frac{\phi(i\nu')d\varepsilon}{\varepsilon^2 + \phi^2(i\nu') + \nu'^2 Z^2(i\nu')} \qquad (28.35)$$

where

$$N(0)\langle U \rangle \equiv \mu^* . \qquad (28.36)$$

Turning now to the phonon kernel (28.23) we express the phonon propagator by the dispersion relation (11.14) in terms of the spectral function $\text{Im}\,\tilde{D}$ and approximate the latter by Eq. (11.36). Thus

$$N(0)\langle K_{\text{ph}}(z) \rangle = -\int_0^\infty \frac{2\Omega d\Omega}{\Omega^2 - z^2} \alpha^2(\Omega)F(\Omega) \equiv -\lambda(z) \qquad (28.37)$$

where

$$\alpha^2(\Omega)F(\Omega) \equiv N(0)V \langle \sum_q |\gamma_q|^2 \delta(\Omega - \omega_q)\delta_{\mathbf{q}, \mathbf{k}-\mathbf{k}'} \rangle \qquad (28.38)$$

is the product of an *averaged coupling function* $\alpha^2(\Omega)$ and an *averaged phonon density of states* $F(\Omega)$ (compare Eq. (76) of Ref. 62).

There are two ways of proceeding from Eqs. (28.29), (28.30) and (28.35), one being a formulation in terms of Matsubara frequencies which will be described later. The other way which is followed in the standard literature[60,62,64)], is to use the dispersion relation (11.10) for the electron propagator (28.4) and to make at the end of the evaluation an analytic continuation $i\nu \to \omega$ to real frequencies. In the notation (28.19), (28.25) this dispersion relation reads

$$\frac{\psi(i\nu')}{\varepsilon^2 + \phi^2(i\nu') + \nu'^2 Z^2(i\nu')} = \frac{1}{\pi} \int_{-\infty}^{+\infty} \frac{d\omega'}{\omega' - i\nu'} \text{Im} \frac{\psi(\omega')}{\varepsilon^2 - R^2(\omega')} \quad (28.39)$$

with $\psi(\omega') = \omega' Z(\omega')$, $\phi(\omega')$ and

$$R^2(\omega') \equiv \omega'^2 Z^2(\omega') - \phi^2(\omega') . \quad (28.40)$$

Insertion into (28.29), (28.30) and (28.35) using (28.37) yields

$$i\nu[Z(i\nu) - 1] = \beta^{-1} \sum_{\nu'} \lambda(i\nu - i\nu') \frac{2}{\pi} \int_0^{\omega_o} d\varepsilon \int_{-\infty}^{+\infty} \frac{d\omega'}{\omega' - i\nu'} \text{Im} \frac{\omega' Z(\omega')}{\varepsilon^2 - R^2(\omega')} , \quad (28.41)$$

$$\phi_{\text{ph}}(i\nu) = \beta^{-1} \sum_{\nu'} \lambda(i\nu - i\nu') \frac{2}{\pi} \int_0^{\omega_o} d\varepsilon \int_{-\infty}^{+\infty} \frac{d\omega'}{\omega' - i\nu'} \text{Im} \frac{\phi(\omega')}{\varepsilon^2 - R^2(\omega')} \quad (28.42)$$

and

$$\phi_C = -\beta^{-1} \sum_{\nu'} \mu^* \frac{2}{\pi} \int_0^{\omega_C} d\varepsilon \int_{-\infty}^{+\infty} \frac{d\omega'}{\omega' - i\nu'} \text{Im} \frac{\phi(\omega')}{\varepsilon^2 - R^2(\omega')} . \quad (28.43)$$

Evaluating next the ε-integration, it is important to remember definition (11.11) of the "imaginary part" which actually is a discontinuity. Since according to (28.24) the numerators under the Im-sign are analytic on the real axis, they do not contribute to the discontinuity and may be taken outside. The remaining integral is elementary but must be done with caution, the result is (Problem 4.13)

$$2 \int_0^{\omega_o} d\varepsilon \, \text{Im} \frac{1}{\varepsilon^2 - R^2(\omega')} = \pi \theta(\tilde{\omega}_o - |\omega'|) \text{Re} \frac{1}{R(\omega')} \quad (28.44)$$

where we have put $\tilde{\omega}_o \equiv \omega_o / Z_C$. Note that for $\omega' > \omega_C$, $R^2(\omega') \simeq \omega'^2 Z_C^2$ so that the square root $R(\omega')$ is defined as odd analytic function of ω'. This is important because with an even $R(\omega')$ one would get $\phi_C = 0$. Introducing the gap function $\Delta(\omega)$ through

$$\phi_{\text{ph}} + \phi_C = \phi = Z\Delta , \quad (28.45)$$

insertion of (28.44) with (28.40) into (28.43) using Eq. (24.1) yields

$$\phi_C = -\mu^* \int_0^{\tilde{\omega}_C} d\omega' \text{Re} \frac{\Delta(\omega')}{\sqrt{\omega'^2 - \Delta^2(\omega')}} \tanh \frac{\beta\omega'}{2} \quad (28.46)$$

where $\tilde{\omega}_C = \omega_C / Z_C$.

As a last step we have to evaluate the ν'-sum in (28.41), (28.42). Inserting (28.37) and combining the frequency denominators, we calculate with the aid of Lindhard functions, Eqs. (10.17) and (10.19), the statistical factor

$$
S(i\nu, \Omega, \omega') \equiv \beta^{-1} \sum_{\nu'} \left[\frac{1}{\Omega - (i\nu - i\nu')} + \frac{1}{\Omega + (i\nu - i\nu')} \right] \frac{1}{\omega' - i\nu'}
$$

$$
= \frac{n_o(-\Omega) + f_o(\omega')}{i\nu - \Omega - \omega'} - \frac{n_o(\Omega) + f_o(\omega')}{i\nu + \Omega - \omega'} , \tag{28.47}
$$

where we used (7.26) to write $f_o(i\nu + \Omega) = -n_o(\Omega)$. In this expression we are now allowed to perform the analytic continuation $i\nu \to \omega$. Inserting (28.44)–(28.47) into Eqs. (28.41) and (28.42) and remembering that $R(\omega')$ defined by (28.40) is an odd function, we finally obtain, in agreement with Eqs. (74) and (75) of Ref. 62

$$
\omega[Z(\omega) - 1] = \int_0^\infty d\Omega \alpha^2(\Omega) F(\Omega) \int_0^{\tilde{\omega}_o} d\omega' \mathrm{Re}\, \frac{\omega'}{\sqrt{\omega'^2 - \Delta^2(\omega')}}
$$
$$
\times \{S(\omega, \Omega, \omega') + S(\omega, \Omega, -\omega')\} \tag{28.48}
$$

and

$$
Z(\omega)\Delta(\omega) = \int_0^\infty d\Omega \alpha^2(\Omega) F(\Omega) \int_0^{\tilde{\omega}_o} d\omega' \mathrm{Re}\, \frac{\Delta(\omega')}{\sqrt{\omega'^2 - \Delta^2(\omega')}}
$$
$$
\times \{S(\omega, \Omega, \omega') - S(\omega, \Omega, -\omega')\}
$$
$$
- \mu^* \int_0^{\tilde{\omega}_C} d\omega' \mathrm{Re}\, \frac{\Delta(\omega')}{\sqrt{\omega'^2 - \Delta^2(\omega')}} \tanh \frac{\beta\omega'}{2} . \tag{28.49}
$$

A more compact form of these equations is obtained following Ref. 60, defining a new phonon kernel

$$
\kappa(\omega, \omega') \equiv \int_0^\infty d\Omega \alpha^2(\Omega) F(\Omega) S(\omega, \Omega, \omega') \theta(\tilde{\omega}_o - |\omega'|)
$$
$$
= \left\{ (\omega - \omega')\gamma(\omega - \omega') + \frac{1}{2}\lambda(\omega - \omega') \tanh \frac{\beta\omega'}{2} \right\} \theta(\tilde{\omega}_o - |\omega'|) \tag{28.50}
$$

where in the second step use was made of Eqs. (7.42), (7.43) and (28.37) and we defined

$$
\gamma(z) \equiv \int_0^\infty \frac{d\Omega \alpha^2(\Omega) F(\Omega)}{\Omega^2 - z^2} \coth \frac{\beta\Omega}{2} . \tag{28.51}
$$

In this notation Eqs. (28.48), (28.49) read

$$\omega[Z(\omega) - 1] = \int_0^\infty d\omega' \{\kappa(\omega, \omega') + \kappa(\omega, -\omega')\} \text{Re} \frac{\omega'}{\sqrt{\omega'^2 - \Delta^2(\omega')}} \quad (28.52)$$

and

$$Z(\omega)\Delta(\omega) = \int_0^\infty d\omega' \left\{ \kappa(\omega, \omega') - \kappa(\omega, -\omega') \right.$$
$$\left. - \mu^* \theta(\tilde{\omega}_C - \omega') \tanh \frac{\beta\omega'}{2} \right\} \text{Re} \frac{\Delta(\omega')}{\sqrt{\omega'^2 - \Delta^2(\omega')}} . \quad (28.53)$$

This form is particularly well suited in the *weak-coupling limit* where $\mu^* \sim 0$ and $\alpha^2(\Omega)$ is small. Since for a simple Debye spectrum $\omega_q = c_\parallel |\mathbf{q}|$, $\alpha^2(\Omega) \propto \Omega$ according to (10.11) and $F(\Omega) \propto \Omega^2$, one finds that $\gamma(0) \propto T^2$. Therefore we may neglect γ in (28.50) and use the approximation[60]

$$\kappa(\omega, \omega') \simeq \frac{1}{2}\lambda \tanh\frac{\beta\omega'}{2}\theta(\tilde{\omega}_o - |\omega'|)\theta(\tilde{\omega}_o - |\omega|) \quad (28.54)$$

where according to (28.37)

$$\lambda \equiv \lambda(0) = 2 \int_0^\infty \frac{d\Omega}{\Omega} \alpha^2(\Omega) F(\Omega) \quad (28.55)$$

is the *mass enhancement factor* of McMillan[66]. Inserting (28.54) one finds from (28.52), $Z = 1$ and hence $\tilde{\omega}_o = \omega_o$. Going close to T_c, so that $\Delta(\omega')$ may be neglected under the root in (28.53), the solution is

$$\Delta(\omega) = \Delta\theta(\omega_o - |\omega|) \quad (28.56)$$

and (28.53) becomes

$$1 = \lambda \int_0^{\omega_o} \frac{d\omega'}{\omega'} \tanh\frac{\beta_c\omega'}{2} . \quad (28.57)$$

But this is Eq. (23.18) so that $\lambda = \Lambda N(0)$. In a more detailed solution keeping $\mu^* \neq 0$ (see Section 2.3.2 of Ref. 60), λ in the exponent of T_c, Eq. (23.21), is replaced by $\lambda - \mu^*$[65]. However, the correlation amplitude at the origin, $\tilde{F}(0)$ of Eq. (22.20), is further reduced by a factor μ^*/μ where $\mu \equiv N(0)\langle V_C \rangle$[67].

It is evident from the above derivation of the real-frequency form of the Eliashberg equations that some quite subtle limiting processes are involved. This sensitivity is likely to be reflected also in numerical calculations. Therefore, the first way of proceeding as mentioned above, namely the Matsubara representation, is also of interest. Going back to Eqs. (28.29), (28.30) and

(28.35) the ε-integration poses no problem since with the symmetry (28.24) and using (28.45), the expression (28.40) has the imaginary root

$$R(i\nu') = iQ(\nu') \equiv iZ(i\nu')\sqrt{\nu'^2 + |\Delta(i\nu')|^2} \qquad (28.58)$$

so that (compare Problem 4.13)

$$\int_0^{\omega_o} \frac{d\varepsilon}{\varepsilon^2 + \phi^2(i\nu') + \nu'^2 Z^2(i\nu')} = \frac{1}{Q(\nu')} \tan^{-1} \frac{\omega_o}{Q(\nu')} . \qquad (28.59)$$

Inserting this with (28.37) into (28.29), (28.30) and (28.35) one finds

$$\nu[Z(i\nu) - 1] = 2\beta^{-1} \sum_{\nu'} \lambda(i\nu - i\nu') \tan^{-1}\left(\frac{\omega_o}{Q(\nu')}\right) \frac{\nu'}{\sqrt{\nu'^2 + |\Delta(i\nu')|^2}} \qquad (28.60)$$

and

$$Z(i\nu)\Delta(i\nu) = 2\beta^{-1} \sum_{\nu'} \left\{ \lambda(i\nu - i\nu') \tan^{-1} \frac{\omega_o}{Q(\nu')} \right.$$
$$\left. - \mu^* \tan^{-1} \frac{\omega_C}{Q(\nu')} \right\} \frac{\Delta(i\nu')}{\sqrt{\nu'^2 + |\Delta(i\nu')|^2}} . \qquad (28.61)$$

These equations become particularly appealing in the linearized approximation in which $\Delta(i\nu')$ is neglected in the root. Introducing the notation

$$\nu Z(i\nu) \equiv \tilde\omega(\nu) ; \quad Z(i\nu)\Delta(i\nu) \equiv \tilde\Delta(\nu) \qquad (28.62)$$

and replacing $(2/\pi)\tan^{-1}$ by a cutoff which we indicate by a prime on the summation sign, we obtain the equations of Bergmann and Rainer[68]

$$\tilde\omega(\nu) = \nu + \pi\beta^{-1} \sum_{\nu'}{}' \lambda(i\nu - i\nu')\,\text{sgn}\,\nu' ,$$

$$\tilde\Delta(\nu) = \pi\beta^{-1} \sum_{\nu'}{}' \{\lambda(i\nu - i\nu') - \mu^*\} \frac{\tilde\Delta(\nu')}{|\tilde\omega(\nu')| + \rho} . \qquad (28.63)$$

The authors have added in the denominator of the second equation a *pair-breaking parameter* ρ which phenomenologically describes the effect of uncorrelated *magnetic impurities* on the superconducting state (see Section 3.2 of Ref. 60 and Section 16 of Ref. 64). They used Eqs. (28.63) in a numerical analysis of the variation $\delta T_c/\delta\alpha^2(\Omega)F(\Omega)$ to assess the contribution of a given region of the phonon spectrum.

It is obvious from looking at Eq. (28.6) that strong-coupling theory may very well be formulated for types of interactions other than the electron-phonon coupling. This question has become of particular interest with regard to the new cuprate superconductors because of their high T_c's and the short coherence lengths which suggest a fairly strong and local pairing[69], as discussed in Section 20. A list of local non-phonon couplings is given in Section I.A of Ref. 69. One of the difficulties of course is the absence of Migdal's theorem (see Section 11) so that vertex corrections cannot be ignored any more. Apart from vertex corrections, one of the simplest strong-coupling models is obtained with an interaction between holes (charge e) centered on neighboring lattice sites and heavy bosons of charge $2e$ localized in between[70]. This model gives rise to "intersite local pairs or Heitler-London bipolarons" (Ref. 69, Section I.A) of the holes and to a transition temperature proportional to the hole concentration which is reminescent of Bose condensation. A new type of superconductivity is obtained in a model of δ-function attraction in a very high external magnetic field[71].

As is evident from Eqs. (28.60), (28.61), a pure Coulomb interaction cannot lead to pairing, at least not in the framework of Eliashberg theory (for particular charge distributions, pairing is of course not excluded, see Refs. 8, 69, 72). Intuitively, one could argue that for extremely strong Coulomb repulsion described, e.g., by a strongly repulsive Hubbard model (see Section 33) the carriers cannot avoid being paired. This seems to be almost the truth since a recent renormalization-group calculation in one dimension shows that the slightest addition of attraction precipitates pairing[73]. Since phonons are ever present there is always some attraction.

Solutions to the Problems of Chapter 4

Problem 4.1: Eq. (20.4).

The canonical transformation S, Eq. (20.1), may be determined by eliminating the interaction $\mathcal{H}_{\text{int}} = \mathcal{H}^{\text{el-ph}}$ to first order,

$$\mathcal{H}^{\text{el-ph}} + [\mathcal{H}_o, S] = 0$$

so that

$$\tilde{\mathcal{H}} = \mathcal{H}_o + \frac{1}{2}[\mathcal{H}^{\text{el-ph}}, S] + \cdots .$$

The condition on S may be written in interaction representation (5.17) in the form

$$\dot{S}^o(t) = -i\mathcal{H}^{\text{el-ph} \, o}(t) ,$$

which integrates to

$$S^o(t) - S^o(t_o) = -i \int_{t_o}^{t} dt' \mathcal{H}^{\text{el-ph} \, o}(t') .$$

Assuming that the interaction was switched on at $t_o = -\infty$ means that $\mathcal{H}^{\text{el-ph}\,o}(t)$ should be multiplied by $e^{\epsilon t}$ with $\epsilon = 0^+$ and that $S^o(-\infty) = 0$; hence

$$S = -i \int_{-\infty}^{0} dt' \mathcal{H}^{\text{el-ph}\,o}(t')e^{\epsilon t'} .$$

Defining the effective interaction as the phonon average of $\tilde{\mathcal{H}} - \mathcal{H}_o$ we have to lowest order

$$\mathcal{H}_{\text{eff}} = -\frac{i}{2} \int_{-\infty}^{0} dt e^{\epsilon t} \langle [\mathcal{H}^{\text{el-ph}}, \mathcal{H}^{\text{el-ph}\,o}(t)] \rangle^{\text{ph}} .$$

Writing the interaction (10.6) as $\sum \gamma_q n(-\mathbf{q})Q_q$, the phonon average in the last expression may be cast into the form

$$\sum_q |\gamma_q|^2 \{ n(-\mathbf{q})n^o(\mathbf{q}; t) \langle [Q_q, Q^o_{-q}(t)] \rangle + [n(-\mathbf{q}), n^o(\mathbf{q}; t)] \langle Q^o_{-q}(t)Q_q \rangle \} .$$

Since, according to (3.22),

$$[n(-\mathbf{q}), n^o(\mathbf{q}; t)] = \sum_{kk'} [a^+_{k'} a_{k'-q}, a^+_k a_{k+q}] e^{i(\varepsilon_k - \varepsilon_{k+q})t}$$

$$= \sum_k (a^+_{k+q} a_{k+q} - a^+_k a_k) e^{i(\varepsilon_k - \varepsilon_{k+q})t} ,$$

one sees that the second term of the phonon average yields, after the t-integration, a trivial term of the form $\sum \alpha_k a^+_k a_k$ and one verifies that

$$\alpha_k = \frac{1}{2} \sum_q |\gamma_q|^2 \left\{ \frac{(1+n_o(\omega_q))(\varepsilon_k - \varepsilon_{k-q})}{(\varepsilon_k - \varepsilon_{k-q})^2 - (\omega_q + i\epsilon)^2} \right.$$

$$\left. + \frac{n_o(\omega_q)(\varepsilon_k - \varepsilon_{k-q})}{(\varepsilon_k - \varepsilon_{k-q})^2 - (\omega_q - i\epsilon)^2} \right\} .$$

Using in the first term of the phonon average the definition (7.15) of $D^{\text{ret}\,o}(q; t)$ one finally obtains

$$\mathcal{H}_{\text{eff}} = \frac{1}{2} \int_{-\infty}^{\infty} dt \, e^{\epsilon t} \sum_q |\gamma_q|^2 n(-\mathbf{q})n^o(\mathbf{q}; t)D^{\text{ret}\,o}(\mathbf{q}; -t) + \sum_k \alpha_k a^+_k a_k .$$

In view of (3.22) and (7.20) the non-trivial term here is exactly Eq. (20.4).

Problem 4.2: Mass independence of Eq. (20.9).

Considering for simplicity a simple cubic crystal with $B = 1$, then the matrix of force constants $\mathbf{C}(\mathbf{R})$ of Eq. (1.23) may be expressed in terms of

$\omega_{\mathbf{q}}$ with the help of Eqs. (1.26), (1.27) and making use of (1.10),

$$\mathbf{C}(\mathbf{R}) = \frac{M}{\hbar^2} \frac{1}{L} \sum_{\mathbf{q} \in Z} \mathbf{D}(\mathbf{q}) e^{i\mathbf{q} \cdot \mathbf{R}} = \frac{Mv}{\hbar^2} \frac{1}{V} \sum_{\mathbf{q} \in Z} \omega_{\mathbf{q}}^2 e^{i\mathbf{q} \cdot \mathbf{R}} \ .$$

Going over to continuous $\mathbf{q}$-vectors one obtains, for $\mathbf{R} = 0$ and with $\omega_{\mathbf{q}} = \hbar c_{\parallel} q$,

$$\mathbf{C}(0) = \frac{M a^3 c_{\parallel}^2}{2\pi^2} \int_0^{\pi/a} dq \, q^4 = \frac{\pi^3}{10 a^2} M c_{\parallel}^2$$

which shows that $M c_{\parallel}^2$ is independent of the mass M, as is well known for the linear harmonic chain.

Problem 4.3: Eq. (20.16).

From the definitions of the Fermi-sea state $|F\rangle$ and of B_0^+ it follows that the energy E defined by (20.16) may be written as

$$E = V^{-1} \sum_{\mathbf{k}\mathbf{k}'} \varphi_{\mathbf{k}'}^* \theta(\varepsilon_{\mathbf{k}'}) \langle 0 | a_{-\mathbf{k}'\downarrow} a_{\mathbf{k}'\uparrow} \mathcal{H} a_{\mathbf{k}\uparrow}^+ a_{-\mathbf{k}\downarrow}^+ | 0 \rangle \varphi_{\mathbf{k}} \theta(\varepsilon_{\mathbf{k}}) \ .$$

Now each of the two terms of $\mathcal{H}$ gives rise to two pairing possibilities resulting in

$$\langle 0 | a_{-\mathbf{k}'\downarrow} a_{\mathbf{k}'\uparrow} \sum_{\sigma} a_{\mathbf{p}\sigma}^+ a_{\mathbf{p}\sigma} a_{\mathbf{k}\uparrow}^+ a_{-\mathbf{k}\downarrow}^+ | 0 \rangle = (\delta_{\mathbf{p}\mathbf{k}} + \delta_{\mathbf{p},-\mathbf{k}}) \delta_{\mathbf{k}'\mathbf{k}}$$

and

$$\langle 0 | a_{-\mathbf{k}'\downarrow} a_{\mathbf{k}'\uparrow} \sum_{\sigma\sigma'} a_{\mathbf{p}'\sigma'}^+ a_{\mathbf{p}\sigma}^+ a_{\mathbf{p}+\mathbf{q},\sigma} a_{\mathbf{p}'-\mathbf{q},\sigma'} a_{\mathbf{k}\uparrow}^+ a_{-\mathbf{k}\downarrow}^+ | 0 \rangle$$

$$= \delta_{\mathbf{p}',-\mathbf{k}'} \delta_{\mathbf{p}\mathbf{k}'} \delta_{\mathbf{p}+\mathbf{q},\mathbf{k}} \delta_{\mathbf{p}'-\mathbf{q},-\mathbf{k}} + \delta_{\mathbf{p}'\mathbf{k}'} \delta_{\mathbf{p},-\mathbf{k}'} \delta_{\mathbf{p}+\mathbf{q},-\mathbf{k}} \delta_{\mathbf{p}'-\mathbf{q},\mathbf{k}} \ .$$

Insertion into the first equation above and taking into account (20.8), (20.10) immediately leads to (20.16).

Problem 4.4: Eqs. (20.25), (20.26).

From the identity

$$e^{B\tau} A - A e^{B\tau} = \tau [B, A] e^{B\tau}$$

derived in Problem 3.8 of Chapter 3 for any operators A and B such that $[B, A]$ commutes with B, we obtain by the substitution $B \to B^+$, $A \to B$, $\tau \to \gamma$

$$B e^{\gamma B^+} = e^{\gamma B^+} B + \gamma e^{\gamma B^+} \ .$$

Acting with this equation on $|0\rangle$ and comparing with (20.24) we find

$$|\gamma\rangle = ce^{\gamma B^+}|0\rangle .$$

Here the normalization constant c may be calculated by expressing $|\gamma\rangle$ in terms of Fock states of constant boson numbers $|n\rangle = (n!)^{-1/2}(B^+)^n|0\rangle$,

$$|\gamma\rangle = c\sum_{n=0}^{\infty} \frac{\gamma^n}{\sqrt{n!}}|n\rangle .$$

Indeed, making use of the orthonormality of the states $|n\rangle$ one finds

$$1 = \langle\gamma|\gamma\rangle = |c|^2 \sum_{n=0}^{\infty} \frac{|\gamma|^{2n}}{n!} = |c|^2 e^{|\gamma|^2}$$

which, up to a phase, gives (20.25).

Applying (20.24), the first relation (20.26) is immediate while the second follows with the help of the identity

$$(B^+B)^2 = B^+B + (B^+)^2(B)^2$$

which is a special case of the Wick ordering (8.10).

Problem 4.5: Eqs. (20.34), (20.35), (21.4).

Since for any $\mathbf{k}$

$$\langle 0|(u_{\mathbf{k}}^* + v_{\mathbf{k}}^* a_{-\mathbf{k}\downarrow} a_{\mathbf{k}\uparrow})(u_{\mathbf{k}} + v_{\mathbf{k}} a_{\mathbf{k}\uparrow}^+ a_{-\mathbf{k}\downarrow}^+)|0\rangle = |u_{\mathbf{k}}|^2 + |v_{\mathbf{k}}|^2 = 1 ,$$

one may write, with $\pm k \equiv (\pm\mathbf{k}, \pm\sigma)$, σ being the sign $+ =\uparrow, \ - =\downarrow$,

$$\langle v|a_k^+ a_k|v\rangle = \langle 0|(u_{\sigma\mathbf{k}}^* + \sigma v_{\sigma\mathbf{k}}^* a_{-k} a_k) a_k^+ a_k (u_{\sigma\mathbf{k}} + \sigma v_{\sigma\mathbf{k}} a_k^+ a_{-k}^+)|0\rangle$$
$$= |v_{\sigma\mathbf{k}}|^2 \langle 0|a_{-k} a_k a_k^+ a_k a_k^+ a_{-k}^+|0\rangle = |v_{\sigma\mathbf{k}}|^2 = |v_{\mathbf{k}}|^2 .$$

This leads immediately to (20.34). Similarly one may write, for $k \neq k'$,

$$\langle v|a_{k'}^+ a_{k'} a_k^+ a_k|v\rangle = \langle 0|(u_{\sigma'\mathbf{k}'}^* + \sigma' v_{\sigma'\mathbf{k}'}^* a_{-k'} a_{k'})(u_{\sigma\mathbf{k}} + \sigma v_{\sigma\mathbf{k}} a_{-k} a_k)$$
$$\times a_{k'}^+ a_{k'} a_k^+ a_k (u_{\sigma'\mathbf{k}'} + \sigma' v_{\sigma'\mathbf{k}'} a_{k'}^+ a_{-k'}^+)$$
$$\times (u_{\sigma\mathbf{k}} + \sigma v_{\sigma\mathbf{k}} a_k^+ a_{-k}^+)|0\rangle$$
$$= |v_{\mathbf{k}'}|^2 |v_{\mathbf{k}}|^2 \langle 0|a_{-k'} a_{k'} a_{-k} a_k a_k^+ a_{k'}^+ a_k^+ a_k a_{k'}^+ a_{-k'}^+ a_k^+ a_{-k}^+|0\rangle$$
$$= |v_{\mathbf{k}'}|^2 |v_{\mathbf{k}}|^2 .$$

On the other hand, for $k = k'$ the exclusion principle implies $(a_k^+ a_k)^2 = a_k^+ a_k$. Therefore

$$\langle v | N_{\text{el}}^2 | v \rangle = \sum_{k\sigma} \sum_{k'\sigma'} (1 - \delta_{\mathbf{kk'}} \delta_{\sigma\sigma'}) |v_{\mathbf{k'}}|^2 |v_{\mathbf{k}}|^2 + \sum_{k\sigma} |v_{\mathbf{k}}|^2$$

$$= \left(2 \sum_{\mathbf{k}} |v_{\mathbf{k}}|^2 \right)^2 - 2 \sum_{\mathbf{k}} |v_{\mathbf{k}}|^4 + 2 \sum_{\mathbf{k}} |v_{\mathbf{k}}|^2$$

which leads to (20.35).

In the same notation one has for $\mathbf{k} \neq \mathbf{k'}$, $\sigma = \sigma'$

$$\langle v | a_k^+ a_{-k}^+ a_{-k'} a_{k'} | v \rangle = \langle 0 | (u_{\sigma k'}^* + \sigma v_{\sigma k'}^* a_{-k'} a_{k'})$$
$$\times (u_{\sigma k}^* + \sigma v_{\sigma k}^* a_{-k} a_k) a_k^+ a_{-k}^+ a_{-k'} a_{k'} (u_{\sigma k'} + \sigma v_{\sigma k'} a_{k'}^+ a_{-k'}^+)$$
$$\times (u_{\sigma k} + \sigma v_{\sigma k} a_k^+ a_{-k}^+) | 0 \rangle = u_{\sigma k'}^* v_{\sigma k}^* v_{\sigma k'} u_{\sigma k}$$

where the u-factors are the result of exclusion-principle cancellations, while for $k = k'$

$$\langle v | a_k^+ a_{-k}^+ a_{-k} a_k | v \rangle = |v_{\mathbf{k}}|^4 .$$

Combining the two cases one finds

$$\sum_{\sigma\sigma'} \langle v | a_k^+ a_{-k}^+ a_{-k'} a_{k'} | v \rangle \delta_{\sigma\sigma'} = \sum_{\sigma} \{ (1 - \delta_{\mathbf{kk'}}) u_{\sigma k'}^* v_{\sigma k}^* v_{\sigma k'} u_{\sigma k} + \delta_{\mathbf{kk'}} |v_{\mathbf{k}}|^4 \}$$

$$= 2 u_{\mathbf{k'}} u_{\mathbf{k}} \text{Re} (v_{\mathbf{k'}} v_{\mathbf{k}}^*) - 2 \delta_{\mathbf{kk'}} |v_{\mathbf{k}}|^2 (|u_{\mathbf{k}}|^2 - |v_{\mathbf{k}}|^2)$$

where the reality conditions $v_{-\mathbf{k}}^* = v_{\mathbf{k}}$, $u_{-\mathbf{k}}^* = u_{\mathbf{k}} = (1 - |v_{\mathbf{k}}|^2)^{1/2}$ were used. This gives (21.4).

Problem 4.6: Eq. (22.3).

Writing in analogy with (21.13) $\theta(-\varepsilon) = (1 - (\varepsilon/|\varepsilon|))/2$ we have

$$\frac{1}{V} \sum_{\mathbf{k}} \varepsilon_{\mathbf{k}} \{ v_{\mathbf{k}}^2 - \theta(-\varepsilon_{\mathbf{k}}) \} = \frac{1}{2V} \sum_{\mathbf{k}} \Delta_{\mathbf{k}}^2 \left(\frac{1}{E_{\mathbf{k}}} - \frac{1}{E_{\mathbf{k}} + |\varepsilon_{\mathbf{k}}|} \right) .$$

Making use of (21.15) and introducing the density of states (10.4) this becomes, with $E_o^2 \equiv \varepsilon^2 + \Delta_o^2$,

$$\frac{1}{2} N(0) \Delta_o^2 \int_{-\omega_o}^{\omega_o} d\varepsilon \left(\frac{1}{E_o} - \frac{E_o - |\varepsilon|}{\Delta_o^2} \right)$$

$$= \frac{\Delta_o^2}{2} \ln \left(\frac{\omega_o}{\Delta_o} + \sqrt{1 + \frac{\omega_o^2}{\Delta_o^2}} \right) - \frac{\omega_o^2}{2} \left(\sqrt{1 + \frac{\omega_o^2}{\Delta_o^2}} - 1 \right) .$$

Using (21.16) and $\omega_o \gg \Delta_o$ this contributes to (22.2)

$$-\frac{2\Delta_o^2}{n\Lambda}\left(1 - \frac{1}{2}\Lambda N(0)\right).$$

Similarly one obtains with (21.8), (21.11)

$$\frac{2}{nV^2}\sum_{\mathbf{k}\mathbf{k}'}\lambda_{\mathbf{k}\mathbf{k}'}u_{\mathbf{k}}v_{\mathbf{k}}u_{\mathbf{k}'}v_{\mathbf{k}'} = \frac{2}{nV}\sum_{\mathbf{k}}\frac{\Delta_{\mathbf{k}}^2}{2E_{\mathbf{k}}} = \frac{2\Delta_o^2}{n\Lambda}$$

which cancels the first term of the previous expression, leaving the first term of (22.3).

For the term $\propto \delta_{\mathbf{k}\mathbf{k}'}$ in (22.2) we may write

$$4(v_{\mathbf{k}}^4 - \theta(-\varepsilon_{\mathbf{k}})) = \left(1 - \frac{\varepsilon_{\mathbf{k}}}{E_{\mathbf{k}}}\right)^2 - \left(1 - \frac{\varepsilon_{\mathbf{k}}}{|\varepsilon_{\mathbf{k}}|}\right)^2$$

$$= \Delta_{\mathbf{k}}^2\frac{\varepsilon_{\mathbf{k}}}{|\varepsilon_{\mathbf{k}}|E_{\mathbf{k}}}\left(\frac{2}{E_{\mathbf{k}} + |\varepsilon_{\mathbf{k}}|} - \frac{\varepsilon_{\mathbf{k}}}{|\varepsilon_{\mathbf{k}}|E_{\mathbf{k}}}\right).$$

Using (21.14), (21.15) this contributes to (22.2)

$$\frac{\Delta_o^2\Lambda N(0)}{nV}\int_0^{\omega_o}\frac{d\varepsilon}{E_o^2}.$$

And since the integral is $\Delta_o^{-1}\tan^{-1}(\omega_o/\Delta_o) \simeq \Delta_o^{-1}\pi/2$ one readily finds the second term of (22.3).

Problem 4.7: Eq. (24.1).

Using the relations of Section 7 we successively obtain with $\epsilon = 0^+$

$$\tanh\frac{\beta\varepsilon_k}{2} = -\mathcal{G}^o(k; \epsilon) - \mathcal{G}^o(k; -\epsilon) = -\beta^{-1}\sum_{\nu}\tilde{\mathcal{G}}^o(k; i\nu)(e^{-i\nu\epsilon} + e^{i\nu\epsilon})$$

$$= \frac{1}{\beta}\sum_{\nu}\cos\nu\epsilon\int_0^{\beta}d\tau\langle a_k^o(-i\tau)a_k^+\rangle_o e^{i\nu\tau} = -\frac{2}{\beta}\sum_{\nu}\frac{\cos\nu\epsilon}{i\nu - \varepsilon_k}.$$

Splitting this fermionic Matsubara sum into two sums with $\nu > 0$ and $\nu' = -\nu$, we obtain a quadratic denominator which makes the sum convergent and allows to let $\epsilon \to 0$. This leads to Eq. (24.1) which may also be written, by reinstating the full Matsubara sum, as

$$\tanh\frac{\beta E}{2} = \frac{2E}{\beta}\sum_{\nu}\frac{1}{\nu^2 + E^2}.$$

This last form is necessary to show the correctness of this formula in reverse with the aid of the identity (10.16). Indeed, the integral over the circle

$|z| \to \infty$ which is needed to close the path Γ in Eq. (10.16) is convergent only if the denominator in the Matsubara sum is at least quadratic in $z = i\nu$. With this formula, Eq. (10.16) reproduces $f_o(-E) - f_o(E)$ as residues of the poles at $z = \pm E$. Note also that $f_o(\pm E)$ taken alone in the above derivation would contain a sum with $\sin \nu\epsilon$, besides the one with $\cos \nu\epsilon$. However, this additional sum is not well defined since Eq. (10.16) cannot be applied. But, fortunately, $f_o(\pm E)$ may be obtained from (24.1) with the aid of Eq. (7.42).

Problem 4.8: Eq. (24.10).

According to (24.8)

$$\sum_{n=0}^{\infty} I_n \cong \frac{1}{4\pi^2} \sum_{n=0}^{\infty} \frac{1}{(2n+1)^3} \; .$$

This sum over all positive odd integers may be obtained from (24.9) by subtraction of the sum over all even integers,

$$\sum_{n=1}^{\infty} \frac{1}{(2n)^z} = 2^{-z} \zeta(z) \; .$$

Thus

$$\sum_{n=0}^{\infty} \frac{1}{(2n+1)^z} = (1 - 2^{-z})\zeta(z)$$

which, inserted with $z = 3$ into the first line above gives (24.10).

Problem 4.9: Eq. (25.11).

Insertion of (25.10) into (25.5) gives, after Bogoljubov transformation (21.23),

$$\tilde{\chi}_{\alpha\gamma}^{S}(\mathbf{q}; i\nu) = \frac{e^2\hbar^2}{m^2 V} \sum_{\mathbf{kk'}} k_\alpha k'_\gamma \int_0^\beta d\tau e^{i\nu\tau} B$$

with the bracket

$$B \equiv \langle (\Phi_{\mathbf{k}-\mathbf{q}/2}^{+} S(\mathbf{k}) \Phi_{\mathbf{k}+\mathbf{q}/2})(-i\tau)(\Phi_{\mathbf{k'}+\mathbf{q}/2}^{+} S^{\mathrm{T}}(\mathbf{k'}) \Phi_{\mathbf{k'}-\mathbf{q}/2}) \rangle$$

where

$$S(\mathbf{k}) \equiv R_{\mathbf{k}-\mathbf{q}/2} R_{\mathbf{k}+\mathbf{q}/2}^{\mathrm{T}} \; .$$

For $\mathbf{q} \neq 0$ and with unperturbed averages (23.3), B factorizes as in (25.6),

$$B = \langle \Phi_{\mathbf{k}-\mathbf{q}/2}^{+}(-i\tau) S(\mathbf{k}) M S^{\mathrm{T}}(\mathbf{k'}) \Phi_{\mathbf{k'}-\mathbf{q}/2} \rangle$$

where M is the 2×2 matrix

$$M \equiv \langle \Phi_{\mathbf{k+q}/2}(-i\tau) \otimes \Phi^+_{\mathbf{k'+q}/2} \rangle \ .$$

Making use of (21.22) and (23.3) one finds

$$M = \begin{pmatrix} q_+, & 0 \\ 0, & p_+ \end{pmatrix} \delta_{\mathbf{kk'}}$$

where

$$p_\pm \equiv f_\pm e^{E_\pm \tau} \ ; \quad q_\pm \equiv (1 - f_\pm)e^{-E_\pm \tau}$$

and the abbreviations (25.12) were used. Applying again (21.22) and (23.3) one also finds

$$B = (p_- N_{11} + q_- N_{22}) \delta_{\mathbf{kk'}}$$

where N_{ij} are the elements of the matrix

$$N \equiv S(\mathbf{k}) \begin{pmatrix} q_+, & 0 \\ 0, & p_+ \end{pmatrix} S^{\mathrm{T}}(\mathbf{k}) \ .$$

With (21.22) one finds

$$S(\mathbf{k}) = \begin{pmatrix} s_+, & s_- \\ -s_-, & s_+ \end{pmatrix} \ ; \quad s_+ \equiv u_- u_+ + v_- v_+ \ ; \quad s_- \equiv u_- v_+ - v_- u_+$$

where $u_\pm \equiv u_{\mathbf{k}\pm\mathbf{q}/2}$, $v_\pm \equiv v_{\mathbf{k}\pm\mathbf{q}/2}$ and

$$N_{11} = q_+ s_+^2 + p_+ s_-^2 \ ; \quad N_{22} = q_+ s_-^2 + p_+ s_+^2 \ ,$$

so that

$$B = \{(p_- q_+ + p_+ q_-)s_+^2 + (p_- p_+ + q_- q_+)s_-^2\}\delta_{\mathbf{kk'}} \ .$$

Use of (21.11) and (21.13) also gives

$$s_\pm^2 = \frac{1}{2}\left(1 \pm \frac{\varepsilon_- \varepsilon_+ + \Delta^2}{E_- E_+}\right) \ .$$

Applying the identity (25.7) in various forms, making use of (7.42), one finds

$$p_- q_+ + p_+ q_- = (f_+ - f_-)\big[n_o(E_- - E_+)$$
$$\times \, e^{(E_- - E_+)\tau} - n_o(E_+ - E_-)e^{(E_+ - E_-)\tau}\big]$$

$$p_- p_+ + q_- q_+ = (1 - f_+ - f_-)\big[n_o(E_- + E_+)$$
$$\times \, e^{(E_- + E_+)\tau} - n_o(-E_- - E_+)e^{(-E_- - E_+)\tau}\big]$$

and with use of (7.26), (7.43)

$$\int_0^\beta d\tau \, e^{i\nu\tau}(p_- q_+ + p_+ q_-) = (f_+ - f_-)\frac{2(E_- - E_+)}{(E_- - E_+)^2 - (i\nu)^2} \ ,$$

$$\int_0^\beta d\tau e^{i\nu\tau}(p_- p_+ + q_- q_+) = (1 - f_+ - f_-)\frac{2(E_- + E_+)}{(E_- + E_+)^2 - (i\nu)^2} \ .$$

Inserting all these ingredients into the above expression for $\tilde{\chi}^S_{\alpha\gamma}$ one readily finds Eq. (25.11).

Problem 4.10: Eq. (25.14), (25.23).

According to (10.1) and (10.4), writing $\varepsilon_{\mathbf{k}} + \varepsilon_F = \hbar^2 \mathbf{k}^2/2m$,

$$n = -\frac{1}{V}\left(\frac{\partial\Omega}{\partial\mu}\right)_T = \frac{2}{V}\sum_{\mathbf{k}} f_o(\varepsilon_{\mathbf{k}}) = 2\int_{-\varepsilon_F}^\infty d\varepsilon N(\varepsilon) f_o(\varepsilon) \ .$$

Introducing the number of states per unit volume with energy $< \varepsilon$,

$$Z(\varepsilon) \equiv \frac{1}{V}\sum_{\mathbf{k}} \theta(\varepsilon - \varepsilon_{\mathbf{k}}) = \frac{1}{6\pi^2}\left[\frac{2m}{\hbar^2}(\varepsilon + \varepsilon_F)\right]^{3/2} \ ; \quad \varepsilon > -\varepsilon_F \ ,$$

so that $N(\varepsilon) = Z'(\varepsilon) = 3Z(\varepsilon)/2(\varepsilon + \varepsilon_F)$, partial integration of the last expression of n gives

$$n = -2\int_{-\varepsilon_F}^\infty d\varepsilon Z(\varepsilon) f_o'(\varepsilon) = -\frac{4}{3}\int_{-\varepsilon_F}^\infty d\varepsilon(\varepsilon + \varepsilon_F) N(\varepsilon) f_o'(\varepsilon) \ .$$

Using (10.4) in reverse this leads to Eq. (25.14).

Combining Eqs. (25.14)–(25.16) we may write

$$n_s = -\frac{2\hbar^2}{3mV}\sum_{\mathbf{k}} \mathbf{k}^2\{f_o'(\varepsilon_{\mathbf{k}}) - f_o'(E_{\mathbf{k}})\}$$

or, since $E - \varepsilon \sim 0$ for $|\varepsilon| > \omega_o$ (Fig. 4.3), using Eq. (14.14) with $D = 3$,

$$\frac{n_s}{n} = -2\int_0^{\omega_o} d\varepsilon\{f_o'(\varepsilon) - f_o'(E)\} \ .$$

For the evaluation of this expression near T_c we use again Eq. (24.1). Remembering (7.42) and noting that the derivative may be taken term by term we have

$$f_o'(\varepsilon) = \frac{2}{\beta}\sum_{\nu>0}\left\{\frac{1}{\nu^2 + \varepsilon^2} - \frac{2\nu^2}{(\nu^2 + \varepsilon^2)^2}\right\} \ .$$

Term-by-term integration of this expression between 0 and ∞, however, yields 0 instead of 1! Therefore, it is absolutely necessary first to calculate the difference $f_o'(\varepsilon) - f_o'(E)$ as power series in $y \equiv \beta\Delta \ll 1$. Since $x_n \equiv \beta\nu = (2n+1)\pi > 1$ we have with $x \equiv \beta\varepsilon$

$$f_o'(\varepsilon) - f_o'(E) = 2\beta y^2 \sum_{n=0}^{\infty} \left\{ \frac{1}{(x^2 + x_n^2)^2} - \frac{4x_n^2}{(x^2 + x_n^2)^3} \right\} + \mathcal{O}(y^4) .$$

Inserting Eq. (24.8) and its derivative,

$$\int_0^{\infty} \frac{dx}{(x^2 + x_n^2)^3} = -\frac{1}{2} \frac{dI_n}{d(x_n^2)} = \frac{3\pi}{16|x_n|^5} ,$$

into the above expression for n_s/n and making use of Eqs. (24.10)–(24.12) one readily obtains the upper line of Eq. (25.23).

Near $T = 0$ the above expression for n_s/n may be rewritten with the help of (25.16) as

$$\frac{n_n}{n} = -2 \int_0^{\infty} d\varepsilon f_o'(E) .$$

Because of $\omega_o/\Delta \gg 1$ we may approximate $f_o'(E)$ by $-\beta \exp(-\beta E)$ and use (24.14) to write

$$\frac{n_n}{n} = 2y \int_0^{\infty} ds\, e^{-y\sqrt{1+s^2}} = -2yK_o'(y) .$$

Applying here the approximation (22.22) then readily yields the second line of (25.23).

Problem 4.11: Eqs. (26.20), (26.25).

With (26.18) the right-hand side of the first equation (26.17) reads

$$-\frac{1}{V^2} \sum_{\mathbf{pp'}} \int d^3u \int d^3v e^{-i(\mathbf{p'}+\mathbf{k})\cdot\mathbf{u}} e^{-i(\mathbf{p}-\mathbf{k})\cdot\mathbf{v}}$$

$$\times \exp\left\{ \frac{\mathbf{u}}{2} \cdot \partial_{\mathbf{R}} + \frac{\mathbf{u}+\mathbf{v}}{2} \cdot \partial_{\mathbf{R''}} + \frac{\mathbf{v}}{2} \cdot \partial_{\mathbf{R'}} \right\}$$

$$\times \tilde{\mathcal{G}}_N(\mathbf{p};\mathbf{R})\Delta(\mathbf{R''})\tilde{\mathcal{F}}^*(\mathbf{p'};\mathbf{R'})\Big|_{\mathbf{R}=\mathbf{R'}=\mathbf{R''}}$$

and similarly for the second equation (26.17). Applying now the identity (26.19), respectively with $\mathbf{q} = -\mathbf{p'}$, $\mathbf{a} = \mathbf{u}$, $\mathbf{b} = (\partial_{\mathbf{R}} + \partial_{\mathbf{R''}})/2$ and $\mathbf{q} = \mathbf{p}$, $\mathbf{a} = -\mathbf{v}$, $\mathbf{b} = (\partial_{\mathbf{R''}} + \partial_{\mathbf{R'}})/2$, this becomes

$$-\frac{1}{V^2} \sum_{\mathbf{pp'}} \int d^3u \int d^3v \exp\left\{ \frac{i}{2}(\partial_{\mathbf{R}} + \partial_{\mathbf{R'}}) \cdot \partial_{\mathbf{p'}} + \frac{i}{2}(\partial_{\mathbf{R''}} + \partial_{\mathbf{R'}}) \cdot \partial_{\mathbf{p}} \right\}$$

$$\times e^{-i(\mathbf{p'}+\mathbf{k})\cdot\mathbf{u}} e^{-i(\mathbf{p}-\mathbf{k})\cdot\mathbf{v}} \tilde{\mathcal{G}}_N(\mathbf{l};\mathbf{R})\Delta(\mathbf{R''})\tilde{\mathcal{F}}^*(\mathbf{l'};\mathbf{R'})\Big|_{\mathbf{R}=\mathbf{R'}=\mathbf{R''},\mathbf{l}=\mathbf{p},\mathbf{l'}=\mathbf{p'}}$$

and similarly for the second equation. Shifting the **p**- and **p**'-derivatives by partial integration to operate on $\tilde{\mathcal{G}}_N$ and $\tilde{\mathcal{F}}^*$, respectively, this changes the sign of the argument of the exponential operator and liberates the plane waves so that they may be integrated over **u** and **v**, respectively. This gives a factor $V^2 \delta_{\mathbf{pk}} \delta_{-\mathbf{p'k}}$, and the result is the first equation (26.20); the second follows similarly.

To obtain Eq. (26.25) we multiply (26.13) from the left with $i\nu - \mathcal{H}(\mathbf{r})$, writing $\nabla = (\partial_{\mathbf{R}}/2) + \partial_{\mathbf{x}}$ and take the Fourier transform (26.15). After a partial integration in **x** one finds, in the gauge (26.24),

$$
\int_V d^3 x\, e^{-i\mathbf{k}\cdot\mathbf{x}} \left\{ i\nu + \mu + \frac{\hbar^2}{2m} \left[\frac{1}{2}\partial_{\mathbf{R}} + i\mathbf{k} + \frac{ie}{\hbar c}\mathbf{A}\left(\mathbf{R} + \frac{\mathbf{x}}{2}\right) \right]^2 \right\}
$$
$$
\times \mathcal{G}_N\left(\mathbf{R} + \frac{\mathbf{x}}{2}, \mathbf{R} - \frac{\mathbf{x}}{2}\right) = 1 \ .
$$

Applying the second assumption we may neglect the **x**-dependence and drop $\partial_{\mathbf{R}}$ in the straight bracket. Using (26.15) again, (26.25), (26.26) readily follow.

Problem 4.12: Eqs. (26.36), (26.37).

With (26.29), application of (24.5) and (23.21) gives

$$
\int_{-\omega_o}^{\omega_o} d\varepsilon \sum_\nu |\tilde{\mathcal{G}}^{(o)}|^2 = 4 \int_0^{\omega_o} d\varepsilon \sum_{\nu>0} \frac{1}{\nu^2 + \varepsilon^2} = \beta\left(\ln \frac{T_c}{T} + \frac{1}{\Lambda N(0)} \right)
$$

while

$$
\int_{-\omega_o}^{\omega_o} d\varepsilon \sum_\nu |\tilde{\mathcal{G}}^{(o)}|^2 \tilde{\mathcal{G}}^{(o)} = \int_{-\omega_o}^{\omega_o} d\varepsilon \sum_\nu \frac{-i\nu - \varepsilon}{(\nu^2 + \varepsilon^2)^2} = 0 \ .
$$

On the other hand, using (24.8), (24.10), (24.11) one obtains

$$
\int_{-\omega_o}^{\omega_o} d\varepsilon \sum_\nu |\tilde{\mathcal{G}}^{(o)}|^4 = 4 \int_0^{\omega_o} d\varepsilon \sum_{\nu>0} \frac{1}{(\nu^2 + \varepsilon^2)^2} = \beta\left(\frac{T_c}{\Delta_1 T} \right)^2
$$

and with

$$
\int_0^\infty \frac{dx}{(x^2 + x_n^2)^3} = \frac{3\pi}{16|x_n|^5}
$$

one also finds

$$\int_{-\omega_o}^{\omega_o} d\varepsilon \sum_\nu |\tilde{\mathcal{G}}^{(o)}|^2 \tilde{\mathcal{G}}^{(o)2}$$

$$= \int_{-\omega_o}^{\omega_o} d\varepsilon \sum_\nu \frac{(i\nu + \varepsilon)^2}{(\nu^2 + \varepsilon^2)^3}$$

$$= 4 \int_0^{\omega_o} d\varepsilon \sum_{\nu > 0} \left(\frac{1}{(\nu^2 + \varepsilon^2)^2} - \frac{2\nu^2}{(\nu^2 + \varepsilon^2)^3} \right) = -\frac{\beta}{2} \left(\frac{T_c}{\Delta_1 T} \right)^2 .$$

Inserting these results into (26.33)–(26.35) and noting that the term with $\varepsilon_{\mathbf{k}}$ in (26.34) does not contribute, one arrives at (26.36), (26.37).

Problem 4.13: Eq. (28.44).

Using the formula

$$2 \int_0^{\omega_o} \frac{d\varepsilon}{\varepsilon^2 - R^2} = \frac{1}{R} \log \frac{R - \omega_o}{R + \omega_o}$$

and the fact that according to (28.24), (28.40) R is analytic on the real axis, definition (11.11) gives

$$I(\omega') \equiv 2 \int_0^{\omega_o} d\varepsilon \mathrm{Im} \, \frac{1}{\varepsilon^2 - R^2(\omega')} = \frac{1}{2iR(\omega')}$$

$$\times \log \left[\frac{R(\omega' + i\epsilon) - \omega_o}{R(\omega' + i\epsilon) + \omega_o} \frac{R(\omega' - i\epsilon) + \omega_o}{R(\omega' - i\epsilon) - \omega_o} \right]_{\epsilon \to 0} .$$

Since we may choose the sign of R such that $R(\omega') \simeq \omega' Z_C > 0$ for $\omega' > \omega_C$, $R(\omega')$ is an odd function with the same sign as ω'. Since according to (28.40), $R^2(0) < 0$, there exists a real ω_m, $\omega_o > \omega_m > 0$, such that $R(\pm \omega_m) = 0$ and

$$I(\omega') = \mathrm{Re}\, R(\omega') = 0 \; ; \quad \omega_m > \omega' > -\omega_m \ .$$

But $I(\omega') = 0$ also for $|R(\omega')| > \omega_o$ or $|\omega'| > \omega_o/Z_C$. The only non-zero values of $I(\omega')$ occur between these two intervals; it is easy to see from the properties of the logarithm that

$$I(\omega') = \frac{\pi}{R(\omega')} \; ; \quad |\omega'| > \omega_m \; ; \quad |\omega'| < \omega_o/Z_C \ .$$

These results for the 3 intervals may be combined to yield (28.44).

It is also interesting to note that integrating illegally under the Im-sign and letting $\omega_o \to \infty$ so that Cauchy's formula may be used, one finds

$$I(\omega') = \mathrm{Im} \int_{-\infty}^{+\infty} \frac{d\varepsilon}{\varepsilon^2 - R^2(\omega')} = \mathrm{Im} \, \frac{i\pi}{R(\omega')} = \pi \mathrm{Re} \, \frac{1}{R(\omega')}$$

which is (28.44) for $\omega_o \to \infty$.

References to Chapter 4

1. **Kamerlingh Onnes, H.**, *Leiden Comm.* **119b**, **122b**, **124c** (1911).
2. **Bednorz, J.G.** and **Müller, K.A.**, *Z. Phys.* **B64**, 189 (1986). See also the Nobel Lectures 1987, *Rev. Mod. Phys.* **60**, 585 (1988).
3. **Rickayzen, G.**, *Theory of Superconductivity* (Wiley-Interscience, New York, 1965).
4. **Meissner, W.** and **Ochsenfeld, R.**, *Naturwiss.* **21**, 787 (1933).
5. **Bardeen, J.**, Cooper, L. and Schrieffer, J. , *Phys. Rev.* **108**, 1175 (1957).
6. **Blatt, J.M.**, *Theory of Superconductivity* (Academic, New York, 1964).
7. **Schafroth, M.R.**, *Phys. Rev.* **96**, 1442 (1954).
8. **Heisenberg, W.**, "The electron theory of superconductivity", in *Two Lectures* (Cambridge Univ. Press, 1949), p. 27 and references therein. Reprinted in *Werner Heisenberg, Collected Works*, ed. Blum, W., Dürr, H.-P. and Rechenberg, H. (Springer, Berlin, 1984), Series B, p. 449.
9. **Wigner, E.** , *Phys. Rev.* **46**, 1002 (1934).
10. **Hirsch, J.E.** and **Tang, S.**, *Phys. Rev. Lett.* **62**, 591 (1989).
11. **Vaknin, D.** *et al.*, *Phys. Rev. Lett.* **58**, 2802 (1987): La_2CuO_{4-y}; **Tranquada, J.M.**, *et al.*, *Phys. Rev. Lett.* **60**, 156 (1988): YBa_2CuO_x.
12. **Wu, M.K.**, **Ashburn, J.R.**, **Torng, C.J.**, **Hor, P.H.**, **Meng, R.L.**, **Gao, L.**, **Huang, Z.J.**, **Wang, Y.Q.**, and **Chu, C.W.**, *Phys. Rev. Lett.* **58**, 908 (1987).
13. **Fröhlich, H.**,, *Proc. Roy. Soc.* **A223**, 296 (1954).
14. **Cooper, L.N.**, *Phys. Rev.* **104**, 1189 (1956).
15. **Gough, C.E.** *et al.*, *Nature* **326**, 855 (1987); **Müller, K.A.**,, "The development of the high-temperature superconductivity field", in *Mechanisms of High Temperature Superconductivity*, ed. Kamimura, H. and Oshiyama, A., Springer Series in Materials Science 11 (Springer, Berlin, 1989), p. 2.
16. **Iye, Y.**, *Int. J. Mod. Phys.* **B3**, 367 (1989); Deutscher, G., *IBM J. Res. Develop.* **33**, 293 (1989).
17. *Physics Today* **44**, no. 6 (June 1991).
18. **Anderson, P.W.**, *Science* **235**, 1196 (1987).
19. **Chen, Y.-H.**, **Wilczek, F.**, **Witten, E.** and **Halperin, B.I.**, *Int. J. Mod. Phys.* **B3**, 1001 (1989). See also *Physics Today* **44**, no. 2 (February 1991), p. 17; *Int. J. Mod. Phys.* **B5** no. 10 (June 1991).
20. **Pauli, W.**, "Selected Topics in Field Quantization", in *Pauli Lectures in Physics*, ed. Enz, C.P. (MIT Press, Cambridge, 1973), vol. 6, p. 144; **Schrieffer, J.R.** and **Wolff, P.A.**, *Phys. Rev.* **149**, 491 (1966).
21. **Leggett, A.J.**, *Rev. Mod. Phys.* **47**, 331 (1975).
22. **Nozières, P.** and **Schmitt-Rink, S.**, *J. Low Temp. Phys.* **59**, 195 (1985); See also **Leggett, A.J.**, "Diatomic molecules and cooper pairs", in *Modern Trends in the Theory of Condensed Matter*, ed. Pekalski, A. and Przystawa, J., Lecture Notes in Physics vol. 115 (Springer, Berlin, 1980), p. 13; **Randeria, M.**, **Duan, J.-M.** and **Shieh, L.-Y.**, *Phys. Rev. Lett.* **62**, 981 (1989); *Phys. Rev.* **B41**, 327 (1990) for 2 dimensions; **Miyake, K.**, *Prog. Theor. Phys.* **69**, 1794 (1983).
23. **Glauber, R.**, *Phys. Rev.* **131**, 2766 (1963).

24. **Schrieffer, J.R.**, *Theory of Superconductivity* (Benjamin, New York, 1964).

25. **Emin, D.**, *Phys. Rev. Lett.* **62**, 1544 (1989).

26. **Shafer, M.W., Penney, T.** and **Olson, B.L.**, *Phys. Rev.* **B36**, 4047 (1987).

27. **Jorgensen, J.D.** *et al.*, *Phys. Rev. Lett.* **58**, 1024 (1987): $La_{1.85}Ba_{0.15}$ CuO_4; **Cava, R.J.** *et al.*, *Phys. Rev. Lett.* **58**, 1676 (1987): $YBa_2Cu_3O_{6.9}$.

28. **Weber, W.**, *Z. Phys.* **B70**, 323 (1988).

29. **Weber, W.**, in *Festkörperprobleme*, Adv. in Solid State Phys., vol. 28, ed. Rössler U. (Vieweg, Braunschweig, 1988), p. 141.

30. **Bogoljubov, N.N.**, *Nuovo Cimento* **7**, 794 (1958); **Valatin, J.G.**, *Nuovo Cimento* **7**, 843 (1958). See also **Galasiewicz, Z.M.**, *Superconductivity and Quantum Fluids* (Pergamon, Oxford and Polish Scientific Publ., Warsaw, 1970).

31. **Nambu, Y.**, *Phys. Rev.* **117**, 648 (1960).

32. **Yosida, K.**, *Phys. Rev.* **111**, 1255 (1958). See also Ref. 6, Section V. 4.

33. **Gradshteyn, I.S.** and **Ryzhik, I.M.**, *Table of Integrals, Series, and Products*, 4th ed. by Geronimus, Yu.V. and Tseytlin, M.Yu; transl. and edited by Jeffrey, A (Academic, New York, 1965).

34. **Anderson, P.W.**, *Phys. Rev.* **112**, 1900 (1958).

35. **Fetter, A.L.** and **Walecka, J.D.**, *Quantum Theory of Many-Particle Systems* (McGraw-Hill, New York, 1971).

36. **Mühlschlegel, B.**, *Z. Phys.* **155**, 313 (1959).

37. **Gor'kov, L.P.**, *Soviet Phys.-JETP* **36**, 1364 (1959).

38. **Abrikosov, A.A., Gor'kov, L.P.** and **Dzyaloshinskii, I.Ye**, *Quantum Field Theoretical Methods in Statistical Physics*, 2nd ed. (Pergamon, Oxford, 1965).

39. **Schafroth, M.R.**, *Helv. Phys. Acta* **24**, 645 (1951).

40. **Gorter, C.J.** and **Casimir, H.B.G.**, *Z. Phys.* **35**, 963 (1934). See also **Enz, C.P.**, *Rev. Mod. Phys.* **46**, 705 (1974).

41. **London, F.** and **London, H.**, *Physica* **2**, 341 (1935); London, F., *Superfluids, Volume 1. Macroscopic Theory of Superconductivity* (Dover, New York, 1961).

42. **Pippard, A.B.**, *Proc. Roy. Soc.* **A216**, 547 (1953); **Faber, T.E.** and **Pippard, A.B.**, *Proc. Roy. Soc.* **A231**, 336 (1955).

43. **Giaever, I.**, *Phys. Rev. Lett.* **5**, 147 (1960); **Giaever, I., Hart, H.R. Jr.** and **Megerle, K.**, *Phys. Rev.* **126**, 941 (1962).

44. **Clogston, A.M.**, *Phys. Rev. Lett.* **9**, 266 (1962); **Chandrasekhar, B.S.**, *Appl. Phys. Lett.* **1**, 7 (1962).

45. **Tinkham, M.**, *Introduction to Superconductivity* (McGraw-Hill, New York, 1975).

46. **Ginzburg, V.L.** and **Landau, L.D.**, *Zh. Eksp. Teor. Fiz.* **20**, 1064 (1950).

47. **Gor'kov, L.P.**, *Soviet Phys.-JETP* **34**, 505 (1958).

48. **Werthamer, N.R.**, *Phys. Rev.* **132**, 663 (1963). See also **Tewordt, L.**, *Phys. Rev.* **132**, 595 (1963).

49. **Itzykson, C.** and **Zuber, J.-B.**, *Quantum Field Theory* (McGraw-Hill, New York, 1980), Sections 7.1 and 8.4.

50. **Tata, J.**, **Cabrera, B.**, **Felch, S.B.** and **Anderson, J.T.**, *Phys. Rev. Lett.* **62**, 845 (1989).
51. **Landau, L.D.** and **Lifshitz, E.M.**, *Statistical Physics* (Pergamon, Oxford, 1958), Section 137.
52. **de Gennes, P.G.**, *Superconductivity of Metals and Alloys*, transl. Pincus, P.A. (Benjamin, New York, 1966).
53. **Bulaevskii, L.N.**, **Ginzburg, V.L.** and **Sobyanin, A.A.**, *Physica* **C152**, 378 (1988); **C156**, 652 (1988).
54. **Abrikosov, A.A.**, *Soviet Phys.-JETP* **5**, 1174 (1957).
55. **Ginzburg, V.L.**, *Soviet Phys.-Solid State* **2**, 1824 (1961).
56. **Lobb, C.J.**, *Phys. Rev.* **B36**, 3930 (1987).
57. **Josephson, B.D.**, *Phys. Lett.* **1**, 251 (1962); *Adv. Phys.* **14**, 419 (1965); *Rev. Mod. Phys.* **46**, 251 (1974).
58. **Clarke, J.** and **Schön, G.**, *Europhys. News* **17**, 94 (1986).
59. **Landau, L.D.**, *J. Phys. (USSR)* **5**, 71 (1941); *Zh. Eksp. Teor. Fiz.* **11**, 592 (1941); *J. Phys. (USSR)* **11**, 91 (1947).
60. **Vonsovsky, S.V.**, **Izyumov, Yu.A.** and **Kurmaev, E.Z.**, *Superconductivity of Transition Metals*, Solid-State Sciences 27 (Springer, Berlin, 1982).
61. **Migdal, A.B.**, *Soviet Phys.-JETP* **7**, 996 (1958).
62. **Scalapino, D.J.**, "The electron-phonon interaction and strong-coupling superconductors", in *Superconductivity*, ed. Parks, R.D. (Marcel-Dekker, New York, 1969), Vol. 1, p. 449.
63. **Eliashberg, G.M.**, *Soviet Phys.-JETP* **11**, 696 (1960); **12**, 1000 (1961).
64. **Crisan, M.**, *Theory of Superconductivity* (World Scientific, Singapore, 1989).
65. **Bogoljubov, N.N.**, **Tolmachev, V.V.** and **Shirkov, D.V.**, *A New Method in the Theory of Superconductivity* (Consultants Bureau, New York, 1959). See also **Morel, P.** and **Anderson, P.W.**, *Phys. Rev.* **125**, 1263 (1962).
66. **McMillan, W.L.**, *Phys. Rev.* **167**, 331 (1968).
67. **Daemen, L.L.** and **Overhauser, A.W.**, *Phys. Rev.* **B38**, 81 (1988).
68. **Bergmann, G.** and **Rainer, D.**, *Z. Phys.* **263**, 59 (1973).
69. **Micnas, R.**, **Ranninger, J.** and **Robaszkiewicz, S.**, *Rev. Mod. Phys.* **62**, 113 (1990).
70. **Enz, C.P.**, *Helv. Phys. Acta* **62**, 122 (1989); see also **Enz, C.P.** and **Galasiewicz, Z.M.**, *Solid State Comm.* **66**, 49 (1988); **Khomskii, D.I.** and **Zvezdin, A.K.**, *Solid State Comm.* **66**, 651 (1988).
71. **Tešanović, Z.** and **Rasolt, M.**, *Phys. Rev.* **B39**, 2718 (1989).
72. **Babichenko, V.S.**, *JETP Lett.* **27**, 531 (1978); **van Dijk, L.G.J.** and **Vertogen, G.**, *Phys. Lett.* **115**, 63 (1986).
73. **Zimanyi, G.T.**, **Kivelson, S.A.** and **Luther, A.**, *Phys. Rev. Lett.* **60**, 2089 (1988).

Chapter 5

MAGNETISM

In contrast to superconductivity, magnetism is a phenomenon occurring in all forms of condensed matter when probed by a magnetic field H and has many diverse manifestations: dia-, para-, ferro-, antiferro-, ferri-, spiral, etc. magnetism. It is therefore not surprising that its most spectacular form, namely the ferromagnetism of magnetite $FeO-Fe_2O_3$, has a long and distinguished history going back to the ancient Greeks (see Chapter 1 of Refs. 1 or 2).

Of course, the symmetrical phenomena of *polarization* in an electric field have an equally distinguished history; indeed, the Greek word for amber is "electron". But in sharp contast to ferromagnets, the first ferroelectric, namely Rochelle salt, was discovered only in the last century[3]. This may be taken as evidence for the fact that the corresponding electric effects have not acquired the importance of *magnetization* phenomena and will serve as a justification for not treating them in this book (for an introduction see Ref. 4, Chapter 13).

It is characteristic of the difficulty of the theory of magnetism that the most common of magnetic materials, namely iron, is still not completely understood (see, e.g. Ref. 5). The reason is that, in distinction to superconductivity, mean- or molecular-field theory is not a satisfactory approximation in the case of ferromagnetism, although the idea of a *molecular field* was first introduced by Pierre Weiss[6] in relation with his hypothesis of ferromagnetic domains[7]. The existence of domains with random orientation of their magnetic moment explained in a simple manner the absence of overall magnetization of a ferromagnet even below its Curie temperature T_C and allowed Weiss to explain *hysteresis* which is a nonequilibrium property common to all ferromagnets.

The difficulty with metallic or *itinerant-electron ferromagnetism* (IEF) as it occurs in iron, cobalt and nickel, may be traced to the nonlocal distribution of the atomic moments . This behavior is in sharp contrast to that of insulating ferromagnets, particularly oxides which have *local moments* whose value μ_s, determined from the saturation magnetization, essentially is an integer number of the Bohr magneton $\mu_B \equiv e\hbar/2mc$ per formula unit, except for corrections due to spin-orbit coupling (see, e.g. Table 2 in Chapter 15 of Ref. 4).

The fact that atoms may possess a magnetic moment is the result of *Hund's rules* which determine the spin, orbital and total angular momenta of the atom, S, L and J, respectively (see, e.g. Chapter 14 of Ref. 4) and hence the *spectroscopic term* $^{2S+1}L_J$. Rule 1 essentially states that the filling of the atomic nl-shell is governed, first, by the exclusion principle which requires the magnetic and spin quantum numbers of any two electrons to differ, $(m_1 s_1^z) \neq (m_2 s_2^z)$ and, second, by the Coulomb repulsion which favors $s_1^z = s_2^z$ so that $S = \max \sum s_i^z$. Rule 2 states that, in addition to Rule 1, $L = \max m_i$ and Rule 3 says that $J = |L - S|$ or $L + S$ if the number of electrons is $<$ or $> 2l + 1$.

Treating the atomic moments as classical local vectors, Langevin concluded in 1905 on thermodynamic grounds that the magnetization M must be an odd function $L(x) = Cx + \mathcal{O}(x^3)$ of $x = H/T$[8,1,2]. At the same time he also had derived an expression for the atomic diamagnetism (see Chapter 14 of Ref. 4). Langevin's formula $M = L(H/T)$ immediately explained the Curie law $\chi_C \equiv (M/H)_{H \to 0} = C/T$, valid in the paramagnetic state $T > T_C$ which Pierre Curie had discovered in 1895[9,1,2]. Adding a molecular field λM to H in Langevin's formula, Weiss obtained not only ferromagnetism but also the generalized *Curie-Weiss law* $\chi_{CW} = C/(T - T_C)$ [6,1,2].

It was Lenz who in 1920 introduced directional quantization of angular momentum[10] which in 1929 was used by Brillouin to generalize Langevin's formula[11]. In 1925 Lenz suggested to his student Ising a model of a linear chain of projected electron spins $S^z = \sigma^z/2$, σ^z being the Pauli matrix of Eq. (2.8), with nearest-neighbor coupling defined by the *Ising Hamiltonian* $\mathcal{H}_I \equiv -J \sum' S_i^z S_j^z$ [12,1,2] (the prime on the sum stands for nearest-neighbor summation). This model was thought to yield a more realistic description of ferromagnetism than the molecular field. Pauli in 1927 was the first to apply Fermi-Dirac statistics to solid state physics deriving his celebrated formula for the *Pauli spin susceptibility* $\chi_P = 2\mu_B^2 N(\mu)$ [13] where $N(\mu)$ is the density of states per spin at the Fermi level.

Although the Ising interaction was a decisive step forward from the molecular field, the physical origin of the coupling constant J remained a mystery. Weiss himself had already been aware of the fact that his molecular field could not be due to magnetic interactions (typically, the dipolar spin-spin interaction is 10^{-4} times smaller than the exchange energy, see Ref. 2, Table 6.3). But it was only Heisenberg who realized in 1928 that the *exchange interaction* he (and independently Dirac) had introduced two years earlier to explain the singlet-triplet splitting in the helium spectrum was the mechanism responsible for ferromagnetism. Heisenberg's derivation was rather complicated[14]; in fact, he never used the compact notation defined by $\mathcal{H}_H \equiv -J\sum' \mathbf{S}_i \cdot \mathbf{S}_j$, $2\mathbf{S}_i$ being the vector of Pauli matrices, which today carries the name *Heisenberg Hamiltonian*. Introduced by Dirac[15], it was van Vleck who was the first to use consistently $\mathcal{H}_H$, calling it first the "Dirac vector model" and later used the name "Heisenberg model"[16]. The form of $\mathcal{H}_H$ establishes the origin of the coupling constant J as the exchange energy.

Solving the 2-dimensional Ising model defined by the above Hamiltonian $\mathcal{H}_I$, Onsager showed in 1944 for the first time that it was possible to describe a second-order phase transition with a finite critical temperature, $T_c > 0$, exactly in the framework of statistical mechanics[17]. His published paper (Ref. 17) contains the solution in the absence of a magnetic field, $H = 0$, which is characterized by a logarithmically divergent specific heat. But Onsager had also indicated the result for the magnetization as function of temperature in a discussion remark in 1942 which remained a puzzle until Yang[18] succeeded in rederiving it 10 years later (see Ref. 1, p. 256, also Ref. 19, p. 91). This result shows that the magnetization plays the role of the *order parameter*, vanishing at T_c as $M \propto (T_c - T)^\beta$ with a *critical exponent* $\beta = 1/8$ (see, e.g. Section 3.10 of Ref. 19). Since mean-field theory gives $\beta = 1/2$, the Onsager-Yang solution shows that the 2-dimensional Ising model is a highly non-trivial example of a second-order or continuous phase transition .

This established the celebrity of the Ising model of which various generalizations involving different observables (Potts model, spherical model, Heisenberg model, plane rotator model, XY model, etc.) on various lattices and more complicated coupling functions (external field, antiferromagnet, eight-vertex model, alloy, lattice gas, spin glass, neural network, etc.) were studied[19,20]. This activity gave rise to a whole new field of statistical mechanics in which a number of highly sophisticated methods (transfer matrix, Bethe ansatz, renormalization group, etc.) were developed[19,20], a particularly important result being the *Mermin-Wagner theorem* which states that 2-dimensional systems with continuous symmetry such as the Heisenberg and the XY models have no long-range order at non-zero temperature[21,19]. However, since in the present text emphasis is on many-body theory, this statistical-mechanics aspect of magnetism will not be systematically pursued here.

From the many-body point of view as well as physically, the problem of non-localized atomic moments giving rise to itinerant-electron magnetism is of fundamental importance. The first magnetic response calculated for itinerant electrons was Landau's diamagnetic susceptibility of free electrons $\chi_L = -\chi_P/3$ [22] which shows that the electron gas is paramagnetic .

Ferromagnetism of itinerant electrons was again first obtained in mean-field approximation. This was done by Stoner in 1939[23] by adding a molecular field λM to H in the spin-shifted chemical potentials $\mu_{\pm} = \mu \pm \mu_B H$ that Pauli had introduced[13]. The result was a Pauli paramagnetism increased by the *Stoner enhancement factor* $(1 - \lambda\chi_P)^{-1}$ which, for sufficiently strong molecular field, gives rise to the ferromagnetic instability expressed by the *Stoner criterion* $1 - \lambda\chi_P = 0$.

Although very tempting, Stoner's theory has several defects, the most important being the impossibility to reproduce the Curie-Weiss law . This shows that, contrary to superconductivity, mean-field theory is not a sufficiently accurate approximation for itinerant-electron ferromagnetism. In other words, order parameter, i.e. magnetization, fluctuations are important and many-body effects cannot be ignored. The question, however, is: what is the interaction giving rise to these many-body effects? The answer lies in Hund's first rule which states that Coulomb repulsion favors electrons with parallel spins. Indeed, this statement may be expressed by an on-site repulsion between electrons with antiparallel spins, $\mathcal{H}_U \equiv U \sum n_{i\uparrow} n_{i\downarrow}$; $U > 0$. Adding this repulsion to the free band Hamiltonian conventionally written in Wannier representation, in simplified form $\mathcal{H}_o \equiv t \sum' c_{i\sigma}^+ c_{j\sigma}$ where t is the hopping energy between nearest-neighbor sites i and j, one arrives at the *Hubbard Hamiltonian* $\mathcal{H}_o + \mathcal{H}_U$ [24].

The Hubbard model has acquired for itinerant-electron ferromagnetism an importance comparable to that of the Ising model in the local-moment case. However, relevant rigorous solutions are scarce. Lieb recently proved that in a bipartite lattice in any dimension defined such that nearest neighbors are on different sublattices A and B containing L_A and L_B lattice sites, respectively, and $N = L_A + L_B$ electrons, corresponding to half-filling, there is a unique groundstate with total spin $|L_B - L_A|/2$ [25]. This gives indeed ferromagnetism for non-identical sublattices such that $L_A \neq L_B$ (e.g., the planar CuO_2-lattice of the cuprate superconductors), but $S = 0$ in the usual case of identical sublattices. Another limiting case was solved by Nagaoka who showed that for certain lattices in dimension $D \geq 2$ and in the limit $U \to \infty$, the groundstate in the presence of one hole, $N = L_A + L_B - 1$ has total spin $(L_A + L_B - 1)/2$ [26,2].

Numerous methods have been devised to treat more realistic systems like Fe, Co or Ni (see the reviews of Refs. 27, 28, 29). The most natural procedure is to generalize the Stoner formalism by including the many-body effects selfconsistently in the form of $\mathcal{H}_U$. It turns out that such an approach is particularly well suited for *weak itinerant-electron ferromagnets* (IEF's) such as $ZrZn_2$ and Sc_3In which are characterized by small atomic *saturation moments*, $\mu_s \ll \mu_B$ and low Curie temperatures T_C as compared to the *strong IEF's* like Fe, Co, Ni for which $\mu_s \geq \mu_B$. This distinction between weak and strong IEF's is well represented in the Rhodes-Wohlfarth plot [30,28] which shows that the strong IEF's are intermediate between local-moment ferromagnets and weak IEF's, and hence are particularly difficult to handle [28].

Localization of the electrons is favored by a narrow conduction band which means small overlaps between the nearest-neighbor Wannier functions. Since a hydrostatic pressure increases these overlaps, one arrives at the important conclusion that, in distinction to superconductivity, pressure reduces band ferromagnetism. But as a generalized Stoner theory is best suited for weak IEF's, this pressure dependence is expected to be particularly pronounced in this case. Indeed, $ZrZn_2$ ceases to be ferromagnetic at a pressure of the order of 10–20 kbar [31].

A generalized Stoner criterion also signals the onset of *itinerant-electron antiferromagnetism* (IEA) . In this case the Pauli susceptibility is replaced by a sublattice susceptibility $\chi_{sl}(\mathbf{Q})$ taken at the wavevector $\mathbf{Q}$ of the underlying *spin density wave* (SDW) which is responsible for the spin polarization. This mechanism was proposed in 1962 by Overhauser to explain the groundstate of chromium [32] which since has become the model case of an itinerant electron antiferromagnet [27].

29. Magnetism of Local Moments. Weiss' Molecular Field Theory

For an atom characterized by a spectroscopic term $^{2S+1}L_J$, directional quantization[10] in a magnetic field H gives rise to a Zeeman energy

$$E_m = -g\mu_B m H ; \quad m = -J, \cdots, +J \tag{29.1}$$

where

$$\mu_B = \frac{e\hbar}{2mc} \tag{29.2}$$

is the Bohr magneton and

$$g = 1 + \frac{J(J+1) + S(S+1) - L(L+1)}{2J(J+1)} \tag{29.3}$$

the spectroscopic or Landé factor. If the atoms form an insulator, the phonons will act as a thermal bath so that the Zeeman levels are canonically distributed with a partition function

$$Z_J(x) = \sum_{m=-J}^{+J} e^{-\beta E_m} = \frac{\sinh(1 + \frac{1}{2J}x)}{\sinh\frac{x}{2J}} = e^{-\beta g_m(x)} \tag{29.4}$$

where

$$x \equiv \beta g\mu_B J H \tag{29.5}$$

and $g_m(x)$ is the *magnetic Gibbs potential* per atom.

The magnetization is then given by

$$M(x) = -n_a \left(\frac{\partial g_m(x)}{\partial H} \right)_T = M_s \frac{Z_J'(x)}{Z_J(x)} \equiv M_s B_J(x) \tag{29.6}$$

where $n_a = L/V$ is the atomic density (L is the number of unit cells, see Section 1),

$$M_s \equiv n_a g\mu_B J \tag{29.7}$$

the *saturation magnetization*, $\mu_s \equiv M_s/n_a$ the *saturation moment* per unit cell and

$$B_J(x) = \left(1 + \frac{1}{2J}\right) \coth\left(\left(1 + \frac{1}{2J}\right)x\right) - \frac{1}{2J} \coth\frac{x}{2J} = B_J(-x) \tag{29.8}$$

the *Brillouin function*[11,19]. The classical limit is obtained by letting the direction of the angular momentum vary continuously which means that $J \to \infty$. In this limit

$$L(x) = B_\infty(x) = \coth x - \frac{1}{x} \tag{29.9}$$

is the *Langevin function*[8] which also applies in the paraelectric case[33].

The Brillouin function determines the whole thermodynamics of the paramagnetic state. From (29.6) the susceptibility is found to follow a *Curie law*[9]

$$\chi \equiv \left(\frac{\partial M}{\partial H}\right)_{H=0} = \beta g \mu_B J M_s B_J'(0) = \frac{C}{T} \tag{29.10}$$

where, using $B_J'(0) = (J+1)/3J$ (Problem 5.1), the Curie constant is found to be

$$C \equiv \frac{n_a g^2 \mu_B^2 J(J+1)}{3 k_B} = \frac{J+1}{J} \frac{n_a}{3 k_B \mu_s^2} . \tag{29.11}$$

The magnetic entropy density s_m follows from the Maxwell relation

$$\left(\frac{\partial s_m}{\partial H}\right)_T = \left(\frac{\partial M}{\partial T}\right)_H = -M_s \frac{x}{T} B_J'(x) \tag{29.12}$$

by integration,

$$-\Delta s_m \equiv s_m(T, 0) - s_m(T, H) = n_a k_B \int_0^x B_J'(y) y \, dy > 0 \tag{29.13}$$

which shows that a magnetic field increases the order. From this expression one deduces the specific heat at constant field

$$\Delta c_H(T) = T \left(\frac{\partial \Delta s_m}{\partial T}\right)_H = -x \frac{d \Delta s_m}{dx} = n_a k_B x^2 B_J'(x) . \tag{29.14}$$

From an experimental determination of $\Delta c_H(T)$ one may deduce the amount ΔS of entropy measuring the frozen-in paramagnetic disorder at constant field (Problem 5.2),

$$\Delta S = V \int_0^\infty \frac{\Delta c_H(T)}{T} dT = L k_B \ln(2J+1) \tag{29.15}$$

which, in principle, yields the value J of the atomic moment. It is this entropy which is reduced by a magnetic field in adiabatic demagnetization (see Ref. 4, p. 446).

Weiss' explanation of ferromagnetism[6] is that a permanent magnetization M contributes a local *molecular field*[19] λM ($\lambda > 0$ is a *phenomenological coupling constant*), which adds to the external field H so that the effective field to be used in Eqs. (29.5), (29.6) is

$$\bar{H} = H + \lambda M . \tag{29.16}$$

Equation (29.6) then becomes nonlinear in M,

$$m \equiv \frac{M}{M_s} = B_J(x + \theta m) \tag{29.17}$$

where

$$\theta \equiv \beta g \mu_B J \lambda M_s \ . \tag{29.18}$$

In the absence of an external field, Eq. (29.17) becomes (Fig. 5.1)

$$\frac{z}{\theta} = m_o \equiv m(H = 0) = B_J(z) \le z B'_J(0) \tag{29.19}$$

where $M_o \equiv M_s m_o$ is the *permanent magnetization*. For $m_o > 0$, Eq. (29.19) implies $T < T_C$ where, using (29.10),

$$T_C = g \mu_B J \lambda M_s B'_J(0)/k_B = \lambda C \tag{29.20}$$

is the *Curie temperature*.

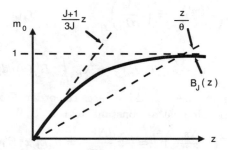

Fig. 5.1: Molecular-field construction of the permanent magnetization $M_o = M_s m_o$ for fixed temperature T, saturation magnetization M_s and phenomenological coupling constant λ according to Eqs. (29.18) and (29.19).

Introducing the dimensionless variables m and

$$t \equiv 1 - \frac{T}{T_C} \ ; \quad h = \frac{H}{\lambda M_s} \ , \tag{29.21}$$

so that $x = \theta h$ and $\theta = 3J/(J+1)(1-t)$, the equation of state (29.17) takes the form

$$m = B_J(\theta(h+m)) \tag{29.22}$$

or, inverting the Brillouin function (Problem 5.1),

$$h = \frac{1}{\theta} B_J^{-1}(m) - m = m \left\{ -t + (1-t)\frac{m^2}{a_J^2}\left(1 + \mathcal{O}(m^2)\right)\right\} \tag{29.23}$$

where

$$a_J \equiv (J+1)\left(\frac{10}{3(2J^2 + 2J + 1)}\right)^{1/2} \ . \tag{29.24}$$

Neglecting the term $\mathcal{O}(m^2)$, Eq. (29.23) yields, for constant t, a linear relation between m^2 and h/m which is called an *Arrott plot*. Its validity is limited to $m \ll 1$, i.e. to the vicinity of T_C where $|t| \ll 1$. For $h = 0$ the solution is

$$\frac{M_o}{M_s} = \begin{cases} 0; & T > T_C \\ a_J t^{1/2}; & T < T_C \, . \end{cases} \tag{29.25}$$

Integrating (29.23) one also determines the *magnetic free energy* F_m. Using the thermodynamic relation

$$H = \frac{1}{V}\left(\frac{\partial F_m}{\partial M}\right)_T \tag{29.26}$$

and Eqs. (29.21) one finds

$$F_m = \frac{\lambda V}{2}\left\{ -tM^2 + \frac{M^4}{2a_J^2 M_s^2} \right\}; \quad |t| \ll 1 \, . \tag{29.27}$$

Since the definition (29.21) of t coincides with that in (24.12), Eqs. (29.25) and (29.27) have the Ginzburg-Landau form of Eqs. (26.38) and (26.39), respectively, which justifies the terminology "mean-field approximation" used there. Hence M_o plays the role of *order parameter* and the transition in zero field is of second order .

We may also calculate the susceptibility and the magnetic specific heat near T_C. Deriving Eq. (29.23) with respect to h using (29.21) one finds

$$1 = \left(-t + \frac{3m_o^2}{a_J^2} \right)\lambda\chi \tag{29.28}$$

and with (29.20), (29.25)

$$\chi = \begin{cases} \chi_{CW}; & T > T_C \\ -2\chi_{CW}; & T < T_C \end{cases} \tag{29.29}$$

where

$$\chi_{CW} = \frac{C}{T - T_C} \tag{29.30}$$

is the *Curie-Weiss susceptibility*[6].

The magnetic specific heat c_m is obtained from Eq. (29.27) by first calculating the magnetic entropy density

$$s_m = -\frac{1}{V}\left(\frac{\partial F_m}{\partial T}\right)_{H=0} = -\frac{\lambda M_s^2}{2}\left\{ \frac{m_o^2}{T_C} + \left(-t + \frac{m_o^2}{2a_J^2} \right)\frac{\partial m_o^2}{\partial T} \right\} . \tag{29.31}$$

Inserting (29.25) and using (29.20) one finds

$$s_m = -\frac{3M_s^2 a_J^2}{4C}t; \quad t \ll 1 \tag{29.32}$$

and the magnetic specific heat

$$c_m = T\left(\frac{\partial s_m}{\partial T}\right)_{H=0} = \frac{3M_s^2 a_J^2}{4C}\frac{T}{T_C} \; ; \quad t \ll 1 \; . \tag{29.33}$$

Near $T = 0$, $m_o \sim 1$, $\theta \sim \infty$ and $z \sim 0$ so that the asymptotic expansion determined in Problem 5.2 has to be used to calculate m_o, χ and c_m from Eqs. (29.19), (29.22) and (29.23), respectively. The result is (Problem 5.3)

$$m_o = 1 - \frac{1}{J}\exp\left(-\frac{3T_C}{(J+1)T}\right) \; ; \quad T \ll T_C \tag{29.34}$$

(see Eq. (10), p. 465 of Ref. 4),

$$\chi = \frac{3C}{J(J+1)T}\exp\left(-\frac{3T_C}{(J+1)T}\right) \; ; \quad T \ll T_C \tag{29.35}$$

and

$$c_m = \frac{3M_s^2}{J(J+1)C}\left(\frac{T_C}{T}\right)^2\exp\left(-\frac{3T_C}{(J+1)T}\right) \; ; \quad T \ll T_C \; . \tag{29.36}$$

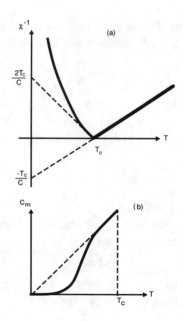

Fig. 5.2: (a) Inverse susceptiblity χ^{-1} according to Eqs. (29.29), (29.30), (29.35). (b) Magnetic specific heat according to Eqs. (29.33), (29.36).

χ and c_m are depicted in Fig. 5.2(a) and (b), respectively. The exponential decrease in these expressions does not agree with experiments (see Ref. 4, p. 465). In particular, m_o decreases much more rapidly, $m_o = 1 - (T/T_C)^{3/2}$ (see Eq. (30.31) below). In fact, the molecular-field construction of Eq. (29.19) and Fig. 5.1 is not very convincing for $T \ll T_C$.

The field dependence of the magnetization near T_C is seen from Eq. (29.23) to be represented by the cubic curve of Fig. 5.3 whose slope at the origin is given by a negative value of the Curie-Weiss susceptibility (29.30) corresponding to the dashed line of Fig. 5.2(a). This, of course, is an unstable situation and the light part of the cubic curve in Fig. 5.3 has to be replaced by the vertical tangents resulting in a *hysteresis loop*. This construction of Weiss[6] is analogous to that of van der Waals in the liquid-gas transition. Note that *saturation* at $T = 0$ is lost if the term $\mathcal{O}(m^2)$ in Eq. (29.23) is neglected (compare Fig. 5.6 below).

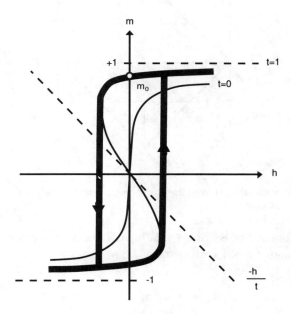

Fig. 5.3: Ferromagnetic equation of state (29.22) resulting in a negative slope $\lambda \chi_{CW}$ at the origin. This instability is corrected by Weiss' construction of the hysteresis loop. At $T = 0$ the permanent magnetization $M_o = m_o M_s$ shows saturation, $m_o = 1$.

In the case of a *local-moment antiferromagnet* the simplest situation is a simple cubic lattice in dimension $D = 2$ or 3, in which the moments in all nearest-neighbor points B of a point A have opposite direction to the moment in A so that the resulting moment in the *antiferromagnetic unit cell* vanishes,

$$\mathbf{m}_{\text{cell}} = \mathbf{m}_A + \mathbf{m}_B = 0 \ . \tag{29.37}$$

A non-zero value of $\mathbf{m}_{\text{cell}}$ leads to *ferrimagnetism*. Since nearest-neighbor points thus become inequivalent , the simple cubic lattice becomes face-centered and the volume of the unit cell doubles . The af unit cell is again a square in $D = 2$, as shown in Fig. 5.4. In $D = 3$ it is the dodecahedron shown in Fig. 14b on p. 225 of Ref. 4.

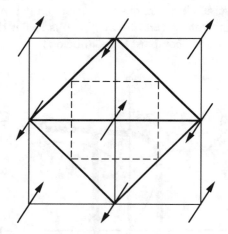

Fig. 5.4: Antiferromagnetic (heavy lines) and nonmagnetic (dashed lines) unit cells in $D = 2$ and arrangement of moments.

In the notation of Section 1 this bipartite lattice is described in terms of the non-magnetic cell vectors $\mathbf{a}_i$ by

$$\mathbf{R}_I = \sum_{i=1}^{D} n_i^I \mathbf{a}_i \ ; \quad I = A, B \tag{29.38}$$

with integers n_i^I such that

$$\sum_{i=1}^{D} n_i^A = \text{even} \ ; \quad \sum_{i=1}^{D} n_i^B = \text{odd} \ . \tag{29.39}$$

We now introduce the vector

$$\mathbf{Q} \equiv \frac{1}{2} \sum_{j=1}^{D} \mathbf{b}_j \qquad (29.40)$$

which in the nonmagnetic (cubic) Brillouin zone defines the R-corner and in the af Brillouin zone the boundary point L shown in Fig. 14a on p. 225 of Ref. 4 (see Ref. 34). From (29.38)–(29.40) then follows, using the orthogonality (1.4), that

$$e^{i\mathbf{Q}\cdot\mathbf{R}_I} = \begin{cases} +1 \; ; & I = A \\ -1 \; ; & I = B \end{cases} . \qquad (29.41)$$

According to (1.11), the Fourier transform of the moment distribution $\mathbf{m}^I = \mathbf{m}(\mathbf{R}^I)$ then simply is

$$\vec{\mu}(\mathbf{k}) = \mathbf{m}_A \delta_{\mathbf{k},\mathbf{Q}} . \qquad (29.42)$$

In $D = 2$ an important example of the cubic bipartite case of local-moment antiferromagnetism is found in the CuO_2-planes of the cuprate superconductors at zero doping[35] (see Section 36). More complicated structures may be described by different bipartite (or even multipartite) lattices with symmetry points on the border of the Brillouin zone, different from the L-point (29.40) (see, e.g. Ref. 36).

In a phenomenological description of the bipartite af case (see, e.g. Ref. 37) one introduces the *sublattice magnetizations* $\mathbf{M}_A$, $\mathbf{M}_B$ such that

$$\mathbf{M} = \mathbf{M}_A + \mathbf{M}_B \qquad (29.43)$$

is the *bulk magnetization* and

$$\mathbf{N} = \mathbf{M}_A - \mathbf{M}_B \qquad (29.44)$$

the *staggered magnetization*. Since in zero field the two sublattices are equivalent except for reflexions along symmetry or moment directions, $|\mathbf{M}_A| = |\mathbf{M}_B|$, and one has the orthogonality relation[37]

$$\mathbf{M} \cdot \mathbf{N} = \mathbf{M}_A^2 - \mathbf{M}_B^2 = 0 . \qquad (29.45)$$

The application of the molecular-field method to this case is due to Néel[38]. He assumed that the molecular field is entirely due to the nearest-neighbor moments and that the vector character of $\mathbf{M}_A$ and $\mathbf{M}_B$ may be neglected. With these assumptions Eq. (29.17) is then generalized to the pair of equations

$$m + p = 2m_A \equiv \frac{2M_A}{M_s} = B_J(x - 2\theta m_B)$$
$$m - p = 2m_B \equiv \frac{2M_B}{M_s} = B_J(x - 2\theta m_A) \ . \tag{29.46}$$

For small fields, $x \ll 1$, it follows from (29.45) that $m \ll 1$ so that the right-hand sides of (29.46) may be developed around $\pm\theta p$. Keeping only first-order terms we thus have

$$p = B_J(\theta p) \tag{29.47}$$

which is the analog of Eq. (29.17) and

$$m = (x - \theta m)B'_J(\theta p) \ . \tag{29.48}$$

Equation (29.47) determines the *Néel temperature* T_N in analogy to (29.20) such that above T_N, $p = 0$ or

$$\theta B'_J(0) = \frac{T_N}{T} \ . \tag{29.49}$$

Inserted into (29.48), using (29.20) and (29.21), this immediately leads to Néel's expression for the susceptibility ,

$$\chi = \frac{C}{T + T_N} \ ; \quad T > T_N \ . \tag{29.50}$$

Below T_N, Eq. (29.47) gives rise to a *permanent staggered magnetization* $p(t)$. Defining an auxiliary temperature[39)]

$$T^*(T) \equiv \frac{T_N}{\theta B'_J(\theta p)} \tag{29.51}$$

we obtain

$$\chi = \frac{C}{T^*(T) + T_N} \ ; \quad T < T_N \ . \tag{29.52}$$

In order to evaluate $B'_J(\theta p)$ we use, for $t \ll 1$, the result of Problem 5.1 together with (29.24) to obtain

$$B'_J(\theta p) \simeq B'_J(0)\left\{1 - \frac{1}{3}\left(\frac{J+1}{Ja_J}\theta p\right)^2\right\} \ ; \quad t \ll 1 \tag{29.53}$$

and, for $T \ll T_N$, the result of Problem 5.3 which gives

$$B'_J(\theta p) \simeq \frac{1}{J^2}e^{-\theta/J} \ ; \quad T \ll T_N \ . \tag{29.54}$$

Using (29.20), (29.25) and (29.49) one then arrives at the result

$$\chi^{-1} = C^{-1} \times \begin{cases} (5T_N - 3T) \; ; & t \ll 1 \\ [J(J+1)T/3] \exp[3T_N/(J+1)T] \; ; & T \ll T_N \end{cases} \quad (29.55)$$

which is depicted in Fig. 5.5. As in the ferromagnetic case of Eq. (29.35) the exponential vanishing of χ at $T = 0$ does not agree with observation. Here it is the result of having ignored the vector character of **M** and **N** which forces M to vanish too strongly at $H = 0$ (see the footnote on p. 116 of Ref. 39).

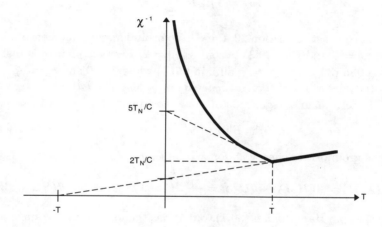

Fig. 5.5: Inverse antiferromagnetic susceptibility χ^{-1}, Eq. (29.55), according to Néel's assumptions.

30. Groundstate and Magnons of the Heisenberg Model

A quite general description of local-moment magnetic systems is given by the anisotropic spin Hamiltonian

$$\mathcal{H}_S(\lambda) = -J \sum_{\langle ij \rangle} \{2(1-\lambda)(S_i^+ S_j^- + S_i^- S_j^+) + (1+\lambda)S_i^z S_j^z\} \quad (30.1)$$

where i runs over the L equivalent lattice sites of the crystal with positions $\mathbf{R}_i$ and $\langle ij \rangle$ is the restriction of j to the z nearest neighbors of i; $\mathbf{S}_i = (S_i^x, S_i^y, S_i^z)$ is the spin vector operator with the properties

$$\mathbf{S}_i^2 = S(S+1) \; ; \quad \text{all } i$$

$$[S_i^z, S_j^\pm] = \pm S_i^\pm \delta_{ij} \; ; \quad [S_i^+, S_j^-] = \frac{1}{2} S_i^z \delta_{ij} \quad (30.2)$$

where, in analogy with Eqs. (2.8),

$$S_i^{\pm} \equiv \frac{1}{2}(S_i^x \pm iS_i^y) \ . \tag{30.3}$$

$\mathcal{H}_{1/2}(0) = \mathcal{H}_H$ describes the *Heisenberg model*, $\mathcal{H}_{1/2}(+1) = \mathcal{H}_I$ the *Ising model* and $\mathcal{H}_{1/2}(-1)$ with $J < 0$ the *XY-model*[1,2]. The groundstate $|0\rangle$ of $\mathcal{H}_S(\lambda)$ is characterized by $S_i^z = \pm S$ which, however, is only a double degeneracy. It is lifted by adding a Zeeman term (29.1) at each site,

$$\mathcal{H}_Z = -\mu_B H \sum_i g_i S_i^z \ . \tag{30.4}$$

Since, in contrast to Section 29, there is no orbital angular momentum here, $L = 0$, $J = S$ and Eq. (29.3) becomes $|g_i| = 2$. *Ferromagnetism* is realized for $J > 0$ in (30.1) and $g_i = 2$, all i, in (30.4) while $J < 0$ and $g_i = -g_j = 2$, $i, j = \langle ij \rangle$, describes *antiferromagnetism* on a *bipartite lattice*.

The ferromagnetic groundstate may now be characterized by

$$S_i^z|0\rangle = S|0\rangle \ ; \quad S_i^+|0\rangle = 0 \ ; \quad \text{all } i \tag{30.5}$$

and the groundstate energy is, for $\lambda \geq 0$ (see Eq. (30.12) below)

$$E_o(\lambda,\, H) = \langle 0|\mathcal{H}_S(\lambda) + \mathcal{H}_Z|0\rangle = -\frac{1}{2}J(1+\lambda)S^2 zL - g\mu_B HSL \ . \tag{30.6}$$

Since the spin Hamiltonian (30.1), (30.4) has the form of lattice sums analogous to the displacement Hamiltonian (1.23), it is therefore to be expected that the excitations are described by operators which have the same form as the phonon operators P_q, Q_q introduced in Eq. (1.24). These are the creation operators of *magnons* or *spin waves*

$$\sigma_{\mathbf{q}}^+ = \sqrt{\frac{2}{SL}} \sum_i e^{i\mathbf{q}\cdot\mathbf{R}_i} S_i^- \ ; \quad \mathbf{q} \in Z \ . \tag{30.7}$$

In order to have propagating modes we must choose periodic boundary conditions (1.7). Then the one-magnon states $\sigma_{\mathbf{q}}^+|0\rangle$ are orthonormal. Indeed, it follows from (30.2), (30.5) and (1.9) that

$$\langle 0|\sigma_{\mathbf{q}'}\sigma_{\mathbf{q}}^+|0\rangle = \delta_{\mathbf{q},\mathbf{q}'} \ . \tag{30.8}$$

These states diagonalize $\mathcal{H}_S(\lambda)$ and $\mathcal{H}_Z$, and one may calculate the one-magnon excitation energy $\omega_{\mathbf{q}}(\lambda) - \mu(H)$ from the eigenvalue equation

$$(\mathcal{H}_S(\lambda) + \mathcal{H}_Z)\sigma_{\mathbf{q}}^+|0\rangle = \{E_o(\lambda,\, H) + \omega_{\mathbf{q}}(\lambda) - \mu(H)\}\sigma_q^+|0\rangle \tag{30.9}$$

where the notation will become clear later (see Eq. (30.21) below). Using the relations

$$S_i^z S_l^- |0\rangle = (S - \delta_{il}) S_l^- |0\rangle$$
$$S_i^+ S_l^- |0\rangle = \frac{1}{2} S \delta_{il} |0\rangle$$
(30.10)

which follow from (30.2), (30.5) one finds in the ferromagnetic case (Problem 5.4)

$$\omega_{\mathbf{q}}(\lambda) = JS\{(1 + \lambda)z - (1 - \lambda)\sum_{\vec{\delta}} e^{i\mathbf{q}\cdot\vec{\delta}}\}$$
(30.11)
$$\mu(H) = -2\mu_B H$$

where $\vec{\delta} \equiv \mathbf{R}_j - \mathbf{R}_i$, $\langle ij \rangle$, are the vectors from any lattice site to its z nearest neighbors. The fact that the low-lying excitations of the Heisenberg Hamiltonian $\mathcal{H}_H$ are wavelike was first recognized by Bloch[40].

The definition of $\vec{\delta}$ just given implies that nearest-neighbor points are equivalent, i.e. the $\vec{\delta}$ are the primitive translations of the lattice. Hence $-\vec{\delta}$ also belongs to the set of these vectors which is not the case, e.g. for the honeycomb lattice. The inversion symmetry implied by this equivalence guarantees the reality of the magnon energy $\omega_{\mathbf{q}}(\lambda)$ since only the real parts of the exponentials contribute. In the long-wavelength limit, development of $\cos(\mathbf{q} \cdot \vec{\delta})$ in powers of q yields

$$\omega_{\mathbf{q}}(\lambda) = JS\{2\lambda z + \frac{1}{2}(1 - \lambda)q^2 \sum_{\vec{\delta}}(\hat{q} \cdot \vec{\delta})^2 + \mathcal{O}(q^4)\}$$
(30.12)

where $\hat{q} = \mathbf{q}/q$. This shows that $\omega_{\mathbf{q}}(\lambda)$ becomes negative for sufficiently small q-values if $\lambda < 0$, so that $|0\rangle$ is no longer the groundstate. This means that the description given above is limited to $\lambda \geq 0$. The antiferromagnetic case is even less favorable since inserting the bounds $-1 \leq \cos(\mathbf{q} \cdot \vec{\delta}) \leq 1$ into (30.11) leads to the inequalities

$$-2zS|J| \leq \omega_{\mathbf{q}}(\lambda) \leq -2\lambda zS|J| \; ; \quad J < 0, \lambda < 1$$
(30.13)

where in the simple cubic lattice in any dimension the lower bound is attained in the corner $\mathbf{q} = (1, \dots, 1)\pi/a$ of the Brillouin zone .

The analogy with phonons suggests that magnons may be treated like bosons. This is indeed what comes out in various representations (see Ref. 1, Chapter 3). The most appropriate among them for many purposes is the

Holstein-Primakov representation[41]

$$S_i^+ = \frac{1}{2}(2S - n_i)^{1/2} a_i$$

$$S_i^- = \frac{1}{2} a_i^+ (2S - n_i)^{1/2} \tag{30.14}$$

$$S_i^z = S - n_i \; ; \quad n_i = 0, 1, \ldots, 2S$$

where a_i and a_i^+ are bosonic operators satisfying

$$[a_i a_j^+] = \delta_{ij} \; ; \quad a_i^+ a_i = n_i \; . \tag{30.15}$$

But, obviously, the representation is not faithful because for $n_i > 2S$ it leads to unphysical values of S_i^z. Within the physical domain of the n_i, however, the square roots in $S_i^\pm$ are perfectly well defined. Indeed, it is possible to construct a polynomial of degree $2S$ in n_i such that the equality

$$\left(1 - \frac{n_i}{2S}\right)^{1/2} = 1 + \sum_{p=1}^{2S} \alpha_p n_i^p \tag{30.16}$$

holds for $n_i = 0, 1, \ldots, 2S$.

In order to verify that the representation (30.14) satisfies the algebra (30.2) we calculate the auxiliary quantities

$$S_i^+ S_i^- = \frac{S}{2}(n_i + 1)\left(1 - \frac{n_i}{2S}\right)$$

$$S_i^- S_i^+ = \frac{S}{2} n_i \left(1 - \frac{n_i - 1}{2S}\right) \tag{30.17}$$

which easily follow from (30.14), (30.15), as does $[S_i^z S_i^+] = S_i^+$. With (30.17) one immediately finds that $2[S_i^+ S_i^-] = S_i^z$ and $2(S_i^+ S_i^- + S_i^- S_i^+) + (S_i^z)^2 = S(S + 1)$. Inserting (30.14) into the two relations (30.5) one also concludes that

$$a_i|0\rangle = 0 \; ; \quad \text{all } i \tag{30.18}$$

or that the boson and spin groundstates coincide. Finally one verifies with (30.7) that the one-magnon states are indeed phonon-like,

$$\sigma_{\mathbf{q}}^+|0\rangle = b_{\mathbf{q}}^+|0\rangle \tag{30.19}$$

where

$$b_{\mathbf{q}}^+ \equiv \frac{1}{\sqrt{L}} \sum_i a_i^+ e^{i\mathbf{q} \cdot \mathbf{R}_i} \tag{30.20}$$

are the exact analogs of the phonon operators defined in (1.24), (1.31). Since because of (30.15), (30.18), Eqs. (30.10) are still true, the calculation of the one-magnon excitation energy is the same as in Problem 5.4 so that the expressions (30.11) for $\omega_{\mathbf{q}}(\lambda)$ and $\mu(H)$ also remain valid. In a lattice with more than one basis atom ($B > 1$, see Section 1) optical magnons also occur (see Fig. 1 of Ref. 42).

Making use of this analogy we may calculate the *magnon distribution function* exactly in the way Eq. (7.41) had been derived,

$$\bar{n}_{\mathbf{q}} \equiv \langle b_{\mathbf{q}}^+ b_{\mathbf{q}} \rangle_o = n_o(\omega_{\mathbf{q}} - \mu) \tag{30.21}$$

where n_o is the Bose distribution function (7.43) and $\omega_{\mathbf{q}} - \mu$ the one-magnon energy defined in (30.9) and given by (30.11). This shows that $\mu = -2\mu_B H \leq 0$ plays the role of *magnetic chemical potential* of the bosons. In terms of $\bar{n}_{\mathbf{q}}$ the magnetic entropy density introduced in (29.31) may be expressed as the Bose analog of Eq. (13.24), namely as (see Ref. 43, Section 54)

$$s_m = k_B V^{-1} \sum_{\mathbf{q} \in Z} \{ (\bar{n}_{\mathbf{q}} + 1) \ln(\bar{n}_{\mathbf{q}} + 1) - \bar{n}_{\mathbf{q}} \ln \bar{n}_{\mathbf{q}} \} . \tag{30.22}$$

Using the Maxwell relation (29.12), the magnetization may now be calculated according to

$$M(H, T) - M_s = \int_0^T \left(\frac{\partial s_m}{\partial H} \right)_{T'} dT' \tag{30.23}$$

where $M_s = M(H, 0)$ is the saturation magnetization which is independent of the field H (see Fig. 5.3). Here the right-hand side is easy to calculate since from definition (7.43) follow the relations

$$-\frac{1}{2\mu_B} \frac{\partial \bar{n}_{\mathbf{q}}}{\partial H} = \frac{T}{\omega_{\mathbf{q}} - \mu} \frac{\partial \bar{n}_{\mathbf{q}}}{\partial T} = \beta \bar{n}_{\mathbf{q}} (\bar{n}_{\mathbf{q}} + 1) ; \quad \ln \frac{\bar{n}_{\mathbf{q}} + 1}{\bar{n}_{\mathbf{q}}} = \beta(\omega_{\mathbf{q}} - \mu) . \tag{30.24}$$

Hence

$$\left(\frac{\partial s_m}{\partial H} \right)_T = -\frac{g\mu_B}{V} \sum_{\mathbf{q} \in Z} \frac{\partial \bar{n}_{\mathbf{q}}}{\partial T} \tag{30.25}$$

and since for $H > 0$, $\bar{n}_{\mathbf{q}} = 0$ at $T = 0$, one finds that the decrease of the magnetization from its saturation value (29.7) is due to thermal magnons,

$$M_s - M(H, T) = 2\mu_B V^{-1} \sum_{\mathbf{q} \in Z} \bar{n}_{\mathbf{q}} ; \quad H > 0 . \tag{30.26}$$

Since $n_o(\omega)$ diverges at $\omega = 0$, small q-values dominate the sum in (30.26) at low temperatures so that we may use (30.12). Concentrating on the gapless spectrum $\lambda = 0$ and assuming the lattice to be simple cubic with lattice constant a, we then have in D dimensions

$$\omega_q = JSa^2q^2 . \tag{30.27}$$

This has the form of an electron energy with $\hbar^2/2m = JSa^2$. We may therefore use Eqs. (14.13), (14.14) to obtain the density of states in dimension D,

$$N_D(\omega) = \frac{\Omega_D}{2(2\pi a)^D (JS)^{D/2}} \omega^{\frac{D}{2}-1} . \tag{30.28}$$

In $D = 3$, Eq. (30.26) then becomes

$$M_s - M(H, T) = 2\mu_B \int_0^\infty d\omega N_3(\omega) n_o(\omega - \mu)$$

$$= \frac{\mu_B}{2\pi^2 a^3} \left(\frac{k_B T}{JS}\right)^{3/2} \int_0^\infty \frac{x^{1/2} dx}{e^{x-y} - 1} \tag{30.29}$$

where $y \equiv \beta\mu \leq 0$. Since the permanent magnetization $M_o(T) \equiv M(\theta, T)$ vanishes at the Curie temperature T_C, the saturation magnetization is found to be

$$M_s = \frac{\mu_B}{2\pi^2 a^3} \left(\frac{k_B T_C}{JS}\right)^{3/2} I_{1/2} \tag{30.30}$$

and

$$\Delta M(T) \equiv M_s - M_o(T) = M_s \left(\frac{T}{T_C}\right)^{3/2} ; \quad T \ll T_C . \tag{30.31}$$

Here (see Eq. (24.9) for the definition of $\zeta(z)$)

$$I_\nu \equiv \int_o^\infty \frac{x^\nu dx}{e^x - 1} = \Gamma(\nu + 1)\zeta(\nu + 1) . \tag{30.32}$$

This is the famous $T^{3/2}$-law derived by Bloch in 1930[40] which agrees much better with experiment than Eq. (29.34). Numerically one finds[44] $I_{1/2} = \Gamma(\frac{3}{2})\zeta(\frac{3}{2}) \simeq 2.315$.

In dimensions $D = 1$ and 2 it follows from (30.28) that, for $y = 0$, the integrand diverges at $x = 0$ as $x^{-3/2}$ and x^{-1}, respectively, so that it becomes impossible to calculate $M_o(T)$ in this way. Note that this difficulty is independent of the fact that, according to the Mermin-Wagner theorem[21], $T_C = 0$ for $D \leq 2$. To see the underlying physical meaning we note that from (30.20) and (1.10) follows

$$\sum_{\mathbf{q} \in Z} \bar{n}_\mathbf{q} = \sum_{\mathbf{q} \in Z} \langle n_\mathbf{q} \rangle_o = \sum_{i=1}^L \langle n_i \rangle_o \tag{30.33}$$

where $n_{\mathbf{q}} \equiv b_{\mathbf{q}}^{+} b_{\mathbf{q}}$. For the second expression the divergence means that there is a Bose condensation into the $q = 0$ state while for the third expression it means that $\langle n_i \rangle = \infty$ for some sites i. But this last fact is incompatible with the restriction $n_i \leq 2S$ imposed on the Holstein-Primakov representation. This incompatibility may be traced to the fixing of the chemical potential , $\mu = 0$, or $H = 0$, which leaves the boson number unspecified . The last remarks point to a remedy which was proposed by Takahashi[45]: Impose the restriction $n_i \leq 2S$ in the average by the requirement

$$\langle n_i \rangle_o = \frac{1}{L} \sum_{\mathbf{q}} \bar{n}_{\mathbf{q}} = S \qquad (30.34)$$

which according to Eq. (30.14) simply means $\langle S_i^z \rangle = 0$, and relax the condition $\mu = 0$ in a selfconsistent way.

In order to implement this idea we first calculate the magnetic Gibbs potential per site introduced in Eq. (29.4) by integrating the entropy. This integration becomes trivial if we write the entropy density (30.22) in the form

$$s_m = k_B V^{-1} \sum_{\mathbf{q}} \left\{ \bar{n}_{\mathbf{q}} \ln \frac{\bar{n}_{\mathbf{q}} + 1}{\bar{n}_{\mathbf{q}}} + \ln(\bar{n}_{\mathbf{q}} + 1) \right\} \qquad (30.35)$$

and transform the first term in the bracket using relations (30.24); the result is

$$g_m(\mu, T) = -\frac{V}{L} \int_o^T dT' s_m(\mu, T') = -\frac{k_B T}{L} \sum_{\mathbf{q}} \ln(\bar{n}_{\mathbf{q}} + 1) . \qquad (30.36)$$

Taking the derivative with respect to $\mu = -2\mu_B H$ one then finds, using again the first relation (30.24) and combining with (30.34),

$$-\frac{\partial g_m}{\partial \mu} = \frac{1}{L} \sum_{\mathbf{q}} \bar{n}_{\mathbf{q}} = S . \qquad (30.37)$$

Thus, the requirement (30.34) appears as stationarity condition on the free energy per site[45],

$$f_m = g_m + S\mu = g_m - \frac{V}{L} M_s H . \qquad (30.38)$$

In order to relax the condition $\mu = 0$ one has to find another variable, in terms of which a second stationarity condition may be formulated. It turns out that there is a natural choice.

Indeed, the thermal decrease of magnetization given by (30.29) may be interpreted as being due to the blurring of the spin alignment which means that neighboring spins cease to be strictly parallel. In other words, the quantity

$$S' \equiv \frac{1}{z} \sum_{\langle i,j\rangle,\, i \text{ fix}} \left(\langle \mathbf{S}_i \cdot \mathbf{S}_j \rangle_o \right)^{1/2} \tag{30.39}$$

becomes smaller than S. We wish to calculate S' at low temperature and for $H > 0$ so that $\bar{n}_\mathbf{q} \ll 1$. This suggests that one may use the development

$$\left(1 - \frac{n_i}{2S} \right)^{1/2} = 1 - \frac{n_i}{4S} + \mathcal{O}(S^{-2}) \tag{30.40}$$

in the operators (30.14). For $i \neq j$ this leads to the expression

$$
\begin{aligned}
2(S_i^+ S_j^- + S_i^- S_j^+) = {}& S(a_i a_j^+ + a_i^+ a_j) \\
& - \frac{1}{4}[n_i a_i a_j^+ + a_i a_j^+ n_j + a_i^+(n_i + n_j)a_j] + \mathcal{O}(S^{-1})
\end{aligned}
\tag{30.41}
$$

which, together with $S_i^z S_j^z = (S-n_i)(S-n_j)$, is most conveniently arranged as[45]

$$
\begin{aligned}
\mathbf{S}_i \cdot \mathbf{S}_j = {}& S^2 - S[a_i^+(a_i - a_j) + a_j^+(a_j - a_i)] \\
& - \frac{1}{4}[a_i^+ a_j^+(a_i - a_j)^2 + (a_i^+ - a_j^+)^2 a_i a_j] \\
& + \mathcal{O}(S^{-1}) \; ; \quad i \neq j \,.
\end{aligned}
\tag{30.42}
$$

Inserting here the inverse of Eq. (30.20),

$$a_i^+ = \frac{1}{\sqrt{L}} \sum_\mathbf{q} b_\mathbf{q}^+ e^{-i\mathbf{q}\cdot\mathbf{R}_i} \,, \tag{30.43}$$

one finds (Problem 5.5)[45]

$$\langle \mathbf{S}_i \cdot \mathbf{S}_j \rangle_o = S^2 \left\{ 1 - \frac{1}{SL} \sum_\mathbf{q} (1 - e^{i\mathbf{q}\cdot\vec{\delta}})\bar{n}_\mathbf{q} + \mathcal{O}(S^{-3}) \right\}^2 \tag{30.44}$$

where $\vec{\delta} \equiv \mathbf{R}_i - \mathbf{R}_j \neq 0$. Summing over the z nearest-neighbor vectors $\vec{\delta}$ one finds for (30.39)

$$S' = \frac{S}{z} \sum_{\vec{\delta}} \left\{ 1 - \frac{1}{SL} \sum_\mathbf{q} (1 - e^{i\mathbf{q}\cdot\vec{\delta}})\bar{n}_\mathbf{q} + \mathcal{O}(S^{-3}) \right\} \tag{30.45}$$

or making use of (30.11) with $\lambda = 0$,[45]

$$S' = S - \frac{1}{JSzL} \sum_\mathbf{q} \omega_\mathbf{q} \bar{n}_\mathbf{q} + \mathcal{O}(S^{-2}) \,. \tag{30.46}$$

This last result suggests that with increasing temperature the magnon energy $\omega_{\mathbf{q}}$ softens which may be described by a renormalization parameter $\nu \equiv S'/S$ such that $\omega_{\mathbf{q}}$ is replaced by $\nu\omega_{\mathbf{q}}$, or[45)]

$$\bar{n}_{\mathbf{q}} = \frac{1}{e^{\beta(\nu\omega_{\mathbf{q}} - \mu)} - 1} \; ; \quad \nu \equiv \frac{S'}{S} < 1 \; . \tag{30.47}$$

This means that g_m now also depends on ν, and we may add an arbitrary function $\gamma(\nu)$ in the integration (30.36). In order to determine $\gamma(\nu)$ we introduce the second stationarity condition

$$0 = \frac{\partial g_m}{\partial \nu} = \frac{1}{L} \sum_{\mathbf{q}} \omega_{\mathbf{q}} \bar{n}_{\mathbf{q}} + \gamma'(\nu) \tag{30.48}$$

and demand that this equation coincides with (30.46). Combined with (30.38) this yields[45)]

$$f_m(\mu, \nu, T) = S\mu - \frac{k_B T}{L} \sum_{\mathbf{q}} \ln(\bar{n}_{\mathbf{q}} + 1) + \frac{1}{2} J S^2 z (\nu - 1)^2 \; . \tag{30.49}$$

In view of Eq. (30.37) the parameters μ and ν are now selfconsistently determined from Eqs. (30.47), (30.49) and the stationarity conditions

$$\frac{\partial f_m}{\partial \mu} = 0, \; \frac{\partial f_m}{\partial \nu} = 0 \; . \tag{30.50}$$

This procedure gives surprisingly good results in dimensions 1 and 2, both for the ferromagnetic[45)] and the antiferromagnetic[46)] cases . For spin 1/2 yet another representation which explicitly takes into account the "hard core" constraint $n_i = 0, 1$ of the bosons has been devised by Zhou[47)].

This renormalization idea may now also be applied to the 3-dimensional result (30.31) by multiplying the magnon energy (30.27) with ν. The result is a renormalized saturation magnetization M_s' which according to (30.30) is $M_s' = \nu^{-3/2} M_s$. Using (30.46) we find

$$\frac{M_s'}{M_s} = \left(\frac{S'}{S}\right)^{-3/2} = 1 + \frac{3u_m}{2JS^2 zL} + \mathcal{O}(S^{-3}) \tag{30.51}$$

where

$$u_m = \frac{1}{V} \sum_{\mathbf{q}} \omega_{\mathbf{q}} \bar{n}_{\mathbf{q}} \tag{30.52}$$

is the magnetic internal energy density. With (30.28) one deduces in 3 dimensions

$$u_m(H, T) = \frac{JS}{4\pi^2 a^3} \left(\frac{k_B T}{JS}\right)^{5/2} \int_0^\infty \frac{x^{3/2} dx}{e^{x-y} - 1} \tag{30.53}$$

and, with $H = 0$ and definition (30.32),

$$\Delta M(T) = M'_s \left(\frac{T}{T_C}\right)^{3/2} = M_s \left(\frac{T}{T_C}\right)^{3/2} \left[1 + \frac{3I_{3/2}}{8\pi^2 a^3 S z L} \left(\frac{k_B T}{JS}\right)^{5/2}\right].$$
(30.54)

This result contains the well-known T^4-correction to the $T^{3/2}$-law first obtained by Dyson[48] and derived here following Ref. 42. From (30.53) we also obtain the magnetic specific heat[42]

$$c_m = \left(\frac{\partial u_m}{\partial T}\right)_{H=0} = \frac{5 k_B I_{3/2}}{8\pi^2 a^3} \left(\frac{k_B T}{JS}\right)^{3/2}.$$
(30.55)

The correction obtained in Eq. (30.54) is the result of interactions among the magnons. Indeed, Eq. (30.42) shows that an expansion in powers of S^{-1} produces 4-magnon terms of order S^0, and higher vertices in $\mathcal{O}(S^{-1})$. The existence of these *dynamic interactions* means that two-magnon states $b_{\mathbf{q}}^+ b_{\mathbf{q}'}^+ |0\rangle$ are not eigenstates of the Heisenberg Hamiltonian $\mathcal{H}_H$, and neither are the states $S_i^- S_{i'}^- |0\rangle$. Diagonalizing $\mathcal{H}_H$ by taking linear combinations of the latter one finds, apart from the two-magnon continuum, also *two-magnon bound states*[1,2]. That *multi-magnon states*

$$|i_1, \ldots, i_n\rangle \equiv S_{i_1}^-, \ldots, S_{i_n}^- |0\rangle$$
(30.56)

have unusual properties follows from the fact that (30.56) must vanish if more than $2S$ of the site indices $i_1, \ldots, i_n$ coincide. This introduces severe restrictions which Dyson calls *kinematical interactions*[48].

For $S = 1/2$ kinematical interactions already occur among two-magnon states, as may be seen by explicitly calculating the general 2-magnon matrix element. Making use of (30.2), (30.5) and (30.56) one finds (Problem 5.6)

$$\langle jj'|ii'\rangle = \frac{1}{4}S^2(\delta_{ij}\delta_{i'j'} + \delta_{ij'}\delta_{i'j}) - \frac{1}{4}S\delta_{ij}\delta_{ii'}\delta_{jj'}$$
(30.57)

which for $i = i'$ becomes $\delta_{ij}\delta_{jj'}S(2S-1)/4$ and hence vanishes for $S = 1/2$. Fourier transformation of (30.57) according to (30.7) leads to (Problem 5.6)

$$\langle 0|\sigma_{\mathbf{p}}\sigma_{\mathbf{p}'}\sigma_{\mathbf{q}}^+\sigma_{\mathbf{q}'}^+|0\rangle = \delta_{\mathbf{qp}}\delta_{\mathbf{q'p'}} + \delta_{\mathbf{qp'}}\delta_{\mathbf{q'p}} - \frac{1}{SL}\sum_{\mathbf{K}}\delta_{\mathbf{q+q'-p-p'},\mathbf{K}}$$
(30.58)

where $\mathbf{K}$ is a vector of the reciprocal lattice. The last term in this expression shows that crystal-momentum conserving initial and final two-magnon states are non-orthogonal to order L^{-1} which is another manifestation of kinematical interactions. Dyson's fundamental work[48] has given rise to a considerable literature[2,49].

31. The Static Magnetic Susceptibility of Band Electrons

Landau showed that for free electrons, 1/3 of the Pauli spin susceptibility χ_P is compensated by orbital diamagnetism[22,50] leaving a net paramagnetism of $(2/3)\chi_P$. However, this proportion may change due to band structure effects. Indeed, many semiconductors, among them Si, Ge, α-Sn and many III–V compounds, are diamagnetic both, in the intrinsic and in the doped regimes[51] which shows that effective masses, bandwidths and more general features of the band structure are of importance. A first systematic study of the magnetism of band electrons was made by Peierls[52,50]. He showed that for this purpose the orbital one-electron Hamiltonian (see Eq. (18.1) for the signs)

$$H_{\text{orb}} = \frac{1}{2m} \left[\mathbf{p} + \frac{e}{c} \mathbf{A}(\mathbf{r}) \right]^2 + U(\mathbf{r}) , \qquad (31.1)$$

$U(\mathbf{r})$ being the periodic potential of Eq. (1.12), may be replaced by an *effective Hamiltonian*[53]

$$H_{\text{eff}} = \varepsilon_n \left(-i\nabla + \frac{e}{c} \mathbf{A}(\mathbf{r}) \right) \qquad (31.2)$$

where $\varepsilon_n(\mathbf{k})$ is the relevant energy band defined in Eq. (1.16). But, of course, this one-band approximation fails if the magnetic energy $\mu_B H$ becomes comparable to the distance between the Fermi level and neighboring bands.

As discussed by Peierls[50], the difficulty of the problem lies in the fact that, for a simple topological reason, the effect of a constant magnetic field on the electrons cannot be treated in perturbation theory. Indeed, the electron orbits projected onto a plane perpendicular to the field are closed with the field present but open without. Formally, this difficulty is reflected in the shape of the vector potential which for a constant field $\mathbf{H} = (0, 0, H)$ may be taken as $\mathbf{A} = (0, Hx, 0)$ and therefore is unbounded in infinite space or, in a finite volume with the periodic boundary condition (2.5), has an infinite slope at the border. Of course, one may gauge away the field contribution in the Hamiltonian at least locally by using the identity

$$e^{i\pi\phi(\mathbf{r})/\phi_o} \left[\mathbf{p} + \frac{e}{c} \mathbf{A}(\mathbf{r}) \right] e^{-i\pi\phi(\mathbf{r})/\phi_o} = \mathbf{p} + \frac{e}{c} \left(\mathbf{A}(\mathbf{r}) - \nabla\phi(\mathbf{r}) \right) \qquad (31.3)$$

where ϕ_o is the flux quantum (27.11) and $\phi = \int \mathbf{A} \cdot d\mathbf{r}$. In this way it is possible to compensate the vector potential in an arbitrary plane $x = x_o$ by $\phi = Hx_o y$[50]. Another useful choice is the one-parameter gauge group $\phi_\lambda = \lambda H x y$ which leads to

$$\mathbf{A}(\mathbf{r}) = (-\lambda y, (1-\lambda)x, 0)H \equiv \frac{mcH}{e} \mathbf{a}(\mathbf{r}) . \qquad (31.4)$$

This form may be used as a simple test of the gauge invariance of a calculation[54,55].

Since susceptibilities are statistical and not dynamical quantities, the difficulty with a perturbative expansion in the field is not as serious as described above. Indeed, ensemble averaging smears out to a certain extent the topological difference between orbits with and without field and makes such an expansion possible, as has been demonstrated by the approximate calculations of the "complete" susceptibility formula[55,56]. According to Ref. 55 the thermodynamic potential per unit volume may be written as the trace (see Eq. (10.1) above),

$$\Omega = \frac{1}{V} \sum_\sigma \mathrm{Tr}\, F(H_\sigma) \qquad (31.5)$$

where

$$F(\varepsilon) \equiv -\beta^{-1} \ln[1 + e^{-\beta(\varepsilon - \mu)}] \qquad (31.6)$$

and

$$H_\sigma = H_{\mathrm{orb}} - \mu_B H \sigma ; \quad \sigma = \pm \qquad (31.7)$$

is the sum of the orbital Hamiltonian (31.1) and the Zeeman term (29.1) for spin 1/2. As seen from Eq. (31.3) the cyclicity of the trace ensures the gauge invariance of Eq. (31.5). In this formulation we neglect the spin-orbit interaction (1.22) in order not to overcharge the formalism. Since $F(w)$ is analytic on the real axis and vanishes exponentially for $w \to +\infty$, we may apply Cauchy's formula to write[54,55]

$$\Omega = \frac{1}{V} \sum_\sigma \mathrm{Tr} \int_\Lambda \frac{dw}{2\pi i} F(w)(w - H_\sigma)^{-1} \qquad (31.8)$$

where the path Λ encloses the spectrum of H_σ in the positive sense.

In Eqs. (31.5) and (31.8) the smearing-out of the topological difference between orbits with and without field is reflected by the invariance of the trace with respect to the representation used. Thus we may expand the resolvent $(w - H_\sigma)^{-1}$ in powers of the field by writing

$$H_{\mathrm{orb}} = H_o + H' \qquad (31.9)$$

where H_o is given by (1.12) and

$$H' = \frac{e}{mc} \mathbf{A} \cdot \mathbf{p} + \frac{e^2}{2mc^2} \mathbf{A}^2 . \qquad (31.10)$$

Note that to obtain the last expression we assumed $\mathrm{div}\mathbf{A} = 0$ which is satisfied by (31.4). Then using the identity $(U - V)^{-1} = (1 - U^{-1}V)^{-1}U^{-1} =$

$\sum_{n=0}^{\infty}(U^{-1}V)^n U^{-1}$ with $U = w - H_o$ we have

$$(w - H_\sigma)^{-1} = (w - H_o)^{-1} - \mu_B H \sigma (w - H_o)^{-2}$$
$$+ \frac{e}{mc}(w - H_o)^{-1}\mathbf{A} \cdot \mathbf{p}\,(w - H_o)^{-1}$$
$$+ \frac{e^2}{2mc^2}(w - H_o)^{-1}\mathbf{A}^2(w - H_o)^{-1} + \mu_B^2 H^2(w - H_o)^{-3}$$
$$+ \frac{e^2}{m^2c^2}(w - H_o)^{-1}\mathbf{A} \cdot \mathbf{p}\,(w - H_o)^{-1}\mathbf{A} \cdot \mathbf{p}\,(w - H_o)^{-1}$$
$$- \frac{e}{mc}\mu_B H \sigma (w - H_o)^{-1}$$
$$\times \left[\mathbf{A} \cdot \mathbf{p}\,(w - H_o)^{-1} + (w - H_o)^{-1}\mathbf{A} \cdot \mathbf{p}\right](w - H_o)^{-1} + \cdots .$$

$$(31.11)$$

Inserting (31.11) with (31.4) into (31.8) and making use of the cyclicity of the trace we may write

$$\Omega = \Omega_o - (M_{\text{spin}} + M_{\text{orb}})H - \frac{1}{2}(\chi_{\text{spin}} + \chi_{\text{orb}})H^2 + \cdots \qquad (31.12)$$

where $\Omega_o = (2/V)\text{Tr}\,F(H_o)$ is the thermodynamic potential without magnetic field,

$$M_{\text{spin}} = \frac{\mu_B}{V}\sum_\sigma \sigma \text{Tr}\int_\Lambda \frac{dw}{2\pi i}F(w)(w - H_o)^{-2}$$

$$M_{\text{orb}} = -\frac{1}{V}\sum_\sigma \text{Tr}\int_\Lambda \frac{dw}{2\pi i}F(w)(w - H_o)^{-2}\mathbf{a} \cdot \mathbf{p}$$

$$(31.13)$$

are the spin and orbital permanent magnetic moments, which of course must vanish, and

$$\chi_{\text{spin}} = -\frac{2\mu_B^2}{V}\sum_\sigma \text{Tr}\int_\Lambda \frac{dw}{2\pi i}F(w)(w - H_o)^{-3}$$

$$\chi_{\text{orb}} = -\frac{1}{V}\sum_\sigma \text{Tr}\int_\Lambda \frac{dw}{2\pi i}F(w)(w - H_o)^{-2}$$

$$\times \left\{ma^2 + 2\mathbf{a} \cdot \mathbf{p}\,(w - H_o)^{-1}\mathbf{a} \cdot \mathbf{p}\right.$$
$$\left. - 2\mu_B \sigma\left[\mathbf{a} \cdot \mathbf{p}\,(w - H_o)^{-1} + (w - H_o)^{-1}\mathbf{a} \cdot \mathbf{p}\right]\right\}$$

$$(31.14)$$

are the spin and orbital susceptibilities .

Evaluating first the spin contributions, it is obvious from the first formula (31.13) that $M_{\text{spin}} = 0$ since the expressions under the trace are all independent of σ. As to the spin susceptibility, the integral in the first formula (31.14) is recognized as $1/2$ the second derivative of Cauchy's formula so that, after summing over σ,

$$\chi_{\text{spin}} = -\frac{2\mu_B^2}{V} \operatorname{Tr} F''(H_o) = -\frac{2\mu_B^2}{V} \sum_k F''(\varepsilon_k) . \tag{31.15}$$

Here we introduced in the second step the representation by eigenstates $|k\rangle$ of H_o where $k = (n, \mathbf{k})$, n being the band index (see Section 1). Since according to (31.6), $F'(\varepsilon)$ is the Fermi function (7.42) and hence, in the limit $T \to 0$, $F''(\varepsilon) = -\delta(\varepsilon - \mu)$, introducing the density of states (10.4) one readily arrives at the *Pauli spin susceptibility*,

$$\chi_{\text{spin}} = \chi_P \equiv 2\mu_B^2 N(\mu) . \tag{31.16}$$

Turning to the orbital terms in Eqs. (31.13) and (31.14), we first show that $M_{\text{orb}} = 0$ by invoking *time-reversal invariance* of H_o (see Section 1)[55]. Choosing some representation Ψ_ν in the crystal volume V we have, after summing over σ and making use of the first derivative of Cauchy's formula,

$$M_{\text{orb}} = -\frac{2}{V} \sum_\nu \int_V d^3r \Psi_\nu^* F'(H_o) \, \mathbf{a} \cdot \mathbf{p} \, \Psi_\nu . \tag{31.17}$$

Now in orbital Hilbert space, time reversal T acts as complex conjugation $*$[57]. Since the Ψ_ν form a basis, it follows that

$$\Psi_\nu^T = \Psi_\nu^* = \sum_\mu U_{\nu\mu} \Psi_\mu ; \quad \Psi_\nu^{*T} = \Psi_\nu = \sum_\mu U_{\nu\mu}^* \Psi_\mu^* \tag{31.18}$$

and

$$\delta_{\nu\nu'} = \int_V d^3r \Psi_\nu^* \Psi_{\nu'} = \sum_{\mu\mu'} U_{\nu\mu} U_{\nu'\mu'}^*$$
$$\times \int_V d^3r \Psi_\mu \Psi_{\mu'}^* = (UU^+)_{\nu\nu'} \tag{31.19}$$

so that U is unitary. In addition, $\mathbf{r}^T = \mathbf{r}^* = \mathbf{r}$, $\mathbf{p}^T = \mathbf{p}^* = -\mathbf{p}$ and hence $H_o^T = H_o^* = H_o$. But these are all hermitean operators and $F(\varepsilon)$ and $\mathbf{a}$ are real so that, taking the complex conjugate of (3.17) one has

$$M_{\text{orb}}^* = -\frac{2}{V} \sum_\nu \int_V d^3r \Psi_\nu^* \mathbf{p} \cdot \mathbf{a} F'(H_o) \Psi_\nu = M_{\text{orb}} \tag{31.20}$$

where in the last step we made use of the cyclicity of the trace. An alternative way of evaluating complex conjugation is to introduce $\mathbf{r}^*$, $\mathbf{p}^*$ and H_o^* and to use Eqs. (31.18), (31.19),

$$M_{\text{orb}}^* = -\frac{2}{V} \sum_\nu \int_V d^3 r \Psi_\nu F'(H_o^*) \, \mathbf{a} \cdot \mathbf{p}^* \Psi_\nu^*$$

$$= +\frac{2}{V} \sum_{\nu\mu\mu'} U_{\nu\mu}^* \int_V d^3 r \Psi_\mu^* F'(H_o) \, \mathbf{a} \cdot \mathbf{p} \, \Psi_{\mu'} U_{\nu\mu'}$$

$$= +\frac{2}{V} \sum_{\mu\mu'} \int_V d^3 r \Psi_\mu^* F'(H_o) \, \mathbf{a} \cdot \mathbf{p} \, \Psi_{\mu'} (U^+ U)_{\mu\mu'} = -M_{\text{orb}} \, .$$

$$(31.21)$$

This proves $M_{\text{orb}} = 0$ without specifying the representation.

The evaluation of the orbital susceptibility χ_{orb}, unfortunately, cannot be done in the same generality; it is a complicated task which has occupied many solid state theorists[58]. Inserting the vector potential (31.4) into (31.14), the expression for χ_{orb} must be independent of λ so that, developing in powers of the gauge parameter λ we obtain after summing over σ,

$$\chi_{\text{orb}} = -\left(\frac{2e}{mc}\right)^2 \frac{1}{V} \int_\Lambda \frac{dw}{2\pi i} F(w) \text{Tr} \left[(w - H_o)^{-2} \right.$$

$$\left. \times \left\{ \frac{m}{2} x^2 + x p_y (w - H_o)^{-1} x p_y \right\} \right] \qquad (31.22)$$

and

$$\int_\Lambda \frac{dw}{2\pi i} F(w) \text{Tr} \left[(w - H_o)^{-2} \{ m x^2 + (x p_y + y p_x)(w - H_o)^{-1} x p_y \right.$$

$$\left. + x p_y (w - H_o)^{-1} (x p_y + y p_x) \} \right] = 0 \, ,$$

$$\int_\Lambda \frac{dw}{2\pi i} F(w) \text{Tr} \left[(w - H_o)^{-2} \left\{ \frac{m}{2} (x^2 + y^2) \right.\right. \qquad (31.23)$$

$$\left.\left. + (x p_y + y p_x)(w - H_o)^{-1} (x p_y + y p_x) \right\} \right] = 0 \, .$$

Making use of the identity

$$[w - H_o, \, xy] = \frac{i\hbar}{m} (x p_y + y p_x) \qquad (31.24)$$

it is straightforward to verify Eqs. (31.23) (Problem 5.7). These identities show the confusing variety of forms one may give to χ_{orb}. Probably the most compact form results from eliminating the term $m x^2$ in the bracket of (31.22) with the help of the first identity (31.23), namely

$$\chi_{\text{orb}} = \left(\frac{2e}{mc}\right)^2 \frac{1}{V} \int_\Lambda \frac{dw}{2\pi i} F(w) \frac{1}{2} (G_{xy}(w) + G_{yx}(w)) \qquad (31.25)$$

with

$$G_{xy}(w) \equiv \mathrm{Tr}\left[(w - H_o)^{-2}xp_y(w - H_o)^{-1}yp_x\right] . \tag{31.26}$$

Equation (31.26) naturally suggests the representation by eigenfunctions of H_o which are the Bloch states $|k\rangle$ mentioned above. The difficulty with this representation is the fact that the periodic boundary conditions give rise to the infinite slopes of the vector potential (31.4) at the border mentioned earlier. This difficulty may be remedied by an approximate treatment of the trace. Physically, the problem is that the vector potential (31.4) does not have the periodicity of the lattice although a constant magnetic field does. This of course is an artefact of the formalism which one is tempted to correct by regauging the vector potential with the help of the transformation (31.3), individually for every one of the L unit cells in the crystal volume V. With (31.4) and with the gauge field

$$\phi_{\mathbf{R}}(\mathbf{r}) = \mathbf{A}(\mathbf{R}) \cdot \mathbf{r} \tag{31.27}$$

the vector potential indeed becomes $\mathbf{A}(\mathbf{r}) - \nabla\phi_{\mathbf{R}}(\mathbf{r}) = \mathbf{A}(\mathbf{r} - \mathbf{R})$. However, with the $\mathbf{R}$-dependent gauge field, the integration over the crystal volume V contained in the trace of Eq. (31.5) or of (31.26) must be broken down into a sum over the L unit cells. Choosing as before a basis Ψ_ν in V and making use of (31.3) with (31.27), this means that we may write

$$\mathrm{Tr}\, F(H_\sigma) = \sum_{\mathbf{R}\in V}\sum_\nu \int_C d^3r \Psi_\nu^*(\mathbf{r} + \mathbf{R})e^{-i\pi\mathbf{A}(\mathbf{R})\cdot\mathbf{r}/\phi_o}$$

$$\times F(H_\sigma)e^{i\pi\mathbf{A}(\mathbf{R})\cdot\mathbf{r}/\phi_o}\Psi_\nu(\mathbf{r} + \mathbf{R}) \tag{31.28}$$

where C is the unit cell at the origin . In Ref. 56 one writes instead (see Eq. (5) of Ref. 56)

$$\mathrm{Tr}\, F(H_\sigma) \simeq L\sum_\nu \int_C d^3r \Psi_\nu^*(\mathbf{r})F(H_\sigma)\Psi_\nu(\mathbf{r}) . \tag{31.29}$$

This is a reasonable approximation since, as observed before, the magnetic field does have the lattice periodicity and, in Bloch representation, a periodic operator does have the property (31.29).

Indeed, for any periodic operator $O(\mathbf{r}+\mathbf{R}) = O(\mathbf{r})$, the matrix element in Bloch representation may be written as

$$\langle n\mathbf{k}|O|n'\mathbf{k}'\rangle = \sum_{\mathbf{R}\in V} e^{i(\mathbf{k}'-\mathbf{k})\cdot\mathbf{R}} \int_C d^3r \psi_{n\mathbf{k}}^* O \psi_{n'\mathbf{k}'} \tag{31.30}$$

$$= \delta_{\mathbf{k}\mathbf{k}'} O_{nn'}(\mathbf{k})$$

where use was made of Bloch's theorem (1.14), of Eq. (1.9) and of the fact that $\mathbf{k}, \mathbf{k}' \in Z$. Here

$$O_{nn'}(\mathbf{k}) \equiv L \int_C d^3 r \psi_{n\mathbf{k}}^* O \psi_{n'\mathbf{k}} \tag{31.31}$$

and we have

$$\operatorname{Tr} O = \sum_{nk} O_{nn}(\mathbf{k}) . \tag{31.32}$$

But, on the other hand, it follows from (31.30) that we may also write

$$\sum_{kk'} \delta_{nn'} \langle k|O|k' \rangle = \sum_{nk} O_{nn}(\mathbf{k}) = \operatorname{Tr} O . \tag{31.33}$$

In Ref. 55 the prescription (31.33) is applied to the almost periodic operator $F(H_\sigma)$ (see Eq. (2.11) of Ref. 55). This approximation is equivalent to (31.29) since, applying Eqs. (31.30), (31.31) we indeed obtain a reduction to the unit cell at the origin,

$$\sum_{nkk'} \langle nk|O|nk' \rangle = \sum_{nk} \int_C d^3 r \psi_{n\mathbf{k}}^* O \psi_{n\mathbf{k}} . \tag{31.34}$$

It must be realized, however, that the prescription (31.33) spoils the cyclicity of the trace and hence also gauge invariance. Therefore, applying (31.33) to Eq. (31.26), the original order (31.11) of the factors $(w - H_o)^{-1}$ must be reinstated. Introducing intermediate states we then obtain

$$G_{xy}(w) = \sum_{kk'k''} \delta_{nn'} (w - \varepsilon_k)^{-1} \langle k|xp_y|k'' \rangle$$
$$\times (w - \varepsilon_{k''})^{-1} \langle k''|yp_x|k' \rangle (w - \varepsilon_{k'})^{-1} . \tag{31.35}$$

The matrix elements of xp_y and yp_x occurring here may be constructed from those of $\mathbf{r}$ and $\mathbf{p}$[55]. While $\mathbf{p}$ is a periodic operator so that Eqs. (31.30), (31.31) give (Problem 5.8),

$$\langle nk|\mathbf{p}|n'k' \rangle = \mathbf{p}_{nn'}(\mathbf{k}) \delta_{kk'} ,$$
$$\mathbf{p}_{nn'}(\mathbf{k}) \equiv L \int_C d^3 r \psi_{n\mathbf{k}}^* \frac{\hbar}{i} \nabla \psi_{n'\mathbf{k}} = \mathbf{p}_{n'n}^*(\mathbf{k}) , \tag{31.36}$$

one finds for the matrix element of $\mathbf{r}$ (Problem 5.8)

$$\langle nk|\mathbf{r}|n'k' \rangle = \delta_{nn'} \frac{\partial}{i\partial \mathbf{k}'} \delta_{kk'} + \mathbf{r}_{nn'}(\mathbf{k}) \delta_{kk'}$$
$$\mathbf{r}_{nn'}(\mathbf{k}) \equiv L \int_C d^3 r u_{n\mathbf{k}}^* \frac{\partial}{i\partial \mathbf{k}} u_{n'\mathbf{k}} = \mathbf{r}_{n'n}^*(\mathbf{k}) . \tag{31.37}$$

With these expressions one obtains

$$\langle n\mathbf{k}|xp_y|n'\mathbf{k}'\rangle = \sum_{k''}\langle n\mathbf{k}|x|k''\rangle\langle k''|p_y|n'\mathbf{k}'\rangle$$

$$= p_{ynn'}(\mathbf{k}')\frac{\partial}{i\partial k'_x}\delta_{\mathbf{kk}'} + (xp_y)_{nn'}(\mathbf{k})\delta_{\mathbf{kk}'} . \qquad (31.38)$$

The derivative acting on $\delta_{\mathbf{kk}'}$ in Eqs. (31.37) and (31.38) is formally well defined since $\delta_{\mathbf{kk}'}$ stands for the orthonormality of the Bloch functions $\psi_{n\mathbf{k}}$ which may be considered analytic in $\mathbf{k}$. The action of the derivative is seen by considering an expression with an arbitrary differentiable function $\varphi(\mathbf{k})$ in a domain D with boundary ∂D and such that $\mathbf{k} \in D \subset Z$,

$$\sum_{\mathbf{k}'\in D} \varphi(\mathbf{k}')\frac{\partial}{\partial \mathbf{k}'}\delta_{\mathbf{kk}'} = \sum_{\mathbf{k}'\in\partial D} \varphi(\mathbf{k}')\delta_{\mathbf{kk}'} - \sum_{\mathbf{k}'\in D} \frac{\partial\varphi(\mathbf{k}')}{\partial\mathbf{k}'}\delta_{\mathbf{kk}'} = -\frac{\partial\varphi(\mathbf{k})}{\partial\mathbf{k}} .$$

$$(31.39)$$

A useful formula is obtained by applying Eq. (31.39) to the matrix element of the identity

$$[H_o\mathbf{r}] = \frac{\hbar}{im}\mathbf{p} . \qquad (31.40)$$

Making use of Eqs. (31.36), (31.37) one finds[55)] (Problem 5.8)

$$\frac{\hbar}{m}\mathbf{p}_{nn'}(\mathbf{k}) = \frac{\partial\varepsilon_{n\mathbf{k}}}{\partial\mathbf{k}}\delta_{nn'} + i(\varepsilon_{n\mathbf{k}} - \varepsilon_{n'\mathbf{k}})\mathbf{r}_{nn'}(\mathbf{k}) . \qquad (31.41)$$

Applying formula (31.38) to (31.35) we first evaluate the k'-sum. Making use of a partial summation (31.39) with $D = Z$ we find

$$\sum_{k'}\delta_{nn'}\langle k''|yp_x|k'\rangle(w - \varepsilon_{k'})^{-1}$$

$$+ \left\{i\frac{\partial}{\partial k''_y}p_{xn''n}(\mathbf{k}'') + (yp_x)_{n''n}(\mathbf{k}'')\right\}(w - \varepsilon_{n\mathbf{k}''})^{-1} \qquad (31.42)$$

where the derivative acts on all factors to the right. Inserting (31.42) and (31.38) into Eq. (31.35) one obtains, using again (31.39) with $D = Z$ (Problem 5.9),

$$G_{xy}(w) = \sum_{nn''\mathbf{k}} \left\{p_{ynn''}(\mathbf{k})\left[-i\frac{\partial}{\partial k_x}(w - \varepsilon_{n\mathbf{k}})^{-1}\right]\right.$$

$$+ (xp_y)_{nn''}(\mathbf{k})(w - \varepsilon_{n\mathbf{k}})^{-1}\bigg\}(w - \varepsilon_{n''\mathbf{k}})^{-1}$$

$$\times \left\{\left[i\frac{\partial}{\partial k_y}p_{xn''n}(\mathbf{k})(w - \varepsilon_{n\mathbf{k}})^{-1}\right] + (yp_x)_{n''n}(\mathbf{k})(w - \varepsilon_{n\mathbf{k}})^{-1}\right\} .$$

$$(31.43)$$

Separating here the terms $n'' = n$ and $n'' \neq n$ we may write $G_{xy} = G^d_{xy} + G^{nd}_{xy}$ where application of (31.41) for $n = n'$ yields, after several partial summations (31.39) (Problem 5.9),

$$
G^d_{xy}(w) = \sum_k (w - \varepsilon)^{-3} \left\{ \frac{1}{12} \left(\frac{m}{\hbar} \right)^2 (\varepsilon_{xx}\varepsilon_{yy} - 2\varepsilon^2_{xy} \right.
$$

$$
+ \varepsilon_y\varepsilon_{xxy} - 2\varepsilon_x\varepsilon_{xyy}) + i\frac{m}{\hbar}(w - \varepsilon)^{-1}\varepsilon_x\varepsilon_y(xp_y - yp_x)_{nn}
$$

$$
\left. + i\frac{m}{\hbar}\varepsilon_{xy}(xp_y)_{nn} + (xp_y)_{nn}(yp_x)_{nn} \right\} \tag{31.44}
$$

where the abbreviated notation $\varepsilon \equiv \varepsilon_{n\mathbf{k}}$ and $\varepsilon_x \equiv \partial\varepsilon_{n\mathbf{k}}/\partial k_x$ etc. is used.

Inserting (31.44) and the corresponding expression for G^{nd}_{xp} obtained from (31.43) into Eq. (31.25), remembering that $\mu_B = e\hbar/2mc$ and making use of the second derivative of Cauchy's formula and of $F'(\varepsilon) = f_o(\varepsilon)$, the final result may be written in the form

$$
\chi_{\text{orb}} = \chi_{LP} + \chi^d + \chi^{nd} . \tag{31.45}
$$

Here

$$
\chi_{LP} \equiv \frac{2}{3}\mu_B^2 \frac{1}{V} \sum_k f'_o(\varepsilon) \left(\frac{m}{\hbar^2} \right)^2 (\varepsilon_{xx}\varepsilon_{yy} - \varepsilon^2_{xy}) \tag{31.46}
$$

is the *Landau-Peierls diamagnetism*[52] which under rotations about the magnetic-field direction transforms as the determinant of the rotation matrix and hence is invariant. χ^d contains the remaining contributions from G^d_{xy},

$$
\chi^d = -\mu_B^2 \frac{1}{V} \sum_k f'_o(\varepsilon) \left\{ \frac{1}{3} \left(\frac{m}{\hbar^2} \right)^2 (2\varepsilon^2_{xy} + \varepsilon_x\varepsilon_{xyy} + \varepsilon_y\varepsilon_{xxy}) \right.
$$

$$
\left. - 4i\frac{m}{\hbar^3}\varepsilon_{xy}(xp_y + yp_x)_{nn} - \frac{8}{\hbar^2}(xp_y)_{nn}(yp_x)_{nn} \right\} \tag{31.47}
$$

and χ^{nd} results from G^{nd}_{xy}. The complication of the expressions for χ^d and for χ^{nd} and the fact that, due to the identities (31.24) and (31.41) there are many ways to write these terms, do not justify going into more detail. The point of interest is to see the form of χ_{orb} in the two limits of nearly free and of tightly bound electrons.

For *nearly free electrons*, only one band is of importance and the $u_{n\mathbf{k}}$ in Eq. (31.36) and (31.37) are nearly independent of $\mathbf{k}$. This means that $\mathbf{p}_{nn'}(\mathbf{k}) \cong \hbar\mathbf{k}\delta_{nn'}$, $\mathbf{r}_{nn'}(\mathbf{k}) \cong 0$ and $\chi^{nd} \cong 0$. From (31.41) it then follows that $\varepsilon_{n\mathbf{k}} \cong \hbar^2\mathbf{k}^2/2m + \text{const.}$ But this implies that $\varepsilon_{xy} \cong 0$ and $\varepsilon_{xx} \cong \varepsilon_{yy} \cong$

$\hbar^2/m$ and, according to (31.47), that $\chi^d \cong 0$. Comparing Eqs. (31.46) and (31.15) one then arrives at Landau's result[22]

$$\chi_{\text{orb}} \cong \chi_{LP} \cong \frac{2}{3}\mu_B^2 \frac{1}{V} \sum_{\mathbf{k}} f_o'(\varepsilon) = -\frac{1}{3}\chi_{\text{spin}} \ . \tag{31.48}$$

In the case of *tightly bound electrons* the **k**-dependence is negligible so that $\varepsilon_x \cong \varepsilon_y \cong 0$ and Eqs. (31.36) and (31.37) become, respectively $\langle n|\mathbf{p}|n'\rangle \cong \mathbf{p}_{nn'}$ and $\langle n|\mathbf{r}|n'\rangle \cong \mathbf{r}_{nn'}$. Hence, it follows from Eqs. (31.46) and (31.47) that $\chi_{LP} \cong 0$ and

$$\chi^d \cong \frac{8\mu_B^2}{\hbar^2 V} \sum_{n} f_o'(\varepsilon_n)(xp_y)_{nn}(yp_x)_{nn} \ . \tag{31.49}$$

For χ^{nd} we find from Eq. (31.25) and from the part $n'' \neq n$ of Eq. (31.43)

$$\chi^{nd} \cong \frac{8\mu_B^2}{\hbar^2 V} \sum_{nn'(n \neq n')} \phi(\varepsilon_n \varepsilon_{n'})$$
$$\times \{(xp_y)_{nn'}(yp_x)_{n'n} + (yp_x)_{nn'}(xp_y)_{n'n}\} \tag{31.50}$$

where

$$\phi(\varepsilon\varepsilon') \equiv \int_\Lambda \frac{dw}{2\pi i} F(w)(w - \varepsilon)^{-2}(w - \varepsilon')^{-1}$$
$$= \frac{F(\varepsilon')}{(\varepsilon' - \varepsilon)^2} + \frac{\partial}{\partial\varepsilon}\left(\frac{F(\varepsilon)}{\varepsilon - \varepsilon'}\right) = \frac{F(\varepsilon') - F(\varepsilon)}{(\varepsilon - \varepsilon')^2} + \frac{F'(\varepsilon)}{\varepsilon - \varepsilon'} \ . \tag{31.51}$$

Making use of the identity (31.24) one finds for the diagonal elements in χ^d

$$(xp_y)_{nn} = -(yp_x)_{nn} = \frac{1}{2}(l_z)_{nn} \tag{31.52}$$

where $\mathbf{l} \equiv \mathbf{r} \times \mathbf{p}$ is the orbital angular momentum. Hence

$$\chi^d \cong \chi_{LD} \equiv -\frac{2\mu_B^2}{\hbar^2 V} \sum_{n} f_o'(\varepsilon_n)|(l_z)_{nn}|^2 \tag{31.53}$$

which is the *Langevin-Debye paramagnetism* [33,55].

To the nondiagonal elements in Eq. (31.50) we apply the identity

$$2\{(xp_y)_{nn'}(yp_x)_{n'n} + (yp_x)_{nn'}(xp_y)_{n'n}\}$$
$$= |(xp_y + yp_x)_{nn'}|^2 - |(l_z)_{nn'}|^2 \tag{31.54}$$

which leads to

$$\chi^{nd} \cong \chi_{vV} + \chi_{at} \tag{31.55}$$

where

$$\chi_{vV} \equiv -\frac{4\mu_B^2}{\hbar^2 V} \sum_{nn'(n\neq n')} \phi(\varepsilon_n \varepsilon_{n'})|(l_z)_{nn'}|^2 \tag{31.56}$$

is the *van Vleck paramagnetism*[33,52,55] and the remaining term χ_{at} may be further transformed with the help of the identity (31.24), writing

$$2|(xp_y + yp_x)_{nn'}|^2 = \frac{im}{\hbar}(\varepsilon_n - \varepsilon_{n'})$$
$$\times \{(xy)_{nn'}(xp_y + yp_x)_{n'n} - (xp_y + yp_x)_{nn'}(xy)_{n'n}\} . \tag{31.57}$$

Inserting this expression back into (31.54) and (31.50) one sees that, according to (31.51), the first term of $\phi(\varepsilon_n \varepsilon_{n'})$ drops out while in the second term we may add the contribution $n' = n$ which is zero. Thus the n'-sum in χ_{at} involves only the large bracket of Eq. (31.57) which gives $([xy, xp_y + yp_x])_{nn} = -(\hbar/i)(x^2 + y^2)_{nn}$. This yields for the remaining term

$$\chi_{at} \equiv -\frac{2\mu_B^2}{\hbar^2 V} m \sum_n f_o(\varepsilon_n)(x^2 + y^2)_{nn} \tag{31.58}$$

which is the *atomic* or *Langevin-Pauli diamagnetism*[33,52,55].

Note that the result (31.53), (31.56), (31.58) could have been obtained directly from Eq. (31.14) with the gauge $\lambda = 1/2$ in the vector potential (31.4). This shows that the approximation (31.33) is exact in this limit. Note also that χ_{LD}, χ_{vV} and χ_{at} have the traditional meaning for gases or liquids where Boltzmann statistics applies[33,55].

32. Itinerant-electron Magnetism. Stoner's Molecular Field Theory

It was Stoner's idea[23] to replace the external magnetic field H in the Zeeman energy (29.1) of a conduction electron by the effective field (29.16) and to add this term to the band energy $\varepsilon_{\mathbf{k}}$ (in this section we drop the band index) writing

$$\varepsilon_{\mathbf{k}\sigma} \equiv \varepsilon_{\mathbf{k}} - \mu_\sigma ; \quad \sigma = \pm \tag{32.1}$$

where

$$\mu_\sigma \equiv \mu + \sigma\mu_B \bar{H} ; \quad \sigma = \pm \tag{32.2}$$

is the spin-shifted chemical potential. As a result, even in the absence of an external field the spectrum has a *Stoner gap* (see Ref. 2, Section 6.15)

$$\Delta^S \equiv [\mu_\uparrow - \mu_\downarrow]_{H=0} = 2\mu_B \lambda M \tag{32.3}$$

and the electrons are divided into *majority* and *minority carriers* with densities $n_\uparrow$ and $n_\downarrow$, respectively. Since in analogy to Eq. (29.7) the magnetization is given by

$$M = \mu_B(n_\uparrow - n_\downarrow) \tag{32.4}$$

and the total electron density is $n = n_\uparrow + n_\downarrow$, one has

$$n_\sigma = \frac{1}{2}\left(n + \sigma\frac{M}{\mu_B}\right) . \tag{32.5}$$

We wish to determine the canonical equation of state, in which all quantities are expressed in terms of n, M and T, by a development in powers of M and T. Fixing the chemical potential such that $\mu_\sigma = \mu = 0$ for $T = 0$ and $\bar{H} = 0$, we may express n_σ in terms of the density of states (10.4) as

$$n_\sigma = \frac{1}{V}\sum_k f_o(\varepsilon_{k\sigma}) = \int_{-\infty}^{+\infty} d\varepsilon N(\varepsilon + \mu_\sigma)f_o(\varepsilon)$$

$$= \sum_{\nu=0}^{\infty} \frac{1}{\nu!}\mu_\sigma^\nu \int_{-\infty}^{+\infty} d\varepsilon N^{(\nu)}(\varepsilon)f_o(\varepsilon) \tag{32.6}$$

where it is understood that $N(\varepsilon) = 0$ below the bottom of the band, i.e. for $\varepsilon < -\varepsilon_F$. In order to develop the integrals in (32.6) into powers of the temperature we observe that, at $T = 0$, $f_o(\varepsilon) = \theta(-\varepsilon) \equiv (1 - \operatorname{sgn}\varepsilon)/2$ and that, according to definition (7.42),

$$f_o(\varepsilon) - \theta(-\varepsilon) = \frac{\varepsilon}{|\varepsilon|}f_o(|\varepsilon|) . \tag{32.7}$$

Since at low temperatures this function is strongly localized around $\varepsilon = 0$, we may develop $N(\varepsilon)$ in the first term of the sum in (32.6) into powers of ε so that

$$\int_{-\infty}^{+\infty} d\varepsilon N(\varepsilon)f_o(\varepsilon) = \int_{-\infty}^{0} d\varepsilon N(\varepsilon) + \sum_{\rho=0}^{\infty} \frac{(k_BT)^{\rho+1}}{\rho!}N^{(\rho)}(0)K_\rho \tag{32.8}$$

where (see Eq. (3.411.3) of Ref. 59)

$$K_\rho \equiv \int_{-\infty}^{+\infty} \frac{x^{\rho+1}dx}{|x|(e^{|x|} + 1)} = \left[1 + (-1)^{\rho+1}\right](1 - 2^{-\rho})\rho!\zeta(\rho + 1) , \tag{32.9}$$

$\zeta(z)$ being defined in (24.9) and $\zeta(2) = \pi^2/6$.

For the remaining terms of the sum in (32.6), a partial integration reduces the integral to an expression as was treated in Problem 2.8 of

Chapter 2,

$$\int_{-\infty}^{+\infty} d\varepsilon N^{(\nu)}(\varepsilon) f_o(\varepsilon) = -\int_{-\infty}^{+\infty} d\varepsilon N^{(\nu-1)}(\varepsilon) f'_o(\varepsilon)$$

$$= \sum_{\rho=0}^{\infty} \frac{(k_B T)^{\rho}}{\rho!} N^{(\nu-1+\rho)}(0) I_{\rho} \qquad (32.10)$$

where (see Eq. (3.527.3) of Ref. 59)

$$I_{\rho} \equiv \int_{-\infty}^{+\infty} \frac{x^{\rho} e^{-x} dx}{(1+e^{-x})^2} = (1+(-1)^{\rho})(1-2^{1-\rho})\rho!\zeta(\rho) = \rho K_{\rho-1} \quad (32.11)$$

and $I_o = 1$, $I_2 = \pi^2/3$, $I_4 = 7\pi^4/15$. Thus

$$n_{\sigma} = \int_{-\infty}^{0} d\varepsilon N(\varepsilon) + \frac{\pi^2}{6}(k_B T)^2 N'(0) + \mathcal{O}(T^4)$$

$$+ \mu_{\sigma}\{N(0) + \frac{\pi^2}{6}(k_B T)^2 N''(0) + \mathcal{O}(T^4)\}$$

$$+ \frac{1}{2}\mu_{\sigma}^2\{N'(0) + \mathcal{O}(T^2)\} + \frac{1}{6}\mu_{\sigma}^3\{N''(0) + \mathcal{O}(T^2)\} + \cdots .$$

$$(32.12)$$

Since the density n is independent of T and $\bar{H}$, subtraction of n, for $T = 0$ and $\bar{H} = 0$, from $n_{\uparrow} + n_{\downarrow}$ given by (32.12), using Eq. (32.2), leads to the condition

$$\frac{\pi}{6}(k_B T)^2 \frac{N'(0)}{N(0)} + \mu\{1 + \mathcal{O}(T^2)\}$$

$$+ \frac{1}{2}(\mu^2 + \mu_B^2 \bar{H}^2)\left\{\frac{N'(0)}{N(0)} + \mathcal{O}(T^2)\right\} + \cdots = 0 \qquad (32.13)$$

which may be solved for μ, giving

$$\mu = -\frac{1}{2}\left\{\frac{\pi^2}{3}(k_B T)^2 + (\mu_B \bar{H})^2\right\}\frac{N'(0)}{N(0)} + \mathcal{O}(T^4, \bar{H}^4) . \qquad (32.14)$$

On the other hand, insertion of (32.12) into (32.4) yields, using Eq. (32.2) and definition (31.16) of the Pauli susceptibility,

$$M = \chi_P \bar{H}\left\{1 + \frac{\pi^2}{6}(k_B T)^2 \frac{N''(0)}{N(0)} + \mu \frac{N'(0)}{N(0)}\right.$$

$$\left. + \frac{1}{6}(\mu_B \bar{H})^2 \frac{N''(0)}{N(0)} + \mathcal{O}(\mu^2, T^4, \bar{H}^4)\right\} \qquad (32.15)$$

or, substituting from (32.14),

$$M = \chi_P \bar{H} \left\{ 1 - \left(\frac{\bar{H}}{H^*} \right)^2 - \left(\frac{T}{T^*} \right)^2 + \mathcal{O}(T^4, \bar{H}^4) \right\} . \tag{32.16}$$

Here we have introduced a characteristic field H^* and a characteristic temperature T^* by the relations

$$(\mu_B H^*)^{-2} \equiv \frac{1}{2} \left\{ \left(\frac{N'(0)}{N(0)} \right)^2 - \frac{1}{3} \frac{N''(0)}{N(0)} \right\} \tag{32.17}$$

and

$$(k_B T^*)^{-2} \equiv \frac{\pi^2}{6} \left\{ \left(\frac{N'(0)}{N(0)} \right)^2 - \frac{N''(0)}{N(0)} \right\} , \tag{32.18}$$

respectively. Note that for the quasi-free band of the form (17.22) one easily deduces from Eq. (17.23) that $(N'(0)/N(0))^2 = -N''(0)/N(0) = (2\varepsilon_F)^{-2}$ so that, in terms of the *Fermi degeneracy temperature* (see Table 1, Chapter 6 of Ref. 4) $T_F \equiv \varepsilon_F/k_B$, one has $T^*/T_F = 2\sqrt{3}/\pi$ and $\mu_B H^* = \sqrt{6}k_B T_F$.

Solving Eq. (32.16) for $\bar{H}$ defined in (29.16), we obtain the *equation of state* (see Eq. (3.7) of Ref. 60)

$$H + \lambda M = \frac{M}{\chi_P} \left\{ 1 + \left(\frac{M}{M^*} \right)^2 + \left(\frac{T}{T^*} \right)^2 + \mathcal{O}(T^4, M^4) \right\} \tag{32.19}$$

where $M^* \equiv \chi_P H^*$. This equation was first derived by Stoner (for $H = 0$ see Eq. (5.8) of paper I and Eq. (3.6) of paper II, Ref. 23). For $H = 0$ and $M = M_o$, Eq. (32.19) has the double solution $M = 0$ and $M = M_o(T)$ analogous to (29.25) where, neglecting higher-order terms,

$$\left(\frac{M_o(T)}{M^*} \right)^2 = \kappa - 1 - \left(\frac{T}{T^*} \right)^2 = \frac{T_S^2 - T^2}{T^{*2}} > 0 \tag{32.20}$$

and

$$\kappa \equiv \lambda \chi_P = 2\lambda \mu_B^2 N(0) . \tag{32.21}$$

This determines the *Stoner temperature* T_S and the saturation magnetization $M_s \equiv M_o(0)$ according to

$$\left(\frac{T_S}{T^*} \right)^2 = \left(\frac{M_s}{M^*} \right)^2 = \kappa - 1 > 0 \tag{32.22}$$

where the last inequality is the *Stoner criterion* for ferromagnetism (see Eq. (5.4) of paper I, Ref. 23).

Introducing definitions analogous to (29.17) and (29.21),

$$h \equiv \frac{H}{H_o} \; ; \quad m \equiv \frac{M}{M_s} \; ; \quad t \equiv 1 - \frac{T}{T_S} \; , \tag{32.23}$$

using Eqs. (32.21), (32.22) and neglecting again higher-order terms, the equation of state (32.19) may be cast into a form analogous to (29.23) (see Eq. (3.15) of Ref. 60),

$$h = m\{-t(2-t) + m^2\} \; ; \quad \kappa > 1 \; , \tag{32.24}$$

and one finds

$$H_o = \frac{\kappa - 1}{\chi_P} M_s = (\kappa - 1)^{3/2} H^* \; . \tag{32.25}$$

Similarly to Eq. (29.23) without the term $\mathcal{O}(m^2)$, Eq. (32.24) does not show saturation as is seen from Fig. 5.6. However, in the case where the band width W is so small that at sufficiently high temperatures $\beta W \ll 1$, the electrons obey Boltzmann statistics within the band so that Eq. (32.6) may be written as $n_\sigma = \int d\varepsilon N(\varepsilon) \exp(-\beta(\varepsilon - \mu_\sigma))$. This means that the electrons

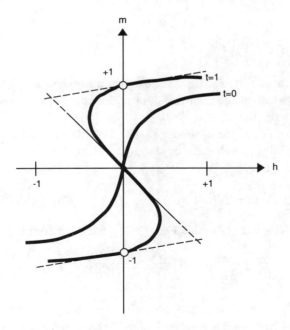

Fig. 5.6: Equation of state (32.24) showing absence of saturation.

are localized on the atoms and we must recover the situation described in Section 29. Indeed, Eq. (32.4) may be expressed in terms of the partition function (29.4) for $J = 1/2$ and of the Zeeman energy (29.1) with $g = 2$. Comparison with Eqs. (29.6), (29.7) then shows that the atomic density n_a is replaced by the electron density n.

The magnetic free energy is obtained with the help of Eqs. (29.26), (32.19), (32.21) and (32.22); it is similar to Eq. (29.27):

$$F_m = \begin{cases} (V/2\chi_P)M^2\{1 - \kappa + (T/T^*)^2 + \frac{1}{2}(M/M^*)^2\} \; ; \; \kappa < 1, > 1 \; ; \\ (\lambda V/2)[1 - (1/\kappa)]M^2\{(T/T_S)^2 - 1 + \frac{1}{2}(M/M_s)^2\} \; ; \; \kappa > 1 \; . \end{cases}$$
(32.26)

Near T_S, $|t| \ll 1$ so that for $H = 0$, Eq. (32.24) takes the form (29.25) of the order parameter with $a_J = \sqrt{2}$.

From Eq. (32.26) the magnetic entropy density s_m and specific heat c_m are obtained according to (29.31) and (29.33). Inserting the expression (32.20), making use of (32.22) and (32.23) one finds

$$s_m = -\lambda\left(1 - \frac{1}{\kappa}\right)M_s^2\frac{T}{T_S^2}\left[1 - \left(\frac{T}{T_S}\right)^2\right] \; ; \; \kappa > 1 \; ; \; T < T_S \qquad (32.27)$$

and

$$c_m = \lambda\left(1 - \frac{1}{\kappa}\right)M_s^2\frac{T}{T_S^2}\left[3\left(\frac{T}{T_S}\right)^2 - 1\right] \; ; \; \kappa > 1 \; ; \; T < T_S \; . \qquad (32.28)$$

For temperatures $T \ll T_S$, Eq. (32.28) gives a negative contribution to the specific heat which is not unphysical, provided that the orbital contribution c_V dominates or that $c_V + c_m \geq 0$.

Using Eq. (10.3) for c_V and Eqs. (32.17), (32.18), (32.21), (32.22) for c_m and describing the density of states (10.4) in the neighborhood of the Fermi level $\varepsilon = 0$ by

$$N(\varepsilon) = N(0)\left\{1 + \frac{\varepsilon}{\varepsilon_o} - \frac{\alpha}{2}\left(\frac{\varepsilon}{\varepsilon_o}\right)^2 + \mathcal{O}(\varepsilon^3)\right\} , \qquad (32.29)$$

one finds for the ratio (see also Eqs. (3.23), (3.24) of Ref. 60)

$$r \equiv -\frac{c_m}{c_V}\bigg|_{T=0} = (\kappa - 1)\frac{(N'(0)/N(0))^2 - (N''(0)/N(0))}{(N'(0)/N(0))^2 - (N''(0)/3N(0))}$$

$$= (\kappa - 1)\frac{1 + \alpha}{1 + \alpha/3} \leq 1 \; . \qquad (32.30)$$

For the quasi-free band considered above, the parenthesis in Eq. (32.29) has the form (17.23), i.e. $(1 + \varepsilon/\varepsilon_F)^{1/2}$ and hence $\varepsilon_o = 2\varepsilon_F$ and $\alpha = 1$. In this case the condition $r \leq 1$ for avoiding a negative specific heat becomes $\kappa \leq 5/3$ which means *weak ferromagnetism*. An analogous calculation for the saturation value of the *relative magnetization* [23,27,60)]

$$\zeta \equiv \frac{M}{n\mu_B} = \frac{n_\uparrow - n_\downarrow}{n_\uparrow + n_\downarrow} \leq 1 \tag{32.31}$$

yields, using Eq. (14.14) with $D = 3$,

$$\zeta_s^2 \equiv \left(\frac{M_s}{n\mu_B}\right)^2 = \frac{8(\kappa - 1)(N(0)/n)^2}{(N'(0)/N(0))^2 - (N''(0)/3N(0))} = \frac{9}{2}\frac{\kappa - 1}{1 + \alpha/3}\left(\frac{\varepsilon_o}{\varepsilon_F}\right)^2$$

$$= \frac{9}{2}\frac{r}{1 + \alpha}\left(\frac{\varepsilon_o}{\varepsilon_F}\right)^2 \leq 1 \tag{32.32}$$

where in the last expression Eq. (32.30) was inserted. This is a second consistency condition for the Stoner theory to be applicable.

In order to obtain the Stoner susceptibility χ_S according to Eq. (29.10), both for the paramagnetic and the ferromagnetic cases $\kappa < 1$ and $\kappa > 1$, we go back to Eq. (32.19) which yields, using (32.21),

$$\chi_S^{-1} = \chi_P^{-1}\left\{1 - \kappa + 3\left(\frac{M_o}{M^*}\right)^2 + \left(\frac{T}{T^*}\right)^2 + \mathcal{O}(T^4, M_o^4)\right\}. \tag{32.33}$$

In the paramagnetic case, $M_o = 0$ so that

$$\chi_S^{-1}(T) = \chi_P^{-1}\left\{1 - \kappa + \left(\frac{T}{T^*}\right)^2\right\} ; \quad \kappa < 1 \tag{32.34}$$

and

$$\chi_S(0) = \frac{\chi_P}{1 - \kappa} \tag{32.35}$$

which shows that the effect of interactions with a coupling constant λ given by (32.21) is to increase the Pauli susceptibility by the *Stoner enhancement factor* $(1 - \kappa)^{-1}$.

In the ferromagnetic case $\kappa > 1$ we have to distinguish the paramagnetic and the ferromagnetic regions $T > T_S$ and $T < T_S$, respectively. For $T > T_S$ we again have $M_o = 0$ so that Eq. (32.33) becomes with (32.22), (32.23)

$$\chi_S^{-1}(T) = \chi_P^{-1}(\kappa - 1)\left\{\left(\frac{T}{T_S}\right)^2 - 1\right\} ; \quad \kappa > 1 ; \quad T > T_S \tag{32.36}$$

while in the case $T < T_S$ application of (32.20), (32.22) yields

$$\chi_S^{-1}(T) = 2\chi_P^{-1}(\kappa - 1)\left\{1 - \left(\frac{T}{T_S}\right)^2\right\} ; \quad \kappa > 1 ; \quad T < T_S \tag{32.37}$$

(compare Eqs. (3.18) of Ref. 60). Equations (32.34), (32.36) and (32.37) are plotted in Fig. 5.7 which shows a weak point of Stoner theory namely the *absence of a Curie-Weiss law* (29.30) as it is shown by the straight lines in Figs. 5.2 and 5.5. This of course is the direct consequence of the fact that Stoner theory is a low-temperature approximation. As mentioned above it is in the opposite limit $\beta W \ll 1$ that the local-moment situation described in Section 29, and hence the Curie-Weiss law is recovered (see Section 35). Experimentally a Curie-Weiss law is indeed observed asymptotically, e.g. in the case of nickel (see Fig. 2 in Chapter 15 of Ref. 4) and in the weak itinerant-electron ferromagnets considered in Section 34.

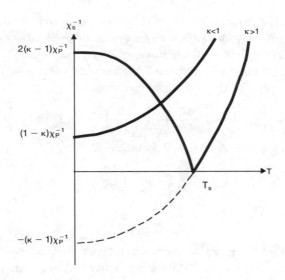

Fig. 5.7: The inverse Stoner susceptibility $\chi_S^{-1}(T)$ for a paramagnet ($\kappa < 1$) according to Eq. (32.34) and for a ferromagnet ($\kappa > 1$) according to Eqs. (32.36) and (32.37). The curves are parabolic in contrast to the linear Curie-Weiss law.

The spin-dependent electron energy (32.1) may be used to define the Stoner Hamiltonian of non-interacting band electrons as

$$\mathcal{H}_o^S = \sum_{\mathbf{k}\sigma} \varepsilon_{\mathbf{k}\sigma} a_{\mathbf{k}\sigma}^+ a_{\mathbf{k}\sigma} \qquad (32.38)$$

where, for $\mu = 0$ and $H = 0$, we have from (32.2), (32.3) and (29.16)

$$\varepsilon_{\mathbf{k}\sigma} = \varepsilon_{\mathbf{k}} - \frac{1}{2}\sigma\Delta^S . \qquad (32.39)$$

Correspondingly, the magnetization (32.4) defines the *order parameter* of itinerant-electron ferromagnetism (IEF) as

$$M = \frac{\mu_B}{V} \sum_{\mathbf{k}\sigma} \sigma \langle a_{\mathbf{k}\sigma}^+ a_{\mathbf{k}\sigma} \rangle_o \tag{32.40}$$

which, on using Eqs. (7.40), (10.4) and (32.39) becomes

$$M = \frac{\mu_B}{V} \sum_{\sigma} \sigma \int d\varepsilon N(\varepsilon) f_o(\varepsilon - \frac{1}{2}\sigma\Delta^S) . \tag{32.41}$$

Developing this expression in powers of the *Stoner gap* Δ^S one obtains, after reducing higher derivatives of f_o to f_o' by partial integrations and using the zero-temperature approximation $f_o'(\varepsilon) = -\delta(\varepsilon)$,

$$M = M_s = \mu_B \Delta^S N(0) \left[1 + \frac{(\Delta^S)^2}{24} \frac{N''(0)}{N(0)} + \mathcal{O}((\Delta^S)^4) \right] . \tag{32.42}$$

Inserting here Eq. (32.3) we may solve for M. With definition (32.21) and $N''(0)/N(0) = -(2\varepsilon_F)^{-2}$ valid for free electrons, one finds

$$M_s = \sqrt{\frac{6(\kappa - 1)}{\kappa^3}} \frac{2\varepsilon_F \chi_P}{\mu_B} [1 + \mathcal{O}(\kappa - 1)] , \tag{32.43}$$

which differs by a factor of 2 from Eq. (32.22) because we have set $\mu = 0$ instead of using (32.14).

From Eq. (32.41) we see that, at $T = 0$, the minority band is less occupied, $n_\downarrow < n_\uparrow$ (see Fig. 5.8). Thus the ferromagnetic groundstate is the polarized Fermi sea $|F\uparrow\rangle$ satisfying

$$\mathcal{H}_o^S |F\uparrow\rangle = E_o |F\uparrow\rangle \tag{32.44}$$

with $E_o = \sum \varepsilon_{\mathbf{k}\sigma} \theta(-\varepsilon_{\mathbf{k}\sigma})$. The *Stoner excitations* above this groundstate then are of the form $a_{\mathbf{k}+\mathbf{q}\downarrow}^+ a_{\mathbf{k}\uparrow} |F\uparrow\rangle$ and their energy is obtained by calculating the commutator $[\mathcal{H}_o^S, a_{\mathbf{k}+\mathbf{q}\downarrow}^+ a_{\mathbf{k}\uparrow}]$, namely,

$$\omega_k^S(q) = \varepsilon_{\mathbf{k}+\mathbf{q}\downarrow} - \varepsilon_{\mathbf{k}\uparrow} = \varepsilon_{\mathbf{k}+\mathbf{q}} - \varepsilon_{\mathbf{k}} + \Delta^S . \tag{32.45}$$

For free electrons this spectrum is limited by the bounds

$$\max \left\{ 0, \Delta^S + \frac{\hbar^2}{2m} q(q - 2k) \right\} \le \omega_k^S(q) \le \Delta^S + \frac{\hbar^2}{2m} q(q + 2k) \tag{32.46}$$

shown in Fig. 5.9 (see Fig. 7.12 in Ref. 1 and Fig. 6.21 in Ref. 2).

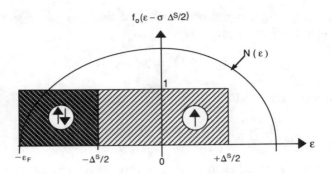

Fig. 5.8: Polarized groundstate.

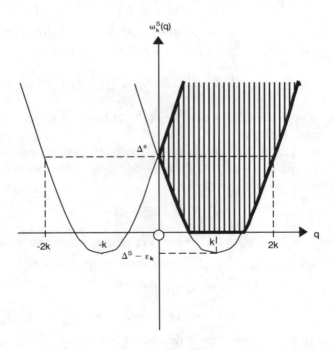

Fig. 5.9: The spectrum of Stoner excitations according to Eq. (32.46).

33. Groundstate, Excitations and Symmetries of the Hubbard Model

Let us now examine the many-body aspect of Stoner's ansatz (32.1), (32.2). Using Eqs. (29.16), (32.4) and setting $H = 0$, we may write the Stoner Hamiltonian (32.38) as

$$\mathcal{H}_o^S \equiv \sum_{\mathbf{k}\sigma} \varepsilon_{\mathbf{k}\sigma} a_{\mathbf{k}\sigma}^+ a_{\mathbf{k}\sigma} = \mathcal{H}_o + \mathcal{H}_{\text{int}} \tag{33.1}$$

where $\mathcal{H}_o = \sum(\varepsilon_{\mathbf{k}} - \mu)a_{\mathbf{k}\sigma}^+ a_{\mathbf{k}\sigma}$ describes free band electrons while

$$\mathcal{H}_{\text{int}} = -V\lambda\mu_B^2(n_\uparrow - n_\downarrow)^2 , \tag{33.2}$$

since

$$n_\sigma = \frac{1}{V} \sum_{\mathbf{k}} a_{\mathbf{k}\sigma}^+ a_{\mathbf{k}\sigma} . \tag{33.3}$$

It is worth noting that Eq. (33.2) is the one-band analog of the *s-d* interaction (17.11) for the particular choice of the coupling functions given by $J_i(\mathbf{q}) = \lambda\mu_B^2\delta_{\mathbf{q},0}\delta_{i,3}$.

A more interesting form of $\mathcal{H}_o^S$ is obtained by going over to local variables with the help of the fermion analog of the transformation (30.20),

$$a_{\mathbf{k}\sigma}^+ = \frac{1}{\sqrt{L}} \sum_i c_{i\sigma}^+ e^{i\mathbf{k}\cdot\mathbf{R}_i} . \tag{33.4}$$

Then the occupation numbers of the unit cells (lattice sites) satisfy the exclusion principle,

$$n_{i\sigma} \equiv c_{i\sigma}^+ c_{i\sigma} = n_{i\sigma}^2 , \tag{33.5}$$

and

$$\mathcal{H}_o = \sum_{ij\sigma} t(\mathbf{R}_i - \mathbf{R}_j)c_{i\sigma}^+ c_{j\sigma} \tag{33.6}$$

where

$$t(\mathbf{R}) \equiv \frac{1}{L} \sum_{\mathbf{k}} (\varepsilon_{\mathbf{k}} - \mu)e^{i\mathbf{k}\cdot\mathbf{R}} . \tag{33.7}$$

Since with (1.10), Eq. (33.3) becomes $n_\sigma = V^{-1}\sum_i n_{i\sigma}$, we may also express Eq. (33.2) in terms of local variables. Using relation (33.5) and defining a new coupling constant by

$$U \equiv 2\lambda\mu_B^2 \frac{L}{V} > 0 , \tag{33.8}$$

one finds, replacing in Eq. (33.2) λ by λL,

$$\mathcal{H}_{\text{int}} = -\frac{U}{2}\sum_{ij}(n_{i\uparrow} - n_{i\downarrow})(n_{j\uparrow} - n_{j\downarrow})$$

$$= -\frac{NU}{2} + U\sum_i n_{i\uparrow}n_{i\downarrow} - 2U\sum_{i \neq j} S_i^z S_j^z \qquad (33.9)$$

where, according to Eqs. (17.12) and (33.4),

$$\mathbf{S}_i \equiv \frac{1}{L}\sum_{\mathbf{q}} \mathbf{s}(\mathbf{q})e^{i\mathbf{q}\cdot\mathbf{R}_i} = \frac{1}{2}\sum_{\sigma\sigma'} c_{i\sigma}^+ \vec{\sigma}_{\sigma\sigma'} c_{i\sigma'} \qquad (33.10)$$

and $N = nV$ is the number of electrons. In the second line of Eq. (33.9), the first term is a trivial constant while the third term is negligible for uncorrelated sites and nonmagnetic electrons, i.e. if $\langle S_i^z \rangle \sim 0$.

Thus the important interaction in the many-body form of the Stoner ansatz is the on-site repulsion

$$\mathcal{H}_U \equiv U\sum_i n_{i\uparrow}n_{i\downarrow} \qquad (33.11)$$

which approximates the Coulomb repulsion between electrons with opposite spin localized in the same unit cell i. Note that the replacement $\lambda \to \lambda L$ in going from (33.2) to (33.9) was necessary to make the diagonal part $i = j$ of $\mathcal{H}_{\text{int}}$ again an *extensive quantity*, i.e. $\mathcal{H}_U \propto L^1$ when $L \to \infty$. This is the interaction of the *Gutzwiller-Hubbard-Kanamori model* $\mathcal{H}_o + \mathcal{H}_U^{24)}$ which is of particular interest for *narrow-band systems*. In this case Eq. (33.7) yields $t(\mathbf{R}) \cong 0$, except for $\mathbf{R} = 0$ and for the z nearest-neighbor vectors $\mathbf{R} = \vec{\delta}$, so that Eq. (33.6) becomes, up to the trivial term $Nt(0)$,

$$\mathcal{H}_o = -t\sum_{\langle ij \rangle \sigma} c_{i\sigma}^+ c_{j\sigma} \qquad (33.12)$$

where $t \equiv -t(\vec{\delta})$ and $\langle ij \rangle$ means summation over nearest neighbors. Unfortunately, it is not sufficient to assume a coupling λ strong enough for the Stoner criterion (32.22) to hold, in order to conclude that the Hubbard model yields ferromagnetism since the latter is favored already by the last term in Eq. (33.9). The question whether and under what conditions a ferromagnetic groundstate exists, has been the central issue with the Hubbard model since its introduction in 1963[1,2,24,27].

Written in the Bloch representation (33.4) the Hubbard interaction becomes, using Eqs. (1.9), (33.5) and neglecting Umklapp processes,

$$\mathcal{H}_U = \frac{U}{L} \sum_{\mathbf{kk'q}} a^+_{\mathbf{k}\uparrow} a^+_{\mathbf{k'}\downarrow} a_{\mathbf{k'+q}\downarrow} a_{\mathbf{k-q}\uparrow} . \tag{33.13}$$

This is of the form (3.38) and hence is represented by Fig. 1.2 with $\sigma = \uparrow$ and $\sigma' = \downarrow$. Note that the spin arrangement is the same as in the reduced BCS Hamiltonian (20.36) which is the reason why the negative-U Hubbard interaction is often used as a model for superconductivity (see Ref. 61). On the other hand, in the case of a *strong on-site repulsion*, $U \gg t$, double occupancy of the lattice sites is very much suppressed so that ordering of the electron spins on the singly occupied sites becomes likely. In this case a perturbative treatment of the Hubbard interaction is not justified and one must resort to a Bethe-Salpeter equation of the form (19.10) as shown in Fig. 3.12. However, in distinction to the impurity potential considered in Section 19, the interaction (33.13) is nonstatic so that *retardation effects* described by frequency variables as in the strong-coupling superconductivity theory of Section 28 must be included.

According to Eqs. (9.15) and (11.16), the connected part of the two-electron Green's function (10.30) may be written as the perturbation expansion

$$\mathcal{G}^2 \equiv \langle T\{a^+_{\mathbf{k}_1\uparrow}(-i\tau_1) a^+_{\mathbf{k}_2\downarrow}(-i\tau_2) a_{\mathbf{k}_3\uparrow}(-i\tau_3) a_{\mathbf{k}_4\downarrow}(0)\}\rangle_{cp}$$

$$= \sum_{N=1}^{\infty} \frac{1}{N!} \langle T\{\left[-\int_o^\beta d\tau \mathcal{H}_U^o(-i\tau)\right]^N a^+_{\mathbf{k}_1\uparrow}(-i\tau_1) \cdots a_{\mathbf{k}_4\downarrow}(0)\}\rangle_{o,cp} \tag{33.14}$$

where time-translation invariance was used to set $\tau_4 = 0$. Taking the time Fourier transform, the t-matrix is defined in analogy to Eqs. (19.11), (19.12) as the series

$$\tilde{\mathcal{G}}^2 \equiv \int_o^\beta d\tau_1 e^{-i\nu_1\tau_1} \int_o^\beta d\tau_2 e^{-i\nu_2\tau_2} \int_o^\beta d\tau_3 e^{i\nu_3\tau_3} \mathcal{G}^2$$

$$= \sum_{N=1}^{\infty} \tilde{\mathcal{G}}_2^{(N)} = \sum_{N=1}^{\infty} \sum_{\nu_4} \left(\prod_{l=1}^4 \tilde{\mathcal{G}}^o(k_l)\right) t^{(N)}(k_1, k_2; k_3, k_4) \tag{33.15}$$

where $k_l \equiv (\mathbf{k}_l, i\nu_l)$. Inserting Eq. (33.13) and using Wick's theorem (9.11) one finds (Problem 5.10)

$$t^{(1)} = \frac{U}{L} \delta_{k_1+k_2, k_3+k_4} \tag{33.16}$$

and

$$t^{(2)} = -\left(\frac{U}{L}\right)^2 \delta_{k_1+k_2, \, k_3+k_4} \beta^{-1} \sum_q \tilde{\mathcal{G}}^o(k_1 + q)\tilde{\mathcal{G}}^o(k_2 - q) + t^{(2)}_{cr} \quad (33.17)$$

where $q \equiv (\mathbf{q}, i\nu)$ and $t^{(N)}_{cr}$, $N \geq 2$, are the contributions from the crossed ladders analogous to Fig. 3.11 (c). $t^{(1)}$ and $t^{(2)}$ are represented in Fig. 5.10 which shows that the q-summation is associated to a closed loop in agreement with the rule formulated in Sections 10 and 11.

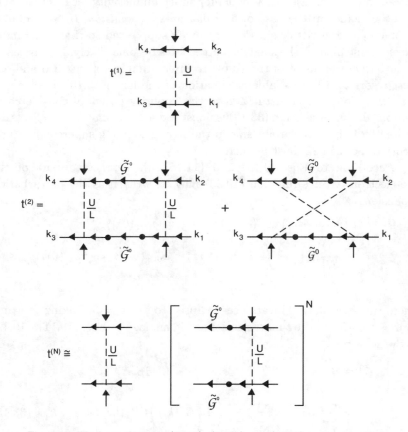

Fig. 5.10: The contributions (33.16), (33.17) and (33.18) to the t-matrix.

Neglecting the crossed ladders, $t^{(1)}$ is the only contribution to the two-electron-irreducible four-point vertex Γ (see Section 12) defined by the Bethe-Salpeter equation analogous to (19.10) and the t-matrix becomes the geometrical series (compare Eq. (10.21) of Ref. 27)

$$\sum_{N=1}^{\infty} t^{(N)} \cong t^{(1)}\{1 - UG_D + (UG_D)^2 - + \cdots\}$$

$$= \frac{U/L}{1 + UG_D} \delta_{k_1+k_2, k_3+k_4} \tag{33.18}$$

with the $t^{(N)}$ shown in Fig. 5.10. Here

$$G_D(k_1 k_2) \equiv \frac{1}{\beta L} \sum_q \tilde{\mathcal{G}}^o(k_1 + q)\tilde{\mathcal{G}}^o(k_2 - q) \tag{33.19}$$

and D is the dimension of the system. Using Eq. (7.44) with $\mu = 0$ for the free propagators $\tilde{\mathcal{G}}^o$, comparison with Eq. (10.17) shows that the ν-sum in (33.19) is the Lindhard function $-\beta L_+(\varepsilon_{k_1+q} - i\nu_1, -\varepsilon_{k_2-q} + i\nu_2)$ so that Eq. (10.19) yields

$$G_D(k_1 k_2) = \frac{1}{L} \sum_q \frac{f_o(-\varepsilon_{k_2-q}) - f_o(\varepsilon_{k_1+q})}{\varepsilon_{k_1+q} + \varepsilon_{k_2-q} - i(\nu_1 + \nu_2)} \tag{33.20}$$

where Eqs. (7.26), (7.42), (7.43) were used to write $n_o(\omega \pm i\nu) = -f_o(\omega)$.

In Eq. (33.20) the analytic continuation $i\nu_1 \to \varepsilon_{k_1}$, $i\nu_2 \to \varepsilon_{k_2}$ may now be executed. Evaluating the resulting expression for the Fermi-sea state $|F\rangle$ at $T = 0$ where $f_o(\varepsilon) = \theta(-\varepsilon)$, we have $\varepsilon_{k_1} < 0$, $\varepsilon_{k_2} < 0$ and $\varepsilon_{k_1+q} > 0$, $\varepsilon_{k_2-q} > 0$ so that the denominator in (33.20) is > 0. Making now the simplifying assumption of a dilute electron gas, the Fermi wavenumber k_F is small and we may neglect k_1 and k_2 relative to q. Then Eq. (33.20) becomes (compare Eq. (15) of Kanamori , Ref. 24 and Eq. (10.23) of Ref. 27)

$$G_D \cong \frac{1}{L} \sum_q \frac{\theta(\varepsilon_q)}{2(\varepsilon_q + \varepsilon_F)} = \frac{V}{2L} \int_o^{W-\varepsilon_F} d\varepsilon \frac{N_D(\varepsilon)}{\varepsilon + \varepsilon_F} \tag{33.21}$$

where $-\varepsilon_F$ is the bottom of the band and W the bandwidth (compare Fig. 3.4), and $N_D(\varepsilon)$ is the density of states (10.4) in dimension D. Note that with the choice $t(\mathbf{R}) = -t\delta_{\mathbf{R}\vec{\delta}}$ the inversion of Eq. (33.7) leads to the band energy

$$\varepsilon_{\mathbf{k}} - \mu = \sum_{\mathbf{R}} t(\mathbf{R})e^{-i\mathbf{k}\cdot\mathbf{R}} = -t\sum_{\vec{\delta}} \cos(\mathbf{k} \cdot \vec{\delta}) \tag{33.22}$$

where, in our usual notation, μ is fixed by requiring the minimum to have the value $-\varepsilon_F$, hence $\mu = zt - \varepsilon_F$. For a simple cubic lattice in dimension D, the number of nearest neighbors, $z = 2D$ and the bandwidth is found to be $W = 4Dt$.

Now the ferromagnetic instability is determined by the Stoner criterion (32.22) which may be expressed in terms of U with the help of Eqs. (32.21) and (33.8). However, since we are interested in the case of strong on-site repulsion $U \gg t$, U should be replaced by the effective coupling determined by the t-matrix (33.18), namely by (see Eq. (16) of Kanamori , Ref. 24 and Eq. (10.24) of Ref. 27)

$$U_D^{\text{eff}} = \frac{U}{1 + U G_D} \ . \tag{33.23}$$

Then the Stoner criterion becomes in dimension D (see Eq. (18) of Kanamori, Ref. 24)

$$\kappa = \kappa_D \equiv U_D^{\text{eff}} \frac{V}{L} N_D(0) > 1 \ . \tag{33.24}$$

For the dilute electron gas considered here $|\mathbf{k} \cdot \vec{\delta}| \ll 1$ so that Eq. (33.22) has the shape of a quasi-free band and the density of states $N_D(\varepsilon)$ has the form (30.28), i.e. using (14.14),

$$N_D(\varepsilon) = \frac{nD}{4\varepsilon_F} \left(1 + \frac{\varepsilon}{\varepsilon_F} \right)^{\frac{D}{2} - 1} . \tag{33.25}$$

With this expression Eq. (33.21) is easily calculated, one finds

$$G_2 = \frac{\rho}{2\varepsilon_F} \ln \frac{W}{\varepsilon_F} \ ; \quad G_3 = \frac{3\rho}{2\varepsilon_F} \left(\sqrt{\frac{W}{\varepsilon_F}} - 1 \right) \tag{33.26}$$

where ρ is the *band-filling factor*,

$$0 \leq \rho \equiv \frac{N}{2L} \leq 1 \ . \tag{33.27}$$

It is convenient to assume that there is just one band state per unit cell and per spin so that the relation

$$\frac{L}{V} = \int_{-\varepsilon_F}^{W - \varepsilon_F} d\varepsilon N_D(\varepsilon) = \frac{n}{2} \left(\frac{W}{\varepsilon_F} \right)^{D/2} \tag{33.28}$$

holds which allows to substitue

$$\varepsilon_F = W \rho^{2/D} \ . \tag{33.29}$$

Making use of Eqs. (33.23)–(33.29) one finds for $D = 2, 3$

$$\kappa_2 \left(\frac{U}{W} \right) = \frac{U/W}{1 + (U/2W) \ln(\rho^{-1})} \ ,$$

$$\kappa_3 \left(\frac{U}{W} \right) = \frac{(3U/2W)\rho^{1/3}}{1 + (3U/2W)(1 - \rho^{1/3})} \ . \tag{33.30}$$

The critical band-filling for the onset of ferromagnetism according to Kanamori[24], ρ_D^K, then is defined by the condition $\kappa_D(\infty) = 1$. This leads in dimensions 2 and 3 to

$$\rho_2^K = e^{-2} = 0.1353 \; ; \quad \rho_s^K = 2^{-3} = 0.125 \; . \tag{33.31}$$

Note that in the first line of Kanamori's Table I[24], $n_{cr} = 2\rho_2^K$.

The t-matrix (33.18) of course is just the *random phase approximation* (RPA, see the remarks leading to Eq. (22.23) above). It is therefore not surprising that the same result is obtained with the help of the equation of motion method used, e.g., in Eq. (32.45). This is indeed the procedure used by Mattis which leads to Eq. (6.105) of Ref. 2. Mattis, however uses a different criterion for the onset of ferromagnetism, namely the difference between the nonmagnetic and the magnetic total energies, E_D^{non} and E_D^{mag}, respectively. The first is, in Hartree-Fock approximation,

$$E_D^{\text{non}} = N\varepsilon_F + \langle F|\mathcal{H}_o + \mathcal{H}_U|F\rangle_{\text{HF}} = \sum_{\mathbf{k}\sigma}(\varepsilon_F + \varepsilon_\mathbf{k})\langle F|a_{\mathbf{k}\sigma}^+ a_{\mathbf{k}\sigma}|F\rangle$$

$$+ \frac{U}{L} \sum_{\mathbf{k}\mathbf{k'}\mathbf{q}} \langle F|a_{\mathbf{k}\uparrow}^+ a_{\mathbf{k}-\mathbf{q}\uparrow}|F\rangle\langle F|a_{\mathbf{k'}\downarrow}^+ a_{\mathbf{k'}+\mathbf{q}\downarrow}|F\rangle \; . \tag{33.32}$$

Making use of Eqs. (33.25), (33.27) and (33.29) one finds

$$E_D^{\text{non}} = 2V \int_{-\varepsilon_F}^{o} d\varepsilon(\varepsilon_F + \varepsilon)N_D(\varepsilon) + \frac{U}{L}\left[V \int_{-\varepsilon_F}^{o} d\varepsilon N_D(\varepsilon)\right]^2$$

$$= LW\left\{\frac{2D}{D+2}\rho^{\frac{D+2}{D}} + \frac{U}{W}\rho^2\right\} \; . \tag{33.33}$$

The magnetic energy E_D^{mag} has the same form (33.32) but taken in the state $|M\rangle$ defined by $\langle M|a_{\mathbf{k}\sigma}^+ a_{\mathbf{k}\sigma}|M\rangle = \delta_{\sigma\uparrow}\theta(\mu_D - \varepsilon_\mathbf{k})$ where the chemical potential μ_D is determined by the condition

$$n = \frac{1}{V}\sum_{\mathbf{k}}\theta(\mu_D - \varepsilon_\mathbf{k}) = \int_{-\varepsilon_F}^{\mu_D} d\varepsilon N_D(\varepsilon) = \frac{n}{2}\left(1 + \frac{\mu_D}{\varepsilon_F}\right)^{D/2} \; . \tag{33.34}$$

Using this condition we obtain in the same way as before

$$E_D^{\text{mag}} = V \int_{-\varepsilon_F}^{\mu_D} d\varepsilon(\varepsilon_F + \varepsilon)N_D(\varepsilon) = LW\frac{D}{D+2}(2\rho)^{\frac{D+2}{D}} \; . \tag{33.35}$$

This leads to an energy difference

$$\Delta E_D\left(\frac{U}{W}\right) \equiv E_D^{\text{non}} - E_D^{\text{mag}} = LW$$

$$\times \left\{-\frac{2D}{D+2}(2^{2/D} - 1)\rho^{\frac{D+2}{D}} + \frac{U_D^{\text{eff}}}{W}\rho^2\right\} \tag{33.36}$$

where we have replaced U by U_D^{eff}. The Mattis criterion for the onset of ferromagnetism, $\Delta E_D(\infty) = 0^{2)}$, then defines another critical band filling ρ_D^M. Using Eqs. (33.23), (33.26) and (33.29) one finds in dimensions 2 and 3

$$\rho_2^M = e^{-2} = \rho_2^K \tag{33.37}$$

and

$$\rho_3^M = \left(\frac{\alpha}{1+\alpha}\right)^3 \; ; \quad \alpha \equiv \frac{9}{5}(2^{2/3} - 1) = 1.0573 \tag{33.38}$$

or $\rho_3^M = 0.1357$, in agreement with Eq. (6.113b) of Ref. 2. The phase diagram U/W versus ρ is shown, for $D = 3$, in Fig. 6.16 of this reference. Note that the diagram is symmetric with respect to electron-hole conjugation, $\rho \leftrightarrow 1 - \rho$.

In view of the approximations made, this phase diagram cannot be the full truth. Indeed, as discussed in the introduction to this chapter, the $U > 0$ Hubbard model on a bipartite lattice is known to have an antiferromagnetic groundstate at half-filling, $\rho = 1/2^{25)}$. This is not entirely surprising since, for $\rho = 1/2$ and $U/W \gg 1$, it can be shown by a canonical transformation (20.1) that the Hubbard Hamiltonian, acting on singly-occupied states where all $n_i \equiv n_{i\uparrow} + n_{i\downarrow} = 1$ (i.e. on the "Gutzwiller-projected" Hilbert space) is equivalent, to second order in the hopping term $\mathcal{H}_o$, with the Heisenberg Hamiltonian $\mathcal{H}_H = \mathcal{H}_{1/2}(0)$, Eq. (30.1) with $J = -t^2/U$. In formulas (see Section 2 of Ref. 62):

$$\mathcal{H} = \mathcal{H}_o + \mathcal{H}_U \to \tilde{\mathcal{H}} = e^S \mathcal{H} e^{-S} \; ; \quad S \propto \frac{t}{U} \; ;$$

$$[S, \mathcal{H}_U] = \mathcal{H}_o' \; ; \quad (\mathcal{H}_o - \mathcal{H}_o')\Big|_{\text{all } n_i = 1} = 0 \; . \tag{33.39}$$

Then

$$\tilde{\mathcal{H}} = \mathcal{H}_o - \mathcal{H}_o' + \mathcal{H}_U + [\mathcal{H}_o - \frac{1}{2}\mathcal{H}_o', S] + \mathcal{O}\left(\frac{t^3}{U^2}\right) \; ;$$

$$[\mathcal{H}_o - \frac{1}{2}\mathcal{H}_o', S] = \frac{4t^2}{U} \sum_{\langle ij \rangle} \left(\mathbf{S}_i \cdot \mathbf{S}_j - \frac{1}{4}n_i n_j\right) \; ; \quad \text{all } n_i = 1 \tag{33.40}$$

where $\mathbf{S}_i$ is given by (33.10). Such a transformation S is far from being unique, and we only assume that it exists. Note that since the condition $n_i = 1$ together with the exclusion principle (33.5) implies $n_{i\uparrow} n_{i\downarrow} = 0$, these singly-occupied states are physically realized for infinite on-site repulsion $U/W = \infty$.

In such a singly-occupied state the only possible motion are spin flips, the system is an insulator. However, this is not an ordinary bandgap insulator but a *Mott insulator*[63] with a *correlation gap* at the Fermi level caused by the strong on-site repulsion U. Hence the energy band (33.22) splits into two narrow *Hubbard bands* separated by $2U$, the lower band being fully occupied by antiferromagnetically aligned electrons and the upper band being empty. This is exactly the situation occurring in the undoped ($\rho = 1/2$) cuprate superconductors where the copper atoms of the CuO_2 planes form a square lattice of ions Cu_σ^{2+} with alternating spin $\sigma = \uparrow, \downarrow$[35].

However, it has been emphasized by Anderson since a long time that in dimension $D = 2$ the antiferromagnetic alignment may not be the true groundstate of the Heisenberg Hamiltonian $\mathcal{H}_H$ with $J < 0$ but that superpositions of products of singlet spin-pairs or *resonating valence bonds* (RVB's) may have lower energy. The reason given is *frustration*, i.e. the existence of non-singlet nearest-neighbor spin-pairs, which is necessarily the case on a triangular lattice while on a square lattice it may result from next-nearest neighbor couplings[64]. The reason why the undoped cuprate superconductors have an antiferromagnetic groundstate then is thought to be the result of couplings among the CuO_2 layers which make these systems 3-dimensional.

Doping of the cuprates induces a *Mott transition*[63] to a metallic (and even superconducting) state with a carrier concentration $\delta \equiv \rho - 1/2 \neq 0$. (Note the difference with the Anderson transition of Section 19 which is due to disorder and not to correlation.) Formally this situation is, in principle, described by the Hamiltonian (33.40) without the restriction that all $n_i = 1$. Since the latter implies simultaneously $\delta = 0$ and $\mathcal{H}_o - \mathcal{H}'_o = 0$, a reasonable substitute for $\mathcal{H}_o - \mathcal{H}'_o$ is $\delta \times \mathcal{H}_o$ (see Section 4 of Ref. 62) which leads to the *t-J model*[64,65].

This is still not the full story of the phase diagram, however, since according to the *Nagaoka-Thouless theorem*[26] already one single hole or electron corresponding to a filling factor $\rho = (L \mp 1)/2L$ flips the system into ferromagnetic order, provided that $U/W = \infty$ and $D \geq 2$ (see the introduction to this chapter). Therefore the true phase diagram is likely to be as shown, for $D = 2$, in Fig. 3 of Ref. 66 (note that in this reference $\rho = N/L$). As to the excitations of the Hubbard model, they are again spin waves and may be calculated again by the equation-of-motion method (see Section 6.15 of Ref. 2).

Recently, Hirsch introduced a generalization of the Hubbard model by adding to $\mathcal{H}_o + \mathcal{H}_U$ an exchange coupling between nearest neighbor cells, of which only the spin-flip part

$$\mathcal{H}_J = +J \sum_{\langle ij \rangle} \sum_\sigma c_{i\sigma}^+ c_{j-\sigma}^+ c_{i-\sigma} c_{j\sigma} \tag{33.41}$$

is important[5]. In mean-field approximation this leads again to a Hamiltonian of the form (33.1) with *spin-split energy bands*,

$$E_{\mathbf{k}\sigma} = R\varepsilon_{\mathbf{k}} - Jzn_\sigma + Un_{-\sigma} - \mu \tag{33.42}$$

where $R \equiv 1 - JI_1/t$ is a renormalization factor (see Eqs. (23)–(30) of Ref. 5). It is interesting that if we set here $U + Jz = 2\mu_B^2\lambda$ we recover again, up to a factor R and the constant term $(U - Jz)n/2$, the Stoner expression (32.1) with (32.2), (32.5) and with $H = 0$ in (29.16). In the case $J = 0$ this correspondence is exact and also agrees with Eq. (33.8), apart from the normalization, which is $L/V = 1$ in Ref. 5.

We close this section with a derivation of some of the symmetries of the Hubbard model since they have turned out to be of considerable importance[67]. First, there is a global unitary group $U(2)$ leaving the full Hubbard Hamiltonian invariant[67] and a local $U(2)$ leaving invariant each term $n_{i\uparrow}n_{i\downarrow}$ of $\mathcal{H}_U$. These groups act in the 2-dimensional spin space and factorize into a $U(1)$ describing an ordinary phase and a $SU(2)$ which has determinant 1 by definition and is parametrized by the 2 directional and 1 rotational Euler angles of angular momentum (called pseudospin symmetry in Ref. 67). In a two-component notation

$$C_i = \begin{pmatrix} c_{i\uparrow} \\ c_{i\downarrow} \end{pmatrix} c \; ; \quad \mathcal{U} = e^{i\varphi} \begin{pmatrix} \alpha & \beta \\ -\beta^* & \alpha^* \end{pmatrix} \; ; \quad \mathcal{U}^+\mathcal{U} = 1 \tag{33.43}$$

where $|\alpha|^2 + |\beta|^2 = 1$ (note the difference with the Nambu representation (21.22) after a Fourier transformation (33.4) of C_i) the group action is

$$TC_iT^+ = \mathcal{U}C_i \; ; \quad T^+T = 1 \; ; \quad T = T(\mathcal{U}) \; . \tag{33.44}$$

Hence the hopping term (33.12) is manifestly invariant since the σ-sum may be written as $C_i^+ C_i$.

To show the invariance of the interaction term (33.11) we use Eqs. (33.10) and (2.8) to write $2\mathbf{S}_i = C_i^+ \vec{\sigma} C_i$, $2S_i^+ = c_{i\uparrow}^+ c_{i\downarrow}$, $2S_i^z = n_{i\uparrow} - n_{i\downarrow}$ and obtain with (30.3) and (33.5)

$$\mathbf{S}_i^2 = 2(S_i^+ S_i^- + S_i^- S_i^+) + (S_i^z)^2 = \frac{3}{4}n_i - \frac{3}{2}n_{i\uparrow}n_{i\downarrow} \; . \tag{33.45}$$

Since according to (33.44), $n_i = C_i^+ C_i$ is invariant, it suffices to show that $\mathbf{S}_i^2$ is. Now

$$TS_iT^+ = \frac{1}{2}C_i^+ \mathcal{U}^+ \vec{\sigma} \, \mathcal{U} C_i ,\qquad (33.46)$$

and since the Pauli matrices form a basis for traceless 2×2 matrices and $\mathrm{Tr}\,\mathcal{U}^+ \vec{\sigma} \mathcal{U} = 0$,

$$\mathcal{U}^+ \sigma^\alpha \mathcal{U} = \sum_{\gamma=x,y,z} R_{\alpha\gamma} \sigma^\gamma ; \quad \alpha = x, y, z ,\qquad (33.47)$$

$R = R(\mathcal{U})$ being a 3×3 matrix acting on the vector components of $\vec{\sigma}$. Since $(\sigma^\alpha)^+ = \sigma^\alpha$, $\alpha = x, y, z$, subtracting the hermitean conjugate of (33.47) from this equation gives $(R - R^*)\vec{\sigma} = 0$, i.e. R is a real matrix. Finally, taking the scalar product of (33.47) with its hermitean conjugate and using $(\sigma^\alpha)^2 = 1$, $\alpha = x, y, z$, we find $RR^T = 1$; i.e. R is a rotation. Hence, combining (33.46) and (33.47) one calculates

$$T(\mathbf{S}_i \cdot \mathbf{S}_j)T^+ = \frac{1}{4}\sum_{\lambda\mu\nu} R_{\lambda\mu}R_{\lambda\nu} C_i^+ \sigma^\mu C_i \cdot C_j^+ \sigma^\nu C_j = \mathbf{S}_i \cdot \mathbf{S}_j .\qquad (33.48)$$

To prove the invariance of $n_{i\uparrow}n_{i\downarrow}$ we actually need Eq. (33.48) only for $i = j$, in which case one may choose $\mathcal{U} = \mathcal{U}_i$ locally and hence $T_i = T(\mathcal{U}_i)$.

In addition, the components of the pseudospin operator $\mathbf{J}$ defined in analogy to Eq. (30.3) on a bipartite lattice of even and odd sites i as

$$J^- = \frac{1}{2}\sum_i (-1)^i c_{i\uparrow} c_{i\downarrow} = (J^+)^+ ; \quad J^z = \frac{1}{2}\sum_i (n_i - 1)\qquad (33.49)$$

satisfy angular-momentum commutation relations analogous to Eq. (30.2)[67],

$$[J^z, J^\pm] = \pm J^\pm ; \quad [J^+, J^-] = \frac{1}{2}J^z ,\qquad (33.50)$$

and hence also have the $SU(2)$ symmetry $\mathcal{U}$ of Eq. (33.43) but with $\varphi = 0$. The total symmetry satisfied by the vectors $\mathbf{S}$ and $\mathbf{J}$ therefore is the group $SO(4) = SU(2) \otimes SU(2)/\mathbf{Z}_2$ where $\mathbf{Z}_2$ is the duality $\mathbf{S} \leftrightarrow \mathbf{J}$. $\varphi = 0$ means that $J^\pm$ also break the $U(1)$ subgroup, as do the singlet-pair operators $\Delta_\pm$ in[67]

$$\Delta_+ = \frac{1}{\sqrt{2}}\sum_i c_{i\uparrow}^+ c_{i\downarrow}^+ = -(\Delta_-)^+ = \frac{1}{\sqrt{2}}\sum_k b_k^+ ; \quad \Delta_o = \frac{1}{2}\sum_i (-1)^i n_i\qquad (33.51)$$

where b_k^+ is defined by Eq. (21.21). The algebra satisfied by the operators (33.49) and (33.51) has important consequences which, in principle, are observable and might be used as a test for the occurrence of the Hubbard model in the cuprate superconductors[67]. In particular, the relations

$$[\mathcal{H}_o, \mathbf{J}] = 0 \; ; \quad [\mathcal{H}_U, J^\pm] = \pm U J^\pm \; ; \quad [\mathcal{H}_U, J^z] = 0 \qquad (33.52)$$

hold for periodic boundary conditions where $\mathcal{H}_o$ and $\mathcal{H}_U$ are defined, respectively, by Eqs. (33.12) and (33.11).

Introducing in addition the total spin projection

$$S^z \equiv \sum_i S_i^z = \frac{1}{2} \sum_i (n_{i\uparrow} - n_{i\downarrow}) \,, \qquad (33.53)$$

it is easy to show that

$$[S^z, \mathbf{J}] = 0 \; ; \quad [S^z, \mathcal{H}_o] = 0 \; ; \quad [S^z, \mathcal{H}_U] = 0 \,. \qquad (33.54)$$

Therefore, one may choose states with definite electron number N and total spin projection S,

$$(\mathcal{H}_o + \mathcal{H}_U)|N, S\rangle = E_{N,S}|N, S\rangle$$
$$J^z|NS\rangle = \frac{N - L}{2}|N, S\rangle \; ; \quad S^z|N, S\rangle = S|N, S\rangle \qquad (33.55)$$

and apply Eqs. (33.52). The result is

$$J^\pm|NS\rangle = |N \pm 2, S\rangle \; ; \quad E_{N\pm2, S} = E_{N, S} \pm U \,. \qquad (33.56)$$

Since there is particle-hole symmetry with respect to half-filling, $N = L$, pair-filling with $J^\pm$ starting at $N = L$, which is known to have $S = 0$[25], is energetically unfavorable for $U \gg t$.

34. Weak Itinerant-electron Magnetism. Paramagnons

With the discovery by Matthias and collaborators of the *weak* itinerant-electron ferromagnets (IEF's) $ZrZn_2$[68], Sc_3In[69] and $TiBe_{2-x}Cu_x$[70] which are all composed of nonmagnetic elements, the theory of itinerant-electron ferromagnetism gained great actuality. The main characteristics of these compounds is a saturation moment per unit cell μ_s of the order of $0.1\mu_B$ which is the reason for their name. With the exception of $TiBe_{2-x}Cu_x$ which follows Stoner theory fairly closely[71], a Curie-Weiss behavior is observed asymptotically above the Curie temperature T_C which is of the order of 10–20 K[72]. While μ_s is determined from an *Arrott plot* of m^2 versus h/m (which to first order in Eq. (29.23) yields straight lines), the Curie constant of Eq. (29.11) determined from the Curie-Weiss law (29.30) gives rise to an *effective moment* μ_C which for weak IEF's is in general much larger than

μ_s (see Table I of Ref. 73 and Fig. 1 of Ref. 30). Pure $TiBe_2$, on the other hand, is an exchange-enhanced spin paramagnet (from Fig. 6 of Ref. 71 one extracts an enhancement factor $(1 - \kappa)^{-1} \sim 65$) like Pd, Ni_3Ga or $HfZn_2$, and there is a critical copper concentration x_{cr} of the order of 0.1 for the onset of ferromagnetism in $TiBe_{2-x}Cu_x$[71].

$ZrZn_2$, $TiBe_2$, $HfZn_2$ and the superconductors ZrV_2 and $ThMg_2$ are all members of the AB_2 family of *cubic Laves phases* (C15) where A is always a transition metal but in general not B (V is an exception) and which have very different Fermi surfaces and hence quite different physical properties[74]. In the magnetic members the spin which is supplied by the A atom, is spread out over much of the unit cell[75] thus being a direct proof of the nonlocal character of this magnetism.

Remarkably, the solid solutions $Zr_{1-x}Ti_x$ and $Zr_{1-x}Hf_x$ are superconducting while $Zr_{1-x}Ti_xZn_2$ and $Zr_{1-x}Hf_xZn_2$ are ferromagnetic. Considering here the limit $x \to 0$, Matthias was led to suspect that the magnetism of $ZrZn_2$ should have the same origin as the superconductivity of Zr. This gave rise to the "p-state pairing" idea that, due to some dominant soft phonon mode ω_q in the propagator (20.5), the effective interaction (20.4) becomes repulsive at short distances (long wavelengths). This inhibits s-state pairing, leaving only the possibility of p- or higher-state superconductivity. But at the same time, this on-site repulsion contributes to the Hubbard interaction (33.11) and hence to the Stoner factor (33.24). Therefore, the speculation was that, under high pressure, the magnetic systems $ZrZn_2$ or $TiBe_2$ could become p-state superconductors and that an *isotope effect* should exist[76]. No such evidence has ever been reported, however. On the side of theory, criticism centered on "Migdal's theorem". In the present situation this criticism stands on weak ground, however, see Section 11.

According to the Stoner criterion (33.24), (33.30), ferromagnetism is favored by a large value of the ratio U/W. Now U is an atomic quantity and hence is essentially independent of variations of the unit-cell volume $v = V/L$. On the other hand, it follows from Eqs. (14.13) and (33.28) for $D = 3$ that $W \propto v^{-2/3}$. This means that $U/W \propto v^{2/3}$, and since v decreases with increasing hydrostatic pressure p, we conclude that applied pressure is detrimental for band ferromagnetism . This is in contrast to superconductivity which in general is favored by applied pressure because the BCS gap (21.17) increases with increasing density of states for which, according to Eqs. (14.13), (14.14) the proportionality $N(0) \propto v^{-1/3}$ holds.

As already mentioned in the introduction to this chapter, $ZrZn_2$ shows a negative slope of the Curie temperature $T_C(p)$ and ceases to be ferromagnetic at a critical pressure $p_{cr} \sim 10\text{–}20$ kbar[31]. Note that the discrepancy in the value of p_{cr} has the same origin as the difference between the saturation and the effective moments μ_s and μ_C obtained, respectively, from magnetization and susceptibility measurements, as mentioned above. Similar negative slopes of $T_C(p)$ were also found in the system $TiBe_{2-x}Cu_x$ from susceptibility measurements[77]. But in addition, the fact that $TiBe_2$ becomes ferromagnetic by sufficient admixture of copper, may also be understood with the same argument. Indeed, since the Cu atom replaces the smaller Be, the lattice parameters increase linearly with x (see Fig. 1 of Ref. 71). In other words, Cu acts as a negative pressure. However, as exceptions seem to be the only general rule in magnetism, Sc_3In exhibits a positive slope of $T_C(p)$ [78]. But as will be explained in the next section, the deeper reasons for Sc_3In to be exceptional are *magnetization fluctuations*.

A many-body treatment of the exchange enhancement of the spin susceptibility may again be based on the Hubbard repulsion (33.13). This had been done very early by Izuyama, Kim and Kubo[79]. Quite generally, the magnetic susceptibility is defined as response function (7.12) connecting the Fourier components of the magnetization density, $\vec{\mu}(\mathbf{q})$, and of an external magnetic field, $\mathbf{H}(\mathbf{q}, t)$. In the case of the spin susceptibility,

$$\vec{\mu}(\mathbf{q}) = \mu_B g \mathbf{s}(\mathbf{q}) \tag{34.1}$$

where the spectroscopic factor (29.3) is, for conduction electrons and neglecting renormalization effects, $g = 2$; $\mathbf{s}(\mathbf{q})$ is the density defined in Eq. (17.12) and the external-field Hamiltonian (7.10) is

$$\delta\mathcal{H}_t = \sum_{\mathbf{q}} \mathbf{H}(\mathbf{q}, t) \cdot \vec{\mu}(-\mathbf{q}) \ . \tag{34.2}$$

Writing $\mathbf{s}(\mathbf{q})$ in Wannier representation with the aid of Eqs. (33.4) and (33.10),

$$\mathbf{s}(\mathbf{q}) = \frac{1}{2} \sum_i \sum_{\sigma\sigma'} c_{i\sigma}^+ \vec{\sigma}_{\sigma\sigma'} c_{i\sigma'} e^{-i\mathbf{q}\cdot\mathbf{R}_i} = \sum_i \mathbf{S}_i e^{-i\mathbf{q}\cdot\mathbf{R}_i} \ , \tag{34.3}$$

one sees that, for a constant magnetic field $\mathbf{H}(\mathbf{q}, t) = (0, 0, H)\delta_{\mathbf{q}0}$, $\delta\mathcal{H}_t$ reduces to the Zeeman Hamiltonian (30.4).

Making use of Eq. (25.5), the spin susceptibility tensor may then be written as

$$\chi_{\alpha\gamma}(q) = \frac{1}{V} \int_0^\beta d\tau e^{i\nu\tau} \langle T\{\mu^\alpha(\mathbf{q}; -i\tau)\mu^\gamma(-\mathbf{q}; 0)\}\rangle \tag{34.4}$$

where $q \equiv (\mathbf{q}, i\nu)$. Inserting (17.12) and (34.1), Eq. (34.4) is again recognized as the two-electron Green's function (33.14) but now with arguments corresponding to the crossed particle-hole channels shown in Figs. 5.11(a) and (b), which are analogous to Fig. 3.6. Written as perturbation expansion as in (33.14), Eq. (34.4) becomes

$$\chi_{\alpha\gamma}(q) = \frac{\mu_B^2}{V} \sum_{\sigma\sigma'} \sum_{\rho\rho'} \sigma_{\sigma\sigma'}^{\alpha} \sigma_{\rho\rho'}^{\gamma} \sum_{N=0}^{\infty} T^{(N)} \tag{34.5}$$

where

$$T^{(N)}(\sigma\sigma', \rho\rho', q) \equiv \frac{1}{N!} \sum_{\mathbf{k}\mathbf{k}'} \int_0^\beta d\tau e^{i\nu\tau}$$

$$\times \langle T\{ \left[-\int_0^\beta d\tau' \mathcal{H}_U^o(-i\tau') \right]^N$$

$$\times (a_{\mathbf{k}\sigma}^+ a_{\mathbf{k}+\mathbf{q},\sigma'})^o(-i\tau)(a_{\mathbf{k}'+\mathbf{q},\rho}^+ a_{\mathbf{k}'\rho'})(0)\} \rangle_{o,cp} . \tag{34.6}$$

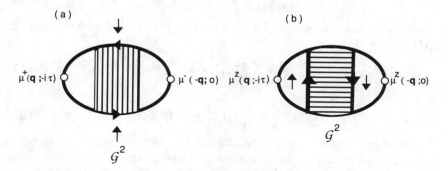

Fig. 5.11: Structure of the response function (34.5). (a) Spin-flip or fluctuation response χ_{+-}. (b) Susceptibility χ_{zz}.

Applying Wick's theorem (9.11) one first observes that the $N = 0$ term cannot have the structure of Fig. 5.11(b) which becomes disconnected in this case while Eq. (34.6) is restricted to connected parts (cp). Using Eqs. (7.25), (7.28) one then obtains with $k \equiv (\mathbf{k}, i\nu_-)$

$$\chi_{\alpha\gamma}^{(0)}(q) = \frac{\mu_B^2}{V} \sum_{\sigma\sigma'} \sigma_{\sigma\sigma'}^{\alpha} \sigma_{\sigma'\sigma}^{\gamma} \Gamma_{\sigma\sigma'}(q) \tag{34.7}$$

where

$$\Gamma_{\sigma\sigma'}(q) \equiv -\beta^{-1} \sum_k \tilde{\mathcal{G}}_\sigma^o(k)\tilde{\mathcal{G}}_{\sigma'}^o(k+q) = \Gamma_{\sigma'\sigma}(-q) \qquad (34.8)$$

and $\tilde{\mathcal{G}}_\sigma^o(k)$ is defined by Eq. (7.44), replacing $\varepsilon_\mathbf{k} - \mu$ by the spin-dependent electron energy $\varepsilon_{\mathbf{k}\sigma}$ of Eq. (32.1). Note the similarity of $\Gamma_{\sigma\sigma'}$ with G_D, Eq. (33.19), the difference being that the arguments of G_D were those of the particle-particle channel (see Fig. 5.10) while the arguments of $\Gamma_{\sigma\sigma'}$ belong to the particle-hole channel as shown in Fig. 5.11. This difference also appears in the different Lindhard function (10.17), (10.19) to which Eq. (34.8) gives rise namely, written for an arbitrary chemical potential μ and making use of Eqs. (7.26), (7.42),

$$\begin{aligned} \Gamma_{\sigma\sigma'}(q) &= -\sum_\mathbf{k} L_-(\varepsilon_{\mathbf{k}\sigma} - \mu, \varepsilon_{\mathbf{k}+\mathbf{q},\sigma'} - \mu - i\nu) \\ &= -\sum_\mathbf{k} \frac{f_o(\varepsilon_{\mathbf{k}\sigma} - \mu) - f_o(\varepsilon_{\mathbf{k}+\mathbf{q},\sigma'} - \mu)}{\varepsilon_{\mathbf{k}\sigma} - \varepsilon_{\mathbf{k}+\mathbf{q},\sigma'} + i\nu} . \end{aligned} \qquad (34.9)$$

Making use of the relations

$$\sigma_{\uparrow\downarrow}^\alpha = \delta_{\alpha+} , \quad \sigma_{\downarrow\uparrow}^\alpha = \delta_{\alpha-} , \quad \sigma_{\uparrow\uparrow}^\alpha = -\sigma_{\downarrow\downarrow}^\alpha = \delta_{\alpha z} \qquad (34.10)$$

which follow from definition (2.8), Eqs. (34.7), (34.8) yield with $\mu^\pm \equiv (\mu^x \pm i\mu^y)/2$

$$\chi_{+-}^{(0)}(q) = \frac{\mu_B^2}{V}\Gamma_{\uparrow\downarrow}(q) = \chi_{-+}^{(0)}(-q) , \qquad (34.11)$$

and

$$\chi_{zz}^{(0)}(\pm q) = \frac{\mu_B^2}{V}(\Gamma_{\uparrow\uparrow}(q) + \Gamma_{\downarrow\downarrow}(q)) . \qquad (34.12)$$

Without magnetic field and interaction we may set $\varepsilon_{\mathbf{k}\sigma} = \varepsilon_\mathbf{k}$ in Eq. (32.1) so that $\Gamma_{\sigma\sigma'}$ becomes independent of the spin indices. Using the relation $\mathrm{Tr}\,(\sigma^\alpha\sigma^\gamma) = 2\delta_{\alpha\gamma}$ which also follows from definition (2.8), the susceptibility tensor (34.7) becomes isotropic,

$$\chi_{\alpha\gamma}^{(0)}(q) = \chi^o\delta_{\alpha\gamma} \qquad (34.13)$$

where, according to (34.9),

$$\chi^o(\pm q) = \frac{2\mu_B^2}{V}\Gamma(\pm q) = -\frac{2\mu_B^2}{V}\sum_\mathbf{k} \frac{f_o(\varepsilon_\mathbf{k} - \mu) - f_o(\varepsilon_{\mathbf{k}+\mathbf{q}} - \mu)}{\varepsilon_\mathbf{k} - \varepsilon_{\mathbf{k}+\mathbf{q}} + i\nu} . \qquad (34.14)$$

Choosing $\mu = \varepsilon_F$ so that $\varepsilon_\mathbf{k} \equiv \hbar^2\mathbf{k}^2/2m$, we have $\varepsilon_{\mathbf{k}+\mathbf{q}} = \varepsilon_\mathbf{k} + 2\zeta\sqrt{\varepsilon_\mathbf{k}\varepsilon_\mathbf{q}} + \varepsilon_\mathbf{q}$ where $\zeta \equiv \mathbf{k}\cdot\mathbf{q}/|\mathbf{k}||\mathbf{q}|$. Introducing the density of states (10.4) and making use of (31.16), Eq. (34.14) then becomes, for $|q_o| < 2\sqrt{\varepsilon_F \varepsilon_{\mathbf{q}}}$ and in the limit of zero temperature (Problem 5.11),

$$\chi^o(\mathbf{q}; q_o) = \chi_P \left\{ 1 + i \frac{\pi q_o}{4\sqrt{\varepsilon_F \varepsilon_{\mathbf{q}}}} - \varphi \frac{\varepsilon_{\mathbf{q}}}{\varepsilon_F} \right\} + \mathcal{O}\left(\left(\frac{\varepsilon_{\mathbf{q}}}{\varepsilon_F} \right)^2, \frac{q_o}{\varepsilon_F}, \frac{q_o^2}{\varepsilon_F \varepsilon_{\mathbf{q}}} \right) \quad (34.15)$$

where

$$\varphi \equiv \frac{\varepsilon_F N'(\varepsilon_F)}{18 N(\varepsilon_F)} + \cdots . \quad (34.16)$$

Thus the isothermal limit $\chi^o(\mathbf{q}; 0)$, $\mathbf{q} \to 0$, coincides with the Pauli susceptibility χ_P as it should.

The fact that the Hubbard interaction (33.13) couples opposite spins implies that the higher-order contributions to $\chi_{\alpha\gamma}$, Eq. (34.5), involve quite different diagrams for the *ordinary susceptibility* χ_{zz} and for the *spin-flip* or *spin fluctuation response* χ_{+-}. The structure of χ_{+-} and χ_{zz} is shown in Figs. 5.11(a) and (b), respectively. In RPA , $\chi_{+-} - \chi_{+-}^{(0)}$ is represented by the *ladders* $T_{\text{ld}}^{(N)}$, $N = 1, 2, \cdots$ of Fig. 5.12(a) while $\chi_{zz} - \chi_{zz}^{(0)}$ consists of the *chains* $T_{\text{ch}}^{(2N-1)}$ of Fig. 5.12(b). Note that $T_{\text{ld}}^{(2)}$ and the crossed diagram $t^{(2)}$ of Fig. 5.10 are related by a relative out-of-plane rotation of the two propagator lines.

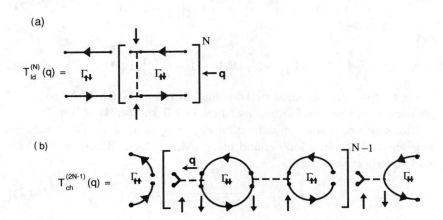

(a)

$$T_{\text{ld}}^{(N)}(q) =$$

(b)

$$T_{\text{ch}}^{(2N-1)}(q) =$$

Fig. 5.12: (a) Ladder diagram representating Eq. (34.19). (b) Chain diagram representing Eq. (34.20). $\Gamma_{\sigma\sigma'}$ is defined by Eq. (34.8).

The ladders and chains are determined by the calculation of the iteration steps of Fig. 5.12(a) and (b) which are already contained in $T^{(1)}$. From

Eq. (34.6) one finds, using definition (34.8) (Problem 5.12),

$$T^{(1)}(\sigma\sigma', \rho\rho'; q) = \frac{U}{L}\Gamma_{\sigma\sigma'}(q)\Gamma_{\rho'\rho}(q)$$
$$\times (\delta_{\uparrow\sigma}\delta_{\downarrow\rho} - \delta_{\downarrow\sigma}\delta_{\uparrow\rho})(\delta_{\downarrow\sigma'}\delta_{\uparrow\rho'} - \delta_{\uparrow\sigma'}\delta_{\downarrow\rho'}) \ . \tag{34.17}$$

Writing

$$T^{(N)} = T^{(N)}_{ld}(q)\delta_{\uparrow\downarrow, \sigma\sigma'}\delta_{\downarrow\uparrow, \rho\rho'} + T^{(N)}_{ld}(-q)\delta_{\downarrow\uparrow, \sigma\sigma'}\delta_{\uparrow\downarrow, \rho\rho'}$$
$$+ T^{(2N-1)}_{ch}(q)(\delta_{\uparrow\uparrow, \sigma\sigma'}\delta_{\downarrow\downarrow, \rho\rho'} + \delta_{\downarrow\downarrow, \sigma\sigma'}\delta_{\uparrow\uparrow, \rho\rho'}) \tag{34.18}$$

where, according to Figs. 5.12(a) and (b),

$$T^{(N)}_{ld}(q) = \Gamma_{\uparrow\downarrow}(q)\left[\frac{U}{L}\Gamma_{\uparrow\downarrow}(q)\right]^N \tag{34.19}$$

and

$$T^{(2N-1)}_{ch}(q) = \Gamma_{\uparrow\uparrow}(q)\left[\frac{U}{L}\Gamma_{\downarrow\downarrow}(q)\frac{U}{L}\Gamma_{\uparrow\uparrow}(q)\right]^{N-1}\frac{U}{L}\Gamma_{\downarrow\downarrow}(q), \tag{34.20}$$

respectively, one verifies that, for $N = 1$, Eq. (34.18) reduces to (34.17). Therefore, taking the sum over N using (34.11), (34.12) and applying the relations (34.10), Eq. (34.5) yields the RPA expressions

$$\chi_{+-}(q) = \frac{\mu_B^2}{V}\Gamma_{\uparrow\downarrow}\left[1 - \frac{U}{L}\Gamma_{\uparrow\downarrow}\right]^{-1} = \chi_{-+}(-q) \tag{34.21}$$

and

$$\chi_{zz}(\pm q) = \frac{\mu_B^2}{V}\left\{\Gamma_{\uparrow\uparrow} + \Gamma_{\downarrow\downarrow} + 2\frac{U}{L}\Gamma_{\uparrow\uparrow}\Gamma_{\downarrow\downarrow}\left[1 - \left(\frac{U}{L}\right)^2\Gamma_{\uparrow\uparrow}\Gamma_{\downarrow\downarrow}\right]^{-1}\right\} . \tag{34.22}$$

These results were obtained by Izuyama , Kim and Kubo with the aid of the equation-of-motion method (see Eqs. (4.16) and (4.24) of Ref. 79).

Making again the approximation $\varepsilon_{k\sigma} = \varepsilon_k$ so that $\Gamma_{\sigma\sigma'}$ becomes independent of the spin indices and using (34.14), Eqs. (34.21) and (34.22) reduce, respectively to

$$\chi_{+-}(\pm q) = \frac{1}{2}\frac{\chi^o(q)}{1 - \lambda\chi^o(q)} = \chi_{-+}(q) \tag{34.23}$$

and

$$\chi_{zz}(\pm q) = \chi^o(q)\left\{1 + \frac{\lambda\chi^o(q)}{1 - (\lambda\chi^o(q))^2}\right\} \simeq \frac{\chi^o(q)}{1 - \lambda\chi^o(q)} \tag{34.24}$$

where λ is defined in Eq. (33.8), and in the last expression a term $\propto \lambda^2$ was neglected in the numerator. Taking the isothermal limit of Eq. (34.15) and making use of Eq. (32.21), one recovers the zero-temperature Stoner expression (32.35),

$$\lim_{\mathbf{q}\to 0}\chi_{zz}(\mathbf{q};\,0) \simeq \lim_{\mathbf{q}\to 0}\big[\chi_{-+}(\mathbf{q};\,0) + \chi_{+-}(\mathbf{q};\,0)\big] = \frac{\chi_P}{1-\kappa}\,. \qquad (34.25)$$

In Section 10 we calculated the effect of the electron-phonon interaction on the specific heat . An analogous calculation may be done with the Hubbard repulsion (33.13). But since we wish to keep the same RPA accuracy as for the magnetic response, a more sophisticated procedure is required. Thus we adopt Gor'kov's idea of a variable coupling constant used in Eqs. (10.27) and (24.17) to calculate the interaction contribution $\Delta\Omega$ to the grand-canonical potential by first deriving Eq. (9.20) with respect to the coupling constant U and then reintegrating the resulting expression. This procedure works as long as no phase transition is traversed in increasing U or κ from 0 to the respective physical value, in other words, as long as we stay in the paramagnetic region $\kappa < 1$. The result of the differentiation is given by Eq. (10.28) which in the present case reads

$$\frac{\partial\Delta\Omega}{\partial U} = -\beta^{-1}\left\langle\frac{\partial S(-i\beta,\,0)}{\partial U}\right\rangle_{o,cp}$$

$$= \frac{1}{U\beta}\int_0^\beta d\tau\langle T\{S(-i\beta,\,0)\mathcal{H}_U^o(-i\tau)\}\rangle_{o,cp} = \sum_{N=0}^\infty E^{(N)} \quad (34.26)$$

where $E^{(N)}$ may be written in a form similar to Eq. (34.6), namely

$$E^{(N)} \equiv \frac{1}{N!}\frac{1}{L\beta}\sum_{\mathbf{kk'q}}\int_o^\beta d\tau$$

$$\times\left\langle T\left\{\left[-\int_o^\beta d\tau'\mathcal{H}_U^o(-i\tau')\right]^N (a_{\mathbf{k}\uparrow}^+ a_{\mathbf{k+q}\uparrow} a_{\mathbf{k'+q}\downarrow}^+ a_{\mathbf{k'}\downarrow})^o(-i\tau)\right\}\right\rangle_{o,cp}. \tag{34.27}$$

The diagrammatic representation of Eq. (34.26) is not unique and may therefore give rise to confusion. Indeed, since the singled-out factor $\mathcal{H}_U/U$ is independent of U, it may be considered not to give rise to an interaction line. Then the four creation and annihilation operators it contains become endings of external lines and the integrand in Eq. (34.26) has the structure of Fig. 5.11 but with equal times τ on both sides. In this case RPA is again given by the ladders or chains of Fig. 5.12 and the effect of the derivation with respect to U is to pull out four legs. This is not very satisfactory, however since, as discussed in Section 10, thermodynamic diagrams have no external lines. And derivation and reintegration with respect to the coupling constant should not alter this fact.

It is more appropriate, therefore to associate an interaction line with the singled-out factor $\mathcal{H}_U$ in Eq. (34.26). In this case the four creation and annihilation operators it contains remain internal and the diagrams have the structure of the Hartree-type diagrams of Fig. 2.4(a) and (b) while the exchange-type diagrams of Fig. 2.4(c) and (d) are excluded by the fact that $\mathcal{H}_U$ couples opposite spins. Figure 2.4(b) generates the ladders of Fig. 5.12(a) if there the upper and lower ends are connected into propagators $\tilde{\mathcal{G}}_\downarrow^o$ and $\tilde{\mathcal{G}}_\uparrow^o$, respectively. This then is the RPA of Eq. (34.26). But it is clear that, if the last interaction to the right of Fig. 5.12(b) is discarded, the remaining chains may be closed in the same way and hence also contribute to Eq. (34.26). However, according to Eq. (34.27) these chains are $E_{\text{ch}}^{(2N+1)} \propto (U/L)^{2N+2}/U$, $N = 0, 1, 2, \cdots$, while the ladders are $E_{\text{ld}}^{(N)} \propto (U/L)^{N+1}/U$. For this reason we shall retain only the ladders.

Applying Wick's theorem (9.11) to Eq. (34.27), the $N = 0$ term is obtained with the help of the prescription (11.24), using (33.3) and $N_\sigma \equiv V \langle n_\sigma \rangle_o$,

$$E^{(0)} = \frac{1}{L} \sum_{\mathbf{k}\mathbf{k}'\mathbf{q}} \langle a_{\mathbf{k}\uparrow}^+ a_{\mathbf{k}+\mathbf{q}\uparrow} \rangle_o \langle a_{\mathbf{k}'+\mathbf{q}\downarrow}^+ a_{\mathbf{k}'\downarrow} \rangle_o = \frac{1}{L} N_\uparrow N_\downarrow \; . \qquad (34.28)$$

For $N = 1$ a comparison with the calculation of Problem 5.11 and with the result (34.17) shows, using definition (34.8), that

$$E^{(1)} = \frac{1}{L\beta} \sum_q T^{(1)}(\uparrow\uparrow, \downarrow\downarrow; q)$$

$$= -\frac{U}{L^2\beta} \sum_q \Gamma_{\uparrow\uparrow}(q)\Gamma_{\downarrow\downarrow}(q) = -\frac{U}{L^2\beta} \sum_p (\Gamma_{\uparrow\downarrow}(p))^2 \; . \qquad (34.29)$$

Here the first and second expressions of the second line are $E_{\text{ch}}^{(1)}$ and $E_{\text{ld}}^{(1)}$, respectively, which shows that the chains start contributing only with $E^{(3)}$. This justifies the neglect of the chains.

For the ladder contributions we conclude from Eq. (34.29) and from Fig. 5.12(a) that

$$E_{\text{ld}}^{(N)} = -\frac{1}{U\beta} \sum_p \left[\frac{U}{L} \Gamma_{\uparrow\downarrow}(p) \right]^{N+1} \; ; \quad N \geq 1 \; . \qquad (34.30)$$

Inserted into Eq. (34.26) together with (34.28) this yields

$$\frac{\partial \Delta \Omega}{\partial U} = \frac{1}{L} N_\uparrow N_\downarrow - \frac{1}{U\beta} \sum_p \left(\frac{U}{L} \Gamma_{\uparrow\downarrow}(p)\right)^2 \left[1 - \frac{U}{L}\Gamma_{\uparrow\downarrow}(p)\right]^{-1} . \qquad (34.31)$$

The U-integration of this expression is straightforward since, with $x \equiv U\Gamma_{\uparrow\downarrow}/L$, the second term involves $\int dx\, x(1-x)^{-1}$. Hence, fixing the integration constant such that $\Delta\Omega = 0$ for $U = 0$, one finds

$$\Delta\Omega = \frac{U}{L} N_\uparrow N_\downarrow + \beta^{-1} \sum_p \left\{\frac{U}{L}\Gamma_{\uparrow\downarrow}(p) + \ln\left[1 - \frac{U}{L}\Gamma_{\uparrow\downarrow}(p)\right]\right\} . \qquad (34.32)$$

In the approximation $\varepsilon_{\mathbf{k}\sigma} = \varepsilon_{\mathbf{k}}$, i.e. $\Gamma_{\sigma\sigma'} = \Gamma$ this formula becomes, using Eq. (34.14) and definition (33.8),

$$\Delta\Omega = \frac{U}{L} N_\uparrow N_\downarrow + \beta^{-1} \sum_p \{\lambda\chi^o(p) + \ln[1 - \lambda\chi^o(p)]\} . \qquad (34.33)$$

Quite generally, interactions affect the specific heat (10.3) in two ways. According to Eq. (14.14), for fixed particle number, $c_V/T \propto m$ so that the first effect is a *mass renormalization* while the second are *corrections to the linear temperature dependence*. As regards the mass renormalization, a comparison of Figs. 2.4 (b) and 2.9 (b) shows that the selfenergy is obtained from $\Delta\Omega$ by cutting out one of the two singled-out propagator lines,

$$\Sigma_\sigma(k) = \beta \frac{\delta\Delta\Omega}{\delta\tilde{\mathcal{G}}_\sigma^o(k)} . \qquad (34.34)$$

That the prefactor is indeed correct may be checked in a special case, e.g. by comparing Eqs. (10.15) and (11.27). Applying this prescription to Eq. (34.32) and using Eqs. (33.8), (34.8) and (34.21) one finds (see Eq. (7) of Ref. 80)

$$\Sigma_\sigma(k) = \frac{4\lambda^2\mu_B^2}{V\beta} \sum_p \chi_{+-}(p)\{\tilde{\mathcal{G}}_\downarrow^o(k+p)\delta_{\uparrow\sigma} + \tilde{\mathcal{G}}_\uparrow^o(k-p)\delta_{\downarrow\sigma}\} . \qquad (34.35)$$

From Eqs. (34.15), (34.23) one deduces the expression (see Eq. (4) of Ref. 80)

$$\chi^o(xk_F; 2tx\varepsilon_F) = \chi_P\{1 - \varphi x^2 + i\frac{\pi}{2}t\theta(1-t)\} , \qquad (34.36)$$

valid for $x < x_1 \simeq 1$. Inserted into Eq. (34.23) this is obviously not sufficient to calculate the sum over $p_o = i\nu$ in Eq. (34.35). It may be achieved, however, by use of a dispersion relation of the form (11.10). Indeed, applying

(7.21), (7.29), (7.31) and (7.43) to Eq. (34.4) one finds, using (34.1) and (34.3),

$$\chi_{+-}(\mathbf{q}; z) = \frac{1}{V} \sum_{nn'} e^{\beta(\Omega - E_n)} \frac{|\langle n|\mu^+(\mathbf{q})|n'\rangle|^2}{n_o(E_n - E_{n'})(E_n - E_{n'} + z)} . \qquad (34.37)$$

But this formula shows that $\chi_{+-}(\mathbf{q}; z)$ is analytic in the z-plane, except for cuts along the real axis. Hence Eq. (11.10) applies,

$$\chi_{+-}(\mathbf{q}; i\nu) = \frac{1}{\pi} \int_{-\infty}^{+\infty} \frac{d\omega}{\omega - i\nu} \text{Im}\, \chi_{+-}(\mathbf{q}; \omega) . \qquad (34.38)$$

Inserted into Eq. (34.35) for $\sigma = \uparrow$, the ν-sum is again a Lindhard function (10.17) namely $-\beta L_+(\omega, \varepsilon_{\mathbf{k+p}\downarrow} - \mu - i\nu_-)$. Applying Eq. (10.19) together with (7.26), (7.42), (7.43) it thus follows that

$$\Sigma_\uparrow(\mathbf{k}; i\nu_-) = 4\lambda^2 \mu_B^2 V^{-1} \sum_{\mathbf{p}} \int_{-\infty}^{+\infty} \frac{d\omega}{\pi} \text{Im}\, \chi_{+-}(|\mathbf{p}|; \omega)$$

$$\times \frac{n_o(\omega) + f_o(\varepsilon_{\mathbf{k+p}\downarrow} - \mu)}{\omega - \varepsilon_{\mathbf{k+p}\downarrow} + \mu + i\nu_-} . \qquad (34.39)$$

We wish to apply this result to the case of an enhanced paramagnet, in which case $0 < 1 - \kappa \ll 1$ and the spin fluctuations are called *paramagnons*. We may neglect the spin dependence and, at low temperatures also $n_o(\omega)$. Setting then $\varepsilon_{\mathbf{k}} = \mu = \varepsilon_F$ and writing as in Eq. (34.15), $\varepsilon_{\mathbf{k+p}} - \mu = 2\zeta\sqrt{\varepsilon_F \varepsilon_{\mathbf{p}}} + \varepsilon_{\mathbf{p}}$, this yields

$$\Sigma(k_F; k_o) = 4\lambda^2 \mu_B^2 \int_o^\infty \frac{p^2 dp}{(2\pi)^2} \int_{-\infty}^{+\infty} \frac{d\omega}{\pi} \text{Im}\, \chi_{+-}(p; \omega)$$

$$\times \int_{-1}^{+1} d\zeta \frac{f_o(2\zeta\sqrt{\varepsilon_F \varepsilon_{\mathbf{p}}} + \varepsilon_{\mathbf{p}})}{\omega - 2\zeta\sqrt{\varepsilon_F \varepsilon_{\mathbf{p}}} - \varepsilon_{\mathbf{p}} + k_o} . \qquad (34.40)$$

Here we assumed that $\text{Im}\, \chi_{+-}$ depends only on $p = |\mathbf{p}|$ and not on ζ. This is the case with the form (34.36) that we shall eventually use and which, in addition, has the symmetry

$$\text{Im}\, \chi_{+-}(p; -\omega) = -\text{Im}\, \chi_{+-}(p; \omega) . \qquad (34.41)$$

Performing a partial integration in the new variable $\varepsilon' \equiv 2\zeta\sqrt{\varepsilon_F \varepsilon_{\mathbf{p}}} + \varepsilon_{\mathbf{p}}$ one then finds (see Eq. (9) of Ref. 80)

$$\Sigma(k_F; k_o) = \frac{2\lambda^2 \mu_B^2 k_F}{\varepsilon_F} \int_o^\infty \frac{pdp}{(2\pi)^2} \int_o^\infty \frac{d\omega}{\pi} \text{Im}\, \chi_{+-}(p; \omega)$$

$$\times \int_{-\infty}^{+\infty} d\varepsilon' f_o'(\varepsilon') \ln\left|\frac{k_o + \omega - \varepsilon'}{k_o - \omega - \varepsilon'}\right| . \qquad (34.42)$$

Inserting now (34.23), (34.36) and (32.21), Eq. (34.42) becomes in the approximation $f'_o(\varepsilon') \simeq -\delta(\varepsilon')$ and replacing in the t-integration the upper limit 1 by ∞ which is justified since $1 - \kappa \ll 1$ and φ may be considered to be small,

$$
\Sigma(k_F; k_o) \cong -\frac{\lambda \mu_B^2 k_F^3}{(2\pi)^2 \kappa} \int_0^{x_1} x^2 \, dx
$$
$$
\times \int_0^\infty \frac{t \, dt}{((1/K) - 1 + \varphi x^2)^2 + (\frac{\pi}{2} t)^2} \ln \left| \frac{k_o + 2tx\varepsilon_F}{k_o - 2tx\varepsilon_F} \right| .
$$

$$(34.43)$$

The mass renormalization may be obtained from Eq. (11.34) by writing $\varepsilon_{\mathbf{k}}^r - \varepsilon_{\mathbf{k}} = (1 - m^*/m)\varepsilon_{\mathbf{k}}^r$ and taking the derivative with respect to $k_o = \varepsilon_{\mathbf{k}}^r$. Applying this to Eq. (34.43) one finds (see Eq. (10) of Ref. 80)

$$
\frac{m^*}{m} - 1 = -\frac{\partial}{\partial k_o} \mathrm{Re}\, \Sigma(k_F; k_o) = \frac{m k_F}{2(2\pi\hbar)^2 N(\varepsilon_F)\varphi}
$$
$$
\times \ln \left[1 + \frac{\kappa\varphi x_1^2}{1 - \kappa} \right] \cong \frac{m k_F x_1^2 \kappa}{2(2\pi\hbar)^2 N(\varepsilon_F)(1 - \kappa)} \qquad (34.44)
$$

where in the last expression we have set $\varphi \simeq 0$. In order to obtain this result it was necessary to introduce the cutoff $x_1\varepsilon_F$ in the p-integration which was a necessary condition for Eq. (34.36) to be valid. This defect is, of course, due to the oversimplified expression used for χ^o. But even in this imperfect form, the result (34.44) is quite remarkable since it predicts an important mass renormalization for strongly enhanced paramagnets which is similar to the Kondo-lattice effect in heavy-fermion systems (see end of Section 17).

In order to calculate the corrections to the linear temperature dependence of the specific heat we first calculate directly the low-temperature expression of $\Delta\Omega$. Since in the paramagnetic region $n_\uparrow = n_\downarrow = n/2 = \text{const}$, the first term in Eq. (34.33) only contributes a constant which we neglect. Since, according to Eq. (10.16), the sum over $p_o = i\nu$ introduces a strong T-dependence, it is important to calculate it before taking T-derivatives. Now the dispersion relation (34.38) is also valid for

$$
\Psi \equiv \lambda\chi^o + \ln[1 - \lambda\chi^o] , \qquad (34.45)
$$

so that we only need $\mathrm{Im}\, \Psi$ which, in the approximation (34.36) is again recognized to satisfy the symmetry (34.41). According to Eq. (10.17) the p_o-sum then leads to a Lindhard function $L_+(\omega, -\omega)$ so that, using (10.19) and (7.43), the T-dependent part of Eq. (34.33) becomes

$$\Delta\Omega(T) = \sum_{\mathbf{p}} \int_o^\infty \frac{d\omega}{\pi} \operatorname{Im} \Psi(\mathbf{p}; \omega)[2n_o(\omega) + 1] . \qquad (34.46)$$

This formula may be interpreted as the contribution to the grand-canonical potential of the bosonic excitations of the system, i.e. the *spin fluctuations*.

Note that this result is obtained by applying formula (10.16) and deforming the path Γ according to Fig. 2.7 which is possible because Ψ has only cuts along the real z-axis and $\operatorname{Im} \Psi = 0$ at $z = 0$ because of the symmetry (34.41). Now for $|z| \gg \varepsilon_F$, Eqs. (34.14), (32.21) show that $|\lambda\chi^o| \sim \kappa\varepsilon_F/|z| \ll 1$ and hence $|\Psi| \sim |\lambda\chi^o|^2 \sim (\kappa\varepsilon_F/|z|)^2$. Therefore the circle with radius $|z| \to \infty$ does not contribute while the integral around the cuts leads to (34.46). This is the meaning of Eq. (5) of Ref. 80 where, however, the factor $2n_o(\omega) + 1$ is missing. This omission corresponds to the zero-temperature limit $n_o(\omega) \sim 0$, the justification of which is not, however, evident *a priori* since both factors, $\operatorname{Im} \Psi(\mathbf{p}; \omega)$ and $2n_o(\omega) + 1$ contribute to the T-dependence of Eq. (34.46).

Applying the thermodynamic relation (10.2) to Eq. (34.46) we therefore write the entropy-density correction as

$$-\frac{1}{V}\left(\frac{\partial\Delta\Omega(T)}{\partial T}\right)_\mu = s_1 + s_2 \qquad (34.47)$$

where

$$\begin{aligned}
s_1 &= \frac{2\lambda^2}{V} \sum_{\mathbf{p}} \int_o^\infty \frac{d\omega}{\pi} \operatorname{Im}\left(\chi_{+-}(\mathbf{p}; \omega)\frac{\partial\chi^o(\mathbf{k}; \omega)}{\partial T}\right)[2n_o(\omega) + 1] \\
&= \frac{2\lambda^2}{V\beta} \sum_p \chi_{+-}(p)\frac{\partial\chi^o(p)}{\partial T} \qquad (34.48)
\end{aligned}$$

and

$$s_2 = -\frac{2\beta}{TV} \sum_{\mathbf{p}} \int_o^\infty \frac{d\omega}{\pi} \operatorname{Im} \Psi(\mathbf{p}; \omega)\omega n_o(\omega)[n_o(\omega) + 1] . \qquad (34.49)$$

In the first expression of Eq. (34.48) the operations $\partial/\partial T$ and Im were interchanged and use was made of (34.23) and (34.45) while the second expression for s_1 is obtained by doing the analog of the step from (34.33) to (34.46) in reverse.

We first calculate s_1 by observing that from Eq. (34.14) and (7.44) it follows that

$$\frac{\partial \chi^o(p)}{\partial T} = -\frac{2\mu_B^2}{V} \sum_{\mathbf{k}} \{\tilde{\mathcal{G}}^o(k+p) + \tilde{\mathcal{G}}^o(k-p)\} \frac{\partial f_o(\varepsilon_{\mathbf{k}} - \mu)}{\partial T} . \qquad (34.50)$$

Making use of (34.35), Eq. (34.48) may therefore be written in the form of Eq. (12) of Ref. 80,

$$s_1 = -\frac{2}{V} \sum_{\mathbf{k}} \Sigma(\mathbf{k}; \varepsilon_{\mathbf{k}} - \mu) \frac{\partial f_o(\varepsilon_{\mathbf{k}} - \mu)}{\partial T} . \qquad (34.51)$$

In the approximation (34.43) one finds at low temperatures, using (32.21) and inserting the density of states (10.4) and $\partial f_o(\varepsilon)/\partial T = -\varepsilon f_o'(\varepsilon)/T$,

$$s_1 = \frac{k_B k_F^3}{(2\pi)^2} \int_{-\infty}^{+\infty} \frac{e^{-y} dy}{(1+e^{-y})^2} \int_o^{x_1} x^2 dx$$

$$\times \int_o^\infty \frac{dt}{(\frac{1}{\kappa}-1)^2 + (\frac{\pi}{2}t)^2} \left[ty \ln \left| \frac{y + 2txT_F/T}{y - 2txT_F/T} \right| \right] \qquad (34.52)$$

where we have set $\varepsilon \equiv k_B T y$, $\varepsilon_F \equiv k_B T_F$, $N(\varepsilon_F + \varepsilon) \simeq N(\varepsilon_F)$ and $\varphi \simeq 0$. For fixed y and x the integrand vanishes as t^2 at $t = 0$ and as t^{-2} at $t = \infty$. Therefore we may replace t by the dominant value $t_o \equiv 2(1-\kappa)/\pi\kappa \ll 1$ in the bracketed part and use the standard result for the remaining t-integral. Replacing x by the new variable $u \equiv xT_o/T$ where $T_o \equiv 2t_o T_F$, the u-integration may then be done explicitly and, making use of definition (32.11), the correction to the specific heat resulting from s_1 is finally found to be (Problem 5.13)

$$c_{V1} \equiv T \left(\frac{\partial s_1}{\partial T} \right)_\mu \simeq \frac{k_B k_F^3 x_1^3}{6\pi} \left\{ \frac{T}{T_1} + \left[\frac{2}{3} + \frac{30}{7\pi^4} K_+ + \ln \frac{T}{T_1} \right] \right.$$

$$\left. \times \frac{14\pi^2}{5} \left(\frac{T}{T_1} \right)^3 + \mathcal{O}\left(\left(\frac{T}{T_1} \right)^5 \right) \right\} \qquad (34.53)$$

where $T_1 \equiv x_1 T_o \simeq T_o$ and

$$K_\pm \equiv \int_o^{+\infty} \frac{y^4 e^{-y} \ln y}{(1 \pm e^{-y})^2} dy = \int_o^\infty \frac{y^3(1 + 3\ln y)}{e^y \pm 1} dy . \qquad (34.54)$$

Note that, actually, we should calculate $(\partial s/\partial T)_n = (\partial s/\partial T)_\mu (\partial \mu/\partial T)_n$. According to Nernst's theorem and to the fact that $c_V/T < \infty$ for $T \to 0$, $(\partial s/\partial \mu)_T \propto T(1 + \alpha \ln T)$ at least while, according to Problem 2.8 of Chapter 2, $(\partial \mu/\partial T)_n \propto T$ so that the correct expression c_{V1} may contain an additional term $\propto T^3(1 + \alpha \ln T)$. In Eq. (34.53) the term $\propto T$ contributes to the mass renormalization; it is easy to check using Eq. (10.3) that it coincides with the last expression (34.44) which gives confidence to the calculation made. On the other hand, the term $\propto T^3$ is a correction to the phonon specific heat (see Ref. 4, Eq. (32) of Chapter 5). The really new term is the one $\propto T^3 \ln T$ which is due to Doniach and Engelsberg (compare Eq. (13) of Ref. 80); it is a paramagnon contribution to the specific heat.

Turning now to s_2, we calculate Eq. (34.49) using (34.36) and (32.21) in (34.45) and setting $\varphi \simeq 0$ as for s_1. Introducing the integration variable $y \equiv \beta\omega = 2txT_F/T$, the result is in the above notation,

$$
s_2 = -\frac{k_B k_F^3}{\pi^3} \int_o^{x_1} x^2 dx \int_o^{2xT_F/T} \frac{y e^y dy}{(e^y - 1)^2} \left\{ (1 - \kappa) \frac{Ty}{T_o x} - \tan^{-1} \frac{Ty}{T_o x} \right\}.
$$
(34.55)

Interchanging the order of the x- and y-integrations and making use of the formula (see Eq. (3.411.1) of Ref. 59 and Eq. (16.9) above)

$$
J_n \equiv \int_o^\infty \frac{y^n e^y dy}{(e^y - 1)^2} = n \int_o^\infty \frac{y^{n-1} dy}{(e^y - 1)^2} = n! \zeta(n)
$$
(34.56)

one finally arrives at the result (Problem 5.14)

$$
c_{V2} \equiv T \left(\frac{\partial s_2}{\partial T} \right)_\mu = \frac{k_B k_F^3 x_1^3}{6\pi} \left\{ \kappa \frac{T}{T_1} + \left[\frac{15}{4\pi^4} K_- + \ln \frac{T}{T_1} \right. \right.
$$
$$
\left. \left. + \mathcal{O}((1 - \kappa)^3) \right] \frac{8\pi^2}{5} \left(\frac{T}{T_1} \right)^3 + \mathcal{O}\left(\left(\frac{T}{T_1} \right)^5 \right) \right\}
$$
(34.57)

where K_- is defined in Eq. (34.54) and the same observation applies as the one made after Eq. (34.54). This expression for c_{V2} is surprisingly similar to the one for c_{V1}. In particular, there is again a term $\propto T^3 \ln T$, for which a derivation is sketched in the Appendix of Ref. 81. As to the term $\propto T$, it also contributes to the mass renormalization and must be contained in the part $\propto n_o(\omega)$ of Eq. (34.39) which was neglected. Note that the chains which were neglected in Eq. (34.32) also contribute in principle[82].

The *paramagnon contribution* $\propto T^3 \ln T$ to the specific heat is best seen in the 2-parameter plot of $\gamma \equiv c_V/T$,

$$
\gamma(T) - \gamma(0) = 2A \left(\frac{T}{T_p} \right)^2 \left[1 + \ln \left(\frac{T}{eT_p} \right) \right] = A \left(\frac{T}{T_p} \right)^2 \ln \left(\frac{T}{T_p} \right)^2
$$
(34.58)

which has the typical shape shown in Fig. 5.13. According to this figure, the minimum of γ occurs at $1/\sqrt{e} = 0.61$ times the "paramagnon temperature"

T_p. Such shapes have been observed in many enhanced paramagnets[83]; for UAl$_2$, $T_p \simeq 12$ K and for TiBe$_2$, $T_p \simeq 21$ K.

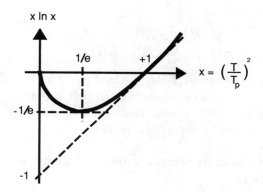

Fig. 5.13: Typical shape (34.58) of the paramagnon contribution to the specific heat.

Since γ and the susceptibility χ are second derivatives of the Gibbs potential density g with respect to temperature T and magnetic field H, respectively, the following thermodynamic relation holds (it also follows from Eq. (29.12) above):

$$-\left(\frac{\partial^4 g}{\partial T^2 \partial H^2}\right)_n = \left(\frac{\partial^2 \gamma}{\partial H^2}\right)_{T,n} = \left(\frac{\partial^2 \chi}{\partial T^2}\right)_{H,n}. \qquad (34.59)$$

Therefore, if A in Eq. (34.58) depends at least quadratically on H, the susceptibility also has a term $\propto T^2 \ln T$. Such terms in χ are controversial; they were reported in Ref. 84 but refuted in Ref. 85. If we write

$$\chi(T, H) = \chi_o(H) + \chi_1(H)T + \chi_2(H)T^2 + \chi_\ell(H)T^2 \ln T$$
$$\gamma(T, H) = \gamma_o(T) + \gamma_1(T)H + \gamma_2(T)H^2 + \gamma_\ell(T)H^2 \ln H \qquad (34.60)$$

then time reversal invariance of a paramagnet implies $\gamma_1(T) = 0$ while from Eq. (34.59) follows that

$$2\chi_2(H) + (3 + 2\ln H)\chi_\ell(H) = 2\gamma_2(T) + (3 + 2\ln H)\gamma_\ell(T). \qquad (34.61)$$

Thus, if $\chi_\ell(0) \neq 0$ and $\gamma_\ell(0) \neq 0$, the limit $H, T \to 0$ of this relation implies the specific form

$$\frac{\chi_\ell(H)}{\ln H} = \frac{\gamma_\ell(T)}{\ln T} = \lambda = \text{const} \neq 0. \qquad (34.62)$$

If such terms do not exist, then $\lambda = 0$ and Eq. (34.61) implies

$$\chi_2(H) = \gamma_2(T) = \alpha = \text{const} \tag{34.63}$$

and hence

$$\chi(T, H) = \chi_o(H) + \chi_1(H)T + \alpha T^2$$
$$\gamma(T, H) = \gamma_o(T) + \alpha H^2 . \tag{34.64}$$

Precision measurements on $TiBe_2$[86] show (see Fig. 3 of Ref. 86) that at small field, $\chi_1 \simeq 0$ and α decreases with H instead of being constant which might be interpreted in the sense that $\lambda \neq 0$. Terms $\propto H^2 \ln H$ in χ have indeed been reported in Ref. 87 but refuted in Ref. 88.

35. Itinerant-electron Ferromagnetism. Magnetization Fluctuations

As observed in Section 32, Stoner theory fails to give the Curie-Weiss law at high temperatures. It turns out that this defect may be corrected by taking into account fluctuations of the local magnetization $M(\mathbf{r})$. In mean-field approximation the magnetic free energy (32.26) may be written as

$$\bar{F}_m = V \frac{m_o^2}{2\chi_o} \tag{35.1}$$

where $m_o = M = \text{const}$ is the $q = 0$ component of the Fourier transform

$$m_{\mathbf{q}} = \frac{1}{V} \int_V d^3r M(\mathbf{r}) e^{-i\mathbf{q}\cdot\mathbf{r}} \tag{35.2}$$

and, using $M^4|_{\text{RPA}} = 6\langle M^2 \rangle M^2$ (there are 6 ways of pairing the factors in M^4),

$$\frac{1}{\chi_o} \equiv \frac{1}{\chi_P}\left\{1 - \kappa + 3\frac{\langle M^2 \rangle}{(M^*)^2}\right\} \equiv a + 3b\langle M^2 \rangle . \tag{35.3}$$

The last expression is the same as Eq. (32.33) where, however, we neglected the term $(T/T^*)^2$, assuming $T_C \ll T^*$ which is certainly satisfied for weak itinerant ferromagnets. In the ferromagnetic case, $\kappa > 1$, this also amounts to formally setting, $T_S \sim \infty$ in Eq. (32.26).

$\bar{F}_m$ may now be considered as the $q = 0$ part of the fluctuation-energy functional[89]

$$\varepsilon[m] = \frac{1}{2}\sum_{\mathbf{q}} \Omega_q |m_{\mathbf{q}}|^2 \tag{35.4}$$

where the factor $1/2$ accounts for double counting due to the fact that $m_{\mathbf{q}}^* = m_{-\mathbf{q}}$ and

$$\frac{\Omega_q}{V} \equiv \frac{1}{\chi_q} = \mu^2 q^2 + \frac{1}{\chi_o} . \tag{35.5}$$

ε is then also the mean-field approximation of the Ginzburg-Landau functional analogous to Eqs. (26.41), (26.27),

$$F_m[M(\mathbf{r})] = \frac{1}{2} \int_V d^3r \left\{ \mu^2 (\nabla M)^2 + aM^2 + \frac{1}{2} bM^4 \right\} . \tag{35.6}$$

From Eq. (35.4) the magnetic-field fluctuation is obtained in analogy with Eq. (29.26) as

$$H_{\mathbf{q}} = \frac{1}{V} \frac{\delta\varepsilon}{\delta m_{-\mathbf{q}}} = \frac{m_{\mathbf{q}}}{\chi_q} \tag{35.7}$$

which shows that χ_q is the corresponding static susceptibility.

The average occurring in Eq. (35.3) may be calculated by introducing the canonical ensemble of the independent fluctuation modes which for $\mathbf{q} \neq 0$ are given by the decomposition into real and imaginary parts, $m_{\mathbf{q}} = m_{\mathbf{q}}' + im_{\mathbf{q}}''$. In terms of the partition function

$$Z_{\mathbf{q}} = \int_{-\infty}^{+\infty} dm\, e^{-\beta\Omega_q m^2} = \left(\frac{\pi}{\beta\Omega_q} \right)^{1/2} \tag{35.8}$$

one finds in the same way as in Problem 2.2 of Chapter 2

$$\langle m_{\mathbf{q}}'^2 \rangle = \langle m_{\mathbf{q}}''^2 \rangle = -\partial \ln Z_{\mathbf{q}} / \partial\beta\Omega_q = \frac{1}{2\beta\Omega_q} . \tag{35.9}$$

On the other hand, $\langle m_{\mathbf{q}}' \rangle = \langle m_{\mathbf{q}}'' \rangle = 0$ and hence $\langle m_{\mathbf{q}}' m_{\mathbf{q}}'' \rangle = 0$ and $\langle m_{\mathbf{q}}^2 \rangle = \langle m_{\mathbf{q}}'^2 \rangle - \langle m_{\mathbf{q}}''^2 \rangle = 0$, but also $\langle m_{\mathbf{p}}^s m_{\mathbf{q}}^t \rangle = 0$ for $\mathbf{p} \neq \mathbf{q}$ and $s, t =' $ or $''$. Therefore,

$$\langle m_{\mathbf{q}} m_{\mathbf{q}'}^* \rangle = \frac{1}{\beta\Omega_q} \delta_{\mathbf{q}\mathbf{q}'} \tag{35.10}$$

and, according to (35.2), (see Eq. (8) of Ref. 89)

$$\langle M^2 \rangle = \sum_{\mathbf{q}\mathbf{q}'} \langle m_{\mathbf{q}} m_{\mathbf{q}'}^* \rangle e^{i(\mathbf{q}-\mathbf{q}')\cdot\mathbf{r}} = k_B T \sum_{\mathbf{q}} \frac{1}{\Omega_q} . \tag{35.11}$$

Since Ω_q grows only as q^2, the $\mathbf{q}$-sum in Eq. (35.11) must be supplemented with a cutoff factor $\theta(q_m - |\mathbf{q}|)$ which is physically reasonable since short-wavelength modes die out quickly. Evaluation with the help of Eqs. (35.3) and (35.5) then gives

$$\frac{\langle M^2 \rangle}{k_B T} = \frac{1}{2\pi^2 \mu^2} \int_0^{q_{max}} \frac{q^2 \, dq}{q^2 + \lambda^2 q_m^2} = \frac{q_m}{2\pi^2 \mu^2} \left[1 - \lambda \tan^{-1} \frac{1}{\lambda} \right] \tag{35.12}$$

where

$$\lambda^2 \equiv \frac{a + 3b\langle M^2 \rangle}{\mu^2 q_{max}^2} = \frac{1}{\mu^2 q_m^2 \chi_o} \, . \tag{35.13}$$

Eliminating $\langle M^2 \rangle$ from the last two relations we find the selfconsistency condition

$$\lambda^2 = \frac{a}{\mu^2 q_m^2} + \frac{T}{T_o} \left[1 - \lambda \tan^{-1} \frac{1}{\lambda} \right] \tag{35.14}$$

where

$$k_B T_o \equiv \frac{2\pi^2 \mu^4 q_m}{3b} \, . \tag{35.15}$$

For a ferromagnet, Eq. (35.3) implies that $a \equiv (1 - \kappa)/\chi_P < 0$. And $\chi_o^{-1} = 0$ or, according to (35.13), $\lambda = 0$ determines the Curie temperature T_C. We thus conclude from Eq. (35.14) that[89]

$$\frac{T_C}{T_o} = \frac{|a|}{\mu^2 q_m^2} \equiv |a| \frac{C}{T_o} \tag{35.16}$$

where the last relation defines the Curie constant C. Near T_C where $|\lambda| \ll 1$, combination of Eqs. (35.14) and (35.16) results in

$$\lambda^2 = \frac{T - T_C}{T_o} - \frac{T}{T_o} \left\{ \frac{\pi}{2} \lambda + \mathcal{O}(\lambda^2) \right\} \, . \tag{35.17}$$

This equation has no real solution for $T < T_C$ and is consistent only for $T_C \ll T_o$ which according to Eqs. (35.6), (35.16) means strong local magnetization fluctuations. In this case the solution of the quadratic approximation of Eq. (35.17) which vanishes at T_C is

$$\lambda \cong \begin{cases} 2|t|/\pi; & \text{if } 0 < -t \ll T_C/T_o \ll 1 \\ \sqrt{|t|T_C/T_o}; & \text{if } T_C/T_o \ll -t \end{cases} \tag{35.18}$$

where, as in Eqs. (29.21) and (32.23), $t \equiv 1 - T/T_C$. According to Eq. (35.13), the last expression gives rise to a *Curie-Weiss law* [89].

In the ferromagnetic state at $T < T_C$ there is a macroscopic occupation of the $q = 0$ mode so that now $M^4|_{RPA} = (6\langle M^2 \rangle + m_o^2)M^2$. Inserted into Eq. (35.6) we have for $M = m_o$

$$\bar{F}_m(m_o) = \frac{V}{2} \left\{ -|a|m_o^2 + b\left(3\langle M^2 \rangle + \frac{1}{2}m_o^2\right)m_o^2 \right\} \tag{35.19}$$

and hence, according to (29.26), $H = \{-|a|+3b\langle M^2\rangle+bm_o^2\}m_o$. Therefore[89]

$$m_o^2 = \frac{|a|}{b} - 3\langle M^2\rangle ; \quad T < T_C \qquad (35.20)$$

and

$$\frac{1}{\chi_o} \equiv \frac{\partial H}{\partial m_o} = 2|a| - 6b\langle M^2\rangle ; \quad T < T_C . \qquad (35.21)$$

Using the last expression in (35.5), Eqs. (35.11), (35.12) yield the selfconsistency condition

$$\lambda^2 = 2\frac{T_C}{T_o} - 2\frac{T}{T_o}\left[1 - \tan^{-1}\frac{1}{\lambda}\right] ; \quad T < T_C \qquad (35.22)$$

where

$$\lambda^2 \equiv \frac{2|a| - 6b\langle M^2\rangle}{\mu^2 q_m^2} = \frac{1}{\mu^2 q_m^2 \chi_o} \qquad (35.23)$$

and the definitions (35.15), (35.16) were used. Again we look for a solution near T_C where $|\lambda| \ll 1$. In the same way as Eq. (35.18) was derived we now find

$$\lambda \cong \begin{cases} -2t/\pi; & \text{if } 0 < t \ll T_C/T_o \ll 1 \\ -\sqrt{2tT_C/T_o}; & \text{if } T_C/T_o \ll t . \end{cases} \qquad (35.24)$$

From this solution it follows, according to Eqs. (35.20), (35.21) and (35.23) that $m_o^2 = 1/2b\chi_o$ vanishes quadratically at T_C but follows a linear behavior as in Eqs. (29.23) and (32.24) in an intermediate temperature range. χ_o^{-1} as obtained from Eqs. (35.18) and (35.24) is shown in Fig. 5.14.

With the help of Eq. (35.8) the total partition function Z and free energy φ of the fluctuations are given by the functional integral

$$e^{-\beta\varphi} = Z = \prod_{\mathbf{q}} Z_{\mathbf{q}} = \int \mathcal{D}m e^{-\beta\varepsilon[m]} . \qquad (35.25)$$

Note that to correct for double counting in the **q**-product, $Z_{\mathbf{q}}^2$ is replaced by $Z_{\mathbf{q}}$. With (35.8) one finds the *fluctuation entropy* [89]

$$\sigma = -\left(\frac{\partial\varphi}{\partial T}\right)_V = \frac{k_B}{2}\sum_{\mathbf{q}}\left(1 + \ln\frac{\pi k_B T}{\Omega_q}\right) \qquad (35.26)$$

and the contribution to the specific heat

$$\Delta c_V = \frac{T}{V}\left(\frac{\partial\sigma}{\partial T}\right)_V = \frac{k_B}{2V}\sum_{\mathbf{q}}\left(1 - \frac{T}{\Omega_q}\frac{\partial\Omega_q}{\partial T}\right) . \qquad (35.27)$$

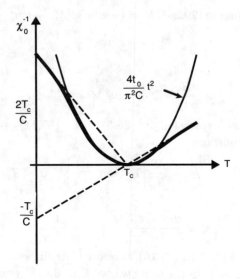

Fig. 5.14: Inverse susceptibility $\chi_o^{-1} = T_o \lambda^2/C$ according to Eqs. (35.18) and (35.24).

Using Eq. (35.5) with (35.3) above T_C and with (35.22) below, we find in view of Eq. (35.11)

$$\Delta c_V = \frac{1}{2}k_B \nu - \frac{1}{2}\langle M^2\rangle \frac{\partial \chi_o^{-1}}{\partial T} \tag{35.28}$$

where νV is the number of fluctuation modes. Making use of Eqs. (35.13), (35.18), (35.23) and (35.24) one finds the cusp-like result

$$\Delta c_V - \frac{1}{2}k_B \nu = \frac{T_o}{6bC^2} \begin{cases} 8t/\pi^2; & |t| \ll T_C/T_o \\ (1+3\,\mathrm{sgn}\,t)T/2T_o; & T_C/T_o \ll |t| \end{cases} \tag{35.29}$$

which is plotted in Fig. 5.15. Such a specific heat anomaly has been observed in Sc_3In (see Fig. 2 of Ref. 90, also Fig. 1 of Ref. 89).

A feature of this material which has not been sufficiently emphasized in the literature is the remarkable fact that ferromagnetism exists only in the narrow interval $0.032 < z < 0.048$ of the system $Sc_{3+z}In_{1-z}$[69] or, in the notation of Ref. 69, of $Sc_{1-x}In_x$ where $x = (1 - z)/4$. The results of this reference were obtained by fitting a Curie law so that Stoner theory cannot be used to explain the data while the Murata-Doniach model[89] described above gives a natural qualitative understanding. In Ref. 69 the Curie constant defined in Eq. (35.16) is written as $C = N\mu_B^2 p_{\mathrm{eff}}^2/3k_B$. Since according to Fig. 2 and Table I of this reference p_{eff} varies not more than 13% in the ferromagnetic range, C may be assumed to be constant.

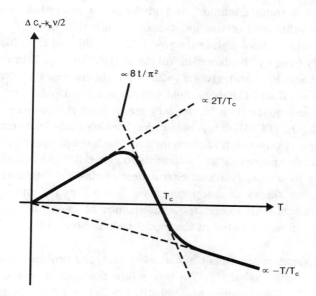

Fig. 5.15: The specific heat anomaly of Eq. (35.29).

Now from Eqs. (32.21), (33.8), (35.3) and (35.16) we have[89]

$$\frac{T_C}{C} = |a| \equiv \frac{\kappa - 1}{\chi_P} = \frac{vUN(\mu) - 1}{2\mu_B^2 N(\mu)} \qquad (35.30)$$

where $v \equiv V/L$ and we introduced the chemical potential μ as a variable. Indeed, the idea is that the carrier concentration and hence μ strongly depend on the chemical composition x while the exchange coupling constant U is less sensitive to x in the narrow range $0.238 < x < 0.242$ of ferromagnetism. From Tables I and II of Ref. 69 (see also Fig. 3 there) one finds that $T_C \simeq 6$ K almost uniformly in this interval while $T_C < 0$ outside. Assuming that $\mu \propto x + const$, the peak structure of $N(\mu)$ introduced in Eq. (32.29) qualitatively explains the data, provided that this peak falls into the ferromagnetic interval such that at the extremities $vUN(\mu) = 1$[89]. Note that this qualitative argument also works in Stoner theory but, as said above, Stoner theory is not adequate in this case.

The above discussion shows that the Murata-Doniach model gives a surprisingly accurate and simple description of itinerant-electron ferromagnetism. Where it fails is in the *critical region* near T_C where long-wavelength fluctuations become dominant and give rise to power-law behaviors of the susceptibility and of the spontaneous magnetization, $\chi_o^{-1} \propto |t|^\gamma$ and $m_o \propto t^\beta$, with *critical exponents* $\gamma \sim 1.22$ to 1.36 and $\beta \sim 0.31$ to 0.36, respectively (see, e.g. Kadanoff in Vol. 5a of Ref. 20). The Murata-Doniach model, on the other hand, gives $\gamma = 2\beta = 2$ if the interval $0 < t \ll T_C/T_o$ is observable at all and the mean-field value $\gamma = 2\beta = 1$ otherwise, both values being inadequate. This inadequacy stems from the fact that according to Eqs. (35.30), (14.14), U/ε_F is of order 1 which makes itinerant-electron magnetism a *strong-coupling phenomenon* which is not properly taken care of in the above approximation. Attempts to generalize the Murata-Doniach model both by a more systematic treatment of classical fluctuations[91] and by a quantum theory of a ferromagnetic Fermi liquid[92] have not led to a deeper insight. An interesting consequence of quantum corrections is that the cutoff wavenumber of the model becomes temperature dependent, $q_m \propto T^{1/3}$ [91].

An alternative treatment of the selfconsistency problem was proposed by Moriya and Kawabata[93]. We here follow the critical review of Gumbs and Griffin[94]. Using definitions (29.10), (29.26) and Eq. (32.33) we may express the renormalization effect on the susceptibility by

$$\chi^{-1} = \frac{1}{V}\left(\frac{\partial^2 F_m}{\partial M^2}\right)_{T,n} = \chi_P^{-1}\{1 - \kappa(T) + \Delta\kappa(T)\} \tag{35.31}$$

where $\kappa(T) \equiv \kappa - (T/T^*)^2$ is the Stoner enhancement factor (including the temperature dependence) and

$$\Delta\kappa(T) \equiv \frac{\chi_P}{V}\left(\frac{\partial^2 \Delta F_m}{\partial M^2}\right)_{n,T} \tag{35.32}$$

the interaction contribution. Here ΔF_m may be replaced by $\Delta\Omega$ since switching on the interaction from 0 to λ does not change the equilibrium carrier density,

$$\left(\frac{\partial\Delta\Omega}{\partial\mu}\right)_{T,M} = \Delta n = 0 \tag{35.33}$$

so that $\Delta\Omega$ is independent of the chemical potential, or $\partial\mu/\partial\lambda = 0$. Hence we may use Eq. (34.32) where the M-dependence is implicitly given by Eqs. (34.9), (32.1), (32.2) and, according to (29.16), $\bar{H} = \lambda M$. A straightforward calculation yields, setting $\bar{H} = 0$ at the end and making use of

Eqs. (34.14) and (32.21), (33.8),

$$\Delta\kappa(T) = -\frac{\lambda^2\kappa^2}{2N(0)V\beta} \sum_p \left\{ \frac{\chi^o(p)}{1-\lambda\chi^o(p)} \left(\frac{2}{V}\frac{\partial^2\Gamma_{\uparrow\downarrow}(p)}{\partial\bar{H}^2} \right)_{\bar{H}=0} \right.$$
$$\left. + \frac{1}{(1-\lambda\chi^o(p))^2} \left(\frac{2\mu_B}{V}\frac{\partial\Gamma_{\uparrow\downarrow}(p)}{\partial\bar{H}} \right)^2_{\bar{H}=0} \right\} \qquad (35.34)$$

where

$$\left(\frac{2\mu_B}{V}\frac{\partial\Gamma_{\uparrow\downarrow}(p)}{\partial\bar{H}} \right)_{\bar{H}=0} = 2\left(\vartheta(p) - \frac{\partial\chi^o(p)}{\partial p_o} \right),$$

$$\left(\frac{2}{V}\frac{\partial^2\Gamma_{\uparrow\downarrow}(p)}{\partial\bar{H}^2} \right)_{\bar{H}=0} = \frac{\partial^2\chi^o(p)}{\partial\mu^2} + 4\frac{\partial^2\chi^o(p)}{\partial p_o^2} + 8\frac{\partial\vartheta(p)}{\partial p_o}; \qquad (35.35)$$

$$\vartheta(p) \equiv \frac{\mu_B^2}{V} \sum_k \frac{f_o'(\varepsilon_k - \mu) + f_o'(\varepsilon_{k+p} - \mu)}{\varepsilon_k - \varepsilon_{k+p} + p_o}.$$

From Eq. (35.34) it is evident that the renormalization effect comes mainly from the *paramagnon pole* $1 - \lambda\chi^o(\mathbf{q}, \omega_q^{\text{para}}) = 0$ which, according to Eq. (34.15), has the q-dependence[94]

$$\omega_q^{\text{para}} \cong i\frac{4}{\pi}\varepsilon_F\frac{1-\kappa}{\kappa}\frac{q}{q_F}; \quad q \ll q_F. \qquad (35.36)$$

Hence it is inconsistent to use Eq. (34.33) in (35.31) to determine the Curie temperature T_C since the dominant renormalization effect is expected to come from fluctuations which become soft at T_C[94]. In order to remedy this inconsistency Moriya and Kawabata[93] propose the recipe of making the substitution

$$\frac{1}{1-\lambda\chi^o(p)} \longrightarrow \frac{1}{1-\lambda\chi^o(p) + \Delta\kappa(T)} \qquad (35.37)$$

in Eq. (35.34). This leads to a selfconsistent and, as a consequence highly nonlinear, determination of $\Delta\kappa(T)$ with the desired property. The procedure does not, however, seem to have a deeper physical basis. It is not surprising, therefore, that other adjustments are necessary, a particularly sensitive point being the wavenumber cutoff q_c in the $\mathbf{p}$-sum of Eq. (35.34) which must guarantee the domination of the long-wavelength modes[94]. An interesting feature is that, after adjustment and under sufficiently restrictive conditions (limitation to the p_o-mode and a q_c-value allowing only thermal fluctuations) Eqs. (35.34), (35.37) may be cast into the form of the selfconsistency condition (35.22) of Murata and Doniach[89].

While the selfconsistent procedures described above are best suited for weak itinerant ferromagnets, no method to deal with strong ferromagnets such as Fe, Co, Ni has yet been mentioned. The main problem with these materials appears to be the fact that, besides being metals, they also show *local-moment behavior*. This circumstance bears some analogy to the Kondo problem mentioned in Section 15 which arises as a result of the interaction between localized f-electrons and band electrons. But while in the latter situation the local moments may rotate with a characteristic "Kondo frequency" so that the groundstate remains a singlet, the Hubbard repulsion among the d-electrons will compete with the s-d interaction in the sense of a "Mott localization" of the *direction* of the local moments. The relevant parameter in this competition then is the ratio between the Hubbard U and the d-level width Γ due to the s-d coupling (see Section 3.6 of Ref. 63).

Since we are interested in the problem of *local-moment formation* we may concentrate on the paramagnetic state and neglect short-range order. It is then sufficient to consider a one-center problem with just one d-electron[95]. But this is the situation of the *Anderson model*[96]: The Hamiltonian of the s-d system consists of the terms

$$\mathcal{H}_s = \sum_{\mathbf{k}\sigma} \varepsilon_{\mathbf{k}} a^+_{\mathbf{k}\sigma} a_{\mathbf{k}\sigma} \; ; \quad \mathcal{H}_d = \sum_\sigma \varepsilon_d n_{d\sigma} \; ;$$
$$\mathcal{H}_{sd} = V^{-1/2} \sum_{\mathbf{k}\sigma} [g(\varepsilon_{\mathbf{k}}) a^+_{\mathbf{k}\sigma} c_{d\sigma} + \text{h.c.}] \tag{35.38}$$

where in analogy to Eq. (33.5), $n_{d\sigma} = c^+_{d\sigma} c_{d\sigma} = n^2_{d\sigma}$. To this is added the d-level repulsion of the Hubbard form (33.11),

$$\mathcal{H}_{dd} = U n_{d\uparrow} n_{d\downarrow} = \frac{1}{2} U \sum_\sigma n_{d\sigma} - 2U(S^z_d)^2 \tag{35.39}$$

where we made use of Eq. (33.10). In the last form of $\mathcal{H}_{dd}$ only the second term represents an interaction while the first may be absorbed into $\mathcal{H}_d$ with the new notation $\bar{\varepsilon}_d \equiv \varepsilon_d + U/2$ and $\bar{\mathcal{H}}^{95)}_d$.

Thus, expressing the total Hamiltonian as a sum of the two terms,

$$\mathcal{H}_o = \mathcal{H}_s + \bar{\mathcal{H}}_d + \mathcal{H}_{sd} \; ; \quad \mathcal{H}_{\text{int}} = -2U(S^z_d)^2 \, , \tag{35.40}$$

we may use Eqs. (5.9), (5.14), (5.25) and (5.28) to write the partition function in the form

$$Z = \text{Tr}\left[e^{-\beta \mathcal{H}_o} T \exp\left\{ -\int_o^\beta d\tau \mathcal{H}^o_{\text{int}}(-i\tau) \right\} \right] . \tag{35.41}$$

Replacing here the integral over $0 < \tau < \beta$ by a Riemann sum over intervals $\Delta\tau$, we may apply the *Hubbard-Stratonovich transformation*[97)]

$$e^{x_\tau^2 \Delta\tau/\beta} \equiv \int_{-\infty}^{+\infty} \sqrt{\frac{\Delta\tau}{\beta}} dy_\tau e^{-(\pi y_\tau^2 + 2\sqrt{\pi} x_\tau y_\tau)\Delta\tau/\beta} \qquad (35.42)$$

to the ordered exponential by setting $x_\tau = (2\beta U)^{1/2} S_d^{zo}(-i\tau)$ and $y_\tau = \tilde{h}(\tau)$ and thereby transform the quadratic interaction term into a linear external-field term,

$$Z = \mathrm{Tr}\left[e^{-\beta\mathcal{H}_o} \int \mathcal{D}\tilde{h}(\tau) T \exp\left\{ -\int_o^\beta d\tau \left(\frac{\pi}{\beta}\tilde{h}^2(\tau) + \mathcal{H}_{Z\tau}^o(-i\tau) \right) \right\} \right]. \qquad (35.43)$$

Here

$$\mathcal{H}_{Z\tau} \equiv m\tilde{h}(\tau)2S_d^z \; ; \quad m \equiv \sqrt{2\pi U/\beta} \qquad (35.44)$$

is the Zeeman term (30.4) with a time-dependent magnetic field $\tilde{h}(\tau)$ and $\int \mathcal{D}\tilde{h}(\tau)$ is the functional integral over the instantaneous field values.

It is convenient to subtract the average field h_o, defining $\bar{h}(\tau) \equiv \tilde{h}(\tau) - h_o$, where

$$h_\mu \equiv \beta^{-1} \int_o^\beta d\tau \tilde{h}(\tau) e^{i\mu\tau} = h_{-\mu}^* \; ; \quad \tilde{h}(\tau) = \sum_\mu h_\mu e^{-i\mu\tau} , \qquad (35.45)$$

μ being a bosonic Matsubara frequency (7.27), and to transfer the contribution of h_o to $\mathcal{H}_o$ by writing

$$\bar{\mathcal{H}}_o(h_o) = \mathcal{H}_o + mh_o \sum_\sigma \sigma n_{d\sigma} \; ; \quad \bar{\mathcal{H}}_{Z\tau} = m\bar{h}(\tau)2S_d^z . \qquad (35.46)$$

Note that this shift appears to lead to complications in Eq. (35.43) since, because of $\mathcal{H}_{sd}$, the shifted term does not commute with $\mathcal{H}_o$. This is not so, however, because $\beta\mathcal{H}_o$ may also be included in the time-ordered exponential. Indeed, in analogy to Eqs. (5.20), (5.25) and (5.28), integration of Eq. (5.4) yields, in the case of any time-dependent Hamiltonian $\mathcal{H}_\tau = \mathcal{H}_o + \mathcal{H}_\tau'$, the general formula

$$U(-i\tau, 0) = T\exp\{-\int_o^\beta d\tau \mathcal{H}_\tau\} = e^{-\beta\mathcal{H}_o} T\exp\{-\int_o^\beta d\tau \mathcal{H}_\tau'\} . \qquad (35.47)$$

Such formulas actually follow even more directly with the aid of Feynman's ordering-label technique[98)] used by Hubbard in Ref. 97.

Expressing the partition function (35.43) in terms of the independent variables h_o and $h_\mu = h_\mu' + ih_\mu''$, $\mu > 0$, using Eq. (7.28) and multiplying

$\mathcal{H}_{Z\tau}$ by an arbitrary coupling parameter λ, we may write $Z \equiv Z_1$ with

$$Z_\lambda \equiv \int_{-\infty}^{+\infty} dh_o \left(\prod_{\mu>0} 2 \int d^2 h_\mu \right) e^{-\pi \sum_\mu |h_\mu|^2} \zeta_\lambda[h] \qquad (35.48)$$

and

$$\zeta_\lambda[h] \equiv \text{Tr} \left[e^{-\beta \bar{\mathcal{H}}_o} T \exp \left\{ -\lambda \int_o^\beta d\tau \bar{\mathcal{H}}_{Z\tau}^o(-i\tau) \right\} \right] \qquad (35.49)$$

where the upper index o on $\bar{\mathcal{H}}_{Z\tau}$ now means the interaction representation (5.17) with respect to $\bar{\mathcal{H}}_o$. Note that the factor 2 in front of the h_μ-integral in Eq. (35.48) is essential since it assures the correct normalization. Indeed, for $U = 0$ one finds $Z_\lambda = \text{Tr} \exp(-\beta \mathcal{H}_o)$ in agreement with Eq. (35.41).

The expression under the trace in Eq. (35.49) may be used to define the statistical average of an operator O by

$$\langle O \rangle_\lambda \equiv \text{Tr} \left[e^{-\beta \bar{\mathcal{H}}_o} T \exp \left\{ -\lambda \int_o^\beta d\tau \bar{\mathcal{H}}_{Z\tau}^o(-i\tau) \right\} O \right] / \zeta_\lambda[h] . \qquad (35.50)$$

Note that this average is not time-translation invariant because of the explicit τ-dependence of the external field $\bar{h}(\tau)$ in $\mathcal{H}_{Z\tau}$. For this reason the one-electron Green's function defined by

$$\bar{\mathcal{G}}_\sigma^\lambda(i\nu, i\nu') \equiv \beta^{-1} \int_o^\beta d\tau e^{-i\nu\tau} \int_o^\beta d\tau' e^{i\nu'\tau'} \langle T\{c_{d\sigma}^{+o}(-i\tau) c_{d\sigma}^o(-i\tau')\} \rangle_\lambda \qquad (35.51)$$

has two frequency arguments. This definition is natural since for $\lambda = 0$ it reduces to the usual unperturbed Green's function (7.34) with respect to $\bar{\mathcal{H}}_o$ (distinguished by a bar instead of the tilde),

$$\bar{\mathcal{G}}_\sigma^o(i\nu, i\nu') = \bar{\mathcal{G}}_\sigma^o(i\nu) \delta_{\nu\nu'} . \qquad (35.52)$$

Physically, the violation of τ-translation invariance shows in the fact that the external field may exchange any amount $i\nu$ of energy with the d-electron. This is analogous to the violation of space-translation invariance in the impurity problem of Section 18 which led to the Green's functions (18.34) depending on two momenta and to a selfenergy whose argument was the arbitrary momentum exchanged between the impurity and the electron. Likewise the selfenergy now depends on the arbitrary energy $i\nu$ exchanged between the external field and the d-electron: $\Sigma_\sigma(i\nu)$. It connects the propagators (35.51) and (35.52) through the Dyson equation (11.5), viz. (see Eq. (5a) of Ref. 95),

$$\bar{\mathcal{G}}_\sigma^\lambda(i\nu, i\nu') = \bar{\mathcal{G}}_\sigma^o(i\nu) \delta_{\nu\nu'} + \lambda \sum_{\nu''} \bar{\mathcal{G}}_\sigma^o(i\nu) \bar{\Sigma}_\sigma(i\nu'' - i\nu) \bar{\mathcal{G}}_\sigma^\lambda(i\nu'', i\nu') , \qquad (35.53)$$

and is easily obtained by calculating $\bar{\mathcal{G}}_\sigma^o(i\nu)\Sigma_\sigma(i\nu' - i\nu)\bar{\mathcal{G}}_\sigma^o(i\nu')$ according to Eqs. (11.17), (35.51). The result is

$$\bar{\Sigma}_\sigma(i\mu) = \sigma m h_\mu (1 - \delta_{\mu o}) . \tag{35.54}$$

On the other hand, making use of the inverse of relation (7.28), i.e. of

$$\beta^{-1} \sum_\nu e^{i\nu(\tau - \tau')} = \delta(\tau - \tau') , \tag{35.55}$$

$\bar{\mathcal{G}}_\sigma^\lambda$ may also be generated from ζ_λ. Indeed (see Eq. (4) of Ref. 95),

$$\frac{\partial}{\partial\lambda} \ln \zeta_\lambda[h] = -\sum_{\sigma\mu\nu} \bar{\Sigma}_\sigma(i\mu)\bar{\mathcal{G}}_\sigma^\lambda(i\nu, i\nu - i\mu) . \tag{35.56}$$

The idea now is to determine $\bar{\mathcal{G}}_\sigma^\lambda$ from Eqs. (35.53), (35.54) and then calculate ζ_1 by integrating Eq. (35.56) with respect to λ.

Multiplying Eq. (35.53) from the right by $\Sigma_\sigma(i\mu - i\nu)$ and defining the (ν, μ)-matrix elements

$$K_{\nu\nu'}^\sigma \equiv \bar{\mathcal{G}}_\sigma^o(i\nu)\bar{\Sigma}_\sigma(i\nu' - i\nu)$$
$$G_{\nu\nu'}^{\lambda\sigma} \equiv \sum_{\nu''} \bar{\mathcal{G}}_\sigma^\lambda(i\nu, i\nu'')\bar{\Sigma}_\sigma(i\nu' - i\nu'') , \tag{35.57}$$

the Dyson equation may be written in the matrix form

$$(1 - \lambda K^\sigma)G^{\lambda\sigma} = K^\sigma \tag{35.58}$$

which is a particular case of a Fredholm integral equation. The existence of a solution amounts formally to assume that there is a linear transformation T which diagonalizes K^σ (which is not entirely trivial since K^σ is neither symmetric nor hermitean). Thus writing

$$TK^\sigma T^{-1} = k^\sigma \text{ (diagonal)} ; \quad TG^{\lambda\sigma}T^{-1} \equiv F^{\lambda\sigma} \tag{35.59}$$

the solution of Eq. (35.58) becomes

$$F_{nn'}^{\lambda\sigma} = \frac{k_n^\sigma}{1 - \lambda k_n^\sigma}\delta_{nn'} . \tag{35.60}$$

Now, in Eq. (35.56) we need $\text{Tr}\, G^{\lambda\sigma} = \text{Tr}\, F^{\lambda\sigma}$ which may be generated from the determinant

$$D^{\lambda\sigma} \equiv \text{Det} |1 - \lambda K^\sigma| = \prod_n (1 - \lambda k_n^\sigma)$$

$$= \exp\left\{ \sum_n \ln(1 - \lambda k_n^\sigma)\right\} = \exp \text{Tr} \ln(1 - \lambda K^\sigma) . \tag{35.61}$$

Indeed,

$$\frac{\partial}{\partial\lambda}\ln D^{\lambda\sigma} = -\sum_n \frac{k_n^\sigma}{1-\lambda k_n^\sigma} = -\sum_n F_{nn}^{\lambda\sigma} = -\text{Tr}\, G^{\lambda\sigma}$$

$$= -\sum_{\mu\nu} \bar{\mathcal{G}}_\sigma^\lambda(i\nu, i\nu - i\mu)\bar{\Sigma}_\sigma(i\mu) \ . \tag{35.62}$$

Therefore, combining Eqs. (35.56), (35.61) and (35.62),

$$\frac{\partial}{\partial\lambda}\ln\zeta_\lambda[h] = \sum_\sigma \frac{\partial}{\partial\lambda}\text{Tr}\,\ln(1-\lambda K^\sigma) \ , \tag{35.63}$$

or integrated over λ from 0 to 1,

$$\zeta_1[h] = \zeta_o(h_o)\exp\Big\{\sum_\sigma \text{Tr}\,\ln(1-K^\sigma)\Big\} \tag{35.64}$$

where $\zeta_o \equiv \text{Tr}\,\exp(-\beta\bar{\mathcal{H}}_o)$. Inserting this into Eq. (35.48) we finally arrive at the result (see Eq. (6) of Ref. 95)

$$Z = Z_1 = \int_{-\infty}^{+\infty} dh_o e^{-\pi h_o^2}\zeta_o(h_o)$$

$$\times \Big(\prod_{\mu>0} 2\int d^2 h_\mu\Big)\exp\Big\{-2\pi\sum_{\mu>0}|h_\mu|^2 + \sum_\sigma \text{Tr}\,\ln(1-K^\sigma)\Big\} \ . \tag{35.65}$$

The evaluation of this complicated formula has to proceed in several steps. We start with $\zeta_o(h_o)$, treating the term $mh_o\sum_\sigma \sigma n_{d\sigma}$ in Eq. (35.46) as perturbation. The slight modification of the chemical potential μ this produces may be neglected so that we may put $\mu = 0$ and write

$$-\beta^{-1}\ln\frac{\zeta_o(h_o)}{\zeta_o(0)} = \Delta\Omega \ . \tag{35.66}$$

This shift in the thermodynamic potential may now be evaluated with the help of Eq. (9.20) together with (5.25) and (5.28),

$$-\beta\Delta\Omega = \sum_{N=1}^{\infty} \frac{(-mh_o)^N}{N!}\Big\langle T\Big[\sum_\sigma \sigma\int_0^\beta d\tau\, n_{d\sigma}^o(-i\tau)\Big]^N\Big\rangle_{o,cp} \ . \tag{35.67}$$

Because of the limitation to connected parts (cp) in this formula it is possible to calculate the general term and do the sum explicitly. The result is (Problem 5.15)

$$-\beta\Delta\Omega = \sum_\nu \log[1 - (mh_o\tilde{\mathcal{G}}^o(i\nu))^2] \tag{35.68}$$

where $\tilde{\mathcal{G}}^o(i\nu)$ is the d-electron propagator for $U = 0$ but in the presence of the s-d hybridization $\mathcal{H}_{sd}$, Eq. (35.38). Making use of Eqs. (11.7), (11.17), (11.30) and (11.34) the renormalization due to $\mathcal{H}_{sd}$ is found to yield (Problem 5.16)

$$\tilde{\mathcal{G}}^o(i\nu) = \frac{1}{i\nu - \bar{\varepsilon}_d + i\Gamma \operatorname{sgn}\nu} \tag{35.69}$$

where $\Gamma = \pi N_s(\bar{\varepsilon}_d)|g(\bar{\varepsilon}_d)|^2$ is the d-level width and $N_s(\varepsilon)$ the s-band density of states (10.4).

With (35.69), Eq. (35.68) may be evaluated further by using the formulas (see Eqs. (8.326.1) and (8.327) of Ref. 59)

$$\left|\frac{\Gamma(x)}{\Gamma(x+iy)}\right| = \prod_{n=0}^{\infty}\left[1 + \left(\frac{y}{n+x}\right)^2\right] \tag{35.70}$$

and

$$\Gamma(z) = z^{z-1/2}e^{-z}\sqrt{2\pi}[1 + \mathcal{O}(z^{-1})] \,, \tag{35.71}$$

provided that we put $\bar{\varepsilon}_d = 0$ and assume $\beta\Gamma \gg 1$. One then finds (Problem 5.17)

$$\Delta\Omega(h_o) \cong \frac{\Gamma}{2\pi} \ln\left[1 + \left(\frac{mh_o}{\Gamma}\right)^2\right] - \frac{mh_o}{\pi}\tan^{-1}\frac{mh_o}{\Gamma} \tag{35.72}$$

(see Eq. (8) of Ref. 95). Adding to this the value $\Omega_o = \pi h_o^2/\beta$ contained in Eq. (35.65), $\Omega \equiv \Omega_o + \Delta\Omega$ has minima determined by $\Omega'(h_o) = 0$ where

$$\Omega'(h_o) = \frac{m}{\pi}\left\{\frac{1}{u}\frac{mh_o}{\Gamma} - \tan^{-1}\frac{mh_o}{\Gamma}\right\} \tag{35.73}$$

and $u \equiv U/\pi\Gamma$. For $u > 1$ one finds two non-zero solutions $\pm\bar{h}_o$ which, for $u \gg 1$ are given by $\bar{h}_o \cong (\beta\Gamma u/8)^{1/2}$ (see Fig. 1 of Ref. 95).

From $\Omega = \Omega_o + \Delta\Omega$ the contribution to the magnetic susceptibility may now be calculated. Indeed, comparison of Eqs. (30.4) and (35.46) shows that an external magnetic field H may be introduced by adding $-\mu_B H$ to mh_o. Thus, according to Eq. (31.12),

$$\Delta\chi = -\frac{1}{V}\frac{\partial^2\Omega}{\partial H^2}\bigg|_{H=0,h_o=h_{av}} = -\frac{1}{V}\left(\frac{\mu_B}{m}\right)^2\Omega''(h_{av}) \tag{35.74}$$

where

$$h_{av}^2 \equiv \langle h_o^2\rangle = \int_{-\infty}^{+\infty} dh_o h_o^2 e^{-\beta\Omega(h_o)} \bigg/ \int_{-\infty}^{+\infty} dh_o e^{-\beta\Omega(h_o)} \,. \tag{35.75}$$

After another derivation of the expression (35.73) one obtains

$$\Delta\chi = \frac{\mu_B^2}{V}\frac{1}{\pi\Gamma}\left[\frac{1}{1+(mh_{av}/\Gamma)^2} - \frac{1}{u}\right]. \tag{35.76}$$

Assuming $u < 1$ and $mh_{av}/\Gamma \ll 1$, the Gaussian approximation may be used in Eq. (35.75) with $\beta\Omega \cong (\pi - \beta m^2/2\pi\Gamma)h_o^2$. The result then is $h_{av}^2 \cong [2\pi(1-u)]^{-1}$ which, inserted into Eq. (35.76) gives

$$\Delta\chi \cong \frac{\mu_B^2}{V}\left[\frac{\beta/\pi}{\beta\Gamma + \pi u/(1-u)} - \frac{1}{U}\right] \tag{35.77}$$

which is quite an interesting result. Indeed, if the diamagnetic term $\propto -U^{-1}$ is compensated by the paramagnetism of the s-electrons (see Section 31), Eq. (35.77) yields a *Curie law* $\chi \cong \mu_B^2(1-u)\beta/\pi^2 uV$ at sufficiently high temperatures and for a sufficiently narrow d-level, $k_B T \gg \Gamma/\pi u$. In the opposite limit, $u \gg 1$, one may put $h_{av}^2 = \bar{h}_o^2 = \beta\Gamma u/8$ in which case the first term in the bracket of Eq. (35.76) becomes $(2/\pi u)^2$ and hence is negligible.

Unfortunately, $\Omega = \Omega_o + \Delta\Omega$ does not yet give the complete dependence on h_o contained in Eq. (35.65) since K^σ also depends on h_o through $\bar{\mathcal{G}}_\sigma^o(i\nu)$. The latter is the propagator obtained by going over from $\mathcal{H}_o$ to $\bar{\mathcal{H}}_o$, Eq. (35.46), which means replacing $\bar{\varepsilon}_d$ in Eq. (35.69) by $\varepsilon_\sigma \equiv \bar{\varepsilon}_d + \sigma m h_o$. Making use of Eqs. (35.54), (35.57) we may calculate $\ln(1 - K^\sigma)$ as power series. Since $\text{Tr}\, K^\sigma = 0$ one easily finds

$$\sum_\sigma \text{Tr}\, \ln(1 - K^\sigma) = m^2 \sum_{\mu>0}\varphi_\mu|h_\mu|^2 + \mathcal{O}(h_\mu h_{\mu'} h_{\mu+\mu'}^*) \tag{35.78}$$

where

$$\varphi_\mu \equiv -\sum_\sigma\sum_\nu \bar{\mathcal{G}}_\sigma^o(i\nu)\bar{\mathcal{G}}_\sigma^o(i\nu + i\mu) = \varphi_{-\mu}. \tag{35.79}$$

Inserting here the expression (35.69) with $\bar{\varepsilon}_d$ replaced by ε_σ one finds (Problem 5.18)

$$\varphi_\mu = \frac{2\Gamma}{\mu(\mu + 2\Gamma)}2\text{Re}\sum_\sigma\sum_{0<\nu<\mu}\frac{1}{\nu + \Gamma + i\varepsilon_\sigma}. \tag{35.80}$$

Again setting $\bar{\varepsilon}_d = 0$ and approximating the ν-sum by a Riemann sum where, according to Eq. (7.27), the step is $\Delta\nu = 2\pi/\beta$, one finds (see Eq. (9b) of Ref. 95)

$$\varphi_\mu \cong \frac{2\beta\Gamma}{\pi\mu(\mu + 2\Gamma)}\ln\left[1 + \frac{\mu(\mu + 2\Gamma)}{\Gamma^2 + m^2 h_o^2}\right]. \tag{35.81}$$

Inserting the expression (35.78) into Eq. (35.65) and integrating out the h_μ with $\mu > 0$ one finds the additional contribution to the thermodynamic potential

$$\Delta\bar{\Omega}(h_o) = \beta^{-1} \sum_{\mu>0} \ln\left[1 - \frac{m^2}{2\pi}\varphi_\mu(h_o)\right]. \tag{35.82}$$

It is possible to estimate an upper limit μ_1 for the Matsubara frequency $\mu = 2k\pi/\beta$, k being an integer, by going back to $\zeta_1[h]$, Eq. (35.49). Indeed, according to Eqs. (35.45), (35.46) and since $2S_d^z$ may take values $\Delta n = 0, \pm 1, \pm 2, \cdots$,

$$\beta^{-1}\int_o^\beta d\tau \bar{\mathcal{H}}_{Z\tau}^o(-i\tau) = m\sum_\mu h_\mu \Delta n\beta^{-1}\int_o^\beta d\tau e^{[-i\mu + \Delta n(\varepsilon_\sigma \pm i\Gamma)]\tau}. \tag{35.83}$$

Here the τ-integration defines a function of k which is vanishingly small for $|k| \gg k_1 \sim x(t \pm 1)$ where $x \equiv \beta\Gamma/\pi$, $t \equiv mh_o/\Gamma$ and $\bar{\varepsilon}_d = 0$ as before. Hence, if $x \ll 1$ and $xt \ll 1$ one concludes that k is essentially limited to the single value $k_1 = 1$ or μ to $\mu_1 = 2\pi/\beta$. But this means that the only relevant Fourier component in Eq. (35.78) is h_{μ_1} and that all higher powers of the h_μ become negligible, leaving as only term $m^2\varphi_{\mu_1}|h_{\mu_1}|^2$. Then Eqs. (35.81), (35.82) lead to the simple formula

$$\beta\Delta\bar{\Omega} \cong \ln\left[1 + \frac{ux^2}{2}\ln\left(\frac{x^2}{4}(1+t^2)\right)\right]. \tag{35.84}$$

Taking $u \equiv U/\pi\Gamma \gg 1$ but such that $ux^2 \ll 1$ and defining $\bar{\Omega} \equiv \Omega_o + \Delta\Omega + \Delta\bar{\Omega}$ where $\Omega_o = \pi h_o^2/\beta$ and $\Delta\Omega$ is given by Eq. (35.72), one finds

$$\beta\bar{\Omega}'(h_o) \cong \left\{\frac{t}{u} - \tan^{-1}t + \frac{uxt}{1+t^2}\right\}\sqrt{2\pi ux}. \tag{35.85}$$

This represents a negligible modification of Eq. (35.73) so that the position of the minima is only little altered, $\bar{t} \cong \pi u/2$. With this value and with Eq. (35.74) one finally arrives at the result

$$\Delta\chi = -\frac{\mu_B^2}{V\beta m^2}\beta\bar{\Omega}'' \cong \frac{\mu_B^2}{V}\left\{\left(\frac{2\Gamma}{\pi U}\right)^2 2\beta - \frac{1}{U}\right\} \tag{35.86}$$

which is a Curie law (compare Fig. 2 of Ref. 95). We note, however, that a more consistent procedure would have been to use Eq. (35.80) restricted to the single term $\nu = \pi/\beta$ instead of Eq. (35.81). Unfortunately, this does not lead to a Curie law.

Higher-order contributions to Eq. (35.78) have been considered by Hertz and Klenin in a systematic analysis of the paramagnetic state[99].

This functional-integral approach has also developed into a major tool in the theory of ferromagnetism[28,29] which justifies the lengthy description given above.

36. Itinerant-electron Antiferromagnetism. Spin-density Waves

Ever since Overhauser proposed the existence of a spin-density wave (SDW) groundstate in chromium[32], this element (see Ref. 100 for a review on Cr) has become the model case of itinerant-electron antiferromagnetism (IEA). And the SDW idea has essentially remained the only explanation of IEA . Quite generally, the idea here is that a spatially inhomogenous distribution of the electrons in the form of a wave may, under certain circumstances, have lower energy than the homogenous Fermi sea. Such a state may be generated by applying one of the one-electron densities (3.22)–(3.25) to the Fermi sea $|F\rangle$. Indeed, using Eqs. (33.4) and (1.10) we have the expression

$$\sum_{\mathbf{k}} \langle a_{\mathbf{k}\sigma}^+ a_{\mathbf{k}-\mathbf{Q}\tau} \rangle = \sum_{j} e^{i\mathbf{Q}\cdot\mathbf{R}_j} \langle c_{j\sigma}^+ c_{j\tau} \rangle \tag{36.1}$$

which exhibits a spatial oscillation with wavelength $\lambda = 2\pi/|\mathbf{Q}|$ in the direction of $\mathbf{Q}$. This oscillation is *commensurate* with the crystal structure if $\mathbf{Q} = \mathbf{Q}_c$ is a high-symmetry point on the boundary of the Brillouin zone such that $2\mathbf{Q}_c = \mathbf{K}$ is a reciprocal lattice vector (see Eqs. (1.3) and (29.40) *et seqq.*) and *incommensurate* for any other point of the Brillouin zone.

Of course, the expression (36.1) can only be an order parameter at sufficiently low temperatures if the interaction among the electrons is such that the exchange of electrons with spin σ and τ between points $\mathbf{k}$ and $\mathbf{k} - \mathbf{Q}$ in the Brillouin zone is energetically favorable, $\mathbf{k}$ and $\mathbf{k} - \mathbf{Q}$ being close to the $\sigma-$ and τ-Fermi surfaces $E_{F\sigma}$ and $E_{F\tau}$, respectively. Here σ and τ may also designate different bands. This exchange may be realized if there exists a portion $\Delta E_{F\sigma}$ of $E_{F\sigma}$ such that, when translated across the Brillouin zone by the vector $-\mathbf{Q}$, it goes over into a portion $\Delta E_\sigma'$ of surface which overlaps, within a narrow domain ϕ, with a portion $\Delta E_{F\tau}$ of $E_{F\tau}$. This condition of *nesting* of portions of surfaces was first recognized intuitively by Lomer to be of importance for the IEA of Cr[101] and was subsequently formalized by Fedders and Martin[102].

To see more explicitly how nesting works, we express the Fermi surfaces $E_{F\sigma}$ and $E_{F\tau}$ by the equations $\varepsilon_{\mathbf{k}\sigma} = 0$ and $\varepsilon_{\mathbf{k}'\tau} = 0$, respectively. Then, if $\mathbf{k}' = \mathbf{k} - \mathbf{Q} \in \phi$, the portions $\Delta E_{F\sigma}, \Delta E_\sigma'$ and $\Delta E_{F\tau}$ are defined, respectively, by $\varepsilon_{k\sigma} = 0$, $\varepsilon_{k'+Q\sigma} = 0$ and $\varepsilon_{k'\tau} = 0$. This situation is shown in Fig. 5.16(a) (see also Fig. 1a of Ref. 103). The reason why nesting favors an order parameter of the form (36.1) is that for *perfect nesting* defined by a vanishing volume of the domain ϕ, there exists a *symmetry relation* (see below). In addition, the density of states develops a *van Hove singularity* (see Section 11) in this case.

The tight-binding band structure (33.22) in dimension $D = 2$ near half-filling describes well the holes in the CuO_2 planes of the cuprates[104],

$$\varepsilon_{\mathbf{k}} = -2t(\cos k_x a + \cos k_y a) \ . \tag{36.2}$$

Calculating the number of states per unit volume $Z(\varepsilon)$ contained in the shaded area of Fig. 5.16(b) one finds the density of states (Problem 5.19)

$$N(\varepsilon) = Z'(\varepsilon) = \frac{1}{8\pi^2 a^2 t}\left[-2 + \ln\frac{\pi^2 t}{\varepsilon} + \mathcal{O}(\varepsilon)\right] \ . \tag{36.3}$$

This logarithmic van Hove singularity (see the end of Section 11) is the result of the fact that, due to the periodicity (1.20), $\varepsilon_{\mathbf{k}}$ has zero slope at $ak_{x,y} = \pm\pi$, as indicated by Eq. (1.21). It favors magnetic ordering, that is, a commensurate SDW order parameter (36.1) with $\sigma = \tau = \uparrow$ and $-\mathbf{Q}_c = (\pi/a, \pi/a)$ or, more generally,

$$\Delta_{\text{SDW}} = \frac{\Lambda}{V}\sum_{\mathbf{k}\in\phi}\langle a_{\mathbf{k}\uparrow}^+ a_{\mathbf{k}-\mathbf{Q}\uparrow}\rangle \tag{36.4}$$

where Λ is a coupling constant. IEA with a Néel temperature (see Section 29) of 100 to 400 K is indeed observed in the cuprate super conductors[35]. It must be emphasized, however, that because of the Mermin-Wagner theorem[21] this is only possible in 3 dimensions which means that antiferromagnetic (af) coupling must also exist between adjacent CuO_2 layers.

Since itinerant-electron ferromagnetism (IEF) may be considered as the special case $\mathbf{Q} = 0$ of IEA, the order parameter of the latter case may be obtained from Eq. (32.40) by the substitution $a_{\mathbf{k}\sigma} \rightarrow a_{\mathbf{k}-\mathbf{Q}\sigma}$. This leads to a *staggered magnetization* analogous to Eq. (29.44),

$$N = \frac{\mu_B}{V}\sum_{\mathbf{k}\in\phi}\sum_{\sigma}\sigma\langle a_{\mathbf{k}\sigma}^+ a_{\mathbf{k}-\mathbf{Q}\sigma}\rangle \ . \tag{36.5}$$

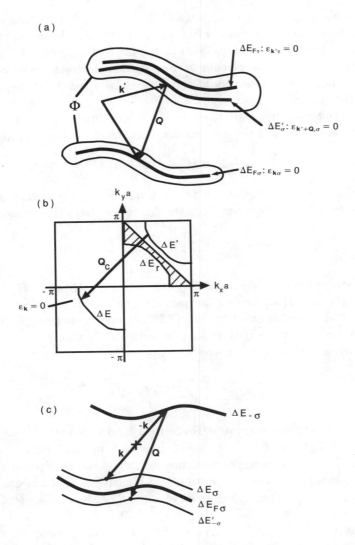

Fig. 5.16: (a) Nesting of the portion $\Delta E'_\sigma$ of surface with the portion $\Delta E_{F\tau}$ of the Fermi surface $E_{F\tau}$ within the narrow domain ϕ. $\Delta E'_\sigma$ is obtained from the portion $\Delta E_{F\sigma}$ of $E_{F\sigma}$ through translation by the nesting vector $-\mathbf{Q}$ from $\mathbf{k}$ to $\mathbf{k}' = \mathbf{k} - \mathbf{Q} \in \phi$. (b) Portion ΔE of the Fermi surface of the band (36.2) near half-filling, its reflection ΔE_τ and its translation by $-\mathbf{Q}_c = (\pi/a, \pi/a)$, $\Delta E'$. (c) Charge-conjugate states $|\mathbf{k}\sigma\rangle$ and $|-\mathbf{k} + \mathbf{Q}, -\sigma\rangle$ according to Eq. (36.14). ΔE_σ and $\Delta E'_{-\sigma}$ are charge-conjugate surfaces.

The difference between the order parameters (36.4) and (36.5) may be expressed in terms of *time-reversal symmetry*, $\mathcal{T}$. On the orbital state $\mathcal{T}$ acts as complex conjugation K (see Eq. (31.18) *et seqq.*) while on the spin state it acts as the Pauli matrix σ^y [57]. Using the explicit form of σ^y given by Eq. (2.8) one finds for any state with labels $\mathbf{k}, \sigma$ that $|\Psi_{\mathbf{k}\sigma}\rangle^{\mathcal{T}} = -i\sigma|\Psi^*_{-\mathbf{k}-\sigma}\rangle$ or, writing $|\Psi_{\mathbf{k}\sigma}\rangle = a^+_{\mathbf{k}\sigma}|\Psi\rangle$, $\langle\Psi_{\mathbf{k}\sigma}| = \langle\Psi|a_{\mathbf{k}\sigma}$,

$$(a^+_{\mathbf{k}\sigma})^{\mathcal{T}} = -i\sigma a^+_{-\mathbf{k}-\sigma}K \ ; \ (a_{\mathbf{k}\sigma})^{\mathcal{T}} = Ki\sigma a_{-\mathbf{k}-\sigma} \ , \tag{36.6}$$

which is true provided that the state $|\Psi\rangle$ is invariant under $\mathcal{T}$. Therefore, it follows for any $\mathcal{T}$-invariant Hamiltonian that, since $K^2 = 1$,

$$\langle a^+_{\mathbf{k}\downarrow}a_{\mathbf{k}-\mathbf{Q}\downarrow}\rangle^{\mathcal{T}} = \langle ia^+_{-\mathbf{k}\uparrow}KK(-i)a_{-\mathbf{k}+\mathbf{Q}\uparrow}\rangle = \langle a^+_{-\mathbf{k}\uparrow}a_{-\mathbf{k}+\mathbf{Q}\uparrow}\rangle \ . \tag{36.7}$$

Now because of $\mathcal{T}$-invariance, $\varepsilon_{-\mathbf{k}} = \varepsilon_{\mathbf{k}}$ (see Section 1) so that $-\mathbf{k} \in \phi$ if $\mathbf{k} \in \phi$. If on the other hand, the nesting vector is commensurate , $\mathbf{Q} = \mathbf{Q}_c$, then $2\mathbf{Q}_c = \mathbf{K}$ is a reciprocal lattice vector (see, e.g. Fig. 5.16(b)). Hence

$$\sum_{\mathbf{k}\in\phi}\langle a^+_{-\mathbf{k}\uparrow}a_{-\mathbf{k}+\mathbf{Q}_c\uparrow}\rangle = \sum_{\mathbf{k}\in\phi}\langle a^+_{\mathbf{k}\uparrow}a_{\mathbf{k}-\mathbf{Q}_c\uparrow}\rangle \tag{36.8}$$

and therefore, setting $\Lambda = \mu_B$,

$$N = \Delta_{\text{SDW}} - \Delta^{\mathcal{T}}_{\text{SDW}} = -N^{\mathcal{T}} \ ; \ \mathbf{Q} = \mathbf{Q}_c \tag{36.9}$$

since Eq. (36.7) implies $\mathcal{T}^2 = 1$. For the same reason the IEF order parameter (32.40) also satisfies

$$M = -M^{\mathcal{T}} \ . \tag{36.10}$$

In other words, M and a commensurate N violate time reversal invariance maximally . But of course, in the ferromagnetic case, $\mathbf{Q} = 0$, the expression (36.4) cannot be used as order parameter since it is non-zero above T_c.

With the aid of Eq. (36.6) it is also straightforward to show that the superconducting order parameter (26.38) is $\mathcal{T}$-invariant provided that the underlying Hamiltonian is. From Eqs. (22.25), (21.21) and (21.14), (21.15) the order parameter is seen to be

$$\Delta = \frac{\Lambda}{V}\sum_{\mathbf{k}\in\phi}\langle a^+_{\mathbf{k}\uparrow}a^+_{-\mathbf{k}\downarrow}\rangle \tag{36.11}$$

where the domain ϕ is defined in Eq. (20.10) or in (20.36). Applying Eq. (36.6) one then finds

$$\langle a^+_{\mathbf{k}\uparrow}a^+_{-\mathbf{k}\downarrow}\rangle^{\mathcal{T}} = \langle(-i)a^+_{-\mathbf{k}\downarrow}Kia^+_{\mathbf{k}\uparrow}K\rangle = -\langle a^+_{-\mathbf{k}\downarrow}a^+_{\mathbf{k}\uparrow}\rangle \tag{36.12}$$

and hence, permuting the two operators in the last expression,

$$\Delta^{\mathcal{T}} = \Delta \ . \tag{36.13}$$

This is *Anderson's theorem* which states that non-magnetic impurities (which preserve the $\mathcal{T}$-invariance of the Hamiltonian) do not affect the thermodynamic properties of a superconductor[105].

Concerning nesting, the important symmetry, however, is *charge conjugation*, $\mathcal{C}$, because it establishes a formal identity between the order parameters (36.4) and (36.11). We define $\mathcal{C}$ by identifying, within the domain ϕ, the creation/annihilation of an electron at $\mathbf{k}$ with spin σ and the annihilation/creation of a hole at $-\mathbf{k} + \mathbf{Q}$ with spin $-\sigma$ (see, Fig. 5.16(c) above),

$$(a_{\pm\mathbf{k}\sigma}^+)^{\mathcal{C}} = a_{\mp(\mathbf{k}-\mathbf{Q}),-\sigma} \; ; \; (a_{\pm\mathbf{k}\sigma})^{\mathcal{C}} = a_{\mp(\mathbf{k}-\mathbf{Q}),-\sigma}^+ \; ; \; \mathbf{k} \in \phi \; . \qquad (36.14)$$

Since this also implies $a_{\pm\mathbf{k}\sigma}^{+\mathcal{C}} = a_{\pm\mathbf{k}\sigma}^{\mathcal{C}+}$ we may write Eq. (36.4) in the form[103]

$$\Delta_{\text{SDW}} = \frac{\Lambda}{V} \sum_{\mathbf{k}\in\phi} \langle a_{\mathbf{k}\uparrow}^+ a_{-\mathbf{k}\downarrow}^{\mathcal{C}+} \rangle \qquad (36.15)$$

which is indeed formally the same as Eq. (36.11). On the other hand, defining in analogy to Eq. (36.5) a *staggered polarization*

$$P = \frac{p_o}{V} \sum_{\mathbf{k}\in\phi} \sum_{\sigma} \langle a_{\mathbf{k}\sigma}^+ a_{\mathbf{k}-\mathbf{Q}\sigma} \rangle \qquad (36.16)$$

where p_o is some unit of electric dipole moment, application of the $\mathcal{C}$ operation (36.14) yields, in the commensurate case $\mathbf{Q} = \mathbf{Q}_c$ and for a $\mathcal{C}$-invariant Hamiltonian, making use of Eq. (36.8) and setting $\Lambda = p_o$,

$$P = \Delta_{\text{SDW}} - \Delta_{\text{SDW}}^{\mathcal{C}} = -P^{\mathcal{C}} \; ; \; \mathbf{Q} = \mathbf{Q}_c \qquad (36.17)$$

since (36.14) implies $\mathcal{C}^2 = 1$. This means that a commensurate P violates charge conjugation maximally . P may therefore be identified as a *charge-density wave* (see, e.g. Refs. 106, 107).

Renormalization effects may slightly shift the wavevector in the order parameter (36.4) to $\mathbf{Q}' = \mathbf{Q} - \mathbf{q}$ relative to the nesting vector $\mathbf{Q}$ defining charge conjugation, Eq. (36.14) so that, with $|\mathbf{q}| \ll |\mathbf{Q}|$,

$$\Delta_{\text{SDW}} = \Delta_q \equiv \frac{\Lambda}{V} \sum_{\mathbf{k}\in\phi} \langle a_{\mathbf{k}\uparrow}^+ a_{-\mathbf{k}-\mathbf{q}\downarrow}^{\mathcal{C}+} \rangle \; . \qquad (36.18)$$

The corresponding superconducting order parameter is well known and describes a superconductor in the presence of a molecular magnetic field[108]. Indeed, since a magnetic field violates $\mathcal{T}$ invariance , Anderson's theorem discussed above implies that some Cooper pairs are broken, giving rise to a normal current $\mathbf{j}_n$ which in the groundstate is compensated by a supercurrent, $\mathbf{j}_s = -\mathbf{j}_n$. According to Eq. (26.45), $\mathbf{j}_S = \mathbf{j}_s$ here is character-ized by a wavevector $\mathbf{q}$ which appears in the order parameter and describes a spatially inhomogenous state of long wavelength. The effect of the inter-nal magnetic field on the superconductor is an *incomplete Meissner effect* (see Section 25).

This analogy with the *Fulde-Ferrell superconductor* allows us to use the same effective interaction as in Eq. (20.36) which also defines the domain ϕ and which for $\mathbf{k} \in \phi$ is nothing else than the Hubbard interaction (33.13) with $U/L = -\Lambda/V$. Translated into the af situation this interaction reads

$$\mathcal{H}_{\text{red}} = -\frac{\Lambda}{V} \sum_{\mathbf{kk'} \in \phi} a_{\mathbf{k}\uparrow}^+ a_{-\mathbf{k}-\mathbf{q}\downarrow}^{\mathcal{C}+} a_{-\mathbf{k'}-\mathbf{q}\downarrow}^{\mathcal{C}} a_{\mathbf{k'}\uparrow} . \tag{36.19}$$

In the case of a superconductor with an internal magnetic field the corre-sponding free Hamiltonian is given by the Stoner expression (32.38) while for the IEA it describes electrons and holes with spin up in the domain ϕ,

$$\mathcal{H}_o = \sum_{\mathbf{k} \in \phi} (\varepsilon_{\mathbf{k}\uparrow} a_{\mathbf{k}\uparrow}^+ a_{\mathbf{k}\uparrow} + \varepsilon_{\mathbf{k}-\mathbf{Q'}\uparrow} a_{\mathbf{k}-\mathbf{Q'}\uparrow}^+ a_{\mathbf{k}-\mathbf{Q'}\uparrow})$$

$$= \sum_{\mathbf{k} \in \phi} \{ \varepsilon_{\mathbf{k}\uparrow} a_{\mathbf{k}\uparrow}^+ a_{\mathbf{k}\uparrow} - \varepsilon_{\mathbf{k}-\mathbf{Q'}\uparrow} (a_{-\mathbf{k}-\mathbf{q}\downarrow}^{\mathcal{C}+} a_{-\mathbf{k}-\mathbf{q}\downarrow}^{\mathcal{C}} - 1) \} . \tag{36.20}$$

Applying the RPA prescription (22.24) to $\mathcal{H} = \mathcal{H}_\varrho + \mathcal{H}_{\text{red}}$ one obtains, in the abbreviated notation $a_{\mathbf{k}\uparrow} \equiv a$, $a_{\mathbf{k}-\mathbf{Q'}\uparrow}^+ = a_{-\mathbf{k}-\mathbf{q}\downarrow}^{\mathcal{C}} \equiv b$, $\varepsilon_{\mathbf{k}\uparrow} \equiv \varepsilon^a$, $-\varepsilon_{\mathbf{k}-\mathbf{Q'}\uparrow} \equiv \varepsilon^b$ [103)]

$$\tilde{\mathcal{H}} \equiv \text{RPA}\{\mathcal{H}\} = \sum_{\mathbf{k}} \{ \varepsilon^a a^+ a + \varepsilon^b b^+ b - \Delta_q ba - \Delta_q^* a^+ b^+ \} . \tag{36.21}$$

Because of the freedom to multiply $a_{\mathbf{k}\uparrow}$ with an arbitrary $\mathbf{k}$-dependent phase factor it is always possible to choose Δ_q real, $\Delta_q^* = \Delta_q \equiv \Delta$. In this case $\tilde{\mathcal{H}}$ is diagonalized by the Bogoljubov tansformation (21.18),

$$a^+ = u\alpha^+ + v\beta ; \quad b = -v\alpha^+ + u\beta \tag{36.22}$$

where, u, v are real and satisfy $u^2 + v^2 = 1$. One finds

$$\tilde{\mathcal{H}} = \sum_{\mathbf{k}} \{ E_+ \alpha^+ \alpha + E_- \beta^+ \beta + (\varepsilon^a + \varepsilon^b)v^2 - 2\Delta uv \} \tag{36.23}$$

where

$$E_+ \equiv \varepsilon^a u^2 - \varepsilon^b v^2 + 2\Delta uv ,$$

$$E_- \equiv \varepsilon^b u^2 - \varepsilon^a v^2 + 2\Delta uv , \tag{36.24}$$

$$0 = (\varepsilon^a + \varepsilon^b)uv - \Delta(u^2 - v^2) .$$

Defining

$$\left.\begin{array}{c}\xi\\\eta\end{array}\right\} \equiv \frac{1}{2}(\varepsilon^a \pm \varepsilon^b) \tag{36.25}$$

the last relation (36.24) has the same form as Eq. (21.9). Hence, using $u^2 + v^2 = 1$, one finds successively Eqs. (21.11) and (21.13), that is, $uv = \Delta/2E$ and

$$\left.\begin{array}{c}u^2\\v^2\end{array}\right\} = \frac{1}{2}\left(1 \pm \frac{\xi}{E}\right) ; \quad E^2 = \xi^2 + \Delta^2 . \tag{36.26}$$

Insertion into the first two relations (36.24) then yields

$$E_\pm = E \pm \eta . \tag{36.27}$$

Applying the Bogoljubov transformation (36.22) to the order parameter (36.18) and using relations (23.3) we finally arrive at the *gap equation*

$$\Delta = \frac{\Lambda}{V}\sum_{\mathbf{k}}\langle a^+ b^+\rangle = \frac{\Lambda}{V}\sum_{\mathbf{k}\in\phi}\frac{\Delta}{2E}[1 - f_o(E_+) - f_o(E_-)] . \tag{36.28}$$

At zero temperature, $f_o(E + |\eta|) = 0$, while $f_o(E - |\eta|) = 1$ defines the *blocking regions* of Fulde and Ferrell[108], i.e. the states which are excluded from pairing or, in other words, which give rise to *pair breaking*. In order to determine these states we have to examine explicitly the dependence on $\mathbf{k}$ and $\mathbf{q}$. Since close to the Fermi surface, $\varepsilon_{-\mathbf{k}+\mathbf{Q}\uparrow} \cong -\varepsilon_{\mathbf{k}\downarrow}$ (see Fig. 5.16(c) above) we have from Eqs. (36.25) and (32.39)

$$\begin{aligned}\xi = \xi_q(\mathbf{k}) &\equiv \frac{1}{2}(\varepsilon_{\mathbf{k}\uparrow} + \varepsilon_{\mathbf{k}+\mathbf{q}\downarrow}) \cong \varepsilon_{\mathbf{k}} + \frac{\hbar}{2}\mathbf{v}_{\mathbf{k}} \cdot \mathbf{q}\\\eta = \eta_q(\mathbf{k}) &\equiv \frac{1}{2}(\varepsilon_{\mathbf{k}\uparrow} - \varepsilon_{\mathbf{k}+\mathbf{q}\downarrow}) \cong -\frac{\hbar}{2}\mathbf{v}_{\mathbf{k}} \cdot \mathbf{q} - \frac{\Delta^S}{2} .\end{aligned} \tag{36.29}$$

Note that in the *af* case the Stoner gap Δ^S may come from a *nesting mismatch* due to electron transfer from a reservoir[103]. Introducing the *pair density of states*[109]

$$N_q(\xi\,\eta) = V^{-1}\sum_{\mathbf{k}\in\phi}\delta(\xi - \xi_q(\mathbf{k}))\delta(\eta - \eta_q(\mathbf{k})) \tag{36.30}$$

the gap equation (36.28) may be written, at $T = 0$ and for $\Delta_q \neq 0$, as

$$\frac{1}{\Lambda} = \int d\xi \int d\eta \frac{N_q(\xi, \eta)}{2\sqrt{\xi^2 + \Delta_q^2}}[1 - \theta(\eta^2 - \xi^2 - \Delta_q^2)] \tag{36.31}$$

where the θ-function describes the excluded states.

From the last equation it is evident that pair breaking is impossible for $|\eta_q(\mathbf{k})| = |\eta| < \Delta_q$ which means, essentially, that $q = 0$ and $\Delta^S = 0$. In this ideal case Eq. (36.30) simply becomes

$$N_o(\xi, \eta) = N(0)\theta(\omega_o - |\xi|)\delta(\eta) \tag{36.32}$$

where $N(0)$ is the density of states (10.4) at the Fermi level and the domain ϕ is defined by $|\varepsilon_\mathbf{k}| < \omega_o$. With this expression the gap equation (36.31) is solved by Eq. (21.16) or, for $\Delta_o \ll \omega_o$ by

$$\frac{1}{\Lambda} = \int d\xi \int d\eta \frac{N_o(\xi, \eta)}{2\sqrt{\xi^2 + \Delta_o^2}} \cong N(0)\ln\frac{2\omega_o}{\Delta_o} . \tag{36.33}$$

In this BCS case the gap Δ_o is determined by the width ω_o of the nesting regions . To see now the effect of pair breaking we neglect the dependence on $\mathbf{q}$ and Δ^S in the pair density of states (36.30) so that we may analytically continue the second equality in Eq. (36.33) from Δ_o to Δ_q and write

$$\int d\xi \int d\eta \frac{N_q(\xi, \eta)}{2\sqrt{\xi^2 + \Delta_q^2}} \cong N(0)\ln\frac{2\omega_o}{\Delta_q} . \tag{36.34}$$

Substituting the last two equations in Eq. (36.31), we explicitly obtain the pair-breaking effect,

$$0 < \int d\xi \int d\eta \frac{N_q(\xi, \eta)}{2\sqrt{\xi^2 + \Delta_q^2}}\theta(\eta^2 - \xi^2 - \Delta_q^2) \cong N(0)\ln\frac{\Delta_o}{\Delta_q} . \tag{36.35}$$

In the case of the Fulde-Ferrell superconductor with spherical energy surfaces it is easy to evaluate the pair density of states (36.30). With $\mathbf{v}_\mathbf{k} \cdot \mathbf{q} = v_F q\zeta$ one finds from Eq. (36.29)

$$N_q(\xi, \eta) = N(0)\int_{-\omega_o}^{\omega_o} d\varepsilon\,\delta\left(\varepsilon - \xi - \eta - \frac{\Delta^S}{2}\right)$$

$$\times \frac{1}{2}\int_{-1}^{+1} d\zeta\,\delta\left(\eta + \frac{\hbar}{2}v_F q\zeta + \frac{\Delta^S}{2}\right)$$

$$= \frac{N(0)}{\hbar v_F q}\theta\left(\omega_o - \left|\xi + \eta + \frac{\Delta^S}{2}\right|\right)\theta\left(\frac{\hbar}{2}v_F q - \left|\eta + \frac{\Delta^S}{2}\right|\right) . \tag{36.36}$$

These two θ-functions imply the bounds $|\xi| < \omega_o + \hbar v_F q/2$ and $|\eta| < (\hbar v_F q + \Delta^S)/2$, respectively. The latter shows that Eq. (36.36) goes over smoothly into (36.32) only if Δ^S vanishes at least as q^1. On the other hand, pair breaking implies that $|\xi| < |\eta|$. Therefore, assuming $\omega_o > \Delta^S/2$, the first θ-function in Eq. (36.36) may be dropped in the integration of Eq. (36.35) which yields Eqs. (8), (9) of Fulde and Ferrell[108], namely

$$2 \ln \frac{\Delta_o}{\Delta_q} \cong \frac{1}{r} \{ G(r+s) + \mathrm{sgn}\,(r-s) G(|r-s|) \} \tag{36.37}$$

where $r \equiv \hbar v_F q / 2\Delta_q$, $s \equiv \Delta^S / 2\Delta_q$ and

$$G(z) \equiv \int_1^z dx \int_0^{\sqrt{x^2-1}} \frac{dy}{\sqrt{y^2+1}} = \int_1^z dx \left| \cosh^{-1} |x| \right|$$

$$= \{ z | \cosh^{-1} z | - \sqrt{z^2 - 1} \} \theta(z-1) \ . \tag{36.38}$$

In the *af* case we take again the two-dimensional energy contours (36.2) as example. Since, apart from the corner regions of the Brillouin zone, these contours are quite straight near half-filling (see Fig. 5.16(b)), we may set $\mathbf{v_k} \cdot \mathbf{q} \cong v_F q$, independent of $\mathbf{k}$. Then follows from Eqs. (36.29), (36.30) that

$$N_q(\xi, \eta) = N(0)\, \theta\!\left(\omega_o - |\xi - \frac{\hbar}{2} v_F q| \right) \delta\!\left(\eta + \frac{\hbar}{2} v_F q + \frac{\Delta^S}{2} \right) \ . \tag{36.39}$$

Here the θ-function implies $|\xi| < \omega_o + \hbar v_F q / 2$ while for calculating the pair breaking contribution (36.35), $|\xi| < |\eta|$. Hence if $\omega_o > \Delta^S / 2$, the θ-function may again be dropped and Eq. (36.35) yields

$$\ln \frac{\Delta_o}{\Delta_q} \cong \left| \cosh^{-1}(r+s) \right| \ ; \ r+s > 1 \tag{36.40}$$

or, taking the inverse relation,

$$\frac{\Delta_q}{\Delta_o} = r + s - \sqrt{(r+s)^2 - 1} \leq 1 \ . \tag{36.41}$$

This is a monotonically decreasing function which vanishes as $(2(r+s))^{-1}$ when Δ_o reaches the value $\hbar v_F q + \Delta^S$. Now since, as observed after Eq. (36.29), the Stoner gap Δ^S may be considered as a measure of doping, Eq. (36.41) gives a qualitative explanation of the vanishing of antiferromagnetism observed in the high-T_c cuprates[110].

As shown by Hirsch[111], nesting may work yet in another way and lead both, to IEF and to IEA, the latter without SDW, provided that the crucial coupling is not the negative-U Hubbard interaction but the exchange interaction (33.41) used by Heisenberg in his original work to explain ferromagnetism[14]. Using the transformation (33.4) this interaction may be written in the form

$$\mathcal{H}_J = \frac{J}{L} \sum_{\mathbf{q}\vec{\delta}\sigma} d_\sigma(\mathbf{q}; \vec{\delta}) d_{-\sigma}(-\mathbf{q}; -\vec{\delta}) \ ; \quad J > 0 \tag{36.42}$$

where we have introduced the one-electron densities

$$d_\sigma(\mathbf{q}; \vec{\delta}) \equiv \sum_{\mathbf{k}} e^{i(\mathbf{k}+\mathbf{q}/2)\cdot\vec{\delta}} a^+_{\mathbf{k}+\mathbf{q}\sigma} a_{\mathbf{k}\sigma} = \sum_{\langle ij \rangle} e^{i\mathbf{q}\cdot(\mathbf{R}_i+\mathbf{R}_j)/2} c^+_{i\sigma} c_{j\sigma} \ . \tag{36.43}$$

Since no spatial oscillation is now invoked it is natural that the $\mathbf{q} = 0$ term in Eq. (36.42) should be crucial for IEF as well as for IAF . Applying RPA to this reduced Hamiltonian $\mathcal{H}_J^{\text{red}}$, the order parameter presents itself naturally as

$$I_\sigma(\vec{\delta}) = \langle d_\sigma(0; \vec{\delta}) \rangle \tag{36.44}$$

and

$$\langle \mathcal{H}_J^{\text{red}} \rangle = \frac{J}{L} \sum_{\vec{\delta}\sigma} \sum_{\mathbf{k}\mathbf{k}'} e^{i(\mathbf{k}-\mathbf{k}')\cdot\vec{\delta}} \langle a^+_{\mathbf{k}\sigma} a_{\mathbf{k}\sigma} \rangle \langle a^+_{\mathbf{k}'-\sigma} a_{\mathbf{k}'-\sigma} \rangle \ . \tag{36.45}$$

If the magnetic instability gives rise to *spin-split Fermi surfaces* then the strictly positive terms $\mathbf{k} = \mathbf{k}'$ are absent from the $\mathbf{k}, \mathbf{k}'$-sum, thus giving rise to a lowering of $\langle \mathcal{H}_J^{\text{red}} \rangle$. But in addition, *nesting* implies that $\mathbf{k} - \mathbf{k}' = \mathbf{Q}$ is such that the exponential factor is essentially -1 (see, e.g. Fig. 5.16(b)) which results in a second lowering of $\langle \mathcal{H}_J^{\text{red}} \rangle$. This makes a magnetic instability quite plausible. But the crucial question of course, is which of the two couplings, $\mathcal{H}_U$ or $\mathcal{H}_J$, is more important.

Solutions to the Problems of Chapter 5

Problem 5.1: Eqs. (29.10), (29.24).

We need a development of $B_J(x)$ in powers of x. Using

$$\alpha \coth \alpha x = \frac{1}{x} + \frac{\alpha^2}{3} x - \frac{\alpha^4}{45} x^3 + \mathcal{O}(x^5)$$

we find from (29.8)

$$B_J(x) = \frac{J+1}{3J} x \left\{ 1 - \left(1 + \frac{1}{J} + \frac{1}{2J^2} \right) \frac{x^2}{15} + \mathcal{O}(x^4) \right\}$$

which gives the value $B_J'(0) = (J+1)/3J$ used in (29.10). Determining $B_J^{-1}(m)$ as a power series we insert the ansatz

$$x = \frac{3J}{J+1} m \left\{ 1 + \frac{m^2}{a_J^2} + \mathcal{O}(m^4) \right\} = B_J^{-1}(m)$$

used in (29.23) into the last equation. This yields

$$\left\{1 + \frac{m^2}{a_J^2} + \cdots\right\}\left\{1 - \left(1 + \frac{1}{J} + \frac{1}{2J^2}\right)\frac{3J^2m^2}{5(J+1)^2} + \cdots\right\} = 1$$

from which (29.24) follows.

Problem 5.2: Eq. (29.15).

With (29.14), Eq. (29.15) may be written, at constant field, as

$$\Delta S = V \int_o^\infty \frac{\Delta c_H(T)}{x}\,dx = Lk_B \int_o^\infty B'_J(x)x\,dx \ .$$

Here partial integration is the obvious thing to do. However, the upper limits being divergent one has to write the result as a limit,

$$\Delta S = Lk_B \lim_{x\to\infty}\left\{xB_J(x) - \int_o^x B'_J(y)y\,dy\right\} \ .$$

Now from (29.6), $B_J(x) = (\ln Z_J(x))'$ so that

$$\frac{\Delta S}{Lk_B} = \lim_{x\to\infty}\left\{xB_J(x) - \ln\frac{Z_J(x)}{Z_J(0)}\right\} \ .$$

Since (29.4) immediately gives $Z_J(0) = 2J + 1$ we have to show that

$$\lim_{x\to\infty}\{xB_J(x) - \ln Z_J(x)\} = 0 \ .$$

For this we need expansions of both functions for $x \gg 1$. In this case Eq. (29.4) may be written

$$Z_J(x) = \frac{e^x - e^{-(1+1/J)x}}{1 - e^{-x/J}} \cong \frac{e^x}{1 - e^{-x-J}} = \sum_{n=0}^\infty e^{(1-n/J)x}$$

which according to (29.6) yields the asymptotic expansion

$$B_J(x) \cong \sum_{n=0}^\infty \left(1 - \frac{n}{J}\right)e^{(1-n/J)x} \Big/ \sum_{n=0}^\infty e^{(1-n/J)x}$$

$$= 1 - \frac{1}{J}e^{-x/J} + \mathcal{O}(e^{-2x/J}) \ ; \quad x \gg 1 \ .$$

On the other hand

$$\ln Z_J(x) \cong x + \ln\sum_{n=0}^\infty e^{-nx/J} = x + e^{-x/J} + \mathcal{O}(e^{-2x/J})$$

which proves the above limit. Hence (29.15).

Problem 5.3: Eqs. (29.34)–(29.36).

Using for B_J the asymptotic expansion of Problem 5.2, Eq. (29.19) becomes

$$\frac{z}{\theta} = m_o = B_J(z) \cong 1 - \frac{1}{J}e^{-z/J} \cong 1 - \frac{1}{J}e^{-\theta/J} \; ; \; \theta/J \gg 1 \; .$$

Inserting $\theta = 3JT_C/(J+1)T$ this is Eq. (29.34).

To obtain χ we derive Eq. (29.22) with respect to h and use (29.10) which gives

$$\lambda\chi = \left(\frac{\partial m}{\partial h}\right)_{h=0} = (1 + \lambda\chi)\theta B'_J(z)$$

or

$$(\lambda\chi)^{-1} = \frac{1}{\theta B'_J(z)} - 1 \cong \frac{J^2}{\theta}e^{\theta/J} - 1 \; .$$

Using (29.20) one finds, to leading order, Eq. (29.35).

To calculate c_m we must first determine F_m near $T = 0$ by integrating (29.26). Using (29.23) and (29.21) the result is

$$F_m = \lambda V M_s^2 \left\{ \frac{1}{\theta}A_J(m) - \frac{1}{2}m^2 \right\}$$

where

$$A_J(m) \equiv \int_o^m B^{-1}(m')dm' \; .$$

According to definition (29.31) the magnetic entropy density then becomes

$$s_m = -\lambda M_s^2 \left\{ \frac{J+1}{3JT_C}A_J(m_o) + \left(\frac{1}{\theta}B_J^{-1}(m_o) - m_o \right)\frac{\partial m_o}{\partial T} \right\} \; .$$

Here the bracket multiplying $\partial m_o/\partial T$ vanishes because of (29.23) so that, with definition (29.33), one finds

$$c_m = -\lambda M_s^2 \frac{J+1}{3J}\frac{T}{T_C}B_J^{-1}(m_o)\frac{\partial m_o}{\partial T} = -\lambda M_s^2 m_o \frac{\partial m_o}{\partial T} \; .$$

Using (29.20) this readily yields (29.36).

Problem 5.4: Eq. (30.11).

With (30.10) we calculate the 3 expressions

$$\sum_i S_i^z S_l^- |0\rangle = \sum_i (S - \delta_{il})S_l^- |0\rangle = (SL - 1)S_l^- |0\rangle \; ,$$

$$\sum_{\langle ij \rangle} S_i^z S_j^z S_l^- |0\rangle = \sum_{\langle ij \rangle} (S - \delta_{il})(S - \delta_{jl}) S_l^- |0\rangle = (\frac{1}{2} S^2 Lz - Sz) S_l^- |0\rangle$$

and

$$\sum_{\langle ij \rangle} (S_i^+ S_j^- + S_i^- S_j^+) \sum_l e^{i\mathbf{q} \cdot \mathbf{R}_l} S_l^- |0\rangle = \frac{1}{2} S \sum_l e^{i\mathbf{q} \cdot \mathbf{R}_l} \sum_{\langle ij \rangle} (S_j^- \delta_{il} + S_i^- \delta_{jl}) |0\rangle$$

$$= \frac{1}{2} S \sum_{i, \vec{\delta}} e^{i\mathbf{q} \cdot (\mathbf{R}_i + \vec{\delta})} S_i^- |0\rangle = \frac{1}{2} S \sum_{\vec{\delta}} e^{i\mathbf{q} \cdot \vec{\delta}} \sum_i e^{i\mathbf{q} \cdot \mathbf{R}_i} S_i^- |0\rangle \ .$$

Inserting these results into the respective terms of (30.1) and (30.4) in Eq. (30.9) we find

$$E_o + \omega_{\mathbf{q}} - \mu = -J\{(1 - \lambda) S \sum_{\vec{\delta}} e^{i\mathbf{q} \cdot \vec{\delta}} + (1 + \lambda)(\frac{1}{2} S^2 Lz - Sz)\}$$

$$- g\mu_B H(SL - 1)$$

and subtracting (30.6) we obtain (30.11).

Problem 5.5: Eq. (30.44).

Insertion of (30.43) into (30.42) yields

$$\langle \mathbf{S}_i \cdot \mathbf{S}_j \rangle_o = S^2 - \frac{S}{L} \sum_{\mathbf{q}\mathbf{q}'} \{ e^{-i\mathbf{q} \cdot \mathbf{R}_i} (e^{i\mathbf{q}' \cdot \mathbf{R}_i} - e^{i\mathbf{q}' \cdot \mathbf{R}_j})$$

$$+ e^{-i\mathbf{q} \cdot \mathbf{R}_j} (e^{i\mathbf{q}' \cdot \mathbf{R}_j} - e^{i\mathbf{q}' \cdot \mathbf{R}_i}) \} \langle b_{\mathbf{q}}^+ b_{\mathbf{q}'} \rangle_o$$

$$- \frac{1}{4L^2} \sum_{\mathbf{q}\mathbf{q}'\mathbf{p}\mathbf{p}'} \{ e^{-i\mathbf{q} \cdot \mathbf{R}_i} e^{-i\mathbf{q}' \cdot \mathbf{R}_j} (e^{i\mathbf{p} \cdot \mathbf{R}_i} - e^{i\mathbf{p} \cdot \mathbf{R}_j})$$

$$\times (e^{i\mathbf{p}' \cdot \mathbf{R}_i} - e^{i\mathbf{p}' \cdot \mathbf{R}_j}) + (e^{-i\mathbf{q} \cdot \mathbf{R}_i} - e^{-i\mathbf{q} \cdot \mathbf{R}_j})(e^{-i\mathbf{q}' \cdot \mathbf{R}_i} - e^{-i\mathbf{q}' \cdot \mathbf{R}_j})$$

$$\times e^{i\mathbf{p} \cdot \mathbf{R}_i} e^{i\mathbf{p}' \cdot \mathbf{R}_j} \} \langle b_{\mathbf{q}}^+ b_{\mathbf{q}'}^+ b_{\mathbf{p}} b_{\mathbf{p}'} \rangle_o + \mathcal{O}(S^{-1}) \ .$$

Now

$$\langle b_{\mathbf{q}}^+ b_{\mathbf{q}'} \rangle_o = \bar{n}_{\mathbf{q}} \delta_{\mathbf{q}\mathbf{q}'}$$

and, if $\mathbf{q} \neq \mathbf{q}'$,

$$\langle b_{\mathbf{q}}^+ b_{\mathbf{q}'}^+ b_{\mathbf{p}} b_{\mathbf{p}'} \rangle_o = (\delta_{\mathbf{q}\mathbf{p}} \delta_{\mathbf{q}'\mathbf{p}'} + \delta_{\mathbf{q}\mathbf{p}'} \delta_{\mathbf{q}'\mathbf{p}}) \bar{n}_{\mathbf{q}} \bar{n}_{\mathbf{q}'} \ .$$

On the other hand, using Eq. (9.1),

$$\langle b_{\mathbf{q}}^{+2} b_{\mathbf{p}} b_{\mathbf{p}'} \rangle_o = \delta_{\mathbf{q}\mathbf{p}} \delta_{\mathbf{p}\mathbf{p}'} \langle b_{\mathbf{q}}^{+2} b_{\mathbf{q}}^2 \rangle_o = \delta_{\mathbf{q}\mathbf{p}} \delta_{\mathbf{p}\mathbf{p}'} 2\bar{n}_{\mathbf{q}}^2 \ .$$

Multiplying the second-to-last relation by $1 - \delta_{\mathbf{qq'}}$ and the last one by $\delta_{\mathbf{qq'}}$ the sum of the two yields

$$\langle b_{\mathbf{q}}^+ b_{\mathbf{q'}}^+ b_{\mathbf{p}} b_{\mathbf{p'}} \rangle_o = (\delta_{\mathbf{qp}} \delta_{\mathbf{q'p'}} + \delta_{\mathbf{qp'}} \delta_{\mathbf{q'p}}) \bar{n}_{\mathbf{q}} \bar{n}_{\mathbf{q'}} .$$

Inserting these two results, one finds with $\vec{\delta} \equiv \mathbf{R}_i - \mathbf{R}_j \neq 0$,

$$\langle \mathbf{S}_i \cdot \mathbf{S}_j \rangle_o = S^2 - \frac{S}{L} \sum_{\mathbf{q}} \{ (1 - e^{-i\mathbf{q}\cdot\vec{\delta}}) + (1 - e^{i\mathbf{q}\cdot\vec{\delta}}) \} \bar{n}_{\mathbf{q}}$$

$$- \frac{1}{4L^2} \sum_{\mathbf{qq'}} \{ (1 - e^{-i\mathbf{q}\cdot\vec{\delta}})(e^{i\mathbf{q'}\cdot\vec{\delta}} - 1) + (e^{i\mathbf{q'}\cdot\vec{\delta}} - 1)(1 - e^{-i\mathbf{q}\cdot\vec{\delta}})$$

$$+ (1 - e^{i\mathbf{q}\cdot\vec{\delta}})(e^{-i\mathbf{q'}\cdot\vec{\delta}} - 1) + (e^{-i\mathbf{q}\cdot\vec{\delta}} - 1)(1 - e^{i\mathbf{q'}\cdot\vec{\delta}}) \}$$

$$\times \bar{n}_{\mathbf{q}} \bar{n}_{\mathbf{q'}} + \mathcal{O}(S^{-1}) .$$

Since $\mathbf{q} \in Z$ in these sums, one may replace $\mathbf{q}$ by $-\mathbf{q}$ where appropriate and since $\bar{n}_{-\mathbf{q}} = \bar{n}_{\mathbf{q}}$ because $\omega_{-\mathbf{q}} = \omega_{\mathbf{q}}$ holds, one readily obtains (30.44).

Problem 5.6: Eqs. (30.57), (30.58).

Using (30.5) and (30.56) we may write

$$\langle jj'|ii' \rangle = \langle 0|S_j^+ [S_{j'}^+, S_i^- S_{i'}^-]|0 \rangle$$

where the commutator may be evaluated with the help of (30.2),

$$[S_{j'}^+, S_i^- S_{i'}^-] = \frac{1}{2} (S_i^- S_{i'}^z \delta_{i'j'} + S_i^z S_{i'}^- \delta_{ij'}) .$$

Substituting here $S_i^z S_{i'}^- = S_{i'}^- S_i^z - S_i^- \delta_{ii'}$ and applying again (30.5) we find

$$\langle jj'|ii' \rangle = \frac{1}{2} \langle 0|[S_j^+, S_i^-]S \delta_{i'j'} + [S_j^+, S_{i'}^-]S \delta_{ij'} - [S_j^+, S_i^-]\delta_{ii'}\delta_{ij'}|0 \rangle .$$

Using (30.2), (30.5) once more, Eq. (30.57) readily follows.

Applying now (30.7) to (30.57) we have

$$\langle 0|\sigma_{\mathbf{p}} \sigma_{\mathbf{p'}} \sigma_{\mathbf{q}}^+ \sigma_{\mathbf{q'}}^+ |0 \rangle = \frac{1}{L^2} \sum_{ii'} \{ e^{-i\mathbf{p}\cdot\mathbf{R}_i} e^{-i\mathbf{p'}\cdot\mathbf{R}_{i'}}$$

$$+ e^{-i\mathbf{p}\cdot\mathbf{R}_{i'}} e^{-i\mathbf{p'}\cdot\mathbf{R}_i} \} e^{i\mathbf{q}\cdot\mathbf{R}_i} e^{i\mathbf{q'}\cdot\mathbf{R}_{i'}}$$

$$- \frac{1}{SL^2} \sum_i e^{i(\mathbf{q}+\mathbf{q'}-\mathbf{p}-\mathbf{p'})\cdot\mathbf{R}_i} .$$

Using (1.9) and remembering that $\mathbf{q}, \mathbf{q'}, \mathbf{p}, \mathbf{p'}$ are all in the reduced zone Z, Eq. (30.58) immediately follows.

Problem 5.7: Eqs. (31.23).

Inserting (31.24), the trace of the second term in the bracket of the first Eq. (31.23) becomes

$$\frac{m}{i\hbar}\text{Tr}\left[(w-H_o)^{-2}\{[w-H_o,\,xy](w-H_o)^{-1}xp_y\right.$$
$$+\,xp_y(w-H_o)^{-1}[w-H_o,\,xy]\}]$$
$$=\frac{m}{i\hbar}\text{Tr}\left[(w-H_o)^{-2}\{(w-H_o)xy(w-H_o)^{-1}xp_y - xyxp_y\right.$$
$$+\,xp_yxy - xp_y(w-H_o)^{-1}xy(w-H_o)\}]$$
$$=\frac{m}{i\hbar}\text{Tr}\,(w-H_o)^{-2}x^2[p_y y] = -m\text{Tr}\,(w-H_o)^{-2}x^2$$

where in the last line use was made of the cyclicity of the trace. This proves the first Eq. (31.23).

To prove the second equation we insert (31.24) once in the first factor of the second term of the bracket and once in the last factor and take half the sum. Then the trace of the second term of the second Eq. (31.23) becomes

$$\frac{m}{2i\hbar}\text{Tr}\left[(w-H_o)^{-2}\{(xp_y+yp_x)(w-H_o)^{-1}[w-H_o,\,xy]\right.$$
$$+\,[w-H_o,\,xy](w-H_o)^{-1}(xp_y+yp_x)\}]$$
$$=\frac{m}{2i\hbar}\text{Tr}\left[(w-H_o)^{-2}\{(xp_y+yp_x)(xy-(w-H_o)^{-1}xy(w-H_o))\right.$$
$$+\,((w-H_o)xy(w-H_o)^{-1}-xy)(xp_y+yp_x)\}]$$
$$=\frac{m}{2i\hbar}\text{Tr}\,(w-H_o)^{-2}[xp_y+yp_x,\,xy] = -\frac{m}{2}\text{Tr}\,(w-H_o)^{-2}(x^2+y^2)$$

where in the last line use was made again of the cyclicity of the trace. This proves the second Eq. (31.23).

Problem 5.8: Eqs. (31.36), (31.37), (31.41).

In (31.36) only the hemiticity of the matrix $\mathbf{p}_{nn'}(\mathbf{k})$ must be proven. It follows with the partial integration

$$\int_C d^3r\nabla(\psi_{n\mathbf{k}}^*\psi_{n'\mathbf{k}}) = \int_{\partial C} d\mathbf{S}\psi_{n\mathbf{k}}^*\psi_{n'\mathbf{k}} = 0$$

where ∂C is the border of C and $d\mathbf{S}$ is the oriented surface element. The vanishing of the surface integral follows from the symmetry of the unit cell

C (see Section 1).

To obtain (31.37) we use (1.14), writing

$$\langle k|\mathbf{r}|k'\rangle = \int_V d^3 r \psi_k^* \mathbf{r} \psi_{k'} = \int_V d^3 r u_k^* u_{k'} \frac{\partial}{i\partial k'} e^{i(\mathbf{k}'-\mathbf{k})\cdot\mathbf{r}} .$$

By a shift of the $\mathbf{k}'$-derivative using $\langle k|k'\rangle = \delta_{kk'}$ and by breaking down the integral into a sum over the L unit cells using the periodicity (1.14) of the $u_k(\mathbf{r})$ one obtains

$$\langle k|\mathbf{r}|k'\rangle = \frac{\partial}{i\partial\mathbf{k}}\delta_{kk'} + \sum_{\mathbf{R}} e^{i(\mathbf{k}'-\mathbf{k})\cdot\mathbf{R}} \int_C d^3 r u_k^* \frac{\partial u_{k'}}{i\partial\mathbf{k}'} e^{i(\mathbf{k}'-\mathbf{k})\cdot\mathbf{r}} .$$

Applying here (1.9) with $\mathbf{k}$ and $\mathbf{k}'$ in Z, (31.37) follows, the hermiticity of the matrix $\mathbf{r}_{nn'}(\mathbf{k})$ being the consequence of the fact that

$$L \int_C d^3 r u_{n\mathbf{k}}^* u_{n'\mathbf{k}} = \delta_{nn'}$$

is independent of $\mathbf{k}$.

(31.41) is obtained by taking the matrix element of (31.40),

$$(\varepsilon_k - \varepsilon_{k'})\langle k|\mathbf{r}|k'\rangle = \frac{\hbar}{im}\langle k|\mathbf{p}|k'\rangle$$

and inserting (31.36), (31.37). This yields

$$(\varepsilon_{n\mathbf{k}} - \varepsilon_{n\mathbf{k}'})\delta_{nn'} \frac{\partial}{i\partial\mathbf{k}'}\delta_{\mathbf{kk}'} + (\varepsilon_{n\mathbf{k}} - \varepsilon_{n'\mathbf{k}})\mathbf{r}_{nn'}(\mathbf{k})\delta_{\mathbf{kk}'} = \frac{\hbar}{im}\mathbf{P}_{nn'}(\mathbf{k})\delta_{\mathbf{kk}'} .$$

Summing over $\mathbf{k}'$ and applying (31.39) to the first term on the left, (31.41) follows immediately.

Problem 5.9: Eqs. (31.43), (31.44).

Insertion of (31.42) and (31.38) into (31.35) yields

$$\begin{aligned}
G_{xy}(w) = \sum_{kk''}(w - \varepsilon_k)^{-1}\Bigg\{ & P_{ynn''}(\mathbf{k}'')\left(\frac{\partial}{i\partial k_x''}\delta_{\mathbf{kk}''}\right) \\
& + (xp_y)_{nn''}(\mathbf{k})\delta_{\mathbf{kk}''}\Bigg\}(w - \varepsilon_{k''})^{-1}\Bigg\{ i\frac{\partial}{\partial k_y''}P_{xn''n}(\mathbf{k}'') \\
& + (yp_x)_{n''n}(\mathbf{k}'')\Bigg\}(w - \varepsilon_{n\mathbf{k}''})^{-1} \\
= \sum_{kk''} & (w - \varepsilon_k)^{-1}\delta_{\mathbf{kk}''}\Bigg\{ i\frac{\partial}{\partial k_x''}P_{ynn''}(\mathbf{k}'') + (xp_y)_{nn''}(\mathbf{k})\Bigg\}
\end{aligned}$$

$$\times (w - \varepsilon_{k''})^{-1} \left\{ i \frac{\partial}{\partial k_y''} P_{xn''n}(\mathbf{k''}) + (yp_x)_{n''n}(\mathbf{k''}) \right\} (w - \varepsilon_{nk''})^{-1}$$

$$= \sum_{kn''} (w - \varepsilon_k)^{-1} \left\{ i \frac{\partial}{\partial k_x} P_{ynn''}(\mathbf{k}) + (xp_y)_{nn''}(\mathbf{k}) \right\} (w - \varepsilon_{n''k})^{-1}$$

$$\times \left\{ i \frac{\partial}{\partial k_y} P_{xn''n}(\mathbf{k}) + (yp_x)_{n''n}(\mathbf{k}) \right\} (w - \varepsilon_k)^{-1}$$

where use was made of (31.39) with $D = Z$ and the derivatives act on all factors to the right. Equation (31.43) follows by applying another partial summation (31.39) to $\partial/\partial k_x$.

Making use of (31.41) for $n' = n$ the terms of (31.43) with $n'' = n$ become, in the notation $\varepsilon \equiv \varepsilon_{nk}$, $\varepsilon_x \equiv \partial \varepsilon_{nk}/\partial k_x$ etc.,

$$G_{xy}^d(w) = \sum_k \left\{ -i(w - \varepsilon)^{-2} \varepsilon_x \frac{m}{\hbar} \varepsilon_y + (w - \varepsilon)^{-1} (xp_y)_{nn} \right\}$$

$$\times (w - \varepsilon)^{-1} \left\{ i \frac{m}{\hbar} \left[(w - \varepsilon)^{-1} \varepsilon_{xy} \right. \right.$$

$$\left. + (w - \varepsilon)^{-2} \varepsilon_y \varepsilon_x \right] + (w - \varepsilon)^{-1} (yp_x)_{nn} \bigg\}$$

$$= \left(\frac{m}{\hbar} \right)^2 \sum_k (w - \varepsilon)^{-3} \left\{ (w - \varepsilon)^{-1} \varepsilon_x \varepsilon_y \varepsilon_{xy} + (w - \varepsilon)^{-2} \varepsilon_x^2 \varepsilon_y^2 \right.$$

$$+ i \frac{\hbar}{m} (w - \varepsilon)^{-1} \varepsilon_x \varepsilon_y (xp_y - yp_x)_{nn} + i \frac{\hbar}{m} \varepsilon_{xy} (xp_y)_{nn}$$

$$+ \left(\frac{\hbar}{m} \right)^2 (xp_y)_{nn} (yp_x)_{nn} \bigg\} .$$

Here the second term in the bracket may be transformed by writing $(w - \varepsilon)^{-5} \varepsilon_x^2 \varepsilon_y^2 = [\partial/4\partial k_x (w - \varepsilon)^{-4}] \varepsilon_x \varepsilon_y^2$. After a partial summation (31.39) this term, combined with the first one, becomes

$$\left(\frac{m}{\hbar} \right)^2 \sum_k (w - \varepsilon)^{-4} \left\{ \varepsilon_x \varepsilon_y \varepsilon_{xy} - \frac{1}{4} (\varepsilon_x \varepsilon_y^2)_x \right\}$$

$$= \frac{1}{4} \left(\frac{m}{\hbar} \right)^2 \sum_k (w - \varepsilon)^{-4} \left\{ 2 \varepsilon_x \varepsilon_y \varepsilon_{xy} - \varepsilon_{xx} \varepsilon_y^2 \right\}$$

$$= \frac{1}{12} \left(\frac{m}{\hbar} \right)^2 \sum_k \left[\frac{\partial}{\partial k_y} (w - \varepsilon)^{-3} \right] \left\{ 2 \varepsilon_x \varepsilon_{xy} - \varepsilon_{xx} \varepsilon_y \right\}$$

$$= -\frac{1}{12} \left(\frac{m}{\hbar} \right)^2 \sum_k (w - \varepsilon)^{-3} \left\{ 2 \varepsilon_x \varepsilon_{xy} - \varepsilon_{xx} \varepsilon_y \right\}_y .$$

Inserted back into $G^d_{xy}(w)$ one arrives at (31.44).

Problem 5.10: Eqs. (33.16), (33.17).

From (33.14), (33.15) we have, inserting (33.13) and using definition (7.21) and Wick's theorem (9.11),

$$\tilde{\mathcal{G}}_2^{(1)} = \int_0^\beta d\tau_1 e^{-i\nu_1\tau_1} \int_0^\beta d\tau_2 e^{-i\nu_2\tau_2} \int_0^\beta d\tau_3 e^{+i\nu_3\tau_3} \int_0^\beta d\tau \sum_{\mathbf{kk'q}} \left(-\frac{U}{L}\right)$$

$$\times \langle T\{(a^+_{\mathbf{k}\uparrow} a^+_{\mathbf{k'}\downarrow} a_{\mathbf{k'+q}\downarrow} a_{\mathbf{k-q}\uparrow})^o (-i\tau) a^{+o}_{\mathbf{k_1}\uparrow}(-i\tau_1)$$

$$\times a^{+o}_{\mathbf{k_2}\downarrow}(-i\tau_2) a^o_{\mathbf{k_3}\uparrow}(-i\tau_3) a_{\mathbf{k_4}\downarrow}\}\rangle_{o,cp}$$

$$= \frac{U}{L} \int_0^\beta d\tau_1 e^{-i\nu_1\tau_1} \int_0^\beta d\tau_2 e^{-i\nu_2\tau_2} \int_0^\beta d\tau_3 e^{+i\nu_3\tau_3}$$

$$\times \int_0^\beta d\tau \mathcal{G}^o(\mathbf{k_1}; \tau - \tau_1) \mathcal{G}^o(\mathbf{k_2}; \tau - \tau_2)$$

$$\times \mathcal{G}^o(\mathbf{k_3}; \tau_3 - \tau) \mathcal{G}^o(\mathbf{k_4}; -\tau) \delta_{\mathbf{k_1+k_2}, \mathbf{k_3+k_4}} .$$

Writing the $\mathcal{G}^o$ as Fourier sums (7.25) and using (7.28) one finds the $N = 1$ term of (33.15) with $t^{(1)}$ given by (33.16).

Similarly we calculate, remembering a combinatorial factor 2,

$$\tilde{\mathcal{G}}_2^{(2)} = \int_0^\beta d\tau_1 e^{-i\nu_1\tau_1} \int_0^\beta d\tau_2 e^{-i\nu_2\tau_2} \int_0^\beta d\tau_3 e^{+i\nu_3\tau_3} \int_0^\beta d\tau \int_0^\beta d\tau'$$

$$\times \sum_{\mathbf{kk'q}} \sum_{\mathbf{pp'q'}} \frac{1}{2}\left(-\frac{U}{L}\right)^2 \langle T\{(a^+_{\mathbf{k}\uparrow} a^+_{\mathbf{k'}\downarrow} a_{\mathbf{k'+q}\downarrow} a_{\mathbf{k-q}\uparrow})^o(-i\tau)$$

$$\times (a^+_{\mathbf{p}\uparrow} a^+_{\mathbf{p'}\downarrow} a_{\mathbf{p'+q'}\downarrow} a_{\mathbf{p-q'}\uparrow})^o(-i\tau') a^{+o}_{\mathbf{k_1}\uparrow}(-i\tau_1) a^{+o}_{\mathbf{k_2}\downarrow}(-i\tau_2)$$

$$\times a^o_{\mathbf{k_3}\uparrow}(-i\tau_3) a_{\mathbf{k_4}\downarrow}\}\rangle_{o,cp}$$

$$= -\left(\frac{U}{L}\right)^2 \int_0^\beta d\tau_1 e^{-i\nu_1\tau_1} \int_0^\beta d\tau_2 e^{-i\nu_2\tau_2} \int_0^\beta d\tau_3 e^{+i\nu_3\tau_3} \int_0^\beta d\tau \int_0^\beta d\tau'$$

$$\times \mathcal{G}^o(\mathbf{k_1}; \tau - \tau_1) \mathcal{G}^o(\mathbf{k_2}; \tau - \tau_2) \mathcal{G}^o(\mathbf{k_3}; \tau_3 - \tau') \mathcal{G}^o(\mathbf{k_4}; -\tau')$$

$$\times \sum_{\mathbf{q}} \mathcal{G}^o(\mathbf{k_1} + \mathbf{q}; \tau' - \tau) \mathcal{G}^o(\mathbf{k_2} - \mathbf{q}; \tau' - \tau) \delta_{\mathbf{k_1+k_2}, \mathbf{k_3+k_4}} .$$

The same manipulations as before lead to the $N = 2$ term of (33.15) with $t^{(2)}$ given by (33.17).

Problem 5.11: Eqs. (34.15), (34.16).

Expanding $f_o(\varepsilon_{\mathbf{k+q}} - \varepsilon_F)$ in powers of $\varepsilon_{\mathbf{k+q}} - \varepsilon_{\mathbf{k}} = 2\zeta\sqrt{\varepsilon_{\mathbf{k}}\varepsilon_{\mathbf{q}}} + \varepsilon_{\mathbf{q}} \equiv \Delta(\varepsilon_{\mathbf{k}})$

and introducing the density of states (10.4), Eq. (34.14) becomes

$$\chi^o(\mathbf{q}; q_o) = -\frac{2\mu_B^2}{V} \sum_{\mathbf{k}} \frac{1}{\Delta - q_o} \left\{ f_o'\Delta + \frac{1}{2} f_o''\Delta^2 + \frac{1}{6} f_o'''\Delta^3 + \cdots \right\}$$

$$= -2\mu_B^2 \int d\varepsilon N(\varepsilon + \varepsilon_F) \int_{-1}^{+1} \frac{d\zeta}{2} \left\{ f_o'(\varepsilon) \left[1 + \frac{q_o}{\Delta(\varepsilon) - q_o} \right] \right.$$

$$+ \frac{1}{2} f_o''(\varepsilon) \left[\Delta(\varepsilon) + q_o + \frac{q_o^2}{\Delta(\varepsilon) - q_o} \right] + \frac{1}{6} f_o'''(\varepsilon)$$

$$\left. \times \left[\Delta^2(\varepsilon) + q_o\Delta(\varepsilon) + q_o^2 + \frac{q_o^3}{\Delta(\varepsilon) - q_o} \right] + \cdots \right\}.$$

The ζ-integration yields

$$\chi^o(\mathbf{q}; q_o) = -2\mu_B^2 \int d\varepsilon N(\varepsilon + \varepsilon_F) \left\{ f_o'(\varepsilon) \right.$$

$$\times \left[1 + \frac{q_o}{4\sqrt{\varepsilon\varepsilon_{\mathbf{q}}}} \log \frac{2\sqrt{\varepsilon\varepsilon_{\mathbf{q}}} + \varepsilon_{\mathbf{q}} - q_o}{-2\sqrt{\varepsilon\varepsilon_{\mathbf{q}}} + \varepsilon_{\mathbf{q}} - q_o} \right]$$

$$+ \frac{1}{2} f_o''(\varepsilon)[\varepsilon_{\mathbf{q}} + q_o + \mathcal{O}(q_o^2)] + \frac{2}{9} f_o'''(\varepsilon)$$

$$\left. \times [\varepsilon\varepsilon_{\mathbf{q}} + \mathcal{O}(\varepsilon_{\mathbf{q}}^2, q_o\varepsilon_{\mathbf{q}}, q_o^2)] + \cdots \right\}.$$

Since at low temperature the derivatives of $f_o(\varepsilon)$ are strongly localized at $\varepsilon \sim 0$, one partial integration of the f_o''-term, two partial integrations of the f_o'''-term, etc. yield

$$\chi^o(\mathbf{q}; q_o) = -2\mu_B^2 \int d\varepsilon f_o'(\varepsilon) \left\{ N(\varepsilon + \varepsilon_F) \right.$$

$$\times \left[1 + \frac{q_o}{4\sqrt{\varepsilon\varepsilon_{\mathbf{q}}}} \log \frac{2\sqrt{\varepsilon\varepsilon_{\mathbf{q}}} + \varepsilon_{\mathbf{q}} - q_o}{-2\sqrt{\varepsilon\varepsilon_{\mathbf{q}}} + \varepsilon_{\mathbf{q}} - q_o} \right]$$

$$- \frac{1}{2} N'(\varepsilon + \varepsilon_F)[\varepsilon_{\mathbf{q}} + q_o + \mathcal{O}(q_o^2)] + \frac{2}{9} (\varepsilon N(\varepsilon + \varepsilon_F))''$$

$$\left. \times [\varepsilon_{\mathbf{q}} + \mathcal{O}(\varepsilon_{\mathbf{q}}^2, q_o\varepsilon_{\mathbf{q}}, q_o^2)] + \cdots \right\}.$$

Using $f_o'(\varepsilon) \simeq -\delta(\varepsilon)$ and Eq. (31.16) this becomes, for $|q_o| < 2\sqrt{\varepsilon_F\varepsilon_{\mathbf{q}}}$,

$$\chi^o(\mathbf{q}; q_o) = \chi_P \left\{ 1 + \frac{q_o}{4\sqrt{\varepsilon_F\varepsilon_{\mathbf{q}}}} \left[i\pi + \mathcal{O}\left(\frac{\varepsilon_{\mathbf{q}} - q_o}{\sqrt{\varepsilon_F\varepsilon_{\mathbf{q}}}} \right) \right] - \frac{1}{2} \frac{N'(\varepsilon_F)}{N(\varepsilon_F)} \right.$$

$$\left. \times [\varepsilon_{\mathbf{q}} + q_o + \mathcal{O}(q_o^2)] + \frac{4}{9} \frac{N'(\varepsilon_F)}{N(\varepsilon_F)} [\varepsilon_{\mathbf{q}} + \mathcal{O}(\varepsilon_{\mathbf{q}}^2, q_o\varepsilon_{\mathbf{q}}, q_o^2)] + \cdots \right\}.$$

From this formula Eqs. (34.15), (34.16) finally follow.

Problem 5.12 : Eq. (34.17).

Inserting (33.13), Eq. (34.6) reads for $N = 1$

$$T^{(1)} = -\frac{U}{L} \sum_{\mathbf{kk'}} \sum_{\mathbf{pp'q'}} \int_0^\beta d\tau e^{i\nu\tau} \int_0^\beta d\tau' \langle T\{(a_{\mathbf{p}\uparrow}^+ a_{\mathbf{p'}\downarrow}^+ a_{\mathbf{p'+q'}\downarrow} a_{\mathbf{p-q'}\uparrow})^o(-i\tau')$$

$$\times (a_{\mathbf{k}\sigma}^+ a_{\mathbf{k+q}\sigma'})^o(-i\tau)(a_{\mathbf{k'+q},\rho}^+ a_{\mathbf{k'}\rho'})(0))\}\rangle_{o,cp} .$$

Applying Wick's theorem (9.11) and definition (7.21) one finds 4 different factorizations which may be combined as follows:

$$T^{(1)} = \frac{U}{L} \sum_{\mathbf{kk'}} \sum_{\mathbf{pp'q'}} \int_0^\beta d\tau e^{i\nu\tau} \int_0^\beta d\tau' \mathcal{G}_\sigma^o(\mathbf{k}; \tau' - \tau) \mathcal{G}_{\sigma'}^o(\mathbf{k} + \mathbf{q}; \tau - \tau')$$

$$\times \mathcal{G}_\rho^o(\mathbf{k'} + \mathbf{q}; \tau') \mathcal{G}_{\rho'}^o(\mathbf{k'}; -\tau')(\delta_{\mathbf{p-q'}\uparrow, \mathbf{k}\sigma} \delta_{\mathbf{p'+q'}\downarrow, \mathbf{k'+q}\rho}$$

$$- \delta_{\mathbf{p'+q'}\downarrow, \mathbf{k}\sigma} \delta_{\mathbf{p-q'}\uparrow, \mathbf{k'+q}\rho})(\delta_{\mathbf{p'}\downarrow, \mathbf{k+q}\sigma'} \delta_{\mathbf{p}\uparrow, \mathbf{k'}\rho'} - \delta_{\mathbf{p}\uparrow, \mathbf{k+q}\sigma'} \delta_{\mathbf{p'}\downarrow, \mathbf{k'}\rho'}) .$$

After eliminating $\mathbf{p}, \mathbf{p'}$ and $\mathbf{q'}$ by summation, writing the $\mathcal{G}^o$ as Fourier sums (7.25) and using (7.28) and definition (34.8), one readily finds (34.17).

Problem 5.13 : Eq. (34.53).

Treating Eq. (34.52) as indicated, one finds

$$s_1 \cong \frac{k_B k_F^3 x_1^3}{2\pi^3} \int_{-\infty}^{+\infty} \frac{ye^{-y}dy}{(1 + e^{-y})^2} F\left(y\frac{T}{T_1}\right)$$

where

$$F\left(y\frac{T}{T_1}\right) \equiv \left(\frac{T}{T_1}\right)^3 \int_0^{T_1/T} u^2 \ln\left|\frac{u+y}{u-y}\right| du$$

$$= \frac{1}{3}\left[\ln\left|\frac{1 + yT/T_1}{1 - yT/T_1}\right| - \left(y\frac{T}{T_1}\right)^3 \ln\left|\frac{1 - (yT/T_1)^2}{(yT/T_1)^2}\right| + y\frac{T}{T_1}\right]$$

$$= y\frac{T}{T_1} + \frac{2}{9}\left(y\frac{T}{T_1}\right)^3\left[1 + 3\ln\left(y\frac{T}{T_1}\right)\right] + \mathcal{O}\left(\left(y\frac{T}{T_1}\right)^5\right) .$$

Hence, with (32.11) and (34.54) follows the expression

$$s_1 \cong \frac{k_B k_F^3 x_1^3}{2\pi^3}\left\{I_2\frac{T}{T_1} + \frac{2}{9}I_4\left(\frac{T}{T_1}\right)^3\left[1 + 3\frac{K_+}{I_4} + 3\ln\frac{T}{T_1}\right] + \mathcal{O}\left(\left(\frac{T}{T_1}\right)^5\right)\right\},$$

from which, using $I_2 = \pi^2/3$ and $I_4 = 7\pi^4/15$, Eq. (34.53) is obtained.

Problem 5.14 : Eq. (34.57).

Interchanging the order of integration, Eq. (34.55) becomes with $y_1 \equiv 2x_1 T_F/T \gg 1$,

$$
\begin{aligned}
s_2 &= -\frac{k_B k_F^3}{\pi^3} \int_0^{y_1} \frac{y e^y \, dy}{(e^y - 1)^2} \int_{yT/2T_F}^{x_1} dx \, x^2 \left\{ (1 - \kappa) \frac{Ty}{T_o x} - \tan^{-1} \frac{Ty}{T_o x} \right\} \\
&= \frac{k_B k_F^3}{\pi^3} \left\{ \frac{1 - \kappa}{2} \frac{T}{T_o} \left[-x_1^2 J_2(y_1) + \left(\frac{T}{2T_F} \right)^2 J_4(y_1) \right] \right. \\
&\quad \left. + \int_0^{y_1} \frac{y e^y \, dy}{(e^y - 1)^2} G\left(y \frac{T}{T_1} \right) \right\}
\end{aligned}
$$

where $J_n(z)$ is defined in Eq. (16.9) and, in the notation introduced for c_{V1},

$$
\begin{aligned}
G\left(y \frac{T}{T_1} \right) &\equiv \int_{yT/2T_F}^{x_1} dx \, x^2 \tan^{-1} \frac{Ty}{T_o x} = \left(\frac{yT}{T_o} \right)^3 \int_{yT/T_1}^{1/t_o} \frac{du}{u^4} \tan^{-1} u \\
&= \frac{x_1^3}{3} \left\{ \tan^{-1} \frac{yT}{T_1} + \frac{1}{2} \frac{yT}{T_1} - \frac{1}{2} \left(\frac{yT}{T_1} \right)^3 \ln \frac{1 + (yT/T_1)^2}{(yT/T_1)^2} \right\} \\
&\quad - \frac{x_1^3}{3} \left(\frac{yT}{T_1} \right)^3 \left\{ t_o^3 \tan^{-1} \frac{1}{t_o} + \frac{1}{2} t_o^2 - \frac{1}{2} \ln(1 + t_o^2) \right\} \\
&= \frac{x_1^3}{2} \left\{ \frac{yT}{T_1} - \left[\frac{2}{9} + \frac{\pi}{3} t_o^3 + \mathcal{O}(t_o^4) - \frac{2}{3} \ln \frac{yT}{T_1} \right] \left(\frac{yT}{T_1} \right)^3 \right. \\
&\quad \left. + \mathcal{O}\left(\left(\frac{yT}{T_1} \right)^5 \right) \right\} .
\end{aligned}
$$

Now for $z \gg 1$

$$
J_n(\infty) - J_n(z) \cong \int_z^\infty y^n e^{-y} \, dy = z^n e^{-z} \left[1 + \mathcal{O}\left(\frac{1}{z} \right) \right] .
$$

Thus, neglecting terms $\propto T^n \exp(-2x_1 T_F/T)$, we may replace the argument y_1 of J_n by ∞ and use definition (34.56) to obtain the expression

$$
\begin{aligned}
s_2 &= \frac{k_B k_F^3 x_1^3}{2\pi^3} \left\{ \kappa J_2 \frac{T}{T_1} + \frac{2}{3} \left[-\frac{1}{3} + \frac{K_-}{J_4} + \ln \frac{T}{T_1} + \mathcal{O}((1 - \kappa)^3) \right] \right. \\
&\quad \left. \times J_4 \left(\frac{T}{T_1} \right)^3 + \mathcal{O}\left(\left(\frac{T}{T_1} \right)^5 \right) \right\} .
\end{aligned}
$$

Comparing Eqs. (32.11) and (34.56) one finds that $J_2 = I_2 = \pi^2/3$ and $J_4 = 4I_4/7 = 4\pi^4/15$. Using these values in the above expression for s_2, Eq. (34.57) follows.

Problem 5.15 : Eq. (35.68).

The average in (35.67) may be written

$$T_N \equiv \sum_{\sigma_1 \cdots \sigma_N} \sigma_1 \cdots \sigma_N \int_0^\beta d\tau_1 \cdots \int_0^\beta d\tau_N \langle T\{c_1^+ c_1 c_2^+ c_2 \cdots c_N^+ c_N\}\rangle_{o,cp}$$

where $c_i \equiv c_{d\sigma_i}^o(-i\tau_i)$. Because of the restriction to connected parts (cp), Wick's theorem (9.11) gives the pairings $c_1 c_2^+, c_2 c_3^+, \cdots, c_{N-1} c_N^+, c_N c_1^+$ with one commutation sign and a combinatorial factor $(N-1)!$. Hence, using definition (7.21),

$$T_N = (N-1)!(-)^{N-1} \sum_{\sigma_1} \sigma_1^N \int_0^\beta d\tau_1 \cdots \int_0^\beta d\tau_N$$
$$\times \mathcal{G}^o(\tau_1 - \tau_2) \cdots \mathcal{G}^o(\tau_{N-1} - \tau_N)\mathcal{G}^o(\tau_N - \tau_1) .$$

Because of the σ_1-sum, $T_N = 0$ for odd N while, using (7.25), (7.28),

$$T_{2N} = -2(2N-1)! \sum_{\nu_1} [\tilde{\mathcal{G}}^o(i\nu_1)]^{2N}$$

and hence

$$-\beta\Delta\Omega = -2 \sum_\nu \sum_{N=1}^\infty \frac{1}{2N} [mh_o\tilde{\mathcal{G}}^o(i\nu)]^{2N}$$

which is the same as (35.68).

Problem 5.16 : Eq. (35.69).

According to (11.17) the selfenergy due to $\mathcal{H}_{sd}$ is determined to lowest order $(N = 2)$ by

$$\Sigma_{sd}(i\nu)[\tilde{\mathcal{G}}_\sigma^o(i\nu)]^2 = -\frac{1}{2} \int_0^\beta d\tau \int_0^\beta d\tau_1 \int_0^\beta d\tau_2 e^{i\nu\tau}$$
$$\times \langle T\{\mathcal{H}_{sd}^o(-i\tau_1)\mathcal{H}_{sd}^o(-i\tau_2)c_{d\sigma}^o(-ir)c_{d\sigma}^+\}\rangle_{o,cp,1-i} .$$

Inserting from (35.38) and using Wick's theorem (9.11) and Eqs. (7.21), (7.25), (7.28) one finds

$$\Sigma_{sd}(i\nu) = V^{-1} \sum_{\mathbf{k}} |g(\varepsilon_{\mathbf{k}})|^2 \tilde{\mathcal{G}}_s^o(\mathbf{k}; i\nu) ,$$

and with (7.44) and the s-band density of states (10.4),

$$\Sigma_{sd}(i\nu) = \int \frac{d\varepsilon}{i\nu - \varepsilon} N_s(\varepsilon)|g(\varepsilon)|^2 \;.$$

The renormalization effect on the d-electrons may be determined by the analytic continuation $i\nu \to \bar{\varepsilon}_d + i\epsilon \, \mathrm{sgn}\,\nu$. Using (11.30) one obtains

$$\Sigma_{sd}(\bar{\varepsilon}_d \pm i\epsilon) = -\int d\varepsilon \frac{P}{\varepsilon - \bar{\varepsilon}_d} N_s(\varepsilon)|g(\varepsilon)|^2 \mp i\pi N_s(\bar{\varepsilon}_d)|g(\bar{\varepsilon}_d)|^2 \;.$$

Absorbing $\mathrm{Re}\,\Sigma_{sd}$ into $\bar{\varepsilon}_d$ according to (11.34) and comparing with (11.7) one finds (35.69) where the superfluous spin label has been dropped.

Problem 5.17 : Eq. (35.72).

Inserting (35.69), using (7.27) to write $\nu = (2n+1)\pi/\beta$, $\mu = 2k\pi/\beta$ and introducing the notation $x \equiv \beta\Gamma/2\pi$, $y \equiv \beta m h_o/2\pi$, $\xi \equiv \beta\bar{\varepsilon}_d/2\pi$, Eq. (35.68) becomes

$$-\beta\Delta\Omega = \log \prod_{n=-\infty}^{+\infty} \left[1 + \left(\frac{y}{n + \frac{1}{2} + x\,\mathrm{sgn}(n + \frac{1}{2}) + i\xi} \right)^2 \right] \;.$$

If $x \gg 1$ and $\xi = 0$ this becomes

$$-\beta\Delta\Omega \cong 2\ln \prod_{n=0}^{\infty} \left[1 + \left(\frac{y}{n + x} \right)^2 \right]$$

and using Eqs. (35.70), (35.71)

$$-\beta\Delta\Omega \cong 2\ln \left| \left(\frac{x}{x + iy} \right)^{x - 1/2} (x + iy)^{-iy} e^{iy}[1 + \mathcal{O}(x^{-1})] \right|$$

$$= 2\mathrm{Re}\,\log \left\{ \left(1 + i\frac{y}{x} \right)^{-x - 1/2} e^{-iy\log(x + iy)}[1 + \mathcal{O}(x^{-1})] \right\}$$

$$= -x\ln \left[1 + \left(\frac{y}{x} \right)^2 \right] + 2y\tan^{-1}\frac{y}{x} + \mathcal{O}(x^{-1})$$

which is (35.72).

Problem 5.18 : Eq. (35.80).

Inserting (35.69) with $\bar{\varepsilon}_d$ replaced by ε_σ, Eq. (35.79) may be written as

$$\varphi_\mu = \sum_\sigma \sum_\nu \left[\mu + \Gamma\big(\mathrm{sgn}\,(\nu + \mu) - \mathrm{sgn}\,\nu\big) \right]^{-1}$$

$$\times \left\{ [\nu + \Gamma\mathrm{sgn}\,\nu + i\varepsilon_\sigma]^{-1} - [\nu + \mu + \Gamma\mathrm{sgn}\,(\nu + \mu) + i\varepsilon_\sigma]^{-1} \right\} \;.$$

Since in the three intervals $\nu > 0$; $0 > \nu > -\mu$ and $-\mu > \nu$ the signs sgn ν, sgn $(\nu + \mu)$ are $+$, $+$; $-$, $+$ and $-$, $-$, respectively, we have

$$\varphi_\mu = \sum_\sigma \left\{ \sum_{\nu>0} \frac{1}{\mu} \left(\frac{1}{\nu + \Gamma + i\varepsilon_\sigma} - \frac{1}{\nu + \mu + \Gamma + i\varepsilon_\sigma} \right) \right.$$
$$+ \sum_{0>\nu>-\mu} \frac{1}{\mu + 2\Gamma} \left(\frac{1}{\nu - \Gamma + i\varepsilon_\sigma} - \frac{1}{\nu + \mu + \Gamma + i\varepsilon_\sigma} \right)$$
$$+ \left. \sum_{-\mu>\nu} \frac{1}{\mu} \left(\frac{1}{\mu - \Gamma + i\varepsilon_\sigma} - \frac{1}{\nu + \mu - \Gamma + i\varepsilon_\sigma} \right) \right\} .$$

Substituting $\nu = \nu' - \mu$ in the second term of each sum leads to cancellations in the first and third sums. Regrouping the remaining sums one finds the expression

$$\varphi_\mu = \sum_\sigma \left\{ \sum_{\mu>\nu>0} \left(\frac{1}{\mu} - \frac{1}{\mu + 2\Gamma} \right) \frac{1}{\nu + \Gamma + i\varepsilon_\sigma} \right.$$
$$\left. - \sum_{0>\nu>-\mu} \left(\frac{1}{\mu} - \frac{1}{\mu + 2\Gamma} \right) \frac{1}{\nu - \Gamma + i\varepsilon_\sigma} \right\}$$

which is easily recognized as Eq. (35.80).

Problem 5.19 : Eq. (36.3).

The hole Fermi surface near half-filling is determined by $\varepsilon_{\mathbf{k}} = -\lambda t$, $0 < \lambda \ll 1$ with $\varepsilon_{\mathbf{k}}$ given by (36.2). Hence the number of states per unit volume in the shaded area of Fig. 5.16(b) is

$$Z(\lambda t) = \frac{1}{V} \sum_{k_x,k_y=0}^{\pi/a} \theta(\varepsilon_{\mathbf{k}} + \lambda t)\theta(-\varepsilon_{\mathbf{k}}) .$$

This expression is best evaluated by taking twice the sum over the area $\pi/2 < k_x a < \pi$, $0 < k_y a < \pi/2$. Writing $k_x a = \pi - u$, $k_y a = v$ and developing the cosines, one finds

$$Z(\lambda t) \cong \frac{2}{(2\pi a)^2} \int_0^{\pi/2} du \int_0^{\pi/a} dv \theta(\lambda - u^2 + v^2)\theta(u - v)$$
$$= \frac{2}{(2\pi a)^2} \int_{\sqrt{\lambda}}^{\pi/2} du \left[u - \sqrt{u^2 - \lambda} \right]$$
$$= \frac{1}{(2\pi a)^2} \left[u^2 - u\sqrt{u^2 - \lambda} + \lambda \ln(u + \sqrt{u^2 - \lambda}) \right]_{\sqrt{\lambda}}^{\pi/2}$$

and therefore

$$Z(\varepsilon) = \frac{\varepsilon}{8\pi^2 a^2 t}\left[-1 + \ln\frac{\pi^2 t}{\varepsilon} + \mathcal{O}(\varepsilon)\right] .$$

This leads to Eq. (36.3).

References to Chapter 5

1. **Mattis, D.C.**, *The Theory of Magnetism* (Harper and Row, New York, 1965).
2. **Mattis, D.C.**, *The Theory of Magnetism I. Statics and Dynamics* (Springer, Berlin, 1981), Solid-State Sciences, Vol. 17.
3. **Busch, G.** , *Ferroelectrics* **74**, 267 (1987).
4. **Kittel, C.**, *Introduction to Solid State Physics, 5th ed.* (Wiley, New York, 1976).
5. **Hirsch, J.E.**, *Phys. Rev.* **B40**, 2354 (1989).
6. **Weiss, P.**, *J. Phys. (Paris)* **6**, 661 (1907); *Phys. Z.* **9**, 358 (1908).
7. **Dillon, J.F.**, "Domains and domain walls", in *Magnetism*, ed. Rado, G.T. and Suhl, H. (Academic, New York, 1963), Vol. III, Chapter 9.
8. **Langevin, P.**, *Ann. Chim. Phys.* **5**, 70 (1905); *J. Phys.* **4**, 678 (1905).
9. **Curie, P.**, *Ann. Chim. Phys.* **5**, 289 (1895).
10. **Lenz, W.**, *Phys. Z.* **21**, 613 (1920).
11. **Brillouin, L.**, *J. Physique Rad.* **8**, 74 (1927).
12. **Ising, E.**, *Z. Phys.* **31**, 253 (1925); see also **Brush, S.G.**, *Rev. Mod. Phys.* **39**, 883 (1967).
13. **Pauli, W.**, *Z. Phys.* **41**, 81 (1927).
14. **Heisenberg, W.**, *Z. Phys.* **49**, 619 (1928).
15. **Dirac, P.A.M.**, *Proc. Roy. Soc.* **A123**, 714 (1929); *The Principles of Quantum Mechanics, 4th ed.* (Oxford University Press, London, 1958), Section 58.
16. **Enz, C.P.**, in *Heisenberg, Collected Works*, ed. Blum, W., Dürr, H.-P. and Rechenberg, H., Series A (Springer, Berlin, 1985), Part I, p. 507.
17. **Onsager, L.**, *Phys. Rev.* **65**, 117 (1944).
18. **Yang, C.N.**, *Phys. Rev.* **85**, 809 (1952).
19. **Mattis, D.C.**, *The Theory of Magnetism II, Thermodynamics and Statistical Mechanics* (Springer, Berlin, 1985), Solid-State Sciences, Vol. 55.
20. **Domb, C.** and **Green, M.S.**, *Phase Transitions and Critical Phenomena* (Academic, London, 1972–1983), Vols. 1–8; **Domb, C.** and **Lebowitz, J.L.** (Academic, London, 1984–1989), Vols. 9–13.
21. **Mermin, N.D.** and **Wagner, H.**, *Phys. Rev. Lett.* **17**, 1133, 1307 (1966).
22. **Landau, L.D.**, *Z. Phys.* **64**, 629 (1930).
23. **Stoner, E.C.**, *Proc. Roy. Soc.* **A165**, 372 (1938), **A169**, 339 (1939).
24. **Hubbard, J.**, *Proc. Roy. Soc.* **A276**, 238 (1963), **Gutzwiller, M.**, *Phys. Rev. Lett.* **10**, 159 (1963); **Kanamori, J.**, *Prog. Theor. Phys.* **30**, 275 (1963). See also *The Hubbard Model*, ed. Rasetti, M., *Int. J. Mod. Phys.* **B5**, nos. 6 & 7 (1991).
25. **Lieb, E.**, *Phys. Rev. Lett.* **62**, 1201, 1927 (1989).

26. Nagaoka, Y., *Phys. Rev.* **147**, 392 (1966); Thouless, D.J., *Proc. Phys. Soc. (London)* **86**, 893 (1965).

27. Herring, C., "Exchange interactions among itinerant electrons", in *Magnetism*, ed. Rado, G.T. and Suhl, H. (Academic, New York, 1966), Vol. IV.

28. Moriya, T., *J. Magn. Magn. Mat.* **14**, 1 (1979).

29. Moriya, T., *Spin Fluctuations in Itinerant Electron Magnetism*, (Springer, Berlin, 1985).

30. Rhodes, P.R. and Wohlfarth, E.P., *Proc. Roy. Soc.* **A273**, 247 (1963).

31. Smith, T.F., Mydosh, J.A. and Wohlfarth, E.P., *Phys. Rev. Lett.* **27**, 1732 (1971); Huber, J.G., Maple, M.B. and Wohlleben, D., *Solid State Commun.* **16**, 211 (1975). See also Lo, I., Mazumdar, S. and Mattocks, P.G., *Phys. Rev. Lett.* **62**, 2555 (1989).

32. Overhauser, A.W., *Phys. Rev.* **128**, 1437 (1962).

33. van Vleck, J.H., *The Theory of Electric and Magnetic Susceptibilities* (Oxford University Press, London, 1952).

34. Lax, M., *Symmetry Principles in Solid State and Molecular Physics* (Wiley, New York, 1974), Apprendix E.

35. Vaknin, D. *et al.*, *Phys. Rev. Lett.* **58**, 2802 (1987) : La_2CuO_{4-y}; Tranquada, J.M. *et al.*, *Phys. Rev. Lett.* **60**, 156 (1988) : $YBa_2Cu_3O_x$; Thurston, T.R. *et al.*, *Phys. Rev. Lett.* **65**, 263 (1990) : $(Nd,Pr)_{2-x}Ce_xCuO_4$.

36. Keffer, F., *Handbuch der Physik*, ed. Flügge, S. (Springer, Berlin, 1966), Vol. XVIII/2, p. 1.

37. Enz, C.P., *Rev. Mod. Phys.* **46**, 705 (1974), Section IV-D.

38. Néel, L., *Ann. Phys. (Paris)* **17**, 64 (1932).

39. Wannier, G.H., *Elements of Solid State Theory* (Cambridge University Press, Cambridge, 1960), Chapter 4.

40. Bloch, F., *Z. Phys.* **61**, 206 (1930).

41. Holstein, T. and Primakoff, H., *Phys. Rev.* **58**, 1098 (1940).

42. Kittel, C., *Quantum Theory of Solids* (Wiley, New York, 1963), Chapter 4.

43. Landau, L.D. and Lifshitz, E.M., *Statistical Physics* (Pergamon, Oxford, 1958).

44. Jahnke-Emde-Lösch, *Tafeln Höherer Funktionen*, ed. Lösch, F. (Teubner, Stuttgart, 1966), Tafel 12.

45. Takahashi, M., *Phys. Rev. Lett.* **58**, 168 (1987); *Prog. Theor. Phys. Suppl.* **87**, 233 (1986); *Phys. Rev.* **B36**, 3791 (1987).

46. Takahashi, M., *Phys. Rev.* **B40**, 2494 (1989); Hirsch, J.E. and Sanyee, T., *Phys. Rev.* **B40**, 4769 (1989). See also Manousakis, E., *Rev. Mod. Phys.* **63**, 1 (1991).

47. Zhou, C. and Enz, C.P., *Physica* **C170**, 119 (1990). See also Barentzen, H., *Phys. Lett.* **A156**, 461 (1991).

48. Dyson, F.J., *Phys. Rev.* **102**, 1217, 1230 (1956).

49. **Zittartz, J.,** Z. Physik **184**, 506 (1965); **van Hemmen, J.L.,** **Brito, A.A.S.** and **Wreszinski, W.F.,** J. Stat. Phys. **37**, 187 (1984).

50. **Peierls, R.E.,** Quantum Theory of Solids (Oxford University Press, London, 1974), Chapter 7.

51. **Busch, G.A.** and **Kern, R.,** Helv. Phys. Acta **32**, 24 (1959); **Kohn, W.,** in Solid State Physics, ed. Seitz, F. and Turnbull, D. (Academic, New York, 1957), Vol. 5, p. 257.

52. **Peierls, R.E.,** Z. Physik **80**, 763 (1933).

53. **Wannier, G.H.,** Rev. Mod. Phys. **34**, 645 (1962).

54. **Enz, C.P.,** Nuovo Cimento **6**, Suppl. 1224 (1957).

55. **Enz, C.P.,** Helv. Phys. Acta **33**, 89 (1960).

56. **Hebborn, J.E.** and **Sondheimer, E.H.,** Phys. Rev. Lett. **2**, 150 (1959); J. Phys. Chem. Solids **13**, 105 (1960).

57. **Wigner, E.P.,** Group Theory and Its Application to the Quantum Mechanics of Atomic Spectra (Academic, New York, 1959), Section 26.

58. **Enz, C.P.,** "Magnetic susceptibility of semiconductors", in Proc. Int. School of Physics "Enrico Fermi", Course XXII, ed. Smith, R.A. (Academic, New York, 1963), p. 458 and references therein; **Wannier, G.H.** and **Upadhyaya, U.N.,** Phys. Rev. **136**, A803 (1964); **Fukuyama, H.,** Prog. Theor. Phys. **45**, 704 (1971); **Misra, S., Tripathi, G.S.** and **Misra, P.K.,** J. Phys. C : Solid State Phys. **19**, 2007 (1986); **Tripathi, G.S.,** Phys. Lett. **A115**, 169 (1986); **Sahu, T.,** Phys. Lett. **A115**, 173 (1986).

59. **Gradshteyn, I.S.** and **Ryzhik, I.M.,** Table of Integrals, Series, and Products, 4th ed. by Geronimus, Yu.V. and Tseytlin, M.Yu., transl. and edited by Jeffrey, A. (Academic, New York, 1965).

60. **Edwards, D.M.** and **Wohlfarth, E.P.,** Proc. Roy. Soc. **A303**, 127 (1968).

61. **Micnas, R., Ranninger, J.** and **Robaszkiewicz, S.,** Rev. Mod. Phys. **62**, 113 (1990).

62. **Balachandran, A.P., Ercolessi, E., Morandi, G.** and **Srivastava, A.,** Int. J. Mod. Phys. **B4**, 2057 (1990).

63. **Mott, N.F.,** Metal-Insulator Transitions (Taylor and Francis, London, 1974).

64. **Anderson, P.W.,** Science **235**, 1196 (1987); and in Frontiers and Borderlines in Many-Particle Physics, Proc. Int. School of Physics "Enrico Fermi", Course 54, ed. Broglia, R.A. and Schrieffer, J.R. (North-Holland, Amsterdam, 1988), p. 1.

65. **Zhang, F.C.** and **Rice, T.M.,** Phys. Rev. **41**, 7243 (1990). See also **Stephan, W.** and **Horsch, P.,** Phys. Rev. Lett. **66**, 2258 (1991).

66. **Hirsch, J.E.,** Phys. Rev. **B31**, 4403 (1985). See also **Mielke, A.,** J. Phys. A : Math. Gen. **24**, L73 (1991).

67. **Zhang, S.C.,** Phys. Rev. Lett. **65**, 120 (1990). See also **Yang, C.N.** and **Zhang, S.C.,** Int. J. Mod. Phys. **B5**, 977 (1991).

68. **Matthias, B.T.** and **Bozorth, R.M.,** Phys. Rev. **109**, 604 (1958).

69. **Matthias, B.T., Clogston, A.M., Williams, H.J., Corenzwit, E.** and **Sherwood, R.C.,** Phys. Rev. Lett. **7**, 7 (1961).

70. Giorgi, A.L., Matthias, B.T., Stewart, G.R., Acker, F. and Smith, J.L., *Solid State Commun.* **32**, 455 (1979).

71. Acker, F., Fisk, Z., Smith, J.L. and Huang, C.Y., *J. Magn. Magn. Mat.* **22**, 250 (1981).

72. Knapp, G.S., Fradin, F.Y. and Culbert, H.V., *J. Appl. Phys.* **42**, 1341 (1971) : $ZrZn_2$; Gardner, W.E., Smith, T.F., Howlett, B.W., Chu, C.W. and Sweedler, A., *Phys. Rev.* **166**, 577 (1968) : Sc_3In.

73. Mishra, S.G., *Mod. Phys. Lett.* **B4**, 83 (1990).

74. Huang, M.-C., Jansen, H.J.F. and Freeman, A.J., *Phys. Rev.* **37**, 3489 (1988).

75. Pickart, S.J., Alperin, H.A., Shirane, G. and Nathans, R., *Phys. Rev. Lett.* **12**, 444 (1964) : $ZrZn_2$; Felcher, G.P., Cable, J.W. and Smith, J.L., *Phys. Rev. Lett.* **45**, 751 (1980) : $TiBe_{1.8}Cu_{0.2}$.

76. Enz, C.P. and Matthias, B.T., *Science* **201**, 828 (1978); *Z. Physik* **B33**, 129 (1979). See also Enz, C.P., in *Superconductivity in d- and f-Band Metals*, ed. Suhl, H. and Maple, B. (Academic, San Francisco, 1980), p. 181.

77. Wu, M.K., Chu, C.W., Smith, J.L., Giorgi, A.L., Huang, C.Y., Matthias, B.T. and Wang, F.E., *Solid State Commun.* **34**, 507 (1980).

78. Grewe, J., Schilling, J.S., Ikeda, K. and Gschneidner, Jr., K.A., *Phys. Rev.* **40**, 9017 (1989).

79. Izuyama, T., Kim, D.J. and Kubo, R., *J. Phys. Soc. Jpn* **18**, 1025 (1963). See also Berk, N.F. and Schrieffer, J.R., *Phys. Rev. Lett.* **17**, 433 (1966).

80. Doniach, S. and Engelsberg, S., *Phys. Rev. Lett.* **17**, 750 (1966). See also Misawa, S., *Prog. Theor. Phys.* **38**, 1207 (1967).

81. Béal-Monod M.T., *J. Physique* **41**, 1109 (1980).

82. Béal-Monod M.T., Ma, S.-K. and Fredkin, D.R., *Phys. Rev. Lett.* **20**, 929 (1968); Brinkman, W.F. and Engelsberg, S., *Phys. Rev.* **169**, 417 (1968).

83. Trainor, R.J., Brodsky, M.B. and Culbert, H.V., *Phys. Rev. Lett.* **34**, 1019 (1975) : UAl_2; Giorgi, A.L., Matthias, B.T., Stewart, G.R., Acker, F. and Smith, J.L., *Solid State Commun.* **32**, 455 (1979) : $TiBe_{2-x}Cu_x$; Stewart, G.R., Matthias, B.T., Giorgi, A.L., Szklarz, E.G. and Smith, J.L., *Solid State Commun.* **30**, 709 (1979) : $TiBe_2$, $ThMg_2$; Stewart, G.R., Smith, J.L., Giorgi, A.L. and Fisk, Z., *Phys. Rev.* **B25**, 5907 (1982) : $TiBe_2$.

84. Misawa, S., *Phys. Lett.* **32A**, 153, 541 (1970); Barnea, G., *J. Phys. C : Solid State Phys.* **8**, L216 (1975).

85. Carneiro, G.M. and Pethick, C.J., *Phys. Rev.* **16**, 1933 (1977); Mishra, S.G. and Ramakrishnan, T.V., *Phys. Rev.* **B18**, 2308 (1978).

86. Acker, F., Huguenin, R., Pelizzone, M. and Smith, J.L., *Phys. Rev.* **B24**, 5404 (1981).

87. Misawa, S., *Phys. Rev. Lett.* **26**, 1632 (1971); Barnea, G. and Edwards, D.M., *J. Phys. F : Metal Phys.* **7**, 1323 (1977).

88. Béal-Monod, M.T., *Physica* **109, 110B**, 1837 (1982).

89. Murata, K.K. and Doniach, S., *Phys. Rev. Lett.* **29**, 285 (1972).

90. Ikeda, K. and Gschneidner, Jr., K.A., J. *Magn. Magn. Mat.* **22**, 207 (1981).
91. Murata, K.K., *Phys. Rev.* **B12**, 282 (1975).
92. Dzyaloshinskii, I.E. and Kondratenko, P.S., *Sov. Phys.-JETP* **43**, 1036 (1976).
93. Moriya, T. and Kawabata, A., J. *Phys. Soc. Jpn* **34**, 639 (1973); **35**, 696 (1973).
94. Gumbs, G. and Griffin, A., *Phys. Rev.* **B13**, 5054 (1976).
95. Wang, S.Q., Evenson, W.E. and Schrieffer, J.R., *Phys. Rev. Lett.* **23**, 92 (1969).
96. Anderson, P.W., *Phys. Rev.* **124**, 41 (1961). See also Blandin, A. and Friedel, J., *J. Physique Rad.* **20**, 160 (1959).
97. Stratonovich, R.L., *Soviet Phys.-Doklady* **2**, 416 (1958); Hubbard, J., *Phys. Rev. Lett.* **3**, 77 (1959).
98. Feynman, R.P., *Phys. Rev.* **84**, 108 (1951).
99. Hertz, J.A. and Klenin, M.A., *Phys. Rev.* **B10**, 1084 (1974).
100. Fawcett, E., *Rev. Mod. Phys.* **60**, 209 (1988).
101. Lomer, W.M., *Proc. Phys. Soc. London* **80**, 489 (1962).
102. Fedders, P.A. and Martin, P.C., *Phys. Rev.* **143**, 245 (1965).
103. Enz, C.P., *Phys. Rev.* **B25**, 6822 (1982).
104. Lee, P.A. and Read, N., *Phys. Rev. Lett.* **58**, 2691 (1987); see also Hirsch, J.E., *Phys. Rev.* **B31**, 4403 (1985).
105. Anderson, P.W., J. *Phys. Chem. Solids* **11**, 26 (1959); Abrikosov, A.A. and Gor'kov, L.P., *Soviet Phys. JETP* **8**, 1090 (1959). See also Maki, K., in *Superconductivity*, ed. Parks, R.D. (Marcel-Dekker, New York, 1969), Vol. 2, p. 1041.
106. Crisan, M., *Theory of Superconductivity* (World Scientific, Singapore, 1989).
107. *Charge Density Waves in Solids*, ed. Hutiray, Gy. and Solyom, J., Lecture Notes in Physics **217** (Springer, Berlin, 1985).
108. Fulde, P. and Ferrell, R.A., *Phys. Rev.* **135**, A550 (1964). See also Larkin, A.I. and Ovchinnikov, Yu.N., *Soviet Phys.-JETP* **20**, 762 (1965); Gruenberg, L.W. and Gunther, L., *Phys. Rev. Lett.* **16**, 996 (1966); Takada, S. and Izuyama, T., *Prog. Theor. Phys.* **41**, 635 (1969).
109. Rice, T.M., *Phys. Rev.* **B2**, 3619 (1970).
110. Birgeneau, R.J. *et al.*, *Phys. Rev.* **B38**, 6614 (1988) : $La_{2-x}Sr_xCuO_4$; Tranquada, J.M. *et al.*, *Phys. Rev.* **B38**, 2477 (1988); **40**, 4503 (1989); Brewer, J.H. *et al.*, *Phys. Rev. Lett.* **60**, 1073 (1988) : $YBa_2Cu_3O_x$.
111. Hirsch, J.E., *Phys. Rev.* **B41**, 6820, 6828 (1990).

AUTHOR INDEX

SUBJECT INDEX